Concrete Structures
Stresses and Deformations

Third Edition

A. Ghali
Professor, The University of Calgary
Canada

R. Favre
Professor, Swiss Federal Institute of Technology (EPFL)
Lausanne, Switzerland

M. Elbadry
Associate Professor, The University of Calgary
Canada

London and New York

First published 1986 by E & FN Spon

Second edition first published 1994

Third edition first published 2002
by Spon Press
11 New Fetter Lane, London EC4P 4EE

Simultaneously published in the USA and Canada
by Spon Press
29 West 35th Street, New York, NY 10001

Spon Press is an imprint of the Taylor & Francis Group

© 1986, 1994 A. Ghali and R. Favre
© 2002 A. Ghali, R. Favre and M. Elbadry

The right of A. Ghali, R Favre and M. Elbadry to be identified as the
Authors of this Work has been asserted by them in accordance with
the Copyright, Designs and Patents Act 1988

Typeset by RefineCatch Limited, Bungay, Suffolk
Printed and bound in Great Britain by
TJ International Ltd, Padstow, Cornwall

British Library Cataloguing in Publication Data
A catalogue record for this book is available from the British Library

Library of Congress Cataloging in Publication Data
A catalog record for this book has been requested

ISBN 0–415–24721–7

Contents

Preface to the third edition

Concrete structures must have adequate safety factor against failure and must also exhibit satisfactory performance in service. This book is concerned with the checks on stresses and deformations that can be done in design to ensure satisfactory serviceability of reinforced concrete structures, with or without prestressing. The following are qualities which are essential for a satisfactory performance:

1   No excessive deflection should occur under the combined effect of prestressing, the self-weight of the structures and the superimposed dead load.
2   Deflections and crack width should not be excessive under the above mentioned loads combined with live and other transitory loads, settlement of support and temperature variations. This makes it necessary to control stress in the reinforcement, which is one of the main parameters affecting width of cracks. Durability of concrete structures is closely linked to the extent of cracking.

Because of creep and shrinkage of concrete and relaxation of prestressed reinforcement, the stresses in the concrete and in the reinforcement vary with time. In addition, when the structure is statically indeterminate the reactions and the internal forces are also time dependent. The strains and consequently the displacement change considerably with time due to the same effects and also due to cracking. The purpose of this text is to present the most effective methods for prediction of the true stresses and deformations during the life of the structure.

The mechanical properties that enter in calculation of stress and strain are the modulus of elasticity, creep and shrinkage of concrete and modulus of elasticity of reinforcements. These properties differ from project to project and from one country to another. The methods of analysis presented in the text allow the designer to account for the effects of variance in these parameters. Appendix A, based on the latest two European codes, British Standards and American Concrete Institute practice, gives guidance on the choice of values of these parameters for use in design. Appendix E, also based on the same sources, deals with crack width and crack spacing.

The methods of analysis of stresses and deformations presented in the chapters of the text are applicable in design of concrete structures regardless of codes. Thus, future code revisions as well as codes of other countries may be employed.

Some of the examples in the text are dimensionless. Some examples are worked out in the SI units and others in the so-called British units, customary to engineers in the USA; the input data and the main results are given in both SI and British Units. It is hoped that the use of both systems of units will make the text equally accessible to readers in all countries. Working out different examples in the two systems of units is considered more useful than the simpler task of working each example in both units.

In the second edition, a chapter discussing control of cracking was added. Four new chapters are added in the third edition. The new Chapter 6 explains how linear computer programs, routinely used by almost all structural engineers, can be employed for analysis of the time-dependent effects of creep, shrinkage and relaxation. Chapter 12 discusses the choice of amount and distribution of prestressed and non-prestressed reinforcements to achieve best serviceability. Fibre-reinforced polymer (FRP) bars and strands are sometimes used as reinforcement of concrete in lieu of steel. Chapter 14 is concerned with serviceability of concrete structures reinforced with these materials. The effect of cracking on the reactions and the internal forces of statically indeterminate reinforced concrete structures requires non-linear analysis discussed in Chapter 13.

The analysis procedures presented in the text can in part be executed using computer programs provided on www.sponpress.com/concretestructures, for use as an optional companion to this book. The new Appendix G describes the programs on the website and how they can be used.

Mr. S. Youakim, doctoral candidate, and Mr. R. Gayed, M.Sc. student, at the University of Calgary prepared the figures and checked the revisions in the third edition; Mrs. K. Knoll-Williams typed the new material. We are grateful to them as well as to those who have helped in the earlier editions.

A. Ghali
R. Favre
M. Elbadry
Calgary, Canada
Lausanne, Switzerland
January, 2002

Acknowledgements

This book was produced through the collaboration of A. Ghali with R. Favre and his research group, mainly during sabbatical leaves spent at the Swiss Federal Institute of Technology, Lausanne. For completion of the work on the first edition, A. Ghali was granted a Killam Resident Fellowship at the University of Calgary for which he is very grateful.

The authors would like to thank those who helped in the preparation of the first edition of the book. In Lausanne, Dr M. Koprna, Research Associate, reviewed parts of the text and collaborated in writing Chapter 8 and Appendix A; Mr J. Trevino, Research Assistant, made a considerable contribution by providing solutions or checking the numerical examples and preparing the manuscript for the publisher; Mr B.-F. Gardel prepared the figures. In Calgary, Mr M. Elbadry and Mr A. Mokhtar, graduate students, checked parts of the text, Mr B. Unterberger prepared by computer the graphs of Appendix F; Miss C. Larkin produced an excellent typescript.

The authors deeply appreciate the work of Dr S. El-Gabalawy of the Department of English at the University of Calgary, who revised the manuscript.

Figures A.1 and A.2 are reproduced with permission of BSI under licence number 2001SK/0331. Complete standards can be obtained from BSI Customer Services, 389 Chiswick High Road, London W4 4AL (tel: 020 8996 9001).

Note

It has been assumed that the design and assessment of structures are entrusted to experienced civil engineers, and that calculations are carried out under the direction of appropriately experienced and qualified supervisors. Users of this book are expected to draw upon other works on the subject including national and international codes of practice, and are expected to verify the appropriateness and content of information they draw from this book.

The SI system of units and British equivalents

Length
metre (m)

$1\,\text{m} = 39.37\,\text{in}$
$1\,\text{m} = 3.281\,\text{ft}$

Area
square metre (m²)

$1\,\text{m}^2 = 1550\,\text{in}^2$
$1\,\text{m}^2 = 10.76\,\text{ft}^2$

Volume
cubic metre (m³)

$1\,\text{m}^3 = 35.32\,\text{ft}^3$

Moment of inertia
metre to the power four (m⁴)

$1\,\text{m}^4 = 2403 \times 10^3\,\text{in}^4$

Force
newton (N)

$1\,\text{N} = 0.2248\,\text{lb}$

Load intensity
newton per metre (N/m)
newton per square metre (N/m²)

$1\,\text{N/m} = 0.06852\,\text{lb/ft}$
$1\,\text{N/m}^2 = 20.88 \times 10^{-3}\,\text{lb/ft}^2$

Moment
newton metre (N-m)

$1\,\text{N-m} = 8.851\,\text{lb-in}$
$1\,\text{N-m} = 0.7376 \times 10^{-3}\,\text{kip-ft}$
$1\,\text{kN-m} = 8.851\,\text{kip-in}$

Stress
newton per square metre (pascal)

$1\,\text{Pa} = 145.0 \times 10^{-6}\,\text{lb/in}^2$
$1\,\text{MPa} = 0.1450\,\text{ksi}$

Curvature
(metre)$^{-1}$

$1\,\text{m}^{-1} = 0.0254\,\text{in}^{-1}$

Temperature change

degree Celsius (°C) $1\,°C = (5/9)\,°Fahrenheit$

Energy and power

joule (J) = 1 N-m $1\,J = 0.7376\,lb\text{-}ft$

watt (W) = 1 J/s $1\,W = 0.7376\,lb\text{-}ft/s$

$1\,W = 3.416\,Btu/h$

Nomenclature for decimal multiples in the SI system

10^9 giga (G)

10^6 mega (M)

10^3 kilo (k)

10^{-3} milli (m)

Notation

The following is a list of symbols which are common in various chapters of the book. All symbols are defined in the text when they first appear and again when they are used in equations which are expected to be frequently applied. The sign convention adopted throughout the text is also indicated where applicable.

A	Cross-sectional area
$\{A\}$	Vector of actions (internal forces or reactions)
$\bar{A}$, $\bar{B}$ and $\bar{I}$	Area, first moment of area and moment of inertia of the age-adjusted transformed section, composed of area of concrete plus $\bar{\alpha}$ times area of reinforcement
B	First moment of area. For $\bar{B}$, see $\bar{A}$
b	Breadth of a rectangular section, or width of the flange of a T-section
c	Depth of compression zone in a fully cracked section
D	Displacement
d	Distance between extreme compressive fibre to the bottom reinforcement layer
E	Modulus of elasticity
$\bar{E}_c$	$= E_c(t_0)/[1 + \chi\varphi(t, t_0)] =$ age-adjusted elasticity modulus of concrete
e	Eccentricity
F	Force
f	Stress related to strength of concrete or steel
$[f]$	Flexibility matrix
f_{ct}	Tensile strength of concrete
h	Height of a cross-section
I	Moment of inertia. For $\bar{I}$, see $\bar{A}$
i, j, m, n	Integers
l	Length of a member
M	Bending moment. In a horizontal beam, a positive moment produces tension at the bottom fibre

M_r and/or N_r	Values of the bending moment and/or the axial force which are just sufficient to produce cracking
N	Normal force, positive when tensile
P	Force
r	Radius of gyration
$r(t, t_0)$	Relaxation function = concrete stress at time t due to a unit strain imposed at time t_0 and sustained to time t
$[S]$	Stiffness matrix
s_r	Spacing between cracks
T	Temperature
t	Time or age (generally in days)
W	Section modulus (length^3)
y	Coordinate defining location of a fibre or a reinforcement layer; y is measured in the downward direction from a specified reference point
α	$= E_s/E_c(t_0)$ = ratio of elasticity modulus of steel to elasticity modulus of concrete at age t_0
$\bar{\alpha}$	$= \alpha[1 + \chi\varphi(t, t_0)] = E_s/\bar{E}_c$ = ratio of elasticity modulus of steel to the age-adjusted elasticity modulus of concrete
α_t	Coefficient of thermal expansion (degree^{-1})
ε	Normal strain, positive for elongation
ζ	Coefficient of interpolation between strain, curvature and deflection values for non-cracked and fully cracked conditions (states 1 and 2, respectively)
η	Dimensionless multiplier for calculation of time-dependent change in axial strain
κ	Dimensionless multiplier for calculation time-dependent change of curvature
ν	Poisson's ratio
ξ	Dimensionless shape function
ρ, ρ'	Ratio of tension and of compression reinforcement to the area (bd); $\rho = A_s/bd$; $\rho' = A'_s/bd$
σ	Normal stress, positive when tensile
τ	Instant of time
$\varphi(t, t_0)$	Creep coefficient of concrete = ratio of creep to the instantaneous strain due to a stress applied at time t_0 and sustained to time t
$\chi(t, t_0)$	Aging coefficient of concrete (generally between 0.6 and 0.9; see Section 1.7 and Figs A.6–45)
$\chi\varphi(t, t_0)$	$= \chi(t, t_0)\,\varphi(t, t_0)$ = aging coefficient × creep coefficient
χ_r	Relaxation reduction coefficient for prestressed steel
ψ	Curvature (length^{-1}). Positive curvature corresponds to positive bending moment
$\{\ \}$	Braces indicate a vector; i.e. a matrix of one column

[ ]	A rectangular or a square matrix
→, ⌒	Single-headed arrows indicate a displacement (translation or rotation) or a force (a concentrated load or a couple)
→→	Double-headed arrow indicates a couple or a rotation; its direction is that of the rotation of a right-hand screw progressing in the direction of the arrow

Subscripts

c	Concrete
cs	Shrinkage
m	Mean
ns	Non-prestressed steel
O	Reference point
0	Initial or instantaneous
pr	Relaxation in prestressed steel
ps	Prestressed steel
s	Steel
st	Total steel, prestressed and non-prestressed
u	Unit force effect, unit displacement effect
φ	Creep effect
1,2	Uncracked or cracked state

Chapter 1

Creep and shrinkage of concrete and relaxation of steel

The 'Saddledome', Olympic Ice Stadium, Calgary, Canada. (Courtesy Genestar Structures Ltd. and J. Bobrowski and Partners Ltd.)

1.1 Introduction

The stress and strain in a reinforced or prestressed concrete structure are subject to change for a long period of time, during which creep and shrinkage of concrete and relaxation of the steel used for prestressing develop gradually. For analysis of the time-dependent stresses and deformations, it is necessary to employ time functions for strain or stress in the materials involved. In this chapter the basic equations necessary for the analysis are presented. The important parameters that affect the stresses or the strains are included in the equations, but it is beyond the scope of this book to examine how these parameters vary with the variations of the material properties.

The modulus of elasticity of concrete increases with its age. A stress applied on concrete produces instantaneous strain; if the stress is sustained the strain will progressively increase with time due to creep. Thus, the magnitude of the instantaneous strain and creep depends upon the age of concrete at loading and the length of the period after loading. Other parameters affecting the magnitude of creep as well as shrinkage are related to the quality of concrete and the environment in which it is kept. Creep and shrinkage are also affected by the shape of the concrete member considered.

Steel subjected to stress higher than 50 per cent of its strength exhibits some creep. In practice, steel used for prestressing may be subjected in service conditions to a stress 0.5 to 0.8 its strength. If a tendon is stretched between two fixed points, constant strain is sustained but the stress will decrease progressively due to creep. This relaxation in tension is of concern in calculation of the time-dependent prestress loss and the associated deformations of prestressed concrete members.

Several equations are available to express the modulus of elasticity of concrete, creep, shrinkage and relaxation of steel as functions of time. Examples of such expressions that are considered most convenient for practical applications are given in Appendix A. However, the equations and the procedures of analysis presented in the chapters of this book do not depend upon the choice of these time functions.

In this chapter the effect of cracking is not included. Combining the effects of creep, shrinkage and relaxation of steel with the effect of cracking on the deformations of concrete structures will be discussed in Chapters 7, 8 9 and 13.

1.2 Creep of concrete

A typical stress–strain curve for concrete is shown in Fig. 1.1. It is common practice to assume that the stress in concrete is proportional to strain in service conditions. The strain occurring during the application of the stress (or within seconds thereafter) is referred to as the instantaneous strain and is expressed as follows:

$$\varepsilon_c(t_0) = \frac{\sigma_c(t_0)}{E_c(t_0)} \qquad (1.1)$$

where $\sigma_c(t_0)$ is the concrete stress and $E_c(t_0)$ is the modulus of elasticity of concrete at age t_0, the time of application of the stress. The value of E_c, the secant modulus defined in Fig. 1.1, depends upon the magnitude of the stress, but this dependence is ignored in practical applications. The value E_c is generally assumed to be proportional to the square or cubic root of concrete

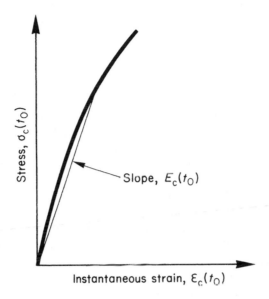

Figure 1.1 Stress–strain curve for concrete. $E_c(t_0)$ = secant modulus of elasticity; t_0 = age of concrete at loading.

strength, which depends on the age of concrete at loading.[1] Expressions for E_c in terms of the strength and age of concrete are given in Appendix A.

Under sustained stress, the strain increases with time due to creep and the total strain – instantaneous plus creep – at time t (see Fig. 1.2) is

$$\varepsilon_c(t) = \frac{\sigma_c(t_0)}{E_c(t_0)} [1 + \varphi(t, t_0)] \qquad (1.2)$$

where $\varphi(t, t_0)$ is a dimensionless coefficient, and is a function of the age at loading, t_0 and the age t for which the strain is calculated. The coefficient φ represents the ratio of creep to the instantaneous strain; its value increases with the decrease of age at loading t_0 and the increase of the length of the period $(t - t_0)$ during which the stress is sustained. When, for example, t_0 is one month and t infinity, the creep coefficient may be between 2 and 4 depending on the quality of concrete, the ambient temperature and humidity as well as the dimensions of the element considered.[2] Appendix A gives expressions and graphs for the creep coefficient according to MC-90, ACI Committee 209 and British Standard BS 8110.[3]

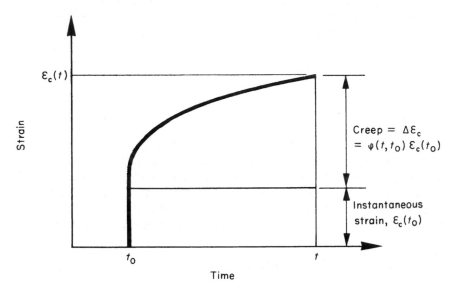

Figure 1.2 Creep of concrete under the effect of sustained stress.

1.3   Shrinkage of concrete

Drying of concrete in air results in shrinkage, while concrete kept under water swells. When the change in volume by shrinkage or by swelling is restrained, stresses develop. In reinforced concrete structures, the restraint may be caused by the reinforcing steel, by the supports or by the difference in volume change of various parts of the structure. We are concerned here with the stresses caused by shrinkage, which is generally larger in absolute value than swelling and occurs more frequently. However, there is no difference in the treatment except in the sign of the term representing the amount of volume change. The symbol ε_{cs} will be used for the free (unrestrained) strain due to shrinkage or swelling. In order to comply with the sign convention for other causes of strain, ε_{cs} is considered positive when it represents elongation. Thus shrinkage of concrete, ε_{cs} is a negative quantity.

Stresses caused by shrinkage are generally reduced by the effect of creep of concrete. Thus the effects of these two simultaneous phenomena must be considered in stress analysis. For this purpose, the amount of free shrinkage and an expression for its variation with time are needed. Shrinkage starts to develop at time t_s when moist curing stops. The strain that develops due to free shrinkage between t_s and a later instant t may be expressed as follows:

$$\varepsilon_{cs}(t, t_s) = \varepsilon_{cs0}\, \beta_s(t - t_s) \tag{1.3}$$

where ε_{cs0} is the total shrinkage that occurs after concrete hardening up to time infinity. The value of ε_{cs0} depends upon the quality of concrete and the ambient air humidity. The function $\beta_s(t - t_s)$ adopted by MC-90 depends upon the size and shape of the element considered (see Appendix A).

The free shrinkage, $\varepsilon_{cs}(t_2, t_1)$ occurring between any two instants t_1 and t_2 can be determined as the difference between the two values obtained by Equation (1.3), substituting t_2 and t_1 for t.

1.4   Relaxation of prestressed steel

The effect of creep on prestressing steel is commonly evaluated by a relaxation test in which a tendon is stretched and maintained at a constant length and temperature and the loss in tension is measured over a long period. The relaxation under constant strain as in a constant-length test is referred to as intrinsic relaxation, $\Delta\sigma_{pr}$. An equation widely used in the US and Canada for the intrinsic relaxation at any time τ of stress-relieved wires or strands is:[4]

$$\frac{\Delta\sigma_{pi}}{\sigma_{p0}} = -\frac{\log(\tau - t_0)}{10}\left(\frac{\sigma_{p0}}{f_{py}} - 0.55\right) \tag{1 4}$$

where f_{py} is the 'yield' stress, defined as the stress at a strain of 0.01. The ratio f_{py} to the characteristic tensile stress f_{ptk} varies between 0.8 and 0.90, with the lower value for prestressing bars and the higher value for low-relaxation strands $((\tau - t_0)$ is the period in hours for which the tendon is stretched).

The amount of intrinsic relaxation depends on the quality of steel. The MC-90[5] refers to three classes of relaxation and represents the relaxation as a fraction of the initial stress σ_{p0}. Steels of the first class include cold-drawn wires and strands, the second class includes quenched and tempered wires and cold-drawn wires and strands which are treated (stabilized) to achieve low relaxation. The third class, of intermediate relaxation, is for bars.

For a given steel and duration of relaxation test, the intrinsic relaxation increases quickly as the initial stress in steel approaches its strength. In the absence of reliable relaxation tests, MC-90 suggests the intrinsic relaxation values shown in Fig. 1.3 for duration of 1000 hours and assumes that the relaxation after 50 years and more is three times these values.

The Eurocode 2-91[6] (EC2–91) allows use of relaxation values differing slightly from MC-90. The values of EC2–91 are given between brackets in the graphs of Fig. 1.3.

The following equation may be employed to give the ratio of the ultimate intrinsic relaxation to the initial stress:

$$\frac{\Delta\sigma_{pr\infty}}{\sigma_{p0}} = -\eta(\lambda - 0.4)^2 \tag{1.5}$$

where

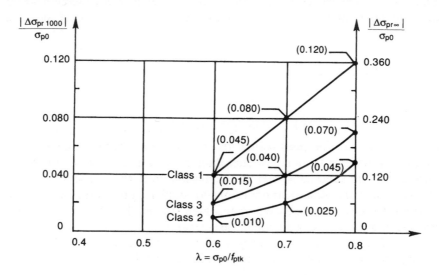

Figure 1.3 Intrinsic relaxation of prestressing steel according to MC-90. The symbols $|\Delta\sigma_{pr\ 1000}|$ and $|\Delta\sigma_{pr\infty}|$ represent respectively absolute values of intrinsic relaxation after 1000 hours and after 50 years or more. σ_{p0} = initial stress; f_{ptk} = characteristic tensile strength. The values indicated between brackets are for 1000 hours relaxation according to EC2–91.

$$\lambda = \frac{\sigma_{p0}}{f_{ptk}} \qquad (1.6)$$

$\Delta\sigma_{pr\infty}$ is the value of intrinsic relaxation of stress in prestressed steel at infinity. The symbol Δ is used throughout this book to indicate an increment. The relaxation represents a reduction in tension; hence it is a negative quantity. σ_{p0} is the initial stress in prestressed steel, f_{ptk} the characteristic tensile strength, and η the dimensionless coefficient depending on the quality of prestressed steel.

Equation (1.5) is applicable only when $\lambda \geq 0.4$; below this value, the intrinsic relaxation is negligible.

When the value of the ultimate intrinsic relaxation is known for a particular initial stress, Equation (1.5) can be solved for the value of η. Subsequent use of the same equation gives the variation of relaxation with the change of σ_{p0}.

Intrinsic relaxation tests are often reported for time equals 1000 h. However, for analysis of the effects of relaxation of steel on stresses and deformations in prestressed concrete structures, it is often necessary to employ expressions that give development of the intrinsic relaxation with time. Such expressions are included in Appendix B.

Relaxation increases rapidly with temperature. The values suggested in

Fig. 1.3 are for normal temperatures (20 °C). With higher temperatures, caused, for example, by steam curing, larger relaxation loss is to be expected.

1.5 Reduced relaxation

The magnitude of the intrinsic relaxation is heavily dependent on the value of the initial stress. Compare two tendons with the same initial stress, one in a constant-length relaxation test and the other in a prestressed concrete member. The force in the latter tendon decreases more rapidly because of the effects of shrinkage and creep. The reduction in tension caused by these two factors has the same effect on the relaxation as if the initial stress were smaller. Thus the relaxation value to be used in prediction of the loss of prestress in a concrete structure should be smaller than the intrinsic relaxation obtained from a constant-length test.

The reduced relaxation value to be used in the calculation of loss of prestress in concrete structures can be expressed as follows:

$$\Delta\sigma_{pr} = \chi_r \Delta\sigma_{pr}$$ (1.7)

where $\Delta\sigma_{pr}$ is the intrinsic relaxation as would occur in a constant-length relaxation test; χ_r is a dimensionless coefficient smaller than unity. The value of χ_r can be obtained from Table 1.1 or Fig. 1.4. The graph gives the value of χ_r as a function of λ, the ratio of the initial tensile stress to the characteristic tensile strength of the prestress steel (Equation (1.6)), and

$$\Omega = -\left(\frac{\Delta\sigma_{ps} - \Delta\sigma_{pr}}{\sigma_{p0}}\right)$$ (1.8)

where σ_{p0} is the initial tensile stress in prestress steel, $\Delta\sigma_{ps}$ is the change in stress in the prestressed steel due to the combined effect of creep, shrinkage

Table 1.1 Relaxation reduction coefficient χ_r

Ω	λ					
	0.55	0.60	0.65	0.70	0.75	0.80
0.0	1.000	1.000	1.000	1.000	1.000	1.000
0.1	0.6492	0.6978	0.7282	0.7490	0.7642	0.7757
0.2	0.4168	0.4820	0.5259	0.5573	0.5806	0.5987
0.3	0.2824	0.3393	0.3832	0.4166	0.4425	0.4630
0.4	0.2118	0.2546	0.2897	0.3188	0.3429	0.3627
0.5	0.1694	0.2037	0.2318	0.2551	0.2748	0.2917

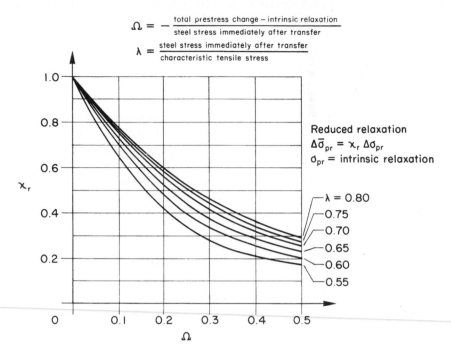

$$\Omega = -\frac{\text{total prestress change} - \text{intrinsic relaxation}}{\text{steel stress immediately after transfer}}$$

$$\lambda = \frac{\text{steel stress immediately after transfer}}{\text{characteristic tensile stress}}$$

Reduced relaxation
$$\Delta\bar{\sigma}_{pr} = \chi_r\,\Delta\sigma_{pr}$$
σ_{pr} = intrinsic relaxation

$\lambda = 0.80$
0.75
0.70
0.65
0.60
0.55

Figure 1.4 Relaxation reduction coefficient χ_r.

and relaxation, and $\Delta\sigma_{pr}$ is the intrinsic relaxation as would occur in a constant-length relaxation test.

The value of the total loss is generally not known *a priori*, because it depends upon the reduced relaxation. Iteration is here required: the total loss is calculated using an estimated value of the reduction factor χ_r (for example 0.7) which is later adjusted if necessary (see Example 3.1).

Appendix B gives the derivation of the relaxation reduction coefficient values in Table 1.1 and the graphs in Fig. 1.4. The values given in the table and the graphs may be approximated by Equation (B.11).

1.6   Creep superposition

Equation (1.2) implies the assumption that the total strain, instantaneous plus creep is proportional to the applied stress. This linear relationship, which is generally true within the range of stresses in service conditions, allows superposition of the strain due to stress increments or decrements and due to shrinkage. Thus, when the magnitude of the applied stress changes with time, the total strain of concrete due to the applied stress and shrinkage is given by (Fig. 1.5):

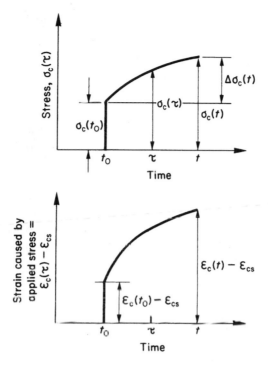

Figure 1.5 Stress versus time and strain versus time for a concrete member subjected to uniaxial stress of magnitude varying with time.

$$\varepsilon_c(t) = \sigma_c(t_0)\frac{1 + \varphi(t, t_0)}{E_c(t_0)} + \int_0^{\Delta\sigma_c(t)}\frac{1 + \varphi(t, \tau)}{E_c(\tau)}\,d\sigma_c(\tau) + \varepsilon_{cs}(t, t_0) \qquad (1.9)$$

where

t_0 and t = ages of concrete when the initial stress is applied and when the strain is considered

τ = an intermediate age between t_0 and t

$\sigma_c(t_0)$ = initial stress applied at age t_0

$d\sigma_c(\tau)$ = an elemental stress (increment or decrement) applied at age τ

$E_c(\tau)$ = modulus of elasticity of concrete at the age τ

$\varphi(t, \tau)$ = coefficient of creep at time t for loading at age τ

$\varepsilon_{cs}(t, t_0)$ = free shrinkage occurring between the ages t_0 and t

Equation (1.9) implies the assumption that a unit stress increment or decrement introduced at the same age and maintained for the same time produces the same absolute value of creep. This equation is the basis of the

methods presented in this book for analysis of the time-dependent stresses
and deformations of concrete structures.

1.7   The aging coefficient χ: definition

The integral in Equation (1.9) represents the instantaneous strain plus creep
due to an increment in concrete stress of magnitude $\Delta\sigma_c$ (Fig. 1.5). This
increment is gradually introduced during the period t_0 to t. A stress introduced gradually in this manner produces creep of smaller magnitude compared to a stress of the same magnitude applied at age t_0 and sustained during
the period $(t - t_0)$. In the following equation the stress increment $\Delta\sigma_c(t)$ is
treated as if it were introduced with its full magnitude at age t_0 and sustained
to age t but the creep coefficient $\varphi(t, t_0)$ is replaced by a reduced value
which equals $\chi\varphi(t, t_0)$ where $\chi = \chi(t, t_0)$ is a dimensionless multiplier (smaller
than 1) which is referred to as the *aging coefficient*. With this important
simplification, the integral in Equation (1.9) can be eliminated as follows:

$$\varepsilon_c(t) = \sigma_c(t_0) \frac{1 + \varphi(t, t_0)}{E_c(t_0)} + \Delta\sigma_c(t) \frac{1 + \chi\varphi(t, t_0)}{E_c(t_0)} + \varepsilon_{cs}(t, t_0) \tag{1.10}$$

Equation (1.10) gives the strain which occurs during a period t_0 to t due to
the combined effect of free shrinkage and a stress which varies in magnitude
during the same period. The first term on the right-hand side of Equation
(1.10) is the instantaneous strain plus creep due to a stress of magnitude $\sigma_c(t_0)$
introduced at time t_0 and sustained without change in magnitude until time t.
The second term is the instantaneous strain plus creep due to a stress increment (or decrement) of a magnitude changing gradually from zero at t_0 to a
value $\Delta\sigma_c(t)$ at time t. The last term is simply the free shrinkage occurring
during the considered period.

For a practical example in which the stress on concrete varies with time as
described above, consider a prestressed concrete cross-section. At time t_0 the
prestressing is introduced causing compression on the concrete which gradually changes with time due to the losses caused by the combined effects of
creep, shrinkage and relaxation of the prestressed steel.

Use of the aging coefficient χ as in Equation (1.10) greatly simplifies the
analysis of strain caused by a gradually introduced stress increment $\Delta\sigma_c$; or
inversely, the magnitude of the stress increment $\Delta\sigma_c$ can be expressed in terms
of the strain it produces. The aging coefficient will be extensively used in this
text for the analysis of the time-dependent stresses and strains in prestressed
and reinforced concrete members.

In practical computations, the aging coefficient can be taken from a table
or a graph (see Appendix A), or simply assumed; its value generally varies
between 0.6 and 0.9. The method of calculating the aging coefficient will be
discussed in Section 1.8; but this may not be of prime concern in practical

design. However it is important that the reader understands at this stage the meaning of the aging coefficient and how it is used in Equation (1.10).

1.8   Equation for the aging coefficient χ

The stress variation between t_0 and t (Fig. 1.5) may be expressed as

$$\xi_1 = \frac{\sigma_c(\tau) - \sigma_c(t_0)}{\Delta\sigma_c(t)} \tag{1.11}$$

where ξ_1 is a dimensionless time function defining the shape of the stress–time curve; the value of ξ_1 at any time τ is equal to the ratio of the stress change between t_0 and τ to the total change during the period $(t - t_0)$. The value of the shape function ξ_1 varies between 0 and 1 as τ changes from t_0 to t.

Differentiation of Equation (1.11) with respect to time gives

$$\frac{d\sigma_c(\tau)}{d\tau} = \Delta\sigma_c(t) \frac{d\xi_1}{d\tau} \tag{1.12}$$

Substitution of Equation (1.12) into (1.9) gives

$$\varepsilon_c(t) = \sigma_c(t_0) \frac{1 + \varphi(t, t_0)}{E_c(t_0)} + \Delta\sigma_c(t) \int_{t_0}^{t} \frac{1 + \varphi(t, \tau)}{E_c(\tau)} \frac{d\xi_1}{d\tau} d\tau$$

$$+ \varepsilon_{cs}(t, t_0) \tag{1.13}$$

Comparison of Equation (1.13) with (1.10) gives the following expression for the aging coefficient:

$$\chi(t, t_0) = \frac{E_c(t_0)}{\varphi(t, t_0)} \int_{t_0}^{t} \frac{1 + \varphi(t, \tau)}{E_c(\tau)} \left(\frac{d\xi_1}{d\tau}\right) d\tau - \frac{1}{\varphi(t, t_0)} \tag{1.14}$$

Three functions of time are included in Equation (1.14): ξ_1, $E_c(\tau)$ and $\varphi(t, \tau)$; of which the last two depend upon the quality of concrete and the ambient air. Examples of expressions that can be used for the variables E_c and φ are given in Appendix A.

In practical applications the actual shape of variation of stress $\sigma_c(\tau)$ is often unknown and the function ξ_1 defining this shape must be assumed. In preparation of the graphs and the table for the aging coefficient χ presented in Appendix A, the time function ξ_1 is assumed to have the same shape as that of the time–relaxation curve for concrete which will be discussed in Section 1.9.

As mentioned in the preceding section, the aging coefficient χ is intended for use in the calculation of strain due to stress which varies with time as for example in the cross-section of a prestressed member made from one or more types of concrete (composite section). Shrinkage, creep and relaxation result in gradual change in stresses in the concrete and the steel. The use of precalculated values of the coefficient χ in the analysis of strain or stress, by Equation (1.10), in such examples implies an assumption of the shape of variation of the stress during the period $(t - t_0)$. The margin of error caused by this approximation is generally small.

We have seen from the equations of this section that χ and φ are functions of t_0 and t, the ages of the concrete at loading and at the time the strain is considered. The product $\chi\varphi$ often occurs in the equations of this book; to simplify the notation, we will use the symbol $\chi\varphi$ to mean:

$$\chi\varphi(t, t_0) \equiv \chi(t, t_0)\,\varphi(t, t_0)$$

1.9   Relaxation of concrete

In the discussion presented in this section, we exclude the effect of shrinkage and consider only the effect of creep.

When a concrete member is subjected at age t_0 to an imposed strain ε_c, the instantaneous stress will be

$$\sigma_c(t_0) = \varepsilon_c E_c(t_0) \tag{1.15}$$

where $E_c(t_0)$ is the modulus of elasticity of concrete at age t_0. If subsequently the length of the member is maintained constant, the strain ε_c will not change, but the stress will gradually decrease because of creep (Fig. 1.6). The value of stress at any time $t > t_0$ may be expressed as follows:

$$\sigma_c(t) = \varepsilon_c r(t, t_0) \tag{1.16}$$

where $r(t, t_0)$ is the relaxation function to be determined in the following section. The value $r(t, t_0)$ is the stress at age t due to a unit strain introduced at age t_0 and sustained constant during the period $(t - t_0)$.

At any instant τ between t_0 and t, the magnitude of the relaxed stress $\Delta\sigma_c(\tau)$ may be expressed as follows:

$$\Delta\sigma_c(\tau) = \xi[\Delta\sigma_c(t)] \tag{1.17}$$

where $\Delta\sigma_c(\tau)$ is the stress increment (the stress relaxed) during the period t_0 to τ:

$$\Delta\sigma_c(\tau) = \sigma_c(\tau) - \sigma_c(t_0) \tag{1.18}$$

Similarly, the stress increment during the period t_0 to t is

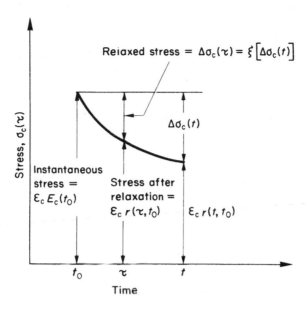

Figure 1.6 Variation of stress with time due to a strain ε_c, imposed at age t_0 and maintained constant thereafter (phenomenon of relaxation).

$$\Delta\sigma_c(t) = \sigma_c(t) - \sigma_c(t_0) \tag{1.19}$$

The symbol ξ is a dimensionless shape function, representing for any value τ the ratio of the stress relaxed during the period $(\tau - t_0)$ to the stress relaxed during the whole period $(t - t_0)$. Thus

$$\xi = \frac{\Delta\sigma_c(\tau)}{\Delta\sigma_c(t)} \tag{1.20}$$

The value of ξ is 0 and 1 when $\tau = t_0$ and t, respectively. The shape function ξ has the same significance as ξ_1 adopted in the preceding section (Equation (1.11)).

Referring to Fig. 1.6, the strain value ε_c which exists at time t may be considered as being the result of: (a) an initial stress $\sigma_c(t_0)$ introduced at age t_0 and maintained constant up to age t; and (b) a stress increment $\Delta\sigma_c(t)$ introduced gradually during the period $(t - t_0)$. Thus, using Equation (1.10):

$$\varepsilon_c = \sigma_c(t_0)\frac{1 + \varphi\,(t, t_0)}{E_c(t_0)} + \Delta\sigma_c(t)\frac{1 + \chi\varphi(t, t_0)}{E_c(t_0)} \tag{1.21}$$

Substitution of Equations (1.15), (1.16) and (1.19) and (1.21) gives

$$\varepsilon_c = \varepsilon_c[1 + \varphi(t, t_0)] + \varepsilon_c[r(t, t_0) - E_c(t_0)]\frac{1 + \chi\varphi(t, t_0)}{E_c(t_0)} \tag{1.22}$$

We recall that the symbol $\chi\varphi(t, t_0)$ indicates the product of two functions χ and φ of the time variables t and t_0. The constant strain value ε_c in Equation (1.22) cancels out and by algebraic manipulation of the remaining terms we can express the aging coefficient χ in terms of $E_c(t_0)$, $r(t, t_0)$ and $\varphi(t, t_0)$:

$$\chi(t, t_0) = \frac{1}{1 - r(t, t_0)/E_c(t_0)} - \frac{1}{\varphi(t, t_0)} \tag{1.23}$$

A step-by-step numerical procedure will be discussed in the following section for the derivation of the relaxation curve in Fig. 1.6. The relaxation function $r(t, t_0)$ obtained in this way can be used to calculate the aging coefficient $\chi(t, t_0)$ by Equation (1.23).

1.10 Step-by-step calculation of the relaxation function for concrete

The step-by-step numerical procedure introduced in this section can be used for the calculation of the time-dependent stresses and deformations in concrete structures. It is intended for computer use and is particularly suitable for structures built or loaded in several stages, as for example in the segmental construction method of prestressed structures. In this section, a step-by-step method will be used to derive the relaxation function $r(\tau, t_0)$. Further development of the method is deferred to Sections 4.6 and 5.8.

The value of the relaxation function $r(t, t_0)$ is defined as the stress at time t due to a unit strain introduced at time t_0 and sustained without change during the period $(t - t_0)$ (see Equation (1.16)).

Consider a concrete member subjected to uniaxial stress and assume that the magnitude of stress varies with time as shown in Fig. 1.7(b). At age t_0 an initial stress value $\sigma_c(t_0)$ is introduced and subsequently increased gradually or step-wise during the period t_0 to t. When the variation of stress with time is known, the step-by-step analysis to be described can be used to find the strain at any time τ between t_0 and t. Alternatively, if the strain is known, the method can be used to determine the time variation of stress.

Divide the period $(t - t_0)$ into intervals (Fig. 1.7(a)) and assume that the stress is introduced in increments at the middle of the intervals. Thus, $(\Delta\sigma_c)_i$ is introduced at the middle of the ith interval. For a sudden increase in stress, consider an increment introduced at an interval of zero length (for example $(\Delta\sigma_c)_1$ and $(\Delta\sigma_c)_k$ in Fig. 1.7(b)). The symbols $t_{j-\frac{1}{2}}$, t_j and $t_{j+\frac{1}{2}}$ are used to refer to the instant (or the age of concrete) at the start, the middle and the end of

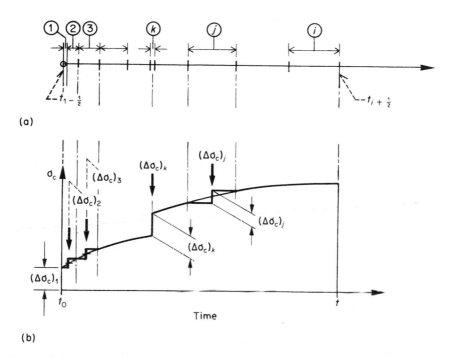

Figure 1.7 Division of: (a) time into intervals and (b) stress into increments for step-by-step analysis.

the jth interval, respectively. The strain at the end of the ith interval can be calculated by Equation (1.9) replacing the first two terms by a summation as follows:

$$\varepsilon_c(t_{i+\frac{1}{2}}) = \sum_{j=1}^{i}\left((\Delta\sigma_c)_j\frac{1+\varphi(t_{i+\frac{1}{2}},\,t_j)}{E_c(t_j)}\right) + \varepsilon_{cs}(t_{i+\frac{1}{2}},\,t_0). \qquad (1.24)$$

The summation represents the superposition of strain caused by stress increments. When the magnitude of the increments is known, the sum gives the strain. In the case when the strain is known, the stress increments can be determined in steps. The stress at the end of the ith interval is

$$\sigma_c(t_{i+\frac{1}{2}}) = \sum_{j=1}^{i}(\Delta\sigma_c)_j. \qquad (1.25)$$

Consider now the case when a strain ε_c is imposed at the time t_0 and sustained constant up to time t. The corresponding stress introduced at t_0 is

$\varepsilon_c E_c(t_0)$ and its value will gradually drop following the relaxation function according to Equation (1.16). Assume that the time after t_0 is divided into intervals as in Fig. 1.7(a) and apply Equation (1.16) at the end of the ith interval:

$$\sigma_c(t_{i+\frac{1}{2}}) = \varepsilon_c r(t_{i+\frac{1}{2}}, t_0). \tag{1.26}$$

Substitution of Equation (1.25) into (1.26) gives the value of the relaxation function at the end of the ith interval:

$$r(t_{i+\frac{1}{2}}, t_0) = \frac{1}{\varepsilon_c} \sum_{j=1}^{i} (\Delta\sigma_c)_j \tag{1.27}$$

Rewrite Equation (1.24) separating the last term of the summation:

$$\varepsilon_c(t_i + \tfrac{1}{2}) = (\Delta\sigma_c)_i \frac{1 + \varphi(t_{i+\frac{1}{2}}, t_i)}{E_c(t_i)} + \sum_{j=1}^{i-1}\left((\Delta\sigma_c)_j \frac{1 + \varphi(t_{i+\frac{1}{2}}, t_j)}{E_c(t_j)}\right)$$

$$+ \varepsilon_{cs}(t_{i+\frac{1}{2}}, t_0). \tag{1.28}$$

Consider that the strain $\varepsilon_c(t_{i+\frac{1}{2}})$ is known at the end of all intervals and it is required to find the stress increments. Values of the modulus of elasticity of concrete, creep coefficients and free shrinkage are also assumed to be known for all intervals as needed in Equation (1.28). In the step-by-step analysis, the stress increment for any interval is determined after the increments of all the preceding intervals have been determined. Thus, Equation (1.28) can be solved for the only unknown stress increment $(\Delta\sigma_c)_i$:

$$(\Delta\sigma_c)_i = \frac{E_c(t_i)}{1 + \varphi(t_{i+\frac{1}{2}}, t_i)}\left[\varepsilon_c(t_{i+\frac{1}{2}}) - \varepsilon_{cs}(t_{i+\frac{1}{2}}, t_0)\right.$$

$$\left. - \sum_{j=1}^{i-1}\left((\Delta\sigma_c)_j \frac{1 + \varphi(t_{i+\frac{1}{2}}, t_j)}{E_c(t_j)}\right)\right] \tag{1.29}$$

Successive application of this equation with $i = 1, 2, \ldots$, gives the stress increments. Equations (1.29) and (1.27) can be employed in this manner to determine the relaxation function, $r(t, t_0)$. For this purpose, $\varepsilon_{cs}(t_{i+\frac{1}{2}}, t_0) = 0$ and $\varepsilon_c(t_{i+\frac{1}{2}}) = \varepsilon_c = $ constant for all i values; ε_c may be conveniently chosen equal to unity. This procedure is employed to calculate $r(t, t_0)$ which is subsequently substituted in Equation (1.23) to determine the aging coefficient $\chi(t, t_0)$ in preparation of the graphs in part (b) of each of Figs A.6 to A.45 and Table A.3 in Appendix A. The same appendix also includes an example plot by computer of the relaxation function (see Fig. A.5).

The aging coefficient $\chi(t, t_0)$ calculated by the above procedure depends mainly upon t_0 and t; other factors affecting χ are the time functions $\varphi(t, \tau)$ and $E_c(\tau)$. The graphs and table presented for χ in Appendix A are based on time functions for φ and E_c in accordance with MC-90 and the ACI Committee 209 report,[7] respectively. Choice of other functions results in small change in the value of χ; but this change may be ignored in practice. Since χ is always used as a multiplier to φ, which is rarely accurately determined, high accuracy in the derivation of χ is hardly justified. Appendix G includes information about computer programs that perform the step-by-step calculations discussed in this section. The programs can be executed on micro-computers using the software provided on the Internet as optional companion of this book (See web address in Appendix G.).

1.11   Age-adjusted elasticity modulus

The three terms in Equation (1.10) represent the strain in concrete at age t due to: a stress $\sigma_c(t_0)$ introduced at age t_0 and sustained during the period $(t - t_0)$, a stress increment of magnitude zero at t_0 increasing gradually to a final value $\Delta\sigma_c(t)$ at age t, and the free shrinkage occurring during the period $(t - t_0)$. This equation may be rewritten as follows:

$$\varepsilon_c(t) = \sigma_c(t_0)\frac{1 + \varphi(t, t_0)}{E_c(t_0)} + \frac{\Delta\sigma_c(t)}{\bar{E}_c(t, t_0)} + \varepsilon_{cs}(t, t_0) \tag{1.30}$$

where

$$\bar{E}_c(t, t_0) = \frac{E_c(t_0)}{1 + \chi\varphi(t, t_0)} \tag{1.31}$$

$\bar{E}_c(t, t_0)$ is the *age-adjusted elasticity modulus* to be used in the calculation of the total strain increment, instantaneous plus creep, due to a stress increment of magnitude developing gradually from zero to a value $\Delta\sigma_c(t)$. Thus, the strain increment in the period $(t - t_0)$ caused by the stress $\Delta\sigma_c(t)$ is given by:

$$\Delta\varepsilon_c(t) = \frac{\Delta\sigma_c(t)}{\bar{E}_c(t, t_0)}. \tag{1.32}$$

1.11.1   Transformed section

In various chapters of this book the term *transformed section* is employed to mean a cross-section of a reinforced concrete member for which the actual area is replaced by a transformed area equal to the area of concrete plus α times the area of steel; where

$$a(t_0) - \frac{E_s(,E_{ps} \text{ or } E_{ns})}{E_c(t_0)} \tag{1.33}$$

where E_s is the modulus of elasticity of the reinforcement. When prestressed or non-prestressed steel are involved, the subscripts ps or ns are employed to refer to the two types of reinforcement. $E_c(t_0)$ is the modulus of elasticity of concrete at age t_0. It thus follows that a is also a function of t_0.

In the analysis of stresses due to forces gradually developed during a period t_0 to t, we will use in Chapter 2 the term *age-adjusted transformed section* to mean a transformed section for which the actual area is replaced by a transformed area composed of the area of concrete plus $\bar{a}$ times the area of steel; where

$$\bar{a}(t, t_0) = \frac{E_s(,E_{ps} \text{ or } E_{ns})}{\bar{E}_c(t, t_0)}. \tag{1.34}$$

1.11.2   Age-adjusted flexibility and stiffness

Similarly, when the age-adjusted modulus of elasticity of concrete is used in the calculation of a flexibility or stiffness of a structure, the result is referred to as an *age-adjusted flexibility* or *age-adjusted stiffness*.

1.12   General

Creep and shrinkage of concrete and relaxation of steel result in deformations and in stresses that vary with time. This chapter presents the basic equations for two methods for the analysis of time-dependent stresses and deformations in reinforced and prestressed concrete structures. The first is suitable for hand computation and requires knowledge of an aging coefficient χ (generally between 0.6 and 0.9) which may be taken from a graph or a table (see Appendix A). The second is a step-by-step numerical procedure intended for computer use. In Chapters 2 to 9 the first method is extensively employed for the analysis of changes of stress and internal forces caused by creep, shrinkage and relaxation of steel in statically determinate and indeterminate structures. The second method, namely the step-by-step procedure, is employed for the same purpose in Sections 4.6 and 5.8. Appendix A gives equations and graphs for the material parameters discussed in this chapter, based upon requirements of codes and technical committee recommendations.

Notes

1 See Neville, A.M. (1997). *Properties of Concrete*, 4th edn. Wiley, New York.
2 See Neville, A.M., Dilger, W.H. and Brooks, J.J. (1983). *Creep of Plain and Structural Concrete*. Construction Press, London.

3 Comité Euro-International du Béton (CEB) – Fédération Internationale de la Précontrainte (FIP) (1990). *Model Code for Concrete Structures*. (MC-90), CEB. Thomas Telford, London, 1993. American Concrete Institute (ACI) Committee 209 (1992). *Prediction of Creep, Shrinkage and Temperature Effects in Concrete Structures*. 209R-92, ACI, Detroit, Michigan, 47 pp. British Standard BS 8110: Part 1: 1997 and Part 2: 1985, *Structural Use of Concrete*, British Standards Institute, 2 Park Street, London W1A 2BS. Part I is reproduced by Deco, 15210 Stagg Street, Van Nuys, Ca. 91405–1092, USA.

4 Based on Magura, D., Sozen, M.A. and Siess, C.P. (1964). A study of stress relaxation in prestressing reinforcement, *PCI Journal*, **9** (2) 13–57.

5 See reference mentioned in note 3 above.

6 Eurocode 2 (1991). *Design of Concrete Structures, Part 1: General Rules and Rules for Buildings*. European Prestandard, ENV 1992–1: 1991E, European Committee for Standardization, rue de Stassart 36, B-1050 Brussels, Belgium.

7 See reference mentioned in note 3 above.

Chapter 2

Stress and strain of uncracked sections

Pre-tensioned element of double tee cross-section at time of cutting of prestressed strands (Courtesy Prestressed Concrete Institute, Chicago).

2.1 Introduction

Cross-sections of concrete frames or beams are often composed of three types of material: concrete, prestressed steel and non-prestressed reinforcement. In some cases, concrete of more than one type is employed in one cross-section, as for example in T-sections where the web is precast and the flanges are cast *in situ*. Concrete exhibits the properties of creep and shrinkage and prestressed steel loses part of its tension due to relaxation. Thus, the components forming one section tend to have different strains. However, because of the bond, the difference in strain is restrained. Thus, the stresses in concrete and the two types of reinforcement change with time as creep, shrinkage and relaxation develop.

This chapter is concerned with the calculation of the time-dependent stresses and the associated strain and curvature in individual cross-sections of reinforced, prestressed or composite members. Cross-sections composed of concrete and structural steel sections are treated in the same way as reinforced concrete members, with the only difference that the steel section has a flexural rigidity which is not ignored.

The cross-sections considered are assumed to have one axis of symmetry and to be subjected to a bending and an axial force caused by prestressing or by other loading. Perfect bond is assumed between concrete and steel; thus, at any fibre the strains in concrete and steel are equal. Plane cross-sections are assumed to remain plane after deformation. No cracking is assumed in the analysis procedures presented in this chapter; analysis of cracked sections is treated in Chapter 7.

Prestressing is generally applied in one of two ways: pre-tensioning or post-tensioning. With pre-tensioning, a tendon is stretched in the form in which the concrete member is cast. After the concrete has attained sufficient strength, the tendon is cut. Because of bond with concrete, the tendon cannot regain its original length and thus a compressive force is transferred to the concrete, causing shortening of the member accompanied by an instant-aneous loss of a part of the prestress in the tendon. We here assume that the change in strain in steel that occurs during transfer is compatible with the concrete strain at the same fibre. The slip that usually occurs at the extremities of the member is ignored.

With post-tensioning, the tendon passes through a duct which is placed in the concrete before casting. After attaining a specified strength, tension is applied on the tendon and it is anchored to the concrete at the two ends and later the duct is grouted with cement mortar. During tensioning of the ten-don, before its anchorage, the strain in steel and concrete are not compatible; concrete shortens without causing instantaneous loss of the prestress force. After transfer, perfect bond is assumed between the tendon, the grout, the duct and the concrete outside the duct. This assumption is not justified when the tendon is left unbonded. However, in most practical calculations the incom-patibility in strain, which may develop after prestress transfer, between the strain in an unbonded tendon and the adjacent concrete is generally ignored.

In this chapter we are concerned with the stress, strain and deformations of a member for which the elongations or end rotation are not restrained by the supports or by continuity with other members. Creep, shrinkage and relaxa-tion of steel change the distribution of stress and strain in the section but do not change the reactions and the induced stress resultants (values of the axial force or bending moment acting on the section). Analysis of the time-dependent effects on continuous beams and other statically indeterminate structures are discussed in Chapters 4 and 5.

Creep and shrinkage of concrete and relaxation of prestressed steel result in prestress loss and thus in time-dependent change of the internal forces (the resultant of stresses) on the concrete cross-section. Generally, in a prestressed section, non-prestressed reinforcement is also present. The time effects of creep, shrinkage and relaxation usually produce a reduction of tension in the prestressed steel and of compression in the concrete and an increase of compression in the non-prestressed steel.

At the time of prestressing, or at a later date, external loads are often introduced, as for example the self-weight. The internal forces due to such loading and the time of their application are assumed to be known. The initial prestressing is assumed to be known but the changes in the stress in the prestressed and non-prestressed steels and the concrete are determined by the analysis.

2.2   Sign convention

The following sign convention will be adopted in all chapters. Axial force N is positive when tensile. In a horizontal beam, a bending moment M that produces tension at the bottom fibre and the corresponding curvature ψ are positive. Tensile stress σ and the corresponding strain ε are positive; thus the value of shrinkage of concrete, ε_{cs}, is generally a negative quantity. The symbol P indicates the absolute value of the prestress force. Δ represents an increment or decrement when positive or negative, respectively. Thus, the loss of tension in the prestressed steel due to creep, shrinkage and relaxation is generally a negative quantity.

2.3   Strain, stress and curvature in composite and homogeneous cross-sections

Fig. 2.1(a) is the cross-section of a member composed of different materials and having an axis of symmetry. For the analysis of stresses due to normal force or moment on the section, we replace the actual section by a *transformed section* for which the actual area of any part i is replaced by a transformed area given by $(E_i/E_{ref})A_i$ where E_{ref} is an arbitrarily chosen value of a reference modulus of elasticity; E_i is the modulus of elasticity of part i of the section. The member is thus considered to have a modulus of elasticity E_{ref} and cross-section properties, for example area and moment of inertia equal to those of the transformed section.

In reinforced and prestressed concrete cross-sections, the reference modulus is taken to be equal to E_c, the modulus of elasticity of concrete of one of the parts, and the reinforcement area, prestressed and non-prestressed, is replaced by a times the actual area; where a is the ratio of the modulus of elasticity of the reinforcement to the modulus of elasticity of concrete (see Equation (1.33)).

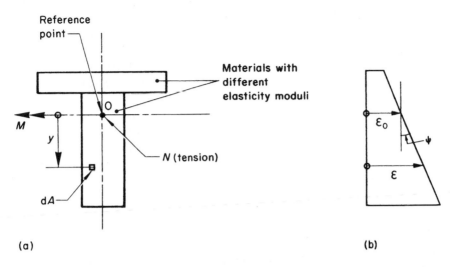

Figure 2.1 Analysis of strain distribution in a composite cross-section by Equation (2.15):
(a) positive M, N and y; (b) strain distribution.

Assume that the cross-section in Fig. 2.1(a) is subjected to a force N normal to the section situated at any point on the symmetry axis. Such a force is statically equivalent to a system composed of a normal force N at a reference point O and a bending moment M as shown in Fig. 2.1(a). The equations most commonly used in calculations of stress, strain and curvature at the cross-section are generally based on the assumption that O is the centroid of the transformed section.

When considering the effects of creep, we shall use for the analysis of the same cross-section different elasticity moduli for concrete and superpose the stresses from several analyses (see Section 2.5). Changing the value of E_c will result in a change of the centroid of the transformed section. To avoid this difficulty, we derive the equations below for the strain, curvature and the stress distribution of a cross-section without requiring that the reference point O be the centroid of the cross-section. Thus, O is an arbitrarily chosen reference point on the axis of symmetry.

The strain distribution is assumed to be linear as shown in Fig. 2.1(b); in other words, a plane cross-section is assumed to remain plane after deformation. At any fibre, at a distance y from the reference point O, the strain is:

$$\varepsilon = \varepsilon_0 + \psi y \qquad (2.1)$$

where ε_0 is the strain at the reference point and ψ is the curvature. The distance y is positive when the point considered is below the reference point.

When the fibre considered is in the ith part of the composite section, the stress at the fibre is

$$\sigma = E_i(\varepsilon_O + \psi y). \tag{2.2}$$

Integration of the stress over the area of the cross-section and taking the moment about an axis through O, gives

$$N = \int \sigma \mathrm{d}A \tag{2.3}$$

$$M = \int \sigma y \mathrm{d}A \tag{2.4}$$

The integral is to be performed for all parts of the cross-section.
Substitution of Equation (2.2) into (2.3) and (2.4) gives

$$N = \varepsilon_O \sum_{i=1}^{m} E_i \int \mathrm{d}A + \psi \sum_{i=1}^{m} E_i \int y \mathrm{d}A \tag{2.5}$$

$$M = \varepsilon_O \sum_{i=1}^{m} E_i \int y \, \mathrm{d}A + \psi \sum_{i=1}^{m} E_i \int y^2 \mathrm{d}A \tag{2.6}$$

Thus summations in Equations (2.5) and (2.6) are to be performed from $i = 1$ to m where m is the number of parts in the cross-section. Equations (2.5) and (2.6) may be rewritten

$$N = E_{\mathrm{ref}}(A\varepsilon_O + B\psi) \tag{2.7}$$

$$M = E_{\mathrm{ref}}(B\varepsilon_O + I\psi) \tag{2.8}$$

where A, B and I are the transformed cross-section area and its first and second moment about an axis through O.
For a composite section, A, B and I are derived by summing up the contribution of the parts:

$$A = \sum_{i=1}^{m} \left(\frac{E_i}{E_{\mathrm{ref}}} A_i \right) \tag{2.9}$$

$$B = \sum_{i=1}^{m} \left(\frac{E_i}{E_{\mathrm{ref}}} B_i \right) \tag{2.10}$$

$$I = \sum_{i=1}^{m} \left(\frac{E_i}{E_{\mathrm{ref}}} I_i \right) \tag{2.11}$$

where A_i, B_i and I_i are respectively the area of the ith part and its first and second moment about an axis through O. A reinforcement layer may be treated as one part.

Equations (2.7) and (2.8) may be rewritten in the matrix form

$$\begin{Bmatrix} N \\ M \end{Bmatrix} = E_{\text{ref}} \begin{bmatrix} A & B \\ B & I \end{bmatrix} \begin{Bmatrix} \varepsilon_O \\ \psi \end{Bmatrix} \tag{2.12}$$

This equation may be used to find N and M when ε_O and ψ are known; or when N and M are known the equation may be solved for the axial strain and curvature:

$$\begin{Bmatrix} \varepsilon_O \\ \psi \end{Bmatrix} = \frac{1}{E_{\text{ref}}} \begin{bmatrix} A & B \\ B & I \end{bmatrix}^{-1} \begin{Bmatrix} N \\ M \end{Bmatrix} \tag{2.13}$$

The inverse of the 2×2 matrix in this equation is

$$\begin{bmatrix} A & B \\ B & I \end{bmatrix}^{-1} = \frac{1}{(AI - B^2)} \begin{bmatrix} I & -B \\ -B & A \end{bmatrix} \tag{2.14}$$

Substitution in Equation (2.13) gives the axial strain at O and the curvature

$$\begin{Bmatrix} \varepsilon_O \\ \psi \end{Bmatrix} = \frac{1}{E_{\text{ref}}(AI - B^2)} \begin{bmatrix} I & -B \\ -B & A \end{bmatrix} \begin{Bmatrix} N \\ M \end{Bmatrix} \tag{2.15}$$

When the reference point O is chosen at the centroid of the transformed section, $B = 0$ and Equation (2.15) takes the more familiar form

$$\begin{Bmatrix} \varepsilon_O \\ \psi \end{Bmatrix} = \frac{1}{E_{\text{ref}}} \begin{Bmatrix} N/A \\ M/I \end{Bmatrix} \tag{2.16}$$

2.3.1   Basic equations

The equations derived above give the stresses and the strains in a cross-section subjected to a normal force and a bending moment (Fig. 2.1). Extensive use of these equations will be made throughout this book in analysis of reinforced composite or non-composite cross-sections. Because of this, the basic equations are summarized below and the symbols defined for easy reference:

$$\varepsilon = \varepsilon_O + \psi y \quad \sigma = E(\varepsilon_O + \psi y) \tag{2.17}$$

$$N = E(A\varepsilon_O + B\psi) \quad M = E(B\varepsilon_O + I\psi) \tag{2.18}$$

$$\varepsilon_O = \frac{IN - BM}{E(AI - B^2)} \quad \psi = \frac{-BN + AM}{E(AI - B^2)} \tag{2.19}$$

$$\sigma_O = \frac{IN - BM}{AI - B^2} \quad \gamma = \frac{-BN + AM}{AI - B^2} \tag{2.20}$$

where

A, B and I	= cross-sectional area and its first and second moment about a horizontal axis through reference point O, respectively
E	= modulus of elasticity
y	= coordinate of any fibre, with respect to a horizontal axis through reference point O; y is measured downward (Fig. 2.2)
N	= normal force
M	= bending moment about a horizontal axis through reference point O
ε and σ	= strain and stress at any fibre
ε_O and σ_O	= strain and stress at reference point O
ψ and γ	= $d\varepsilon/dy$ (the curvature) and $d\sigma/dy$, respectively.

When the section is composed of more than one material (e.g., concrete parts of different age, prestressed, non-prestressed steel, structural steel), E in Equation (2.17) is the modulus of elasticity of the material for which the stress is calculated; A, B and I are properties of a transformed section composed of the cross-section areas of the individual materials, each multiplied by its modulus of elasticity divided by a reference modulus, whose value is to be used in Equations (2.18) and (2.19).

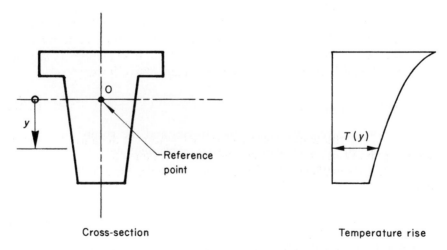

Cross-section   Temperature rise

Figure 2.2 Cross-section of a member subjected to a rise of temperature which varies non-linearly over the depth.

2.4   Strain and stress due to non-linear temperature variation

Analysis of the change in stresses due to creep, shrinkage of concrete and relaxation of prestressed steel in concrete structures can be done in the same way as the analysis of stresses due to temperature (as will be shown in Sections 2.5, 5.4 to 5.6 and 10.7). For this reason, we shall consider here the strain and stress in a cross-section subjected to a temperature rise of magnitude $T(y)$, which varies over the depth of the section in an arbitrary fashion (Fig. 2.2).

In a *statically determinate* frame, uniform or linearly varying temperature over the depth of the cross-section of a member produces no stresses. When the temperature variation is non-linear (Fig. 2.2), stresses are produced because each fibre being attached to adjacent fibres is not free to acquire the full expansion due to temperature. The stresses produced in this way in an individual cross-section must be self-equilibrating; in other words the temperature stress in a statically determinate structure corresponds to no change in the stress resultants (the internal forces). We shall discuss below the analysis of the stresses produced by a rise of temperature which varies non-linearly over the depth of a member of a statically determinate framed structure.

The self-equilibrating stresses caused by non-linear temperature variation over the cross-sections of statically determinate frame are sometimes referred to as the eigenstresses. If the structure is *statically indeterminate*, the elongations and/or the rotations of the joints of the members are restrained or prevented, resulting in a statically indeterminate set of reactions which are also self-equilibrating, but these will produce statically indeterminate internal forces and corresponding stresses. Statically indeterminate forces produced by temperature will be discussed in Section 10.8. The present section is concerned with the axial strain, the curvature and the self-equilibrating stresses in a cross-section of a statically determinate structure subjected to a rise of temperature which varies non-linearly over the depth of the section (Fig. 2.2).

The hypothetical strain that would occur at any fibre if it were free is:

$$\varepsilon_f = a_t T \tag{2.21}$$

where $T = T(y)$, the temperature rise at any fibre at a distance y below a reference point O and a_t = coefficient of thermal expansion.

If this strain is artificially prevented, the stress in the restrained condition will be

$$\sigma_{restrained} = -E\varepsilon_f \tag{2.22}$$

where E is the modulus of elasticity, which is considered, for simplicity, to be constant over the whole depth of the section.

The resultant of this stress may be represented by an axial force ΔN at a reference point O and a bending moment ΔM given by:

$$\Delta N = \int \sigma_{\text{restrained}} \, dA \qquad (2.23)$$

$$\Delta M = \int \sigma_{\text{restrained}} y \, dA \qquad (2.24)$$

Substitution of Equation (2.22) into (2.23) and (2.24) gives:

$$\Delta N = -\int E\varepsilon_f \, dA \qquad (2.25)$$

$$\Delta M = -\int E\varepsilon_f y \, dA \qquad (2.26)$$

The artificial restraint is now released by the application of a force $-\Delta N$ at O and a bending moment $-\Delta M$; the resulting axial strain and curvature are obtained by Equation (2.19) and the corresponding stress by Equation (2.17):

$$\begin{Bmatrix} \Delta\varepsilon_O \\ \Delta\psi \end{Bmatrix} = \frac{1}{E(AI - B^2)} \begin{bmatrix} I & -B \\ -B & A \end{bmatrix} \begin{Bmatrix} -\Delta N \\ -\Delta M \end{Bmatrix} \qquad (2.27)$$

$$\Delta\sigma = E[\Delta\varepsilon_O + (\Delta\psi)y] \qquad (2.28)$$

where A, B and I are the area and its first and second moment about an axis through the reference point O. When O is at the centroid of the section, $B = 0$ and Equation (2.27) becomes

$$\begin{Bmatrix} \Delta\varepsilon_O \\ \Delta\psi \end{Bmatrix} = \frac{1}{E} \begin{Bmatrix} -\Delta N/A \\ -\Delta M/I \end{Bmatrix} \qquad (2.29)$$

The actual stress due to temperature is the sum of $\sigma_{\text{restrained}}$ and $\Delta\sigma$; thus (Equations (2.22) and (2.28))

$$\sigma = E[-\varepsilon_f + \Delta\varepsilon_O + (\Delta\psi)y] \qquad (2.30)$$

The equations of the present section are applicable for composite cross-sections having more than one material; in this case A, B and I are properties of a transformed section with modulus of elasticity of $E = E_{\text{ref}}$. The transformed section is composed of parts of cross-section area a times the actual areas of individual parts, where a is the ratio of the modulus of elasticity of the part considered to E_{ref} (see Equations (1.33) and (2.9)–(2.11)). When the change in temperature occurs at age t_0 and takes a short time to develop, such that creep may be ignored, $E_{\text{ref}} = E_c(t_0)$; where $E_c(t_0)$ is the modulus of elasticity at age t_0 of one of the concrete parts chosen as reference. When the change in temperature develops gradually during a period t_0 to t, a is replaced by $\bar{a}$ and $E_c(t_0)$ by the age-adjusted modulus of elasticity $\bar{E}_c(t, t_0)$ as discussed in Section 1.11.1 (see Equation (1.34)). The analysis in this way accounts for the fact that creep of concrete alleviates the stresses due to temperature.

Example 2.1   Rectangular section with parabolic temperature variation

Calculate the axial strain, the curvature and the stress distribution in a member of a rectangular section subjected to a rise of temperature which varies over the depth in the form of a parabola of the mth degree (Figs. 2.3(a) and (b)). The elongation and rotation at the member

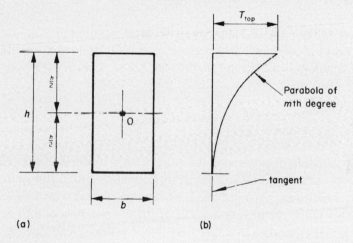

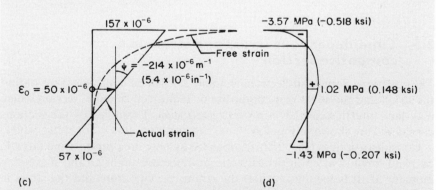

Figure 2.3 Temperature stresses in a statically determinate member (Example 2.1): (a) cross-section; (b) variation of the magnitude of rise of temperature over depth; (c) strain; (d) stress (self-equilibrating).

ends are assumed to occur freely (structure statically determinate externally).

Choose the reference point at the middle of the depth. Equations (2.25) and (2.26) give

$$\begin{Bmatrix} \Delta N \\ \Delta M \end{Bmatrix} = a_t T_{top} E \begin{Bmatrix} -\dfrac{bh}{m+1} \\[2ex] bh^2 \dfrac{m}{2(m+1)(m+2)} \end{Bmatrix}$$

With $A = bh$, $I = bh^3/12$, Equation (2.29) gives

$$\varepsilon_O = \frac{a_t T_{top}}{m+1}$$

$$\psi = -\frac{a_t T_{top}}{h} \frac{6m}{(m+1)(m+2)}$$

The variation of strain over the cross-section is shown in Fig. 2.3(c). The corresponding stress calculated by Equation (2.30) is shown in Fig. 2.3(d). The values given in Figs. 2.3(c) and (d) are calculated assuming the temperature rise to vary over the depth as a parabola of fifth degree ($m = 5$) and other data as follows: $b = 1\,m$; $h = 1\,m$; $a_t = 10^{-5}$ per °C and $T_{top} = 30\,°C$; $E = 25\,GPa$. Or, in British units, $b = 40\,in$; $h = 40\,in$; $a_t = 5.6 \times 10^{-6}$ per °F and $T_{top} = 54\,°F$; $E = 3600\,ksi$.

2.5   Time-dependent stress and strain in a composite section

The equations derived in Sections 2.3 and 2.4 will be employed here to find the strain and the stress in a composite or reinforced concrete section which may have prestressed and non-prestressed steel. Examples of the sections considered are shown in Fig. 2.4.

Consider a section (Fig. 2.5(a)) subjected at age t_0 to a prestressing force P, an axial force N at an arbitrarily chosen reference point O and a bending moment M. It is required to find the strain, the curvature and the stress in concrete and steel at age t_0, immediately after prestressing at age t; where t is greater than t_0. Assumed to be known are: the cross-section dimensions and the reinforcement areas, the magnitudes of P, N and M, the modulus of elasticity of concrete $E_c(t_0)$ at age t_0, the shrinkage $\varepsilon_{cs}(t, t_0)$ that would occur at

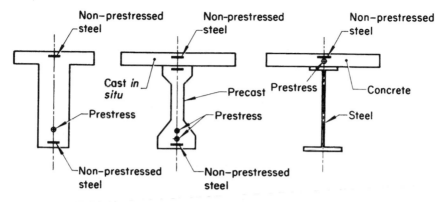

Figure 2.4 Examples of cross-sections treated in Section 2.5.

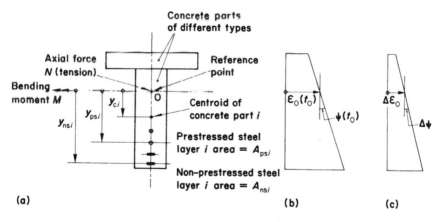

Figure 2.5 Analysis of time-dependent stress and strain in a composite section. All variables are shown in their positive directions: (a) cross-section; (b) strain at t_0; (c) change in strain during the period $t - t_0$.

any fibre if it were free, the creep coefficient $\varphi(t, t_0)$ and the aging coefficient $\chi(t, t_0)$.

The intrinsic relaxation $\Delta\sigma_{pr}$ that occurs during the period $(t - t_0)$ is also assumed to be known. A reduced relaxation value $\Delta\bar{\sigma}_{pr} = \chi_r(\Delta\sigma_{pr})$ will be used in the analysis. The reduction factor χ_r must be assumed at the start of the analysis and adjusted later if necessary as will be further discussed in Section 3.2; here we assume that the reduced relaxation value $\Delta\bar{\sigma}_{pr}$ is known.

2.5.1 Instantaneous stress and strain at age t_0

Before we can apply Equation (2.19), we must combine N and M with the prestressing forces into an equivalent normal force at O and a moment:

$$\begin{Bmatrix} N_{\text{equivalent}} \\ M_{\text{equivalent}} \end{Bmatrix} = \begin{Bmatrix} N - \Sigma P_i \\ M - \Sigma P_i y_{\text{psi}} \end{Bmatrix} \tag{2.31}$$

where the subscript i refers to the ith prestressed steel layer and y_{psi} is its distance below the reference point O. The summation in this equation is to be performed for the prestressed steel layers. Here we assume that the prestress is introduced in one stage; multi-stage prestressing will be discussed in Section 3.7. P is the absolute value of the prestressing force.

The instantaneous axial strain and curvature immediately after prestressing (Equation 2.15) are given by

$$\begin{Bmatrix} \varepsilon_O(t_0) \\ \psi(t_0) \end{Bmatrix} = \frac{1}{E_{\text{ref}}(AI - B^2)} \begin{bmatrix} I & -B \\ -B & A \end{bmatrix} \begin{Bmatrix} N \\ M \end{Bmatrix}_{\text{equivalent}} \tag{2.32}$$

where A, B and I are respectively the area and its first and second moment of the transformed section at time t_0 (see Section 1.11.1); the modulus of elasticity of concrete to be used here is $E_c(t_0)$ for the individual parts of the section; E_{ref} is a reference modulus of elasticity which may be chosen equal to $E_{c1}(t_0)$, the modulus at age t_0 for concrete of part 1 (see Equations (2.9) to (2.11)).

When the reference point O is at the centroid of the transformed section at time t_0, $B = 0$ and Equation (2.32) becomes

$$\begin{Bmatrix} \varepsilon_O(t_0) \\ \psi(t_0) \end{Bmatrix} = \frac{1}{E_{\text{ref}}} \begin{Bmatrix} \dfrac{N_{\text{equivalent}}}{A} \\ \dfrac{M_{\text{equivalent}}}{I} \end{Bmatrix} \tag{2.33}$$

With post-tensioning, the area of prestressed duct should be deducted from the area of concrete and the area of the prestressed steel excluded when calculating the properties of the transformed section to be used in Equation (2.32) or (2.33).

The instantaneous strain and stress in concrete at any fibre (Equation (2.17)) are

$$\varepsilon_c(t_0) = \varepsilon_O(t_0) + \psi(t_0)y \tag{2.34}$$

$$\sigma_c(t_0) = [E_c(t_0)]_i[\varepsilon_O(t_0) + \psi(t_0)y] \tag{2.35}$$

where y is the distance below the reference point O of the layer considered and the subscript i refers to the number of the concrete part of the fibre considered.

The instantaneous stress in the non-prestressed steel is

$$\sigma_{ns}(t_0) = E_{ns}[\varepsilon_O(t_0) + \psi(t_0)y_{ns}] \tag{2.36}$$

In the case of pretensioning, the stress in the prestressed steel immediately after transfer is

$$\sigma_{ps}(t_0) = (\sigma_{ps})_{initial} + E_{ps}[\varepsilon_O(t_0) + \psi(t_0)y_{ps}] \tag{2.37}$$

where $(\sigma_{ps})_{initial}$ is the stress in prestressed steel before transfer. The second term in this equation represents the instantaneous change in stress (generally a loss of tension due to shortening of concrete). Thus, the instantaneous prestress change (the loss) in pretensioned tendon at the time of transfer is

$$(\Delta\sigma_{ps})_{inst} = E_{ps}[\varepsilon_O(t_0) + \psi(t_0)y]. \tag{2.38}$$

With post-tensioning, compatibility of strain in the tendon does not take place at this stage and thus no instantaneous loss occurs.[1] The stresses in post-tensioned tendon immediately before and after transfer are the same:

$$\sigma_{ps}(t_0) = (\sigma_{ps})_{initial} \tag{2.39}$$

where $(\sigma_{ps})_{initial}$ is the initial stress in prestressed steel given by the prestressed force P divided by the cross-section area of prestressed steel.

2.5.2   Changes in stress and strain during the period t_0 to t

In this step of the analysis we deal with a cross-section for which the initial stress and strain are known. Creep, shrinkage and relaxation of steel result in stress redistribution between the various materials involved. The analysis to be presented here gives the stress changes in each material occurring during a specified period of time. In some cases, the cross-section of the member is changed at the beginning of the period, for example, by the addition of a part cast *in situ* to a precast section (see Fig. 2.4). In such a case the initial stress in the added part is known to be zero. Assuming perfect bond, the two parts behave as one cross-section; thus creep, shrinkage and relaxation of any part will affect both parts.

The change in strain during the period t_0 to t (Fig. 2.5(c)) is defined by the increments $\Delta\varepsilon_O$ and $\Delta\psi$ in the axial strain and curvature. To determine these, we follow a similar procedure to that in Section 2.4. The change of strain due to creep and shrinkage of concrete and relaxation of prestressed steel is first artificially restrained by application of an axial force ΔN at O and a bending moment ΔM. Subsequently, these restraining forces are removed, by the application of equal and opposite forces on the composite section, resulting in the following changes in axial strain and in curvature (Equation (2.19)):

$$\begin{Bmatrix} \Delta\varepsilon_O \\ \Delta\psi \end{Bmatrix} = \frac{1}{\bar{E}_c(\bar{A}\bar{I} - \bar{B}^2)} \begin{bmatrix} \bar{I} & -\bar{B} \\ -\bar{B} & \bar{A} \end{bmatrix} \begin{Bmatrix} -\Delta N \\ -\Delta M \end{Bmatrix} \qquad (2.40)$$

where $\bar{A}$, $\bar{B}$ and $\bar{I}$ are, respectively, the area of the age-adjusted transformed section and its first and second moment about an axis through the reference point O (see Section 1.11.1); $\bar{E}_c = E_{ref} = \bar{E}_c(t, t_0)$ is the age-adjusted elasticity modulus of one of the concrete types chosen as reference material (Equation (1.31)). The restraining forces are calculated as a sum of three terms:

$$\begin{Bmatrix} \Delta N \\ \Delta M \end{Bmatrix} = \begin{Bmatrix} \Delta N \\ \Delta M \end{Bmatrix}_{creep} + \begin{Bmatrix} \Delta N \\ \Delta M \end{Bmatrix}_{shrinkage} + \begin{Bmatrix} \Delta N \\ \Delta M \end{Bmatrix}_{relaxation} \qquad (2.41)$$

If creep were free to occur, the axial strain and curvature would increase during the period t_0 to t by the amounts $\varphi(t, t_0)\, \varepsilon(t_0)$ and $\varphi(t, t_0)\, \psi(t_0)$. The forces necessary to prevent these deformations may be determined by Equation (2.18):

$$\begin{Bmatrix} \Delta N \\ \Delta M \end{Bmatrix}_{creep} = -\sum_{i=1}^{m} \left\{ \bar{E}_c \varphi \begin{bmatrix} A_c & B_c \\ B_c & I_c \end{bmatrix} \begin{Bmatrix} \varepsilon_O(t_0) \\ \psi(t_0) \end{Bmatrix} \right\}_i \qquad (2.42)$$

The subscript i refers to the ith part of the section and m is the total number of concrete parts. A_{ci}, B_{ci} and I_{ci} are respectively the area of concrete of the ith part and its first and second moment about an axis through the reference point O; $\bar{E}_{ci} = [\bar{E}_c(t, t_0)]_i$ and $\varphi_i = [\varphi(t, t_0)]_i$ are the age-adjusted modulus of elasticity and creep coefficient for concrete in the ith part.

When applying Equation (2.42), it should be noted that $[\varepsilon_O(t_0)]_i$ and $[\psi(t_0)]_i$ are two quantities defining a *straight line* of the strain distribution on the ith part and the value $[\varepsilon_O(t_0)]_i$ is the strain at the reference point O (which may not be situated in the ith part (see Example 2.4)).

In Equation (2.42), it is assumed that all loads are applied at age t_0; in case there are other loads introduced at an earlier age, the vector $\varphi\{\varepsilon_O, \psi\}$ must be replaced by a vector of two values equal to the total axial strain and curvature due to creep if it were free. This is equal to the sum of products of instantaneous strains and curvatures by appropriate creep coefficients (see part (d) of solution of Example 2.5).

The forces required to prevent shrinkage are

$$\begin{Bmatrix} \Delta N \\ \Delta M \end{Bmatrix}_{shrinkage} = -\sum_{i=1}^{m} \left\{ \bar{E}_c \varepsilon_{cs} \begin{Bmatrix} A_c \\ B_c \end{Bmatrix} \right\}_i \qquad (2.43)$$

where $\varepsilon_{cs} = \varepsilon_{cs}(t, t_0)$ is the free shrinkage for the period t_0 to t.

The age-adjusted moduli of elasticity are used in Equations (2.40), (2.42) and (2.43) (indicated by a bar as superscript) because the forces ΔN and ΔM are gradually developed between the instants t_0 and t.

The forces necessary to prevent the strain due to relaxation of prestressed steel are

$$\left\{\begin{matrix}\Delta N \\ \Delta M\end{matrix}\right\}_{\text{relaxation}} = \sum \left\{\begin{matrix}A_{\text{ps}}\Delta\bar{\sigma}_{\text{pr}} \\ A_{\text{ps}}y_{\text{ps}}\Delta\bar{\sigma}_{\text{pr}}\end{matrix}\right\}_i \tag{2.44}$$

The subscript i in this equation refers to a prestressed steel layer; A_{ps} is its cross-section area and y_{ps} its distance below the reference point O and $\Delta\bar{\sigma}_{\text{pr}}$ is the reduced relaxation during the period t_0 to t.

The stress in concrete required to prevent creep and shrinkage at any fibre is

$$\sigma_{\text{restrained}} = -\bar{E}_c(t, t_0)[\varphi(t, t_0)\varepsilon_c(t_0) + \varepsilon_{\text{cs}}] \tag{2.45}$$

where $\varepsilon_c(t_0)$ is the instantaneous strain determined earlier (Equation (2.34)). In Equation (2.45), we assume that all loads are applied at t_0; in the case when other loads are introduced earlier, the quantity $(\varphi\varepsilon_c)$ must be replaced by the sum of products of instantaneous strains by appropriate creep coefficient (see part (d) of solution of Example 2.5).

The stress increments that develop during the period $(t - t_0)$ are as follows. In concrete, at any fibre in the ith part

$$\Delta\sigma_c = \sigma_{\text{restrained}} + \bar{E}_c(t, t_0)(\Delta\varepsilon_O + y\Delta\psi); \tag{2.46}$$

in non-prestressed steel

$$\Delta\sigma_{\text{ns}} = E_{\text{ns}}(\Delta\varepsilon_O + y_{\text{ns}}\Delta\psi); \tag{2.47}$$

and in prestressed steel

$$\Delta\sigma_{\text{ps}} = \Delta\bar{\sigma}_{\text{pr}} + E_{\text{ps}}(\Delta\varepsilon_O + y_{\text{ps}}\Delta\psi). \tag{2.48}$$

The last equation gives the change in prestress due to creep, shrinkage and relaxation. Multiplication of $\Delta\sigma_{\text{ps}}$ by A_{ps} gives the loss of tension in the prestressed steel.

The procedure of analysis presented in this section is demonstrated by the following examples. The input data and the main results are given in all examples throughout this book in both SI and British units; however, the examples are worked out either in SI units or in British units.

Example 2.2 Post-tensioned section

A prestress force $P = 1400 \times 10^3 N$ (315 kip) and a bending moment $M = 390 \times 10^3$ N-m (3450 kip-in) are applied at age t_0 on the rectangular post-tensioned concrete section shown in Fig. 2.6(a). Calculate the stresses, the axial strain and curvature at age t_0 and t given the following data: $E_c(t_0) = 30.0$ GPa (4350 ksi); $E_{ns} = E_{ps} = 200$ GPa (29 $\times$ 10³ ksi); uniform free shrinkage value $\varepsilon_{cs}(t, t_0) = -240 \times 10^{-6}$; $\varphi(t, t_0) = 3$; $\chi = 0.8$; reduced relaxation, $\Delta\bar{\sigma}_{pr} = -80$ MPa (−12 ksi). The dimensions of the section and cross-section areas of the reinforcement and the prestress duct are indicated in Fig. 2.6(a).

(a)   Stress and strain at age t_0
Calculation of the properties of the transformed section at time t_0 is done in Table 2.1. The reference modulus of elasticity, $E_{ref} = E_c(t_0) = 30.0$ GPa. The forces introduced at age t_0 are equivalent to Equation (2.31) is

$$\begin{Bmatrix} N \\ M \end{Bmatrix}_{equivalent} = \begin{Bmatrix} -1400 \times 10^3 \\ 390 \times 10^3 - 1400 \times 10^3 \times 0.45 \end{Bmatrix} = \begin{Bmatrix} -1400 \times 10^3 \, \text{N} \\ -240 \times 10^3 \, \text{N-m} \end{Bmatrix}$$

The instantaneous axial strain at O and curvature (Equation (2.32)) is

$$\begin{Bmatrix} \varepsilon_O(t_0) \\ \psi(t_0) \end{Bmatrix} = \frac{1}{30 \times 10^9[0.3712 \times 46.88 \times 10^{-3} - (0.208 \times 10^{-3})^2]}$$
$$\times \begin{bmatrix} 46.88 \times 10^{-3} & -0.208 \times 10^{-3} \\ -0.208 \times 10^{-3} & 0.3712 \end{bmatrix} \begin{Bmatrix} -1400 \times 10^3 \\ -240 \times 10^3 \end{Bmatrix}$$
$$= 10^{-6}\begin{Bmatrix} -126 \\ -170 \, \text{m}^{-1} \end{Bmatrix}$$

The concrete stress at top and bottom fibres (Equation (2.35)) is

$$(\sigma_c(t_0))_{top} = 30 \times 10^9[-126 + (-0.6)(-170)] \, 10^{-6}$$
$$= -0.706 \, \text{MPa} \, (-0.102 \, \text{ksi})$$
$$(\sigma_c(t_0))_{bot} = 30 \times 10^9[-126 + 0.6(-170)] \, 10^{-6}$$
$$= -6.830 \, \text{MPa} \, (-0.991 \, \text{ksi})$$

The stress distribution is shown in Fig. 2.6(b).

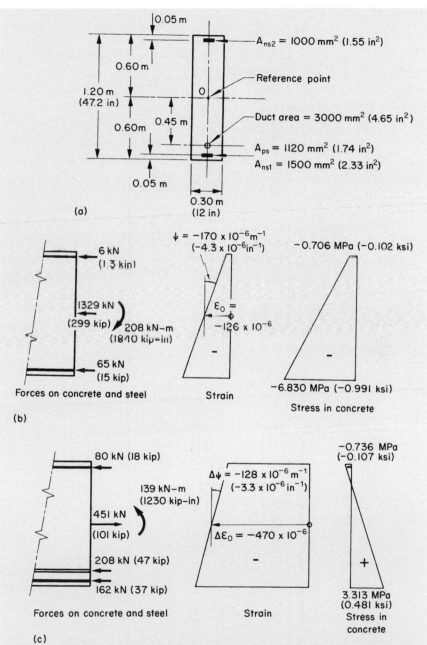

Figure 2.6 Analysis of stress and strain in the cross-section of a post-tensioned member (Example 2.2): (a) cross-section dimensions; (b) condition at age t_0 immediately after prestress; (c) changes caused by creep, shrinkage and relaxation.

Table 2.1 Calculation of A, B and I of transformed section at time t_0

	Properties of area			Properties of transformed area		
	A (m^2)	B (m^3)	I (m^4)	AE/E_{ref} (m^2)	BE/E_{ref} (m^3)	IE/E_{ref} (m^4)
Concrete	0.3545	-1.625×10^{-3}	41.84×10^{-3}	0.3545	-1.625×10^{-3}	41.84×10^{-3}
Non-prestressed steel	2500×10^{-6}	0.275×10^{-3}	0.756×10^{-3}	0.0167	1.833×10^{-3}	5.04×10^{-3}
Prestressed steel	—	—	—			
Properties of transformed section				0.3712 A	0.208×10^{-3} B	46.88×10^{-3} I

*(b)   Changes in stress and strain due to creep, shrinkage
        and relaxation*

The age-adjusted elasticity modulus of concrete (Equation (1.31)) is

$$\bar{E}_c(t, t_0) = \frac{30 \times 10^9}{1 + 0.8 \times 3} = 8.82\,\text{GPa}.$$

The stress in concrete at the top and bottom fibres when the strain
due to creep and shrinkage is artificially restrained (Equations (2.34)
and (2.45)) is:

$$(\sigma_{c\ \text{restrained}})_{\text{top}} = -8.82 \times 10^9[3 \times 10^{-6}(-126 + 170 \times 0.6) - 240 \times 10^{-6}]$$

$$= 2.741\,\text{MPa} \ (0.398\,\text{ksi})$$

$$(\sigma_{c\ \text{restrained}})_{\text{bot}} = -8.82 \times 10^9[3 \times 10^{-6}(-126 - 170 \times 0.6) - 240 \times 10^{-6}]$$

$$= 8.145\,\text{MPa} \ (1.181\,\text{ksi}).$$

The restraining forces (Equations (2.41) to (2.44)) are:

$$\left\{ \begin{matrix} \Delta N \\ \Delta M \end{matrix} \right\}_{\text{creep}} = -8.82 \times 10^9 \times 3 \begin{bmatrix} 0.3545 & -1.625 \times 10^{-3} \\ -1.625 \times 10^{-3} & 41.84 \times 10^{-3} \end{bmatrix}$$

$$\times \left\{ \begin{matrix} -126 \\ -170 \end{matrix} \right\} 10^{-6} = 10^6 \left\{ \begin{matrix} 1.175\,\text{N} \\ 0.1828\,\text{N-m} \end{matrix} \right\}$$

$$\left\{ \begin{matrix} \Delta N \\ \Delta M \end{matrix} \right\}_{\text{shrinkage}} = -8.82 \times 10^9(-240 \times 10^{-6}) \left\{ \begin{matrix} 0.3545 \\ -1.625 \times 10^{-3} \end{matrix} \right\}$$

$$= 10^6 \left\{ \begin{matrix} 0.750\,\text{N} \\ -0.0034\,\text{N-m} \end{matrix} \right\}$$

$$\left\{ \begin{matrix} \Delta N \\ \Delta M \end{matrix} \right\}_{\text{relaxation}} = \left\{ \begin{matrix} 1120 \times 10^{-6}(-80 \times 10^6) \\ 1120 \times 10^{-6} \times 0.45 \ (-80 \times 10^6) \end{matrix} \right\}$$

$$= 10^6 \left\{ \begin{matrix} -0.090\,\text{N} \\ -0.0403\,\text{N-m} \end{matrix} \right\}$$

$$\left\{ \begin{matrix} \Delta N \\ \Delta M \end{matrix} \right\} = 10^6 \left\{ \begin{matrix} 1.175 + 0.750 - 0.090 \\ 0.1828 - 0.0034 - 0.0403 \end{matrix} \right\} = 10^6 \left\{ \begin{matrix} 1.835\,\text{N} \\ 0.139\,\text{N-m} \end{matrix} \right\}.$$

Calculation of the properties of the age-adjusted transformed section is
performed in Table 2.2 using $E_{\text{ref}} = \bar{E}_c(t, t_0) = 8.82\,\text{GPa}$ and $\bar{\alpha}(t, t_0) = 22.68$ (Equation (1.31)).

Table 2.2 Calculation of $\bar{A}$, $\bar{B}$ and $\bar{I}$ of the age-adjusted transformed section

	Properties of area			Properties of transformed area		
	A (m²)	B (m³)	I (m⁴)	AE/E_{ref} (m²)	BE/E_{ref} (m³)	IE/E_{ref} (m⁴)
Concrete	0.3545	-1.625×10^{-3}	41.84×10^{-3}	0.3545	-1.625×10^{-3}	41.84×10^{-3}
Non-prestressed steel	2500×10^{-6}	0.275×10^{-3}	0.756×10^{-3}	0.0567	6.236×10^{-3}	17.24×10^{-3}
Prestressed steel	1120×10^{-6}	0.504×10^{-3}	0.227×10^{-3}	0.0254	11.429×10^{-3}	5.15×10^{-3}
Properties of age-adjusted transformed section				$\underset{\bar{A}}{0.4366}$	$\underset{\bar{B}}{16.040 \times 10^{-3}}$	$\underset{\bar{I}}{64.12 \times 10^{-3}}$

The prestress duct is usually grouted shortly after the prestress; hence, its area may be included in Table 2.2, but this is ignored here.

$$\left\{ \begin{matrix} \Delta\varepsilon_O \\ \Delta\psi \end{matrix} \right\} = \frac{1}{8.82 \times 10^9 [0.4366 \times 64.12 \times 10^{-3} - (16.04 \times 10^{-3})^2]}$$

$$\times \begin{bmatrix} 64.12 \times 10^{-3} & -16.040 \times 10^{-3} \\ -16.040 \times 10^{-3} & 0.4366 \end{bmatrix} \left\{ \begin{matrix} -1.835 \\ -0.139 \end{matrix} \right\} 10^6$$

$$= 10^{-6} \left\{ \begin{matrix} -470 \\ -128\,\mathrm{m}^{-1} \end{matrix} \right\}$$

Increments of stress that will develop during the period $(t - t_0)$ in concrete, non-prestressed steel and prestressed steel are (Equations (2.46–48)):

$$(\Delta\sigma_c)_{\mathrm{top}} = 2.741 \times 10^6 + 8.82[-471 + (-0.6)(-128)]10^3$$

$$= -0.736\,\mathrm{MPa}\ (-0.107\,\mathrm{ksi})$$

$$(\Delta\sigma_c)_{\mathrm{bot}} = 8.145 \times 10^6 + 8.82[-471 + 0.6(-128)]10^3$$

$$= 3.313\,\mathrm{MPa}\ (0.481\,\mathrm{ksi})$$

$$\Delta\sigma_{\mathrm{ns2}} = 200[-471 + (-0.55)(-128)]10^3$$

$$= -80.1\,\mathrm{MPa}\ (-11.6\,\mathrm{ksi})$$

$$\Delta\sigma_{\mathrm{ns1}} = 200[-471 + 0.55(-128)]10^3$$

$$= -108.3\,\mathrm{MPa}\ (-15.7\,\mathrm{ksi})$$

$$\Delta\sigma_{\mathrm{ps}} = -80 \times 10^6 + 200[-471 + 0.45(-128)]10^3$$

$$= -185.7\,\mathrm{MPa}\ (-26.9\,\mathrm{ksi})$$

The last value is the loss of prestress in the tendon. Fig. 2.6(b) shows the distributions of stress and strain on the concrete and the resultants of forces on concrete and steel at age t_0. The changes in these variables caused by creep, shrinkage and relaxation are shown in Fig. 2.6(c). From these figures it is seen that the loss in tension in the prestressed steel due to these effects, is 208 kN or 15% of the original tension (1400 kN). The resultant compressive force on the concrete at age t_0 is 1329 kN and the difference $(1400 - 1329 = 71\,\mathrm{kN})$ represents the compression in the non-prestressed steel. The loss in compression in concrete due to creep, shrinkage and relaxation amounts to 451 kN which is

32% of the initial compression in the concrete (1329 kN). The higher percentage is caused by the compression picked up by the non-prestressed steel as creep and shrinkage develop.

The results of this example may be checked by verifying that the sum of the changes of the resultants of stress in concrete and steel is zero. Thus the system of forces shown in Fig. 2.6(c) is in equilibrium.

A check on compatibility can be made by verifying that the change in strain in prestress steel caused by $(\Delta \sigma_{ps} - \Delta \bar{\sigma}_{pr})$ is equal to the change in strain in concrete at the prestressed steel level.

In Fig. 2.7, we assumed that the cross-section analysed in this

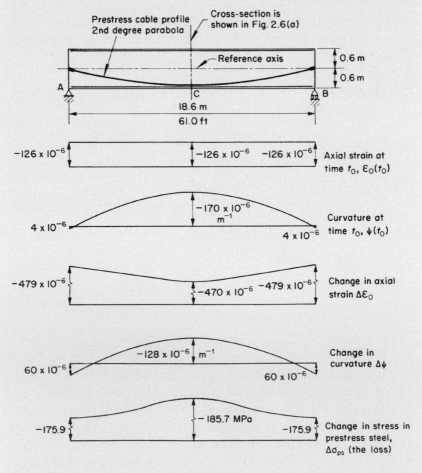

Figure 2.7 Axial strain, curvature and prestress loss in a post-tensioned span (beam of Example 2.2).

example is at the centre of span of a simply supported beam. The absolute value P of the prestressing force at time t_0 is assumed constant at all sections, while the dead load bending moment, M, is assumed to vary as a parabola. The profile of the prestress tendon is assumed a parabola, as shown. The graphs in this figure show the variation over the span of $\varepsilon_O(t_0)$, $\psi(t_0)$, $\Delta\varepsilon_O$, $\Delta\psi$, $\Delta\sigma_{ps}$ which are respectively the axial strain and curvature at t_0 and the changes during the period $(t - t_0)$ in axial strain, in curvature and in tension in prestress steel due to the combined effects of creep, shrinkage and relaxation. The values of $(\varepsilon_O + \Delta\varepsilon_O)$ and $(\psi + \Delta\psi)$ will be used in Example 3.5 to calculate displacement values at time t.

Example 2.3 Pre-tensioned section

Solve the same problem as in Example 2.2 assuming that pre-tensioning is employed (the duct shown in Fig. 2.6(a) is eliminated).

(a) Stress and strain at age t_0

The prestressed steel must now be included in the calculation of the properties of the transformed section at t_0. With this modification and considering that there is no prestress duct in this case, calculation of the area properties of the transformed section in the same way as in Table 2.1 gives: $A = 0.3805 \, \text{m}^2$; $B = 4.413 \times 10^{-3} \, \text{m}^3$; $I = 48.77 \times 10^{-3} \, \text{m}^4$.

The forces applied on the section at t_0 are the same as in Example 2.2. Equation (2.32) gives the strain and the curvature at the reference point immediately after prestress transfer:

$$\varepsilon_O(t_0) = -120 \times 10^{-6}; \qquad \psi(t_0) = -153 \times 10^{-6} \, \text{m}^{-1}.$$

The change in stress in the prestressed steel at transfer (Equation (2.38)) are

$$(\Delta\sigma_{ps})_{inst} = 200[-120 + 0.45(-153)]10^3 = -37.8 \, \text{MPa}.$$

Multiplying this value by the area of the prestressed steel gives the instantaneous prestress loss (−43 kN).

The stresses and strain introduced at transfer and the corresponding resultants of stresses are shown in Fig. 2.8(a).

44   Concrete Structures

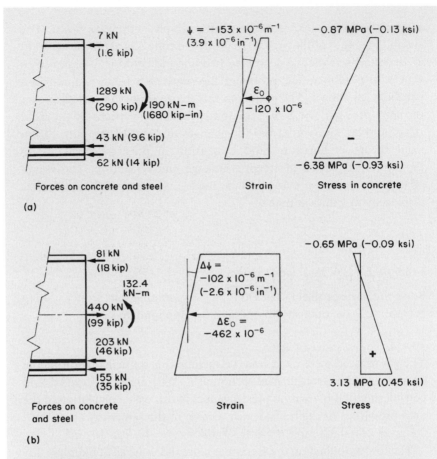

(a)

(b)

Figure 2.8 Stress and strain distribution in the section of Fig. 2.5(a) assuming that pre-tensioning is used (Example 2.3): (a) condition at age t_0 immediately after prestress transfer; (b) changes caused by creep, shrinkage and relaxation.

Using these results and following the same procedure as in Example 2.2, the time-dependent changes in stress and strain are calculated and the results shown in Fig. 2.8(b).

Example 2.4 Composite section: steel and post-tensioned concrete

Figure 2.9(a) shows the cross-section of a composite simply supported beam made of steel plate girder and a prestressed post-tensioned concrete slab. The plate girder is placed first in position and shored. Then

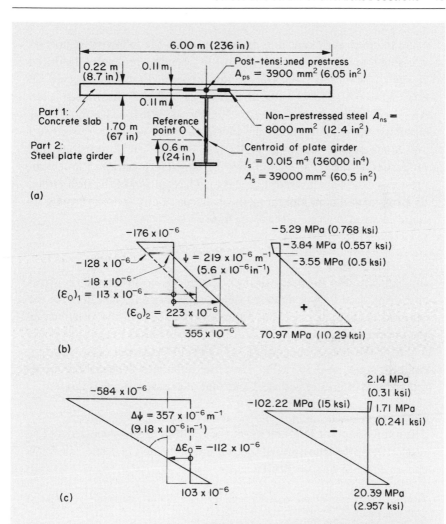

Figure 2.9 Analysis of stress and strain in a composite cross-section (Example 2.4):
(a) cross-section properties; (b) stress and strain immediately after
removal of shores; (c) changes caused by creep, shrinkage and relaxation.

the deck slab is cast *in situ*, but its connection to the steel girder is
delayed, by means of pockets left out around the anchor studs. The
pockets are cast only after the application of the prestress. Assume that
at age t_0 the prestress is applied; shortly after, the anchorage of the deck
to the steel girder is realized and the shoring removed. It is required to
find the stress and strain distribution occurring immediately after
removal of the shores (age t_0) and the changes in these values at time

t due to creep, shrinkage and relaxation using the following data: prestressing force $P = 4500 \times 10^3$ N(1010 kip); bending moment introduced at age t_0, $M = 2800 \times 10^3$ N-m. (24 800 kip-in); $\varphi(t, t_0) = 2.5$; $\chi = 0.75$; $\varepsilon_{cs}(t, t_0) = -350 \times 10^{-6}$; reduced relaxation of the prestressed steel $\Delta\bar{\sigma}_{pr} = -90$ MPa (-13 ksi); $E_c(t_0) = 30$ GPa (4350 ksi). The moduli of elasticity of the plate girder, the prestressed and non-prestressed steel are equal; $E_s = E_{ns} = E_{ps} = 200$ GPa (29 000 ksi). The dimensions and properties of the cross-section area of concrete prestressed and non-prestressed steel are given in Fig. 2.9(a). The centroid of the steel girder, its cross-section area and moment of inertia about an axis through its centroid are also given in the same figure.

(a)   Stress and strain at age t_0, before connection of slab to steel girder

Immediately after prestress, the steel girder has no stress and the stress and strain need to be calculated only in the concrete slab. Because the centroid of the reinforcement coincides with the centroid of concrete, the prestress produces no curvature and the strain is uniform over the depth of the slab of magnitude $= -110 \times 10^{-6}$ and the corresponding concrete stress $= -3.305$ MPa. Here the difference between the cross-section area of prestressed steel and that of prestress ducts is ignored.

(b)   Stress and strain immediately after removal of shores (age t_0)

The reference point O is chosen at the centroid of the steel girder. The properties of the transformed section are calculated in Table 2.3; E_{ref} is chosen equal to $E_c(t_0) = 30$ GPa.

Table 2.3 Properties of the transformed section used in calculation of stress at time t_0

	Properties of areas			Properties of transformed area		
	A (m^2)	B (m^3)	I (m^4)	AE/E_{ref} (m^2)	BE/E_{ref} (m^3)	IE/E_{ref} (m^4)
Concrete	1.3081	−1.5828	1.9205	1.3081	−1.5828	1.9205
Non-prestressed steel	8000×10^{-6}	−0.0097	0.0117	0.0533	−0.0645	0.0781
Prestressed steel	3900×10^{-6}	−0.0047	0.0057	0.0260	−0.0315	0.0381
Steel girder	39000×10^{-6}	0	0.0150	0.2600	0	0.1000
Properties of transformed section				1.6474 A	−1.6788 B	2.1367 I

Axial force at O and bending moment introduced at removal of shores is:

$$\left\{ \begin{matrix} N \\ M \end{matrix} \right\} = \left\{ \begin{matrix} 0 \\ 2800 \times 10^3 \text{ N-m} \end{matrix} \right\}$$

The axial strain at O and the curvature caused by these forces (Equation (2.32)) is

$$\left\{ \begin{matrix} \varepsilon_O(t_0) \\ \psi(t_0) \end{matrix} \right\} = \frac{1}{30 \times 10^9 (1.6474 \times 2.1367 - 1.6788^2)}$$

$$\times \begin{bmatrix} 2.1367 & 1.6788 \\ 1.6788 & 1.6474 \end{bmatrix} \left\{ \begin{matrix} 0 \\ 2800 \times 10^3 \end{matrix} \right\}$$

$$= 10^{-6} \left\{ \begin{matrix} 223 \\ 219 \text{ m}^{-1} \end{matrix} \right\}$$

The values of $\varepsilon_O(t_0)$ and $\psi(t_0)$ are used to find the strain at any fibre and hence the corresponding stress. Superposition of these stresses and strains and of the values determined in (a) above gives the stress and strain distributions shown in Fig. 2.9(b).

(c)   Changes in stress and strain due to creep, shrinkage and relaxation
Age-adjusted elasticity modulus is

$$\bar{E}_c(t, t_0) = \frac{30 \times 10^9}{1 + 0.75 \times 2.5} = 10.435 \text{ GPa.}$$

In the restrained condition, stress in concrete is (Equation (2.45)):

$$(\sigma_{c \text{ restrained}})_{\text{top}} = -10.435 \times 10^9 [2.5(-176 \times 10^{-6}) - 350 \times 10^{-6}]$$
$$= 8.24 \text{ MPa}$$

$$(\sigma_{c \text{ restrained}})_{\text{bot}} = -10.435 \times 10^9 [2.5(-128 \times 10^6) - 350 \times 10^{-6}]$$
$$= 6.99 \text{ MPa}$$

To calculate the axial force at O and the bending moment necessary to prevent creep by Equation (2.42), we must find $(\varepsilon_O)_1$ and ψ_1 defining the straight-line distribution of strain in part 1, the deck slab (Fig. 2.9(b)). These values are: $(\varepsilon_O)_1 = 113 \times 10^{-6}$; $\psi_1 = 219 \times 10^{-6}\,\text{m}^{-1}$.

$$\left\{\begin{matrix} \Delta N \\ \Delta M \end{matrix}\right\}_{\text{creep}} = -10.435 \times 10^9 \times 2.5$$

$$\times \begin{bmatrix} 1.3081 & -1.5828 \\ -1.5828 & 1.9205 \end{bmatrix} \begin{Bmatrix} 113 \\ 219 \end{Bmatrix} 10^{-6}$$

$$= 10^6 \left\{\begin{matrix} 5.187\ \text{N} \\ -6.306\ \text{N-m} \end{matrix}\right\}$$

The forces required to prevent strain due to shrinkage and relaxation (Equation (2.43) and (2.44)) are:

$$\left\{\begin{matrix} \Delta N \\ \Delta M \end{matrix}\right\}_{\text{shrinkage}} = -10.435 \times 10^9 (-350 \times 10^{-6}) \left\{\begin{matrix} 1.3081 \\ -1.5828 \end{matrix}\right\}$$

$$= 10^6 \left\{\begin{matrix} 4.777\ \text{N} \\ -5.781\ \text{N-m} \end{matrix}\right\}$$

$$\left\{\begin{matrix} \Delta N \\ \Delta M \end{matrix}\right\}_{\text{relaxation}} = \left\{\begin{matrix} 3900 \times 10^{-6}(-90 \times 10^6) \\ 3900 \times 10^{-6})(-1.21)(-90 \times 10^6) \end{matrix}\right\}$$

$$= 10^6 \left\{\begin{matrix} -0.351\ \text{N} \\ 0.425\ \text{N-m} \end{matrix}\right\}$$

The total restraining forces are

$$\left\{\begin{matrix} \Delta N \\ \Delta M \end{matrix}\right\} = 10^6 \left\{\begin{matrix} 5.187 + 4.777 - 0.351 \\ -6.306 - 5.781 + 0.425 \end{matrix}\right\} = 10^6 \left\{\begin{matrix} 9.613\ \text{N} \\ -11.662\ \text{N-m} \end{matrix}\right\}$$

Properties of the age-adjusted transformed section are calculated in a similar way as in Table 2.3 giving:

$$\bar{A} = 2.284\,\text{m}^2; \quad \bar{B} = -1.859\,\text{m}^3; \quad \bar{I} = 2.542\,\text{m}^4.$$

E_{ref} used in the calculation of the above values is:

$$E_{\text{ref}} = \bar{E}_c(t, t_0) = 10.435\,\text{GPa}.$$

Increments in axial strain and curvature when the restraining forces are removed (Equation (2.40)) are:

$$\begin{Bmatrix} \Delta\varepsilon_O \\ \Delta\psi \end{Bmatrix} = \frac{10^6}{10.435 \times 10^9(2.284 \times 2.542 - 1.859^2)} \begin{bmatrix} 2.542 & 1.859 \\ 1.859 & 2.284 \end{bmatrix}$$

$$\begin{Bmatrix} -9.613 \\ 11.662 \end{Bmatrix}$$

$$= 10^{-6} \begin{Bmatrix} -112 \\ 357\,\text{m}^{-1} \end{Bmatrix}$$

The corresponding stress and strain distributions are shown in Fig. 2.9(c). The stresses are calculated by Equation (2.46).

Example 2.5 Composite section: pre-tensioned and cast-in-situ parts

The cross-section shown in Fig. 2.10 is composed of a precast pre-tensioned beam (part 1) and a slab cast *in situ* (part 2). It is required to find the stress and strain distribution in the section immediately after prestressing, and the changes in these values occurring between pre-stressing and casting of the deck slab and after a long period using the following data.

Ages of precast beam at the time of prestress, $t_1 = 3$ days and at the

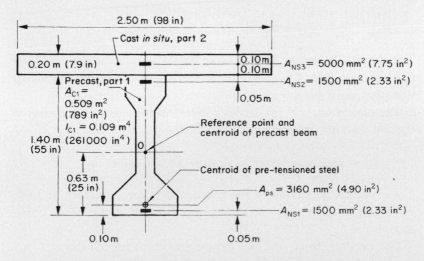

Figure 2.10 Analysis of stress and strain in a cross-section composed of precast and cast *in situ* parts (Example 2.5).

time of casting of the deck slab, $t_2 = 60$ days; the final stress and strain are required at age $t_3 = \infty$. The prestress force, $P = 4100 \times 10^3$ N; (920 kip); the bending moment due to self-weight of the prestress beam (which is introduced at the same time as the prestress), $M_1 = 1400 \times 10^3$ N-m (12400 kip-in); additional bending moment introduced at age t_2 (representing the effect of the weight of the slab plus superimposed dead load), $M_2 = 1850 \times 10^3$ N-m (16400 kip-in). The modulus of elasticity of concrete of the precast beam $E_{c1}(3) = 25$ GPa (3600 ksi) and $E_{c1}(60) = 37$ GPa (5400 ksi).

Soon after hardening of the concrete, the composite action starts to develop gradually. Here we will ignore the small composite action occurring during the first three days. Consider that age $t_2 = 60$ days for the precast beam corresponds to age = 3 days of the deck at which time the modulus of elasticity of the deck $E_{c2}(3) = 23$ GPa (3300 ksi).

Creep and aging coefficients and the free shrinkage values to be used are

Concrete part 1:
$[\varphi(60, 3)]_1 = 1.20$ $[\varphi(\infty, 3)]_1 = 2.30$ $[\varphi(\infty, 60)]_1 = 2.27;$
$[\chi(60, 3)]_1 = 0.86$ $[\chi(\infty, 60)]_1 = 0.80$
$[\varepsilon_{cs}(60, 3)]_1 = -57 \times 10^{-6}$ $[\varepsilon_{cs}(\infty, 60)]_1 = -205 \times 10^{-6}$

Concrete part 2:
$[\varphi(\infty, 3)]_2 = 2.40$ $[\chi(\infty, 3)]_2 = 0.78$
$[\varepsilon_{cs}(\infty, 3)]_2 = -269 \times 10^{-6}$

Reduced relaxation $\Delta\bar{\sigma}_{pr} = -85$ MPa (12 ksi) of which -15 MPa (2.2 ksi) in the first 57 days. Modulus of elasticity of the prestressed and non-prestressed steels = 200 GPa.

The dimensions and properties of areas of the concrete and steel in the two parts are given in Fig. 2.10.

(a)   Stress and strain immediately after prestressing of the precast beam

The geometric properties of the precast beam are calculated in Table 2.4, with the reference point O chosen at the centroid of concrete cross-section and $E_{ref} = E_{c1}(3) = 25$ GPa.

The prestress force and the bending moment introduced at t_1 are equivalent to an axial force at O plus a bending moment given by Equation (2.31)

Table 2.4 Properties of the precast section employed in calculation of stress and strain at time $t_1 = 3$ days

	Properties of area			Properties of transformed area		
	A (m^2)	B (m^3)	I (m^4)	AE/E_{ref} (m^2)	BE/E_{ref} (m^3)	IE/E_{ref} (m^4)
Concrete	0.5090	0.0	0.1090	0.5090	0.0	0.1090
Non-prestressed steel	3000×10^{-6}	-210×10^{-6}	1282×10^{-6}	0.0240	−0.0017	0.0103
Prestressed steel	3160×10^{-6}	1675×10^{-6}	888×10^{-6}	0.0253	0.0134	0.0071
Properties of transformed section				0.5583 A	0.0117 B	0.1264 I

$$\begin{Bmatrix} N \\ M \end{Bmatrix}_{equivalent} = \begin{Bmatrix} -4100 \times 10^3 \\ 1400 \times 10^3 - 4100 \times 10^3 \times 0.53 \end{Bmatrix} = \begin{Bmatrix} -4100 \times 10^3 \text{ N} \\ -773 \times 10^3 \text{ N-m} \end{Bmatrix}$$

Instantaneous axial strain and curvature at $t_1 = 3$ days (Equation (2.32)) are,

$$\begin{Bmatrix} \varepsilon_O(t_1) \\ \psi(t_1) \end{Bmatrix} = \frac{1}{25 \times 10^9 (0.5583 \times 0.1264 - 0.0117^2)} \begin{bmatrix} 0.1264 & -0.0117 \\ -0.0117 & 0.5583 \end{bmatrix}$$
$$\times \begin{Bmatrix} -4100 \times 10^3 \\ -773 \times 10^3 \end{Bmatrix} = 10^{-6} \begin{Bmatrix} -289 \\ -218 \text{ m}^{-1} \end{Bmatrix}$$

The above values of ε_O and ψ are used to calculate the strain at any fibre and the corresponding stress (Fig. 2.11(a)). The strain at the level of prestress tendon is -405×10^{-6}. The instantaneous prestress loss is -256 kN (6.2% of the initial force).

(b) Change in stress and strain occurring between $t = 3$ days and $t = 60$ days
The age-adjusted elasticity modulus of concrete (Equation (1.31)) is:

$$\bar{E}_c(60, 3) = \frac{25 \times 10^9}{1 + 0.86 \times 1.20} = 12.30 \text{ GPa (1780 ksi)}.$$

The stress in concrete required to artificially restrain creep and shrinkage (Equation (2.45)) is:

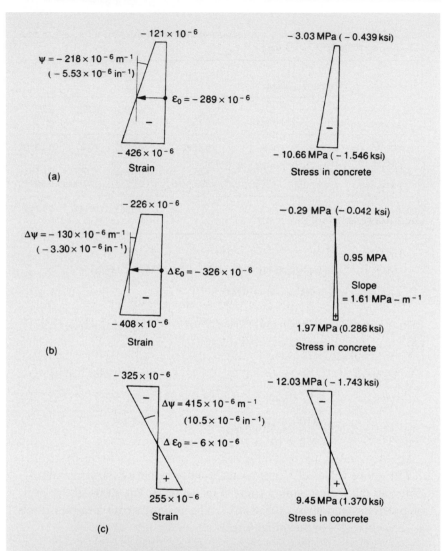

Figure 2.11 Stress and strain in the precast beam of Example 2.5: (a) conditions at age $t_1 = 3$ days; (b) changes caused by creep, shrinkage and relaxation occurring between $t_1 = 3$ days and $t_2 = 60$ days; (c) instantaneous changes at t_2 caused by introduction of moment $M_2 = 1850 \times 10^3$ N-m.

$$(\sigma_{c\ \text{restrained}})_{\text{top}} = -12.30 \times 10^9 [1.2(-121 \times 10^{-6}) - 57 \times 10^{-6}] = 2.487\,\text{MPa}$$

$$(\sigma_{c\ \text{restrained}})_{\text{bot}} = -12.30 \times 10^9 [1.2(-426 \times 10^{-6}) - 57 \times 10^{-6}] = 6.989\,\text{MPa}$$

Strain due to creep, shrinkage and relaxation can be restrained by the following forces (Equations (2.42), (2.43) and (2.44)):

$$\left\{\begin{matrix}\Delta N \\ \Delta M\end{matrix}\right\}_{creep} = -12.30 \times 10^9 \times 1.2 \begin{bmatrix} 0.5090 & 0 \\ 0 & 0.1090 \end{bmatrix} \left\{\begin{matrix}-289 \\ -218\end{matrix}\right\} 10^{-6}$$

$$= 10^6 \left\{\begin{matrix}2.171\ \text{N} \\ 0.351\ \text{N-m}\end{matrix}\right\}$$

$$\left\{\begin{matrix}\Delta N \\ \Delta M\end{matrix}\right\}_{shrinkage} = -12.3 \times 10^9 (-57 \times 10^{-6}) \left\{\begin{matrix}0.5090 \\ 0\end{matrix}\right\} = 10^6 \left\{\begin{matrix}0.357\ \text{N} \\ 0\end{matrix}\right\}$$

$$\left\{\begin{matrix}\Delta N \\ \Delta M\end{matrix}\right\}_{relaxation} = \left\{\begin{matrix}3160 \times 10^{-6}(-15 \times 10^6 \\ 3160 \times 10^{-6} \times 0.53(-15 \times 10^6)\end{matrix}\right\} = 10^6 \left\{\begin{matrix}-0.047\ \text{N} \\ -0.025\ \text{N-m}\end{matrix}\right\}$$

The total restraining forces are

$$\left\{\begin{matrix}\Delta N \\ \Delta M\end{matrix}\right\} = 10^6 \left\{\begin{matrix}2.171 + 0.357 - 0.047 \\ 0.0351 + 0 - 0.025\end{matrix}\right\} = 10^6 \left\{\begin{matrix}2.481\ \text{N} \\ 0.326\ \text{N-m}\end{matrix}\right\}$$

With $\bar{E}_{ref} = \bar{E}_c(60, 3)$ the properties of the age-adjusted transformed section are calculated in the same way as in Table 2.4 giving: $\bar{A} = 0.6092\,\text{m}^2$; $\bar{B} = 0.0238\,\text{m}^3$; $\bar{I} = 0.1443\,\text{m}^4$.

Removal of the restraining forces results in the following increments of axial strain and curvature during the period t_1 to t_2 (Equation (2.40)):

$$\left\{\begin{matrix}\Delta\varepsilon_O(t_2, t_1) \\ \Delta\psi(t_2, t_1)\end{matrix}\right\} = 10^{-6} \left\{\begin{matrix}-326 \\ -130\,\text{m}^{-1}\end{matrix}\right\}$$

The corresponding incremental stress and strain distributions are shown in Fig. 2.11(b). (The stresses are calculated by Equation (2.46).)

The stress at $t_2 = 60$ days may be obtained by superposition of the diagrams in Fig. 2.11(a) and (b).

(c)   Instantaneous increments of stress and strain at $t_2 = 60$ days
The bending moment $M = 1850 \times 10^3$ N-m is resisted only by the prestressed beam. The properties of the transformed section are calculated in the same way in Table 2.4 using $E_{ref} = E_c(60) = 37\,\text{GPa}$, giving: $A = 0.5423\,\text{m}^2$; $B = 0.0079\,\text{m}^3$; $I = 0.1207\,\text{m}^4$. Substitution in Equation (2.32) gives the instantaneous increments in axial strain and curvature occurring at t_2:

$$\left\{\begin{matrix}\Delta\varepsilon_O(t_2) \\ \Delta\psi(t_2)\end{matrix}\right\} = 10^{-6} \left\{\begin{matrix}-6 \\ 415\,\text{m}^{-1}\end{matrix}\right\}$$

The corresponding stress and strain distributions are shown in Fig. 2.11(c).

(d)   Changes in stress and strain due to creep, shrinkage and relaxation during the period $t_2 = 60$ days to $t_3 = \infty$.

The age-adjusted moduli of elasticity for the precast beam and the deck slab are

$$\bar{E}_{c1}\,(\infty,\,60) = \frac{37 \times 10^9}{1 + 0.8 \times 2.27} = 13.14\,\text{GPa}\,(1900\,\text{ksi})$$

$$\bar{E}_{c2}\,(\infty,\,3) = \frac{23 \times 10^9}{1 + 0.78 \times 2.40} = 8.01\,\text{GPa}\,(1160\,\text{ksi})$$

The stresses shown in Figs 2.11(a), (b) and (c) are introduced at various ages and thus have different coefficients for creep occurring during the period considered. In the following, the stresses in Figs 2.11(a) and (b) are combined and treated as if the combined stress were introduced when the age of the precast beam is 3 days; thus the creep coefficient to be used is $\varphi(\infty,\,3) - \varphi(60,\,3) = 2.30 - 1.20 = 1.10$. The stress in Fig. 2.11(c) is introduced when the precast beam is 60 days old; thus the coefficient of creep for the period considered is $\varphi(\infty,\,60) = 2.27$.

For more accuracy, the stress in Fig. 2.11(b) which is gradually intro-duced between the age 3 days and 60 days may be treated as if it were introduced at some intermediate time $\bar{t}$, such that:

$$\frac{1}{E_c(\bar{t})}\,[1 + \varphi(60,\,\bar{t})] = \frac{1}{E_c(3)}\,[1 + \chi(60,\,3)\,\varphi(60,\,3)].$$

Using this approach would result in a slightly smaller coefficient than 1.10 adopted above.

The stresses in the precast beam necessary to artificially restrain creep and shrinkage (Equation (2.45)) are:

$$(\Delta\sigma_{c\ \text{restrained}})_{\text{top}} = -13.14 \times 10^9[1.10\left(-121 \times 10^{-6} - \frac{0.29 \times 10^6}{25 \times 10^9}\right)$$

$$+ 2.27\,(-325 \times 10^{-6}) + (-205 \times 10^{-6})]$$

$$= 14.304\,\text{MPa}$$

$$(\Delta\sigma_{\text{c restrained}})_{\text{bot}} = -13.14 \times 10^9[1.10\left(-426 \times 10^{-6} + \frac{1.97 \times 10^6}{25 \times 10^9}\right)$$

$$+ 2.27(255 \times 10^{-6}) + (-205 \times 10^{-6})]$$

$$= 0.106\,\text{MPa}.$$

The stress in the restrained condition in the deck slab is constant over its thickness and is equal to Equation (2.45).

$$\sigma_{\text{c restrained}} = -8.01 \times 10^9 \, (-269 \times 10^{-6}) = 2.155\,\text{MPa}.$$

The properties of the age-adjusted transformed section for the period t_2 to t_3 are calculated in Table 2.5, using $E_{\text{ref}} = \bar{E}_{\text{cl}} \, (\infty, 60) = 13.14\,\text{GPa}$.

The forces necessary to restrain creep, shrinkage and relaxation during the period t_2 to t_3 are (Equations (2.41) to (2.44)):

$$\begin{Bmatrix} \Delta N \\ \Delta M \end{Bmatrix}_{\text{creep}} = -13.14 \times 10^9 \begin{bmatrix} 0.509 & 0 \\ 0 & 0.109 \end{bmatrix}$$

$$\times \begin{Bmatrix} 1.10\left(-289 + \dfrac{0.95 \times 10^{12}}{25 \times 10^9}\right) + 2.27(-6) \\[2mm] 1.10\left(-218 + \dfrac{1.61 \times 10^{12}}{25 \times 10^9}\right) + 2.27(415) \end{Bmatrix} 10^{-6}$$

$$= 10^6 \begin{Bmatrix} 1.937\,\text{N} \\ -1.108\,\text{N-m} \end{Bmatrix}$$

The term between the curly brackets represents the changes in axial strain and curvature that would occur due to creep if it were unrestrained. The deck slab is not included in this equation because no stress is applied on the slab before the period considered.

$$\begin{Bmatrix} \Delta N \\ \Delta M \end{Bmatrix}_{\text{shrinkage}} = -13.14 \times 10^9(-205 \times 10^{-6}) \begin{Bmatrix} 0.509 \\ 0 \end{Bmatrix}$$

$$-8.01 \times 10^9(-269 \times 10^{-6}) \begin{Bmatrix} 0.495 \\ -0.4307 \end{Bmatrix}$$

$$= 10^6 \begin{Bmatrix} 2.437\,\text{N} \\ -0.928\,\text{N-m} \end{Bmatrix}$$

Table 2.5 Properties of the composite age-adjusted transformed section used in calculation of the changes of stress and strain between $t_2 = 60$ days and $t_3 = \infty$.

	Properties of area			Properties of transformed area		
	A (m^2)	B (m^3)	I (m^4)	AE/E_{ref} (m^2)	BE/E_{ref} (m^3)	IE/E_{ref} (m^4)
Concrete of deck slab	0.495	−0.4307	0.3763	0.3017	−0.2625	0.2294
Non-prestressed steel in deck slab	5000×10^{-6}	-4350×10^{-6}	3785×10^{-6}	0.0761	−0.0662	0.05763
Concrete in beam	0.5090	0.0	0.1090	0.5090	0.0	0.1090
Non-prestressed steel in beam	3000×10^{-6}	-210×10^{-6}	1282×10^{-6}	0.0457	−0.0032	0.0195
Prestressed steel	3160×10^{-6}	1675×10^{-6}	887.6×10^{-6}	0.0481	0.0255	0.0135
Properties of the age-adjusted transformed section				0.9806 $\bar{A}$	−0.3064 $\bar{B}$	0.4290 $\bar{I}$

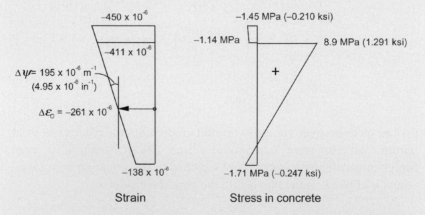

Figure 2.12 Changes in stress and strain in the composite section of Example 2.5 due to creep, shrinkage and relaxation occurring between casting of the deck slab, $t_2 = 60$ days and $t_3 = \infty$.

$$\begin{Bmatrix} \Delta N \\ \Delta M \end{Bmatrix}_{relaxation} = \begin{Bmatrix} 3160 \times 10^{-6}\,(-70 \times 10^{6}) \\ 3160 \times 10^{-6} \times 0.53\,(-70 \times 10^{6}) \end{Bmatrix} = 10^{6} \begin{Bmatrix} -0.221\ \text{N} \\ -0.117\,\text{N-m} \end{Bmatrix}$$

The total restraining forces

$$\begin{Bmatrix} \Delta N \\ \Delta M \end{Bmatrix} = 10^6 \begin{Bmatrix} 1.937 + 2.437 - 0.221 \\ -1.108 - 0.928 - 0.117 \end{Bmatrix} = 10^6 \begin{Bmatrix} 4.153 \text{ N} \\ -2.153 \text{ N-m} \end{Bmatrix}$$

The increments of axial strain and curvature during the period t_2 to t_3 are obtained by substitution in Equation (2.40) and are plotted in Fig. 2.12:

$$\begin{Bmatrix} \Delta\varepsilon_O(t_3, t_2) \\ \Delta\psi(t_3, t_2) \end{Bmatrix} = 10^{-6} \begin{Bmatrix} -261 \\ 195 \text{ m}^{-1} \end{Bmatrix}$$

The corresponding change in stress is calculated by Equation (2.46) and plotted also in Fig. 2.12.

2.6   Summary of analysis of time-dependent strain and stress

The procedure of analysis given in this chapter can be performed in four steps. Figure 2.13 outlines the four steps to determine the instantaneous and the time-dependent changes in strain and stress in a non-cracked prestressed section. For quick reference, the symbols used are defined again below and the four steps summarized.

Notation

A area
B first moment of area
E modulus of elasticity
$\bar{E}_c$ age-adjusted elasticity modulus of concrete
I second moment of area
M bending moment about an axis through O
N normal force at O
P absolute value of prestressing
t time or age of concrete
y coordinate (Fig. 2.5)
σ stress
a ratio of modulus of steel to that of concrete at time t_0
$\bar{a}$ ratio of modulus of elasticity of steel to $\bar{E}_c$
χ aging coefficient of concrete
Δ increment

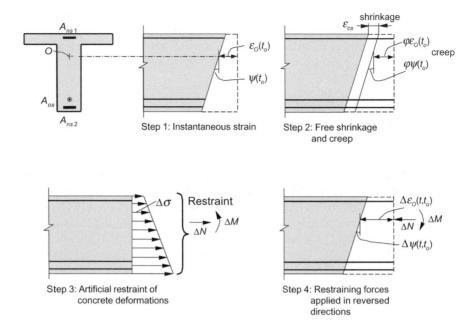

Figure 2.13 Steps of analysis of time-dependent strain and stress.

$\Delta\bar{\sigma}_{\mathrm{pr}}$ reduced relaxation of prestressed steel
ε strain
φ creep coefficient
ψ curvature (slope of strain diagram = $d\varepsilon/dy$)
γ slope of stress diagram (= $d\sigma/dy$)

Subscripts

c concrete
cs shrinkage
ns non-prestressed steel
O arbitrary reference point
0 time of prestressing
ps prestressed steel

Four analysis steps

Step 1   Apply the initial prestressing force and the dead load or other bend-ing moment, which becomes effective at the time of prestressing t_0, on a transformed section composed of A_{c} plus $(\alpha_{\mathrm{ps}}A_{\mathrm{ps}} + A_{\mathrm{ns}}\alpha_{\mathrm{ns}})$. Here the trans-formed section includes only the prestressed and the non-prestressed steel bonded to the concrete at the prestress transfer. Thus, A_{ps} should be included when pre-tentioning is used, but when all the prestressing is post-tensioned in one stage, A_{ps} and the area of the duct should be excluded.

When the structure is statistically indeterminate, the indeterminate normal force and moment should be included in the forces on the section. Determine the resultants N and M of all forces on the section.

Apply Equation (2.19) to determine $\varepsilon_O(t_0)$ and $\psi(t_0)$ which define distribution of the instantaneous strain. Multiplication by $E_c(t_0)$ or application of Equation (2.20) gives $\sigma_O(t_0)$ and $\gamma(t_0)$, which define the instantaneous stress distribution.

Step 2 Determine the hypothetical change, in the period t_0 to t, in strain distribution due to creep and shrinkage if they were free to occur. The strain change at O is equal to $[\varphi(t, t_0) \varepsilon_O(t_0) + \varepsilon_{cs}]$ and the change in curvature is $[\varphi(t, t_0) \psi(t_0)]$.

Step 3 Calculate artificial stress which, when gradually introduced on the concrete during the period t_0 to t, will prevent occurrence of the strain determined in step 2. The restraining stress at any fibre y is (Equations (2.34) and (2.45)):

$$\Delta\sigma_{restrained} = -\bar{E}_c \{\varphi(t, t_0)[\varepsilon_O(t_0) + \psi(t_0)y] + \varepsilon_{cs}\} \tag{2.49}$$

Step 4 Determine by Equation (2.18) a force at O and a moment, which are the resultants of $\Delta\sigma_{restrained}$. The change in concrete strain due to relaxation of the prestressed steel can be artificially prevented by the application, at the level of the prestressed steel, of a force equal to $A_{ps} \Delta\bar{\sigma}_{pr}$; substitute this force by a force of the same magnitude at O plus a couple. Summing up gives $\Delta N_{restrained}$ and $\Delta M_{restrained}$ the restraining normal force and the couple required to prevent artificially the strain change due to creep, shrinkage and relaxation.

To eliminate the artificial restraint, apply $\{\Delta N, \Delta M\}_{restrained}$ in reversed directions on an age-adjusted transformed section composed of A_c plus $(\bar{a}A_{ps} + \bar{a}A_{ns})$; calculate the corresponding changes in strains and stresses by Equations (2.19) and (2.17).

The strain distribution at time t is the sum of the strains determined in steps 1 and 4, while the corresponding stress is the sum of the stresses at t_0 calculated in step 1 and the time-dependent changes calculated in steps 3 and 4 (Equations (2.46)–(2.48)).

Commentary

1   The flow chart in Fig. 2.14 shows how the four steps of analysis can be applied in a general case to determine the instantaneous and

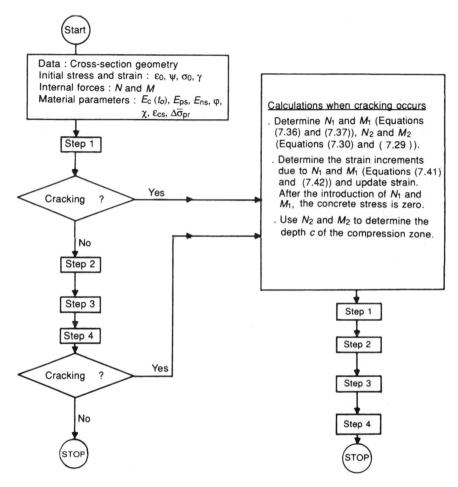

Figure 2.14 Flow chart for calculating stress and strain increments in a section due to a normal force N and a bending moment M introduced at time t_0 and sustained to time t.

time-dependent increments of stress and strain, due to the application at time t_0 of a normal force N and a moment M on a section for which the initial values of stresses and strains are known.

If after step 1, the stress at the extreme fibre exceeds the tensile strength of concrete, the calculation in step 1 must be repeated, using A, B and I of a cracked section, in which the concrete in tension is ignored. The depth of the compression zone c must be determined prior to applying steps 2, 3 and 4 to a cracked section (see Chapter 7). The flow chart also outlines the analysis in a less common case, in which cracking occurs during the period t_0 to t, and thus will be detected only at the end of step 4.

2   Superposition of strains or stresses in the various steps can be done by summing up the increments $\Delta\varepsilon_O$, $\Delta\psi$, $\Delta\sigma_O$ or $\Delta\gamma$. This is possible because of the use of the same reference point O in all steps.

3   The four steps give the stress and strain at time t, without preceding the analysis by an estimate of the loss in tension in the prestressed steel. No empirical equation is involved for loss calculation.

4   The analysis satisfies the requirements of compatibility and equilibrium: the strain changes in concrete and steel are equal at all reinforcement layers; the time-dependent effect changes the distribution of stresses between the concrete and the reinforcements, but does not change the stress resultants.

5   The same four-step analysis applies to reinforced concrete sections without prestressing, simply by setting $A_{ps} = 0$; the effects of cracking will be discussed in Chapter 7.

6   The same procedure can be used for analysis of composite sections, made of more than one type of concrete cast or prestressed in stages, or made of concrete and structural steel.

2.7   Examples worked out in British units

Example 2.6 Stresses and strains in a pre-tensioned section

The pretensioned cross-section shown in Fig. 2.15 is subjected at time t_0 to a prestressing force 600 kip (2700 kN) and a bending moment 10 560 kip-in. (1193 kN-m). Find the extreme fibre stresses at time t_0 immediately after prestressing, and at time t after occurrence of creep, shrinkage and relaxation. The following data are given: $E_c(t_0) = 3600$ ksi (25 GPa); $E_{ns} = 29\,000$ ksi (200 GPa); $E_{ps} = 27\,000$ ksi (190 GPa); $\varphi(t, t_0) = 3$; $\chi = 0.8$; $\Delta\bar\sigma_{pr}(t, t_0) = -13$ ksi (90 MPa); $\varepsilon_{cs}(t, t_0) = -300 \times 10^{-6}$.

Step 1   The reference point O is chosen at the top fibre. The presenting and the bending moment introduced at t_0 are equivalent to ΔN at O and a moment ΔM about an axis through O, calculated by Equation (2.31):

$$\Delta N = -600 \text{ kip}; \qquad \Delta M = -9840 \text{ kip-in.}$$

The ratios of the elasticity moduli E_{ns} and E_{ps} to E_c (t_0) are (Equation (1.33)):

$$\alpha_{ns} = 8.06; \qquad \alpha_{ps} = 7.50.$$

Use of these values to calculate the area properties of the transformed section at time t_0 gives:

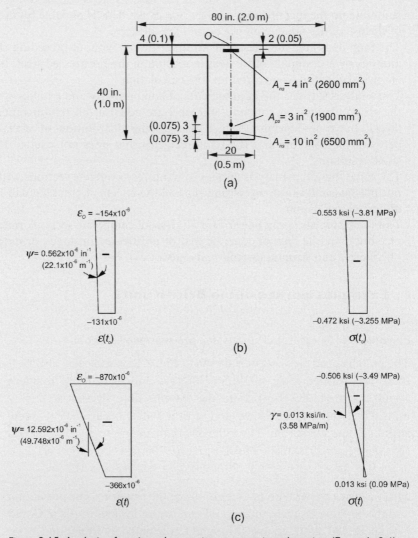

Figure 2.15 Analysis of strain and stress in a pre-tensioned section (Example 2.6): (a) cross-section dimensions; (b) strain and stress at t_0; (c) strain and stress at t.

$A = 1158 \, \text{in}^2$; $B = 19\,819 \, \text{in}^3$; $I = 547\,200 \, \text{in}^4$.

Substitution in Equations (2.19) and (2.17) gives (Fig. 2.15(b)):

$$\Delta\varepsilon_O(t_0) = -154 \times 10^{-6}; \qquad \Delta\psi(t_0) = 0.562 \times 10^{-6} \, \text{in}^{-1}$$

$$\{\Delta\sigma_c(t_0)\}_{\text{top, bot}} = \{-0.553, -0.472\} \, \text{ksi}.$$

Step 2   Hypothetical changes in strain at O and in curvature if creep and shrinkage were free to occur are:

$$(\Delta\varepsilon_O)_{free} = 3(-154 \times 10^{-6}) - 300 \times 10^6 = -762 \times 10^{-6}$$

$$(\Delta\psi)_{free} = 3(0.562 \times 10^{-6}) = 1.69 \times 10^{-6} \text{ in}^{-1}.$$

Step 3   The age-adjusted elasticity modulus (Equation (1.31)):

$$\bar{E}_c(t, t_0) = 1059 \text{ ksi}.$$

The area properties of concrete cross-section are:

$$A_c - 1023 \text{ in}^2; \qquad B_c = 16000 \text{ in}^3; \qquad I_c = 410800 \text{ in}^4.$$

Artificial stress to prevent strain changes due to creep and shrinkage (Equation (2.45)):

$$\{\Delta\sigma_{restrained}\}_{top, bot} = \{0.807, 0.734\} \text{ ksi}.$$

Step 4   Substitution of A_c, B_c, I_c, $\bar{E}_c$, $(-\Delta\varepsilon_O)_{free}$ and $(-\psi)_{free}$ in Equation (2.18) gives the forces necessary to restrain creep and shrinkage:

$$\Delta N_{creep + shrinkage} = 795 \text{ kip} \qquad \Delta M_{creep + shrinkage} = 12151 \text{ kip-in}.$$

Forces necessary to prevent strain change due to relaxation of prestressed steel are (Equation (2.44)):

$$\Delta N_{relaxation} = -39 \text{ kip}; \qquad \Delta M_{relaxation} = -1326 \text{ kip-in}.$$

Summing (Equation (2.41)):

$$\Delta N_{restrained} = 756 \text{ kip}; \qquad \Delta M_{restrained} = 10825 \text{ kip-in}.$$

The ratios of the elasticity moduli E_{ns} and E_{ps} to $\bar{E}_c (t, t_0)$ are (Equation (1.34)):

$$\bar{a}_{ns} = 27.39; \qquad \bar{a}_{ps} = 25.50.$$

The area properties of the age-adjusted section are:

$$\bar{A} = 1483\,\text{in}^2; \qquad \bar{B} = 28\,950\,\text{in}^3; \qquad \bar{I} = 874\,600\,\text{in}^4.$$

Substitution of these three values, $\bar{E}_c$ and $\{-\Delta N,\ -\Delta M\}_{\text{restrained}}$ in Equation (2.19) gives the changes in strain in the period t_0 to t:

$$\Delta\varepsilon_O(t,\ t_0) = -716 \times 10^{-6}; \qquad \Delta\psi(t,\ t_0) = 12.03 \times 10^{-6}\,\text{in}^{-1}.$$

Substitution of these two values in Equation (2.17) and addition of $\{\Delta\sigma_{\text{restrained}}\}$ gives the changes in stress in the period t_0 to t:

$$\{\Delta\sigma(t,\ t_0)\}_{\text{top, bot}} = \{0.047,\ 0.485\}\ \text{ksi}.$$

Addition of these two stress values to the stresses determined in step 1 gives the stresses at time t:

$$\{\Delta\sigma(t)\}_{\text{top, bot}} = \{-0.506,\ 0.013\}\ \text{ksi}.$$

The strain and stress distributions at t_0 and t are shown in Fig. 2.15.

Example 2.7 Bridge section: steel box and post-tensioned slab

Figure 2.16 shows the cross-section of a simply supported bridge of span 144 ft (43.9 m). The deck is made out of precast rectangular segments assembled in their final position, above a structural steel U-shaped section, by straight longitudinal post-tensioned tendons. Each precast segment covers the full width of the bridge. In the longitudinal direction, each segments covers a fraction of the span. At completion of installation of the precast elements, the structural steel section carries, without shoring, a uniform load = 5.4 kip/ft (79 kN/m), representing the weight of concrete and structural steel. Shortly after prestressing, the bridge section is made composite by connecting the deck slab to the structural steel section. This is achieved by the casting of concrete to fill holes in the precast deck at the locations of protruding steel studs welded to the top flanges of the structural steel section. Determine the strain and stress distributions in concrete and structural steel at the mid-span section at time t_0, shortly after prestressing and at time t after occurrence of creep, shrinkage and relaxation.

Consider that the post-tensioning and the connection of concrete to

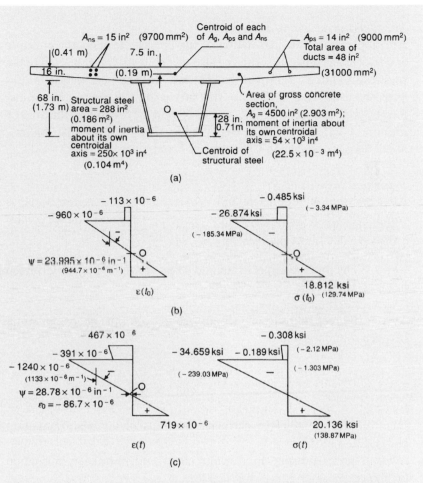

Figure 2.16 Composite cross-section of a bridge (Example 2.7): (a) cross-section dimensions; (b) strain and stress at t_0; (c) strain and stress at t.

structural steel occur at the same time t_0. Assume that during prestressing the deck slides freely over the structural steel section. The following data are given. Initial total prestressing force, excluding loss by friction and anchor set = 2200 kip (9800 kN); creep coefficient $\varphi(t, t_0) = 2.2$; aging coefficient $\chi(t, t_0) = 0.8$; free shrinkage $\varepsilon_{cs}(t, t_0) = -220 \times 10^{-6}$; reduced relaxation $\Delta\bar{\sigma}_{pr} = -6.0$ ksi (−48 MPa); modulus of elasticity of concrete $E_c(t_0) = 4300$ ksi (30 GPa); modulus of elasticity of prestressed steel, non-prestressed steel or structural steel = 28 000 ksi (190 GPa).

Strain and stress at time t_0
At completion of the installation of the precast elements the concrete
and steel act as separate sections; the concrete section is subjected to the
prestressing force 2200 kip at the centroid and the steel section is subjected to a bending moment = 168 000 kip-in. The area of the transformed concrete section composed of A_c and a_{ns} (excluding the area of
ducts) = A = 4535 in^2; with a_{ns} = 6.51. The strain and stress distributions
at this stage are shown in Fig. 2.16(b).

Strain and stress at time t
After connection of concrete and steel at time t_0, the section becomes
composite. Select the reference point O at the centroid of the structural
steel and follow the four analysis steps outlined in Section 2.6:

Step 1   The instantaneous strain and stress at t_0 have been determined
in Fig. 2.16(b).

Step 2   If creep and shrinkage were free to occur, the change in strain
between t_0 and t would be:

$$(\Delta\varepsilon_O)_{free} = -220 \times 10^{-6} - 2.2(112.8 \times 10^{-6})$$
$$= -468.2 \times 10^{-6}; \qquad (\Delta\psi)_{free} = 0.$$

Step 3   The age-adjusted elasticity modulus of concrete (Equation
(1.31)), $\bar{E}_c$ = 1558 ksi.
   The stress necessary to restrain creep and shrinkage (Equation
(2.45)) is:

$$(\sigma_c)_{restraint} = -1558 (-468.2 \times 10^6) = 0.729 \text{ ksi}.$$

Step 4   The area properties of concrete are:

$$A_c = 4471 \text{ in}^2; \qquad B_c = -216 800 \text{ in}^3; \qquad I_c = 10.57 \times 10^6 \text{ in}^4.$$

   The forces necessary to restrain creep and shrinkage (Equation
(2.18)) are:

$$(\Delta N)_{creep + shrinkage} = 3261 \text{ kip}; \qquad (\Delta M)_{creep + shrinkage}$$
$$= -158.2 \times 10^3 \text{ kip-in}.$$

Forces necessary to restrain relaxation (Equation 2.44) are:

$$(\Delta N)_{\text{relaxation}} = -84 \, \text{kip}; \qquad (\Delta M)_{\text{relaxation}} = 4074 \, \text{kip-in.}$$

The total restraining forces are:

$$\Delta N = 3177 \, \text{kip}; \qquad \Delta M = -154.1 \times 10^3 \, \text{kip-in.}$$

Properties of the age-adjusted transformed section are:

$$\bar{A} = 10\,170 \, \text{in}^2; \qquad \bar{B} = -242.1 \times 10^3 \, \text{in}^3; \qquad \bar{I} = 16.29 \times 10^6 \, \text{in}^4.$$

Apply $-\Delta N$ and $-\Delta M$ on the age-adjusted section and use Equation (2.19) to calculate the changes in strain between t_0 and t:

$$\Delta \varepsilon_O(t, t_0) = -86.66 \times 10^{-6}; \qquad \Delta \psi(t, t_0) = 4.784 \times 10^{-6} \, \text{in}^{-1}.$$

Adding the change in strain to the initial strain in Fig. 2.16(b) gives the total strain at time t, shown in Fig. 2.16(c).

The time-dependent change in stress in concrete is calculated by Equation (2.46):

$$[\Delta \sigma_c(t, t_0)]_{\text{top}} = 0.729 + 1558[-86.7 \times 10^{-6} + 4.784 \times 10^{-6} \, (-56)] = 0.177 \, \text{ksi}.$$

$$[\Delta \sigma_c(t, t_0)]_{\text{bot}} = 0.729 + 1558[-86.7 \times 10^{-6} + 4.784 \times 10^{-6} (-40)] = 0.296 \, \text{ksi}.$$

Adding these stresses to the initial stress (Fig. 2.16(b)) gives the total stress at time t shown in Fig. 2.16(c). It is interesting to note the change in the resultant force on the concrete (the area of the concrete cross-section multiplied by the stress at its centroid). The values of the resultants are -2180 and -1130 kip at time t_0 and t respectively. The substantial drop in compressive force is due to the fact that the time-dependent shortening of the concrete is restrained after its attachment to a relatively stiff structural steel section.

2.8   General

A general procedure is presented in Section 2.5 which gives the stress and strain distribution at any time in a composite cross-section accounting for the effects of creep, shrinkage and relaxation of prestress. The analysis employs the aging coefficient χ to calculate the instantaneous strain and creep due to a stress increment which is gradually introduced in the same way as if it were introduced all at once. The analysis employs equations which can be easily programmed on desk calculators or small computers.

We have seen that the axial strain and curvature and the corresponding strain are calculated in two steps with single stage prestress or in more steps when the prestress is applied in more than one step. If a computer is employed, more accuracy can be achieved if the time is divided into increments and a step-by-step calculation is performed to determine the time development of stress and strain (see Sections 4.6 and 5.8). In this case, the aging coefficient χ is not needed and the approximation involved in the assumption used for its derivation is eliminated (see Sections 1.7 and 1.10).

When the equations of Section 2.5 are used, the loss of prestress due to creep, shrinkage and relaxation is accounted for and the effects of the loss on the strain and stress distributions are directly obtained. Among the time-dependent variables obtained by the analysis is the loss of tension in the prestressed steel (Equation 2.48)). However, of more interest in design is the loss of compression in the concrete, because it is this value which governs the possibility of cracking when the strength of concrete in tension is approached. The loss of tension in the prestressed steel is equal in absolute value to the loss of compression on the concrete only in a concrete section without non-prestressed reinforcement. In general, the loss in compression is greater in absolute value than the loss in tension in prestressed steel. The difference represents the compression picked up gradually by the non-prestressed steel as creep, shrinkage and relaxation develop. This will be further discussed in the following chapter, where the time-dependent effects will be considered for sections without non-prestressed steel or with one or more layers of this reinforcement.

Note

1   The loss due to friction or anchor setting are excluded in this discussion; the prestress force P is the force in the tendon excluding the losses due to these effects.

Chapter 3

Special cases of uncracked sections and calculation of displacements

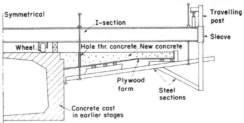

Bow River, Calgary, Canada. Continuous bridge over 430 m (1410 ft). Cantilever slabs are cast on forms moving on box girder cast in earlier stages. (Courtesy KVN, Heavy Construction Division of the Foundation Co. of Canada Ltd. and Stanley Associates Engineering Ltd., Calgary.)

3.1   Introduction

In the preceding chapter, we presented a method of analysis of the time-dependent stresses and strains in composite sections composed of more than one type of concrete or of concrete and structural steel sections with or without prestressed or non-prestressed reinforcement. In the special case when the section is composed of one type of concrete and the prestressed and non-prestressed steel are situated (or approximately considered to be) in one layer, the analysis leads to simplified equations which are presented in this chapter. Another special case which is also examined in this chapter, is a cross-section which has reinforcement without prestressing and we will consider the effects of creep and shrinkage. However, discussion of the effects of cracking is excluded from the present chapter and deferred to Chapters 7 and 8.

The procedures of analysis presented in Chapter 2 and in the present chapter give the values of the axial strain and the curvature at any section of a framed structure at any time after loading. These can be used to calculate the displacements (the translation and the rotation) at any section or at a joint. This is a geometry problem generally treated in books of structural analysis. In Section 3.8 two methods, which will be employed in the chapters to follow, are reviewed: the unit load theory based on the principle of virtual work and the method of elastic weights. The two methods are applicable for cracked or uncracked structures.

3.2   Prestress loss in a section with one layer of reinforcement

The method of analysis in the preceding chapter gives the loss of prestress among other values of stress and strain in composite cross-sections with a number of layers of reinforcements. When the total reinforcement, prestressed and non-prestressed, are closely located, such that it is possible to assume that the total reinforcement is concentrated at one fibre, it may be expedient to calculate the loss of prestress by an equation – to be given below – then find the time-dependent strain and curvature by superposing the effect of the initial forces and the prestress loss.

Consider a prestressed concrete member with a cross-section shown in Fig. 3.1. The section has a total reinforcement area

$$A_{st} = A_{ns} + A_{ps} \tag{3.1}$$

where A_{ns} and A_{ps} are the areas of the non-prestressed reinforcement and the prestress steel, respectively. A reference point O is chosen at the *centroid of the concrete section*. The total reinforcement A_{st} is assumed to be concentrated in one fibre at coordinate y_{st}. The moduli of elasticity of the two types of

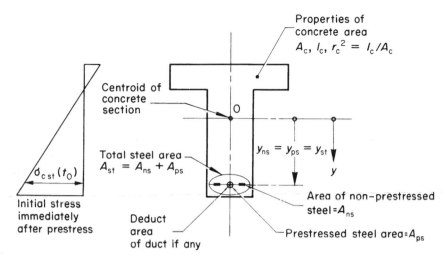

Figure 3.1 Definition of symbols used in Equations (3.1) to (3.14).

reinforcement are assumed to be the same; thus, one symbol E_{st} is used for the modulus of elasticity of the total steel:

$$E_{st} = E_{ns} = E_{ps} \tag{3.2}$$

The prestress force P is applied at age t_0 at the same time as a bending moment and an axial force. It is required to calculate the prestress loss and calculate the changes in axial strain and curvature and in stresses in steel and concrete due to creep, shrinkage and relaxation.

Creep, shrinkage and relaxation cause changes in the distribution of stress in concrete and the two steel types, but any time the sum of the total changes in the forces in the three materials must be zero; thus,

$$\Delta P_c = -\Delta P_{ns} - \Delta P_{ps} \tag{3.3}$$

where ΔP_c is the change in the resultant force on the concrete, ΔP_{ps} is the change in the force in the prestress tendon, and ΔP_{ns} is the change in the force in the non-prestressed reinforcement.

We recall, according to our sign convention (see Section 2.2), that a positive ΔP means an increase in tension. Thus, generally ΔP_c is a positive value while ΔP_{ps} is negative.

The loss in tension in the tendon is equal to the loss of the compressive force on concrete ($\Delta P_c = -\Delta P_{ps}$) only in the absence of non-prestressed reinforcement.

The change of the resultant force on concrete due to creep, shrinkage and relaxation can be calculated by the following equation, which is applicable for post-tensioned and pre-tensioned members:

$$\Delta P_c = -\frac{\varphi(t, t_0)\sigma_{cst}(t_0)A_{st}[E_{st}/E_c(t_0)] + \varepsilon_{cs}(t, t_0)E_{st}A_{st} + \Delta\bar{\sigma}_{pr}A_{ps}}{1 + \dfrac{A_{st}}{A_c}\dfrac{E_{st}}{\bar{E}_c(t, t_0)}\left(1 + \dfrac{y_{st}^2}{r_c^2}\right)} \tag{3.4}$$

This equation can of course be used when the section has only one type of steel (substituting A_{ns} or $A_{ps} = 0$ in Equations (3.1) and (3.4)). When used for a reinforced concrete section, without prestressing Equation (3.4) gives the change in the resultant force in concrete due to creep and shrinkage. We are here assuming that no cracking occurs.

The symbols used in Equation (3.4) are defined below:

A_c = cross-section of concrete
$r_c^2 = I_c/A_c$; where I_c is the moment of inertia of concrete section about an axis through its centroid
A_{st}, E_{st} = the total cross-section area of reinforcement and its modulus of elasticity; one modulus of elasticity is assumed for the two types of steel
y_{st} = the y-coordinate of a fibre at which the total reinforcement is assumed to be concentrated; y is measured downwards from point O, the centroid of concrete area
$E_c(t_0)$ = modulus of elasticity of concrete at age t_0
$\bar{E}_c(t, t_0)$ = age-adjusted elasticity modulus of concrete given by Equation (1.31)
$\varphi(t, t_0)$ = creep coefficient at time t for age at loading t_0.
$\varepsilon_{cs}(t, t_0)$ = the shrinkage that would occur during the period $(t - t_0)$ in free (unrestrained) concrete
$\Delta\bar{\sigma}_{pr}$ = the intrinsic relaxation of the prestressed steel multiplied by the reduction coefficient χ_r (see Fig. 1.4 or Table 1.1)
$\sigma_{cst}(t_0)$ = stress of concrete at age t_0 at the same fibre as the centroid of the total steel reinforcement. This is the instantaneous stress existing immediately after application of prestress (if any) and other simultaneous loading, for example the member self-weight.

Post-tensioned and pre-tensioned members differ only in the calculation of σ_{cst}. With post-tensioning, the area of the cross-section to be used in the calculation of σ_{cst} includes the cross-section areas of the non-prestressed steel and of concrete, excluding the area of prestress duct. With pre-tensioning, the cross-section to be employed is composed of the areas of concrete and prestressed and non-prestressed steel (see Examples 2.2 and 2.3).

The procedure of analysis adopted in Section 2.5 can be employed to calculate ΔP_c and the same result as by Equation (3.4) should be obtained. However, in the special case considered here, Equation (3.4) can be derived more easily as follows.

During the period $(t - t_0)$, the changes of the forces in the prestressed steel and the non-prestressed reinforcement are:

$$\Delta P_{ps} = A_{ps}\Delta\sigma_{ps} \tag{3.5}$$

$$\Delta P_{ns} = A_{ns}\Delta\sigma_{ns} \tag{3.6}$$

The change of resultant force on the concrete during the same period (Equation (3.3)) is

$$\Delta P_c = -(A_{ps}\Delta\sigma_{ps} + A_{ns}\Delta\sigma_{ns}) \tag{3.7}$$

For compatibility, the strain in the non-prestressed steel, in the prestressed steel and in the concrete at the fibre with $y = y_{st}$ must be equal. Thus,

$$\frac{\Delta\sigma_{ns}}{E_{st}} = \frac{\Delta\sigma_{ps} - \Delta\bar{\sigma}_{pr}}{E_{st}}$$

$$= -\frac{\sigma_{cst}(t_0)\varphi(t, t_0)}{E_c(t_0)} + \varepsilon_{cs}(t, t_0) + \frac{1}{\bar{E}_c(t, t_0)}\left(\frac{\Delta P_c}{A_c} + \frac{\Delta P_c y_{st}^2}{I_c}\right) \tag{3.8}$$

The relaxation is deducted from the total change in prestress steel because the relaxation represents a change in stress without associated strain. The first and second terms on the second line are the strains in concrete due to creep and shrinkage. The last term is the instantaneous strain plus creep due to a force ΔP_c. This term represents the strain recovery associated with prestress loss. Solution of Equations (3.5–8) for ΔP_c gives Equation (3.4).

The last term in Equation (3.8) can be presented in this form only when y_{st} is measured from point O the centroid of A_c and I_c is the moment of inertia of the area of concrete about an axis through its centroid. Thus, in determination of the values y_{st} and r_c^2 to be substituted in Equation (3.4), point O must be chosen at the centroid of A_c.

The reduced relaxation $\Delta\bar{\sigma}_{pr}$ to be used in Equation (3.4) is given by Equation (1.7), which is repeated here:

$$\Delta\bar{\sigma}_{pr} = \chi_r\Delta\sigma_{pr} \tag{3.9}$$

where $\Delta\sigma_{pr}$ is the intrinsic relaxation that would develop during a period $(t - t_0)$ in a tendon stretched between two fixed points; χ_r is a reduction factor (see Section 1.5 and Appendix B).

The relaxation reduction factor χ_r may be taken from Table 1.1 or Fig. 1.4.

The value χ_r depends upon the magnitude of the total loss $\Delta\sigma_{ps}$ which is generally not known. Thus for calculation of the total loss due to creep, shrinkage and relaxation, an assumed value of $\Delta\bar{\sigma}_{pr}$ is substituted in Equation (3.4) to give a first estimate of ΔP_c. This answer is used to obtain an improved reduced relaxation value and Equation (3.4) is used again to calculate a better estimate of ΔP_c. In most cases, a first estimate of $\chi_r = 0.7$ followed by one iteration gives sufficient accuracy.

3.2.1 Changes in strain, in curvature and in stress due to creep, shrinkage and relaxation

The changes in axial strain at O or in curvature during the period $(t - t_0)$ may be expressed as the sum of the free shrinkage, the creep due to the prestress and external applied loads plus the instantaneous strain (or curvature) plus creep produced by the force ΔP_c which acts at a distance y_{st} below O. Thus,

$$\Delta\varepsilon_O = \varepsilon_{cs}(t, t_0) + \varphi(t, t_0)\, \varepsilon_O\,(t_0) + \frac{\Delta P_c}{\bar{E}_c(t, t_0)A_c} \tag{3.10}$$

$$\Delta\psi = \varphi(t, t_0)\psi(t_0) + \frac{\Delta P_c y_{st}}{\bar{E}_c(t, t_0)I_c}. \tag{3.11}$$

The change in concrete stress at any fibre due to creep, shrinkage and relaxation is

$$\Delta\sigma_c = \frac{\Delta P_c}{A_c} + \frac{\Delta P_c y_{st}}{I_c}\, y$$

where y is the coordinate of the fibre considered; y is measured downwards from the centroid of concrete area. Substitution of $I_c = A_c r_c^2$ in the last equation gives

$$\Delta\sigma_c = \frac{\Delta P_c}{A_c}\left(1 + \frac{y y_{st}}{r_c^2}\right) \tag{3.12}$$

The changes in stress in the prestressed steel and in the non-prestressed reinforcement caused by creep, shrinkage and relaxation are:

$$\Delta\sigma_{ns} = E_{st}(\Delta\varepsilon_O + y_{ns}\Delta\psi) \tag{3.13}$$

$$\Delta\sigma_{ps} = E_{st}(\Delta\varepsilon_O + y_{ps}\Delta\psi) + \Delta\bar{\sigma}_{pr} \tag{3.14}$$

where y_{ns} and y_{ps} are the y coordinates of a non-prestressed and prestressed steel layer, respectively.

Equation (3.4) may be used to calculate ΔP_c also in the case when the cross-section has more than one layer of reinforcement and when the centroid of the prestressed and the non-prestressed steels do not coincide. In this case, ΔP_c must be considered to act at the centroid of the total steel area and the equations of this section may be used to calculate the changes in strain, in curvature and in stress. The solution in this way involves approximation, acceptable in most practical calculations; the compatibility relations (Equation 3.8) are not satisfied exactly at all layers of reinforcement.

Example 3.1 Post-tensioned section without non-prestressed steel

Calculate the prestress loss due to creep, shrinkage and relaxation in the post-tensioned cross-section of Example 2.2 (Fig. 2.6(a)), ignoring the non-prestressed steel. Assume that the intrinsic relaxation, $\Delta\sigma_{pr\infty} = -115\,\text{MPa}\,(-16.7\,\text{ksi})$. The reduced relaxation value is to be calculated employing the graph in Fig. 1.4, assuming that the characteristic tensile strength of the prestressed steel, $f_{ptk} = 1770\,\text{MPa}\,(257\,\text{ksi})$. Use the calculated prestress loss to find the axial strain and curvature at $t = \infty$.

In this example $A_{ns} = 0$; $A_{st} = A_{ps}$ and $y_{st} = y_{ps}$. Because of the prestress duct, the centroid of concrete section is slightly shifted upwards from centre (Fig. 3.2(a)). With this shift, we have $y_{st}^2 = 0.2059\,\text{m}^2$; $I_c = 42.588 \times 10^{-3}\,\text{m}^4$; $A_c = 0.357\,\text{m}^2$; $r_c^2 = 0.1193\,\text{m}^2$.

For calculation of the stress and the strain at age t_0 immediately after prestressing, consider a plain concrete section subjected to a compressive force $P = 1400\,\text{kN}$ at eccentricity $y_{st} = 0.454\,\text{m}$, plus a bending moment $M = 390\,\text{kN-m}$. Calculation of stress and strain distributions in a conventional way, using a modulus of elasticity equal to $E_c(t_0) = 30\,\text{GPa}$ gives the results shown in Fig. 3.2(b). From this figure, the stress in concrete at age t_0 at the level of the prestress steel is

$$\sigma_{cps}(t_0) = -6.533\,\text{MPa}\quad(-0.9475\,\text{ksi}).$$

As first estimate, assume the relaxation reduction factor $\chi_r = 0.7$; thus the reduced relaxation is

$$\Delta\bar{\sigma}_{pr\infty} = 0.7(-115) = -80.5^1\,\text{MPa}\quad(-11.6\,\text{ksi}).$$

The age-adjusted elasticity modulus of concrete is calculated by Equation (1.31) giving: $\bar{E}_c(t, t_0) = 8.824\,\text{GPa}$. Substitution in Equation (3.4) gives

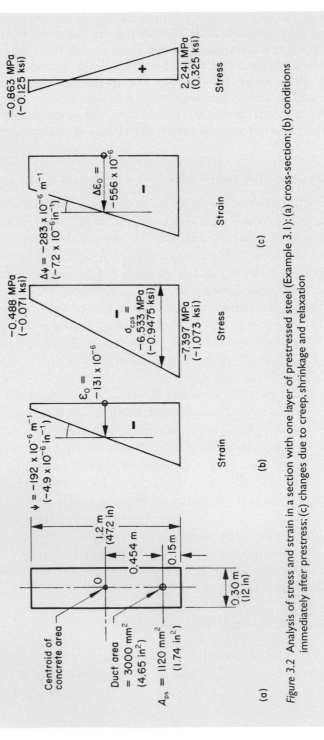

Figure 3.2 Analysis of stress and strain in a section with one layer of prestressed steel (Example 3.1): (a) cross-section; (b) conditions immediately after prestress; (c) changes due to creep, shrinkage and relaxation

$$\Delta P_{\mathrm{c}} = - \left[3(-6.533 \times 10^6)(200/30)1120 \times 10^{-6} + (-240 \times 10^{-6})200 \times 10^9 \right.$$

$$\left. \times 1120 \times 10^{-6} + (-80 \times 10^6)1120 \times 10^{-6}\right]$$

$$\times \left[1 + \frac{1120 \times 10^{-6}}{0.357} \frac{200}{8.824}\left(1 + \frac{0.2059}{0.1193}\right)\right]^{-1}$$

$$= 242.7\,\mathrm{kN} \ (54.56\,\mathrm{kip}).$$

In the absence of non-prestressed steel, $\Delta P_{\mathrm{c}} = -\Delta P_{\mathrm{ps}}$. Thus, the change in stress in the tendon is

$$\Delta\sigma_{\mathrm{ps}} = \frac{\Delta P_{\mathrm{ps}}}{A_{\mathrm{ps}}} = -\frac{\Delta P_{\mathrm{c}}}{A_{\mathrm{ps}}} = -\frac{242.7 \times 10^3}{1120 \times 10^{-6}}\,\mathrm{Pa} = -216.7\,\mathrm{MPa} \ \ (-31.43\,\mathrm{ksi}).$$

We can now find an improved estimate of χ_{r}. The initial stress in the tendon is

$$\sigma_{\mathrm{p0}} = \frac{1400 \times 10^3}{1120 \times 10^{-6}}\,\mathrm{Pa} = 1250\,\mathrm{MPa} \ \ (181\,\mathrm{ksi}).$$

The coefficients λ and Ω are (see Equation (1.7)):

$$\lambda = \frac{\sigma_{\mathrm{p0}}}{f_{\mathrm{ptk}}} = \frac{1250}{1770} = 0.706; \qquad \Omega = \frac{216.7 - 115}{1250} = 0.081.$$

Entering these values in the graph of Fig. 1.4, we obtain $\chi_{\mathrm{r}} = 0.80$.
    An improved estimate of the reduced relaxation is

$$\Delta\overline{\sigma}_{\mathrm{pr\infty}} = 0.80(-115) = -92\,\mathrm{MPa} \ \ (-13.3\,\mathrm{ksi}).$$

Equation (3.5) may be used again to obtain a more accurate value of the prestress loss ($\Delta\sigma_{\mathrm{ps}} = -225.9\,\mathrm{MPa}$). Further iteration is hardly necessary in practical calculations.
    Application of Equations (3.12), (3.10) and (3.11)^2 gives the changes in concrete stress, the axial strain and curvature due to creep, shrinkage and relaxation as follows (Fig. 3.2(c)):

$$(\Delta\sigma_{\mathrm{c}})_{\mathrm{top}} = \frac{242.7 \times 10^3}{0.357}\left(1 + \frac{(-0.596)0.454}{0.1193}\right)\mathrm{Pa}$$

$$= -0.863\,\mathrm{MPa} \ \ (-0.125\,\mathrm{ksi})$$

$$(\Delta\sigma_c)_{bot} = \frac{242.7 \times 10^3}{0.357}\left(1 + \frac{0.604 \times 0.454}{0.1193}\right) Pa = 2.241 \, MPa \, (0.325 \, ksi)$$

$$\Delta\varepsilon_O = -240 \times 10^{-6} + 3(-131 \times 10^{-6}) + \frac{242.7 \times 10^3}{8.824 \times 10^9 \times 0.357}$$

$$= -556 \times 10^{-6}$$

$$\Delta\psi = 3(-192 \times 10^{-6}) + \frac{242.7 \times 10^3 \times 0.454}{8.824 \times 10^9 \times 42.588 \times 10^{-3}}$$

$$= -283 \times 10^{-6} \, m^{-1}(-7.19 \times 10^{-6} \, in^{-1}).$$

Solution of the above problem employing the equations of Section 2.5 would give identical results.

3.3   Effects of presence of non-prestressed steel

Presence of non-prestressed reinforcement in a prestressed member reduces the loss in tension in the prestressed steel. A part of the compressive force introduced by prestressing will be taken by the non-prestressed steel at the time of prestressing and the magnitude of the compressive force in this reinforcement substantially increases with time. As a result, at $t = \infty$ the remaining compressive force in the concrete in a member with non-prestressed steel is much smaller compared with the compressive force on a member without such reinforcement.

The loss of tension in the prestressed steel is equal to the loss of compression in the concrete only when there is no non-prestressed reinforcement. Comparing absolute values, the loss in compression in concrete is generally larger than the reduction in tension in prestressed steel; the difference is the compression picked up by the non-prestressed steel during the period of loss.

The axial strain and curvature are also much affected. Presence of non-prestressed steel substantially decreases the axial strain and curvature at $t = \infty$. Thus, the non-prestressed reinforcement should be accounted for in calculations to predict the displacements, as will be further discussed in Chapter 8.

A comparison is made in Table 3.1 of the results of Examples 2.2 and 3.1, in which two sections are analysed. The data for the two examples are identical, with the only difference being the absence of non-prestressed reinforcement in Example 3.1 (see Figs 2.6(a) and 3.2(a)).

Table 3.1 Comparison of strains, curvatures and losses of prestress in two identical cross-sections with and without non-prestressed reinforcement (Examples 2.2 and 3.1).

	Symbol used	Without non-prestressed reinforcement	With non-prestressed reinforcement
Axial strain immediately after prestress	ε_O	-131×10^{-6}	-126×10^{-6}
Curvature immediately after prestress	ψ	$-192 \times 10^{-6}\,\mathrm{m}^{-1}$	$-170 \times 10^{-6}\,\mathrm{m}^{-1}$
Change in axial strain due to creep, shrinkage and relaxation	$\Delta\varepsilon_O$	-556×10^{-6}	-470×10^{-6}
Change in curvature due to creep, shrinkage and relaxation	$\Delta\psi$	$-283 \times 10^{-6}\,\mathrm{m}^{-1}$	$-128 \times 10^{-6}\,\mathrm{m}^{-1}$
Axial strain at time $t = \infty$	$\varepsilon_O + \Delta\varepsilon_O$	-687×10^{-6}	-596×10^{-6}
Curvature at time $t = \infty$	$\psi + \Delta\psi$	$-475 \times 10^{-6}\,\mathrm{m}^{-1}$	$-298 \times 10^{-6}\,\mathrm{m}^{-1}$
Change in force in prestressed steel (the loss)	$A_{ps}\Delta\sigma_{ps}$	$-243\,\mathrm{kN}$	$-208\,\mathrm{kN}$
Axial force on concrete immediately after prestress	$\int\sigma_c(t_0)dA_c$	$-1400\,\mathrm{kN}$	$-1329\,\mathrm{kN}$
Axial force on concrete at $t = \infty$	$\int\sigma_c(t)dA_c$	$-1157\,\mathrm{kN}$	$-878\,\mathrm{kN}$
Change in force on concrete, ΔP_c	$\int[\sigma_c(t_0) - \sigma_c(t)]dA_c$	$243\,\mathrm{kN}$	$451\,\mathrm{kN}$

3.4 Reinforced concrete section without prestress: effects of creep and shrinkage

The procedure of analysis of Section 2.5 when applied to a reinforced concrete section without prestress can be simplified as shown below.

Consider a reinforced concrete section with several layers of reinforcement (Fig. 3.3), subjected to a normal force N and a bending moment M that produce no cracking. The equations presented in this section give the changes due to creep and shrinkage in axial strain, in curvature and in stress in concrete and steel during a period $(t - t_0)$; where $t > t_0$ and t_0 is the age of concrete at the time of application of N and M. The force N is assumed to act at reference point O chosen at the *centroid of the age-adjusted transformed section* of area A_c plus $[\bar{a}(t, t_0)A_s]$; where $\bar{a}(t, t_0)$ is a ratio of elasticity moduli, given by Equation (1.34).

Following the procedure of analysis in Section 2.5, two equations may be derived for the changes in axial strain and in curvature during the period t_0 to t (the derivation is given at the end of this section):

$$\Delta\varepsilon_O = \eta\{\varphi(t, t_0)[\varepsilon_O(t_0) + \psi(t_0)y_c] + \varepsilon_{cs}(t, t_0)\} \tag{3.15}$$

$$\Delta\psi = \kappa\left[\varphi(t, t_0)\left(\psi(t_0) + \varepsilon_O(t_0)\frac{y_c}{r_c^2}\right) + \varepsilon_{cs}(t, t_0)\frac{y_c}{r_c^2}\right] \tag{3.16}$$

where $\varepsilon_O(t_0)$ and $\psi(t_0)$ are instantaneous axial strain at O and curvature at age

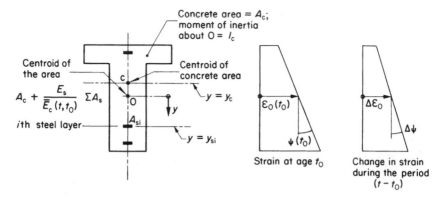

Figure 3.3 Definition of symbols in Equations (3.15) to (3.23) for analysis of effects of creep and shrinkage in a reinforced concrete uncracked section.

t_0; η and κ are the ratios of the area and moment of inertia of the concrete section to the area and moment of inertia of the age-adjusted transformed section (see Section 1.11.1); thus,

$$\eta = A_c/\bar{A} \qquad (3.17)$$

$$\kappa = I_c/\bar{I} \qquad (3.18)$$

A_c and $\bar{A}$ are areas of the concrete section and of the age-adjusted transformed section, I_c and $\bar{I}$ are moments of inertia of the concrete area and of the age-adjusted transformed section about an axis through O the centroid of the age-adjusted transformed section.

The values η and κ, smaller than unity, represent the effect of the reinforcement in reducing the absolute value of the change in axial strain and in curvature due to creep and shrinkage or applied forces. For this reason, η and κ will be referred to as axial strain and curvature reduction coefficients.

$r_c^2 = I_c/A_c$ is the radius of gyration of the concrete area. y_c is the y-coordinate of the centroid c of the concrete area. y is measured in the downward direction from the reference point O; thus in Fig. 3.3, y_c is a negative value.

The change in stress in concrete at any fibre during the period t_0 to t (see Equations (2.45) and (2.46)) is

$$\Delta\sigma_c = \bar{E}_c(t, t_0)\{-[\varepsilon_0(t_0) + \psi(t_0)y]\varphi(t, t_0) - \varepsilon_{cs}(t, t_0) + \Delta\varepsilon_0 + \Delta\psi y\} \qquad (3.19)$$

where $\bar{E}_c$ is the age-adjusted modulus of elasticity of concrete (see Equation (1.31)).

The change in steel stress may be calculated by Equation (2.47).

For the derivation of Equations (3.15), (3.16) and (3.19), apply Equations (2.42) and (2.43) to calculate the forces necessary to artificially prevent deformations due to creep and shrinkage:

$$
\left\{ \begin{matrix} \Delta N \\ \Delta M \end{matrix} \right\}_{\text{creep}} = -\bar{E}_c(t, t_0)\varphi(t, t_0) \begin{bmatrix} A_c & A_c y_c \\ A_c y_c & I_c \end{bmatrix} \left\{ \begin{matrix} \varepsilon_O(t_0) \\ \psi(t_0) \end{matrix} \right\}
\tag{3.20}
$$

$$
\left\{ \begin{matrix} \Delta N \\ \Delta M \end{matrix} \right\}_{\text{shrinkage}} = -\bar{E}_c(t, t_0)\varepsilon_{cs}(t, t_0) \left\{ \begin{matrix} A_c \\ A_c y_c \end{matrix} \right\}
\tag{3.21}
$$

The sum of Equations (3.20) and (3.21) gives the forces necessary to restrain creep and shrinkage:

$$
\left\{ \begin{matrix} \Delta N \\ \Delta M \end{matrix} \right\} = -\bar{E}_c(t, t_0)
$$

$$
\times \left\{ \begin{matrix} \{\varphi(t, t_0)[\varepsilon_O(t_0) + \psi(t_0)y_c] + \varepsilon_{cs}(t, t_0)\}A_c \\ \left[\varphi(t, t_0)\left(\psi(t_0) + \varepsilon_O(t_0)\dfrac{y_c}{r_c^2} \right) + \varepsilon_{cs}(t, t_0)\dfrac{y_c}{r_c^2} \right]A_c r_c^2 \end{matrix} \right\}
\tag{3.22}
$$

The artificial restraint may now be eliminated by application of $-\Delta N$ and $-\Delta M$ on the age-adjusted transformed section. With the reference point O chosen at the centroid of this section, the first moment of area $\bar{B}$ must be zero. The axial strain and curvature due to $-\Delta N$ and $-\Delta M$ can be calculated by Equation (2.29) giving

$$
\left\{ \begin{matrix} \Delta \varepsilon_O \\ \Delta \psi \end{matrix} \right\} = \frac{1}{\bar{E}_c(t, t_0)} \left\{ \begin{matrix} -\Delta N/\bar{A} \\ -\Delta M/\bar{I} \end{matrix} \right\}
\tag{3.23}
$$

Substitution of Equation (3.22) into (3.23) gives Equations (3.15) and (3.16). Equation (3.19) can be obtained by substitution of Equation (2.45) into (2.46).

Example 3.2 Section subjected to uniform shrinkage

Find the stress and strain distribution in the cross-section in Fig. 3.4(a) due to uniform free shrinkage $\varepsilon_{cs}(t, t_0) = -300 \times 10^{-6}$, using the following data: $E_c(t_0) = 30\,\text{GPa}$ (4350 ksi); $E_s = 200\,\text{GPa}$ (29 000 ksi); $\varphi(t, t_0) = 3$; $\chi = 0.8$. The section dimensions and reinforcement areas are given in Fig. 3.4(a).

The age-adjusted modulus of elasticity of concrete and the corresponding modular ratio are (Equations (1.31) and (1.34)):

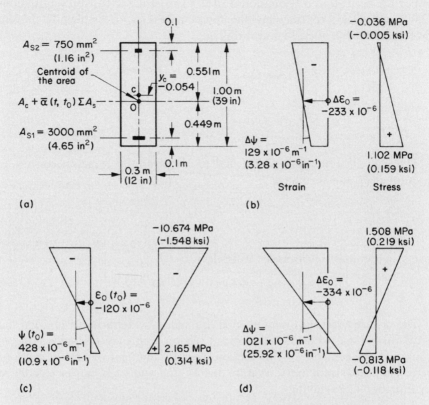

Figure 3.4 Analysis of changes in stress and in strain due to shrinkage and creep in a reinforced concrete section (Examples 3.2 and 3.3): (a) cross-section dimensions; (b) changes in stress and strain due to shrinkage; (c) stress and strain at age t_0 due to axial force of $-1300\,\mathrm{kN}$ ($-292\,\mathrm{kip}$) at midheight and a bending moment of $350\,\mathrm{kN\text{-}m}$ ($3100\,\mathrm{kip\text{-}in}$); (d) changes in stress and strain due to creep.

$$\bar{E}_c(t,\, t_0) = \frac{30 \times 10^9}{1 + 0.8 \times 3} = 8.824\,\mathrm{GPa}\ (1280\,\mathrm{ksi})$$

$$\bar{a}(t,\, t_0) = \frac{200}{8.824} = 22.665.$$

The age-adjusted transformed section composed of A_c plus $\bar{a}A_s$ has a centroid O at a distance $0.551\,\mathrm{m}$ below the top fibre. The centroid of the concrete area is at $0.497\,\mathrm{m}$ below top; thus, $y_c = -0.054$. The area and moment of inertia of concrete section about an axis through O are:

$$A_c = 0.2963 \, \text{m}^2 \quad I_c = 25.26 \times 10^{-3} \, \text{m}^4 \quad r_c^2 = I_c/A_c = 84.75 \times 10^{-3} \, \text{m}^2.$$

The area and moment of inertia of the age-adjusted transformed section about an axis through O are:

$$\overline{A} = 0.3811 \, \text{m}^2 \qquad \overline{I} = 37.50 \times 10^{-3} \, \text{m}^4.$$

The axial strain and curvature reduction coefficient (Equations (3.17) and (3.18)) are:

$$\eta = \frac{0.2963}{0.3811} = 0.777 \qquad \kappa = \frac{25.26}{37.50} = 0.674.$$

Substitution in Equations (3.15) and (3.16) gives the changes in axial strain and in curvature due to shrinkage:

$$\Delta\varepsilon_O = 0.777(-300 \times 10^{-6}) = -233 \times 10^{-6}$$

$$\Delta\psi = 0.674 \, (-300 \times 10^{-6}) \frac{-0.054}{84.45 \times 10^{-3}}$$

$$= 129 \times 10^{-6} \, \text{m}^{-1} \, (3.23 \, \text{in}^{-1}).$$

The changes in concrete stress due to shrinkage (Equation (3.19)) are:

$$(\Delta\sigma_c)_{\text{top}} = 8.824 \times 10^9 [- (-300) + (-233) + 129(-0.551)]10^{-6} \, \text{Pa}$$

$$= -0.036 \, \text{MPa} \, (-0.005 \, \text{ksi}).$$

$$(\Delta\sigma_c)_{\text{bot}} = 8.824 \times 10^9 [- (-300) + (-233) + 129(0.449)]10^{-6} \, \text{Pa}$$

$$= 1.102 \, \text{MPa} \, (0.159 \, \text{ksi}).$$

The changes in stress and strain distributions caused by shrinkage are shown in Fig. 3.4(b).

Example 3.3 Section subjected to normal force and moment

The same cross-section of Example 3.2 (Fig. 3.4(a)) is subjected at age t_0 to an axial force $= -1300 \, \text{kN}$ at mid-height and a bending moment of

350 kN-m. It is required to find the changes during the period $(t - t_0)$ in axial strain, curvature and in concrete stress due to creep. Use the same data as in Example 3.2 but do not consider shrinkage. Assume no cracking.

The applied forces are a bending moment of 350 kN-m and an axial force of -1300 kN at mid-height. Replacing these by equivalent couple and axial force at the reference point O, gives (see Fig. 3.4(a)):

$$N = -1300 \text{kN} \quad (-292 \text{kip}) \qquad M = 350 + 1300(0.051)$$

$$= 416.3 \text{kN-m} \ (3685 \text{kip-in}).$$

These two values are substituted in Equation (2.32) to give the instantaneous axial strain and curvature:

$$\varepsilon_O(t_0) = -120 \times 10^{-6} \qquad \psi(t_0) = 428 \times 10^{-6} \text{m}^{-1} \quad (10.9 \text{in}^{-1}).$$

The stress and strain distributions at age t_0 are shown in Fig. 3.4(c). The modulus of elasticity of concrete used for calculating the values of this figure is $E_c(t_0) = 30 \text{GPa}$.

The values $\bar{E}_c(t, t_0)$, $\bar{a}(t, t_0)$, η and κ are the same as in Example 3.2.

Substitution in Equations (3.15) and (3.16) gives the changes in axial strain and curvature due to creep (Fig. 3.4(d)):

$$\Delta\varepsilon_O = 0.777\{3[-120 + 428(-0.054)]10^{-6}\} = -334 \times 10^{-6}$$

$$\Delta\psi = 0.674 \left[3\left(428 + (-120)\frac{-0.054}{84.45 \times 10^{-3}} \right)10^{-6} \right]$$

$$= 1021 \times 10^{-6} \text{m}^{-1} \quad (25.92 \times 10^{-6} \text{in}^{-1}).$$

The corresponding changes in concrete stress (Equation (3.19)) are

$$(\Delta\sigma_c)_{\text{top}} = 8.824 \times 10^9 \{-[-120 + 428(-0.551)]3$$

$$+ (-334) + 1021(-0.551)\}$$

$$= 1.508 \text{MPa} \ (0.219 \text{ksi})$$

$$(\Delta\sigma_c)_{\text{bot}} = 8.824 \times 10^9 \{-[-120 + 428(0.449)]3 + (-334)$$

$$+ 1021(0.449)\} = -0.813 \text{MPa} \ (-0.118 \text{ksi}).$$

3.5 Approximate equations for axial strain and curvature due to creep

The changes in axial strain and curvature due to creep and shrinkage in a reinforced concrete section without prestressing subjected to a normal force and a bending moment are given by Equations (3.15) and (3.16). When considering only the effect of creep the equations become:

$$\Delta\varepsilon_O = \eta\{\varphi(t, t_0)[\varepsilon_O(t_0) + \psi(t_0)y_c])\} \tag{3.24}$$

$$\Delta\psi = \kappa\left[\varphi(t, t_0)\left(\psi(t_0) + \varepsilon_O(t_0)\frac{y_c}{r_c^2}\right)\right] \tag{3.25}$$

where $\varepsilon_O = \varepsilon_O(t_0)$ is the instantaneous strain at the reference point chosen at the *centroid of the age-adjusted transformed section*; $\psi(t_0)$ is the instantaneous curvature; y_c is the y-coordinate of point c, the centroid of concrete area; $r_c^2 = I_c/A_c$ with A_c and I_c being the area of concrete and its moment of inertia about an axis through O; $\varphi(t, t_0)$ are creep coefficients; η and κ are axial strain and curvature reduction coefficients (see Equations (3.17) and (3.18)).

When the section is subjected only to an axial force at O, or to a bending moment without axial force, Equations (3.24) and (3.25) may be approximated as follows:

(a) *Creep due to axial force*: The change in axial strain due to creep in a reinforced section subjected to axial force

$$\Delta\varepsilon_O \simeq \eta\varepsilon_O(t_0)\varphi(t, t_0). \tag{3.26}$$

(b) *Creep due to bending moment*: The change in curvature due to creep in a reinforced concrete section subjected to bending moment

$$\Delta\psi \simeq \kappa\psi(t_0)\varphi(t, t_0). \tag{3.27}$$

Equation (3.26) is derived from Equation (3.24) ignoring the term $[\psi(t_0)y_c]$ because it is small compared to $\varepsilon_O(t_0)$. Similarly, ignoring the term $[\varepsilon_O(t_0)y_c/r_c^2]$ in Equation (3.25), leads to Equation (3.27).

If the section is without reinforcement, $\Delta\varepsilon_O$ and $\Delta\psi$ due to creep would simply be equal to φ times the instantaneous values. In a section with reinforcement, we need to multiply further by the reduction coefficients η and κ.

3.6 Graphs for rectangular sections

The graphs in Fig. 3.5, for rectangular non-cracked sections, can be employed to determine the position of the centroid O and moment of inertia I (or $\bar{I}$)

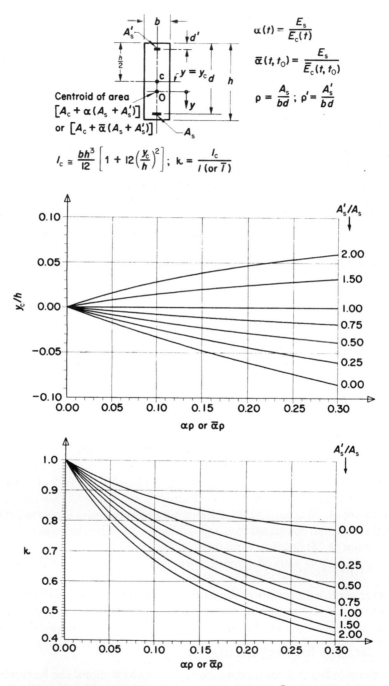

$$\alpha(t) = \frac{E_s}{E_c(t)}$$

$$\bar{\alpha}(t, t_0) = \frac{E_s}{\bar{E}_c(t, t_0)}$$

$$\rho = \frac{A_s}{bd}\;;\; \rho' = \frac{A_s'}{bd}$$

$$I_c \cong \frac{bh^3}{12}\left[1 + 12\left(\frac{y_c}{h}\right)^2\right];\; k = \frac{I_c}{I \text{ (or } \bar{I})}$$

Figure 3.5 Position of the centroid and moment of inertia I (or $\bar{I}$) of transformed (or age-adjusted transformed) non-cracked rectangular section about an axis through the centroid ($d' = 0.1\ h; d = 0.9\ h$).

about an axis through O of the transformed (or age-adjusted transformed) section. The transformed section is composed of the area of concrete $A_c \simeq bh$ plus aA_s (or $\bar{a}A_s$), where (see Equations (1.31) and (1.34))

$$a = a(t_0) = E_s/E_c(t_0) \tag{3.28}$$

$$\bar{a} = \bar{a}(t, t_0) = E_s[1 + \chi\varphi(t, t_0)]/E_c(t_0). \tag{3.29}$$

b and h are breadth and height of the section.

The values of I (or $\bar{I}$) may be used in the calculations for the instantaneous curvature by Equation (2.16) or the change in curvature due to creep and shrinkage by Equation (2.40) (setting $\bar{B} = 0$).

The top graph in Fig. 3.5 gives the coordinate y_c of the centroid of the concrete area (mid-height of the section) with respect to point O. It is to be noted that in the common case when A_s is larger than A_s', y_c has a negative value. A_s and A_s' are the cross-section areas of the bottom and top reinforcement (Fig. 3.5).

The bottom graph in Fig. 3.5 gives the curvature reduction coefficient:

$$\kappa = \frac{I_c}{I \, (\text{or } \bar{I})} \tag{3.30}$$

where I_c is the moment of inertia of the concrete area, A_c about an axis through O. I_c is given by

$$I_c \simeq \frac{bh^3}{12}\left(1 + 12\frac{y_c^2}{h^2}\right). \tag{3.31}$$

In this equation, the area A_c is considered equal to bh and its centroid at mid-height. In other words, the space occupied by the reinforcement is ignored. The graphs in Fig. 3.5 are calculated assuming that the distance between the centroid of the top or bottom reinforcement and the nearby extreme fibre is equal to $0.1\,h$. A small error results when the graphs are used with this distance between $0.05\,h$ and $0.15\,h$.

3.7 Multi-stage prestressing

Consider a cross-section with a number of prestress tendons which are pre-stressed at different stages of construction. This is often used in bridge construction where ducts are left in the concrete for the prestress cables to be inserted and prestressed in stages to suit the development of forces due to the structure self-weight as the construction proceeds.

In the procedure presented in Section 2.5, with one-stage prestressing, the axial strain and curvature were calculated in two steps: one for the instantaneous values occurring at time t_0 and the other for the increments developing during the period t_0 to t due to creep, shrinkage and relaxation. With multi-stage prestress, the two steps are repeated for each prestress stage.

Assume that the prestress is applied at age t_0 and t_1 and we are interested in the stress and strain at these two ages and at a later age t_2. The analysis is to be done in four steps to calculate the following:

1   $\varepsilon_O(t_0)$ and $\psi(t_0)$ are the instantaneous strain at reference point O and the curvature immediately after application of the first prestress.
2   $\Delta\varepsilon_O(t_1, t_0)$ and $\Delta\psi(t_1, t_0)$ are the changes in strain at reference point O and in curvature during the period t_0 to t_1.
3   $\Delta\varepsilon_O(t_1)$ and $\Delta\psi(t_1)$ are the additional instantaneous strain at reference point O and curvature immediately after second prestress.
4   $\Delta\varepsilon_O(t_2, t_1)$ and $\Delta\psi(t_2, t_1)$ are the additional change in strain at reference point O and curvature during the period t_1 to t_2.

In each of the four steps, appropriate values must be used for the properties of the cross-section, the modulus of elasticity of concrete, the shrinkage and creep coefficients and the relaxation; all these values vary according to the age or the ages considered in each step.

It is to be noted that when the prestress is introduced in stage 2, instantaneous prestress loss occurs in the tendons prestressed in stage 1. This is accounted for in the increments calculated in step 3.

3.8   Calculation of displacements

In various sections of Chapters 2 and 3 equations are given for calculation of the axial strain and the curvature and the changes in these values caused by temperature, creep and shrinkage of concrete and relaxation of prestressed steel.

The present section is concerned with the methods of calculation of displacements in a framed structure for which the axial strain and curvature are known at various sections. The term 'displacement' is used throughout this book to mean a translation or a rotation at a coordinate. A coordinate is simply an arrow drawn at a section or a joint of a structure to indicate the location and the positive direction of a displacement.

Once the axial strain and curvature are known, calculation of the displacement is a problem of geometry and thus the methods of calculation are the same whether the material of the structure is linear or non-linear and whether cracking has occurred or not.

3.8.1   Unit load theory

The most effective method to find the displacement at a coordinate j is the unit load theory, based on the principle of virtual work.[3] For this purpose, a fictitious virtual system of forces in equilibrium is related to the actual displacements and strains in the structure. The virtual system of forces is composed of a single force $F_j = 1$ and the corresponding reactions at the supports, where the displacements in actual structure are known to be zero. When shear deformations are ignored the displacement at any coordinate j on a plane frame is given by:

$$D_j = \int \varepsilon_O N_{uj}\, dl + \int \psi M_{uj}\, dl \qquad (3.32)$$

where ε_O and ψ are the axial strain at a reference point O and the curvature ψ in any cross-section of the frame; N_{uj} and M_{uj} are the axial normal force and bending moment at any section due to unit virtual force at coordinate j. The cross-section is assumed to have a principal axis in the plane of the frame and the reference point O is arbitrarily chosen on this principal axis. The axial force N_{uj} acts at O and M_{uj} is a bending moment about an axis through O. The integral in Equation (3.32) is to be performed over the length of all members of the frame.

The principle of virtual work relates the deformations of the actual structure to any virtual system of forces in equilibrium. Thus, in a statically indeterminate structure, the unit virtual load may be applied on a released *statically determinate* structure obtained by removal of redundants. This results in an important simplification of the calculation of N_{uj} and M_{uj} and in the evaluation of the integrals in Equation (3.32). For example, consider the transverse deflection at a section of a continuous beam of several spans. The unit virtual load may be applied at the section considered on a released structure composed of simple beams. Thus, M_{uj} will be zero for all spans except one while N_{uj} is zero everywhere.

Only the second integral in Equation (3.32) needs to be evaluated and the value of the integral is zero for all spans except one.

3.8.2   Method of elastic weights

The rotation and the deflection in a beam may be calculated respectively as the shearing force and the bending moment in a *conjugate beam* subjected to a transverse load of intensity numerically equal to the curvature ψ for the actual beam. This load is referred to as elastic load (Fig. 3.6).

The method of elastic weights is applicable for continuous beams. The conjugate beam is of the same length as the actual beam, but the conditions of the supports are changed,[4] whereas for a simple beam, the conjugate and actual beam are the same (Fig. 3.6(a) and (b)).

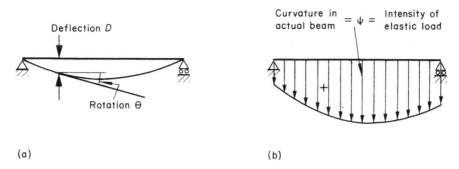

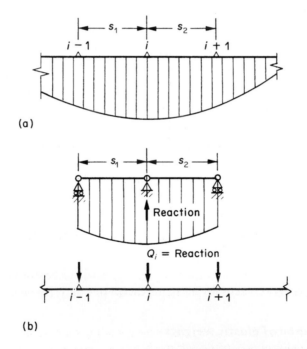

Figure 3.6 Actual and conjugate beams: (a) deflection of actual beam; (b) elastic load on conjugate beam.

Figure 3.7 Equivalent concentrated loads which produce the same bending moment at the nodes and reactions at the supports of a statically determinate beam subjected to variable load: (a) variable load intensity; (b) equivalent concentrated load at *i*.

The ψ-diagram in the actual beam is treated as the load on the conjugate beam as shown in Fig. 3.6(b). Positive curvature is positive (downward) load. It can be shown that the shear V and the moment M in the conjugate beam are equal respectively to the rotation θ and the deflection D at the corresponding point of the actual beam. The calculation of the reactions and bending

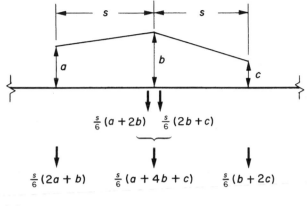

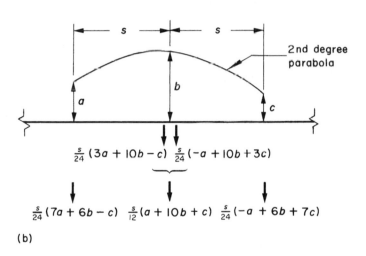

(a)

(b)

Figure 3.8 Equivalent concentrated load for: (a) straight-line; and (b) parabolic varying load.

moments in a beam due to irregular elastic loading may be simplified by the use of equivalent concentrated elastic loads applied at chosen node points (Fig. 3.7(a), (b)). At any node i the equivalent concentrated load Q_i is equal and opposite to the sum of the reactions at i of two simply supported beams between $i-1$, i and $i+1$, carrying the same elastic load as the conjugate beam between the nodes considered (Fig. 3.7(b)).

The equivalent concentrated loads for straight-line and second-degree parabolic variations are given in Fig. 3.8(a) and (b). The formulae for the parabolic variation can, of course, be used to approximate other curves.

The method of elastic weights and the equivalent concentrated loads are used to derive a set of equations presented in Appendix C for the elongation, end rotations and central deflection of a beam in terms of the values of axial strain and curvature at a number of equally spaced sections of a simple beam (Fig. C.1). The same equations are applicable to a general member of a plane frame, but the equations in this case give deflections and rotations measured from a straight line joining the two displaced ends of the member (Line A 'B' in Fig. C.2). Appendix C also includes equations for the displacements of a cantilever.

Example 3.4 Simple beam: derivation of equations for displacements

Express the displacements D_1 to D_4 in the simple beam in Fig. C.1(a) in terms of the axial strain $\{\varepsilon\}$ and the curvature $\{\psi\}$ at three sections (Fig. C.1(b)).

Assume parabolic variation of ε_O and ψ between the three sections.

Equivalent concentrated elastic loads at the three nodes are (see Fig. 3.8(b)):

$$\{Q\} = \begin{bmatrix} 7l/48 & l/8 & -l/48 \\ l/24 & 5l/12 & l/24 \\ -l/48 & l/8 & 7l/48 \end{bmatrix} \{\psi\} \tag{a}$$

The first column of the 3×3 matrix represents the equivalent concentrated forces at the three nodes when $\psi_1 = 1$, while $\psi_2 = \psi_3 = 0$. The other two columns are derived in a similar way.

The displacements may be expressed in terms of $\{Q\}$:

$$D_2 = [0 - 0.5 - 1]\{Q\} \tag{b}$$

$$D_3 = [1\ 0.5\ 0]\{Q\} \tag{c}$$

$$D_4 = l[0\ 0.25\ 0]\ \{Q\} \tag{d}$$

The first element in each of the 1×3 matrices in the last three equations represents the shear at one of the two ends or the bending moment at the centre of the conjugate simple beam, subjected to $Q_1 = 1$, while $Q_2 = Q_3 = 0$. The other two elements of each matrix are derived in a similar way.

Substitution of Equation (a) in each of equations (b), (c) and (d) respectively gives Equations (C.6), (C.7) and (C.8).

The sum of the elements of the first column in the 3×3 matrix in Equation (a) is $l/6$; this is equal to the total elastic load on the beam which is the integral $\int \psi \, dl$ when $\psi_1 = 1$, while $\psi_2 = \psi_3 = 0$. The sum of the elements in the second and third column of the matrix is equal to similar integrals.

The displacement at coordinate 1 in Fig. C1(a) is equal to the change in length of the beam; thus

$$D_1 = \int \varepsilon_O \, dl. \tag{e}$$

This integral is to be evaluated over the length of the beam for the variable ε_O when $(\varepsilon_O)_1 = 1$ while $(\varepsilon_O)_2 = (\varepsilon_O)_3 = 0$, and this procedure has to be repeated two more times, each time setting one of the ε_o values equal to unity and the others zero. Thus, summing the elements in each column of the matrix in Equation (a) and changing the variable ψ to ε_O, we can express the displacement D_1 in terms of the axial strain at the three nodes:

$$D_1 = \left[\frac{l}{6} \; \frac{2l}{3} \; \frac{l}{6} \right] \{ \varepsilon_O \}.$$

Appendix C lists a series of expressions derived by the same procedure as Example 3.4. The variation of ε_O and ψ is assumed either linear or parabolic and the number of nodes either 3 or 5.

Example 3.5 Simplified calculation of displacements

Use the values of the axial strain and curvature at mid-span and at the ends of the post-tensioned simple beam in Fig. 2.7 to calculate the vertical deflection at point C, the centre of the span and the horizontal movement of the roller at B at time t, after occurrence of creep, shrinkage and relaxation. Assume parabolic variation of the axial strain and curvature between the three sections.

We prepare the vectors $\{ \varepsilon_O \}$ and $\{ \psi \}$ to be used in the equations of Appendix C (see table p. 94):

The deflection at the centre (Equation (C.8)) is given by

$$\frac{(18.6)^2}{96} [1 \; 10 \; 1] \begin{Bmatrix} 64 \\ -298 \\ 64 \end{Bmatrix} 10^{-6} = -0.0103 \, \text{m} = -10.3 \, \text{mm}.$$

The minus sign indicates upward deflection.

	Left end	Mid-span	Right end
Axial strain at t_0, $\varepsilon_O(t_0)$	-126×10^{-6}	-126×10^{-6}	-126×10^{-6}
Change in axial strain $\Delta\varepsilon_O$	-479×10^{-6}	-470×10^{-6}	-479×10^{-6}
Total axial strain at time t	-605×10^{-6}	-596×10^{-6}	-605×10^{-6}
Curvature at t_0, $\psi(t_0)$	4×10^{-6}	$-170 \times 10^{-6}\,\mathrm{m}^{-1}$	4×10^{-6}
Change in curvature, $\Delta\psi$	60×10^{-6}	$-128 \times 10^{-6}\,\mathrm{m}^{-1}$	60×10^{-6}
Total curvature at time t	64×10^{-6}	$-298 \times 10^{-6}\,\mathrm{m}^{-1}$	64×10^{-6}

The change of length at the level of the reference axis (Equation (C.5)) is

$$\frac{18.6}{6}[1\ 4\ 1]\begin{Bmatrix} -605 \\ -596 \\ -605 \end{Bmatrix}10^{-6} = -0.0111\,\mathrm{m} = -11.1\,\mathrm{mm}.$$

(Here Equation (C.1) could have been used, assuming straight-line variation between the section at mid-span and the two ends, but the answer will not change within the significant figures employed.)

Rotation at the left end A (Equation (C.7)) is

$$\frac{18.6}{6}[1\ 2\ 0]\begin{Bmatrix} 64 \\ -298 \\ 64 \end{Bmatrix}10^{-6} = -1.65 \times 10^{-3}\ \text{radian}.$$

The minus sign means an anticlockwise rotation.

The same rotation but opposite sign occurs at B. The change in length of AB (on the bottom fibre) is

$$-0.0111 + 2 \times 0.6(-1.65 \times 10^{-3}) = -0.0131\,\mathrm{m}.$$

Horizontal movement of the roller at B is $-0.0131\,\mathrm{m} = -13.1\,\mathrm{mm}$. The minus indicates shortening of AB, and hence B moves to the left.

3.9 Example worked out in British units

Example 3.6 Parametric study.

The structure shown in Fig. 3.9(a) represents a 1 ft wide (305 mm) strip of a post-tensioned, simply supported solid slab. At time t_0, the structure is subjected to dead load $q = 0.40$ kip/ft (5.8 kN/m) and an initial pre-stressing force $P = 290$ kip (1300 kN), which is assumed constant over the length. The objectives of this example are to study the effects of the presence of the non-prestressed steel on the stress distributions between concrete and the reinforcement and on the mid-span deflection at time t after occurrence of creep, shrinkage and relaxation. Non-prestressed steel of equal cross-section area A_{ns} is provided at top and bottom. The steel ratio $\rho_{ns} = A_{ns}/bh$, is considered variable between zero and 1 per cent.

The modulus of elasticity of concrete $E_c(t_0) = 4350$ ksi (30 GPa); the change in E_c with time is ignored. The modulus of elasticity of the prestressed and the non-prestressed steel $E_s = 29\,000$ ksi (200 GPa). Other data are:

$$\varphi(t, t_0) = 3.0; \qquad \varepsilon_{cs}(t, t_0) = -300 \times 10^{-6};$$
$$\Delta\bar{\sigma}_{pr} = -9.3\,\text{ksi}\,(-64\,\text{MPa}).$$

The effects of varying the values of φ and ε_{cs} on the results will also be discussed.

The dead load q produces a bending moment at mid-span = 1500 kip-in (169 kN-m).

Only the results of the analyses are given and discussed below. For ease in verifying the results, the simplest cross-section is selected. Also the variation of the initial prestressing force P because of friction is ignored and the difference in the cross-section area of the tendon and the area of the prestressing duct is neglected.

Table 3.2 gives the concrete stresses at midspan at time t after occurrence of creep, shrinkage and relaxation. It can be seen that the stress at the bottom fibre varies between -1026 and -502 psi (-7.08 and -3.46 MPa) as the non-prestressed steel ratio, ρ_{ns} is increased from zero to 1%.

In other words, ignoring the non-prestressed steel substantially overestimates the compressive stress provided by prestressing to prevent or to control cracking by subsequent live load; the overestimation is of the same order of magnitude as the tensile strength of concrete. The

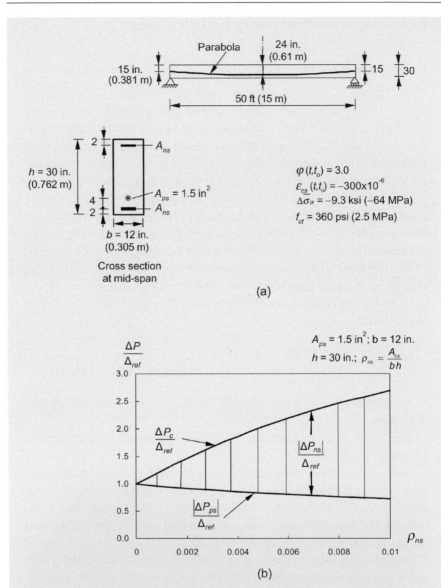

Figure 3.9 Post-tensioned slab (Example 3.6): (a) slab dimensions and material parameters; (b) relative time-dependent change in forces in concrete, prestressed steel and non-prestressed steel at mid-span cross-section.

compressive stress reserve, commonly intended to counteract the tension due to live load, is substantially eroded as a result of the presence of the non-prestressed steel. On the other hand, the non-prestressed steel is beneficial in controlling the width of cracks (see Example 7.6).

Table 3.2 Stress and deflection at mid-span in non-cracked slab, Example 3.6

Non-prestressed steel ratio ρ_{ns} (percent)		0	0.2	0.4	0.6	0.8	1.0	0.4 with reduced φ & ε_{cs}
Concrete stresses at time t (psi)	σ_{top}	−302	−276	−246	−215	−184	−155	−250
	σ_{bot}	−1026	−879	−759	−659	−574	−502	−969
Change of force in the three materials between t_0 and t (kips) — Concrete	ΔP_c	52	76	97	114	128	140	59
Non-prestressed steel	ΔP_{ns}	0	−28	−52	−72	−88	−102	−28
Prestressed steel	ΔP_{ps}	−52	−48	−45	−42	−40	−38	−30
Deflection at time t before application of the live load; $(10^{-3}\,in)$		−923	−794	−696	−621	−560	−510	−553
Ratio of deflection at time t, before application of live load to the instantaneous deflection			2.56	2.32	2.13	2.00	1.88	1 78	1.69
Steel stresses at time t, before live load application (ksi)	σ_{ns} (bot)	−36	−33	−30	−28	−26	−24	−20
	σ_{ps}	159	161	163	165	167	168	173

Table 3.2 also gives the force changes ΔP_c, ΔP_{ns} and ΔP_{ps} in the concrete, the non-prestressed and the prestressed steel due to creep, shrinkage and relaxation. The sign of ΔP_c is positive, indicating a decrease of the initial compressive force in concrete. The negative ΔP_{ns} indicates an increase in compression. Also, the negative ΔP_{ps} indicates loss of tension in the prestressing tendon.

The non-dimensional graphs in Fig. 3.9(b) represent the variation of ρ_{ns} versus $(|\Delta P_{ps}|/\Delta_{ref})$ or $(\Delta P_c/\Delta_{ref})$; where Δ_{ref} is a reference force equal to $|\Delta P_{ps}|$ when $\rho_{ns} = 0$, in which case $|\Delta P_{ps}| = \Delta P_c$. The difference between the ordinates of the two curves in Fig. 3.9(b) represents the relative increase in compressive force in the non-prestressed steel.

Unless $\rho_{ns} = 0$, the absolute value of the tension loss in prestressing steel, $|\Delta P_{ps}|$ should not be considered as equal to the compression loss in concrete, because this will overestimate the compression remaining in concrete after losses.

Table 3.2 also gives the deflection at the centre of span with varying

ρ_{ns}. The negative sign indicates camber. It is clear that the camber will be overestimated if non-prestressed steel is ignored. Also it can be seen that the deflection after creep, shrinkage and relaxation cannot be accurately predicted by multiplying the instantaneous deflection by a constant number, because such a number must vary with ρ_{ns} and with the creep, shrinkage and relaxation parameters.

Effects of varying creep and shrinkage parameters
It is sometimes argued that the effort required for an accurate analysis of the strain and the stress is not justified because accurate values of the creep coefficient φ and the free shrinkage ε_{cs} are not commonly available. A more rational approach for important structures is to perform accurate analyses using upper and lower bounds of the parameters φ and ε_{cs}.

The analysis is repeated in the above example for the case $\rho_{ns} = 0.4$ with $\varphi = 1.5$ and $\varepsilon_{cs} = -150 \times 10^{-6}$ (instead of 3.0 and -300×10^{-6}). The results, shown in the last column of Table 3.2, indicate that reducing φ and ε_{cs} by a factor of 2 has some effect, but the effect is not as important as the effect of ignoring the non-prestressed steel.

3.10   General

The loss in tension in prestressed steel, ΔP_{ps} caused by creep, shrinkage and relaxation is equal in absolute value to the loss in compression on the concrete, ΔP_c only in a cross-section without non-prestressed reinforcement. In general the value of ΔP_c is greater in absolute value than ΔP_{ps}; the difference depends on several variables one of which, of course, is the amount of non-prestressed reinforcement (in Example 2.2, $\Delta P_{ps} = -208\,\text{kN}$; and $\Delta P_c = 451\,\text{kN}$; see Fig. 2.6). The presence of non-prestressed reinforcement may substantially reduce the instantaneous strains and to a greater extent the time-dependent strains. Thus, the non-prestressed steel must be taken into consideration for accurate prediction of deformations of prestressed structures.

Equation (3.4) gives the value of ΔP_c when the prestressed steel and the non-prestressed reinforcement are at one level, and the force ΔP_c is situated at this level. Once ΔP_c is known, it may be used to calculate the changes in stresses and in strain variation over the section. The same procedure may also be employed involving approximation, when the section has more than one layer of reinforcement.

The methods discussed in Section 3.8 can be used to determine the displacements when the axial strain ε_O and the curvature ψ are known at all sections (or at a number of chosen sections). Here the calculation represents a solution of a problem of geometry and thus the same methods are equally applicable in structures with or without cracking.

Notes

1 The value of the reduced relaxation = 80 MPa is used below in order to compare the results with those of Example 2.2.
2 The value $\Delta\sigma_{ps} = -216.7$ MPa (not the slightly improved value obtained after iteration) is substituted in these equations in order to be able to compare the results with Example 2.2 where the reduced relaxation was -80 MPa (see section 3.4).
3 See Ghali, A. and Neville, A.M. (1997). *Structural Analysis: A Unified Classical and Matrix Approach*, 4th edn. E & FN Spon, London (Sections 6.5, 6.6, 7.2 and 7.3).
4 See p. 187 of the reference mentioned in note 3, above.

Chapter 4

Time-dependent internal forces in uncracked structures: analysis by the force method

Pasco-Kennewick Intercity Bridge, Wa., USA. Segmentally assembled concrete cable-stayed bridge. (Courtesy A. Grant and Associates, Olympia.)

4.1   Introduction

The preceding two chapters were concerned with the analysis of stress and strain in an uncracked reinforced or prestressed concrete section subjected to internal forces for which the magnitude and the time of application are known. Creep and shrinkage of concrete and relaxation of steel were considered to affect the distribution of stress and strain and the magnitude of the prestressing force in a prestressed member, but it was assumed that the elongation or curvature occur without restraint by the supports or by continuity with other members, which is the case in a statically determinate structure. The present chapter is concerned with the analysis of changes in internal forces due to creep, shrinkage and relaxation of steel in statically indeterminate structures.

Consider the effect of creep on a statically indeterminate structure made up of concrete as a homogeneous material, neglecting the presence of reinforcement. A sustained load of given magnitude produces strains and displacements that increase with time, but this is not accompanied by any change in the internal forces or in the reactions at the supports. Creep effect on displacements in such a case can be accounted for simply by using – for the modulus of elasticity of the material – a reduced value equal to $E/(1 + \varphi)$; where φ is the creep coefficient.

On the other hand, if the structure is composed of parts that have different creep coefficients, or if its boundary conditions change, the internal forces will, of course, be affected by creep. Concrete structures are generally constructed in stages, thus made up of concrete of different ages and hence different creep coefficients. Precast parts are often made continuous with other members by casting joints or by prestressing and hence the boundary conditions for the members change during construction. For all these cases, statically indeterminate forces gradually develop with time.

Change in the length of members due to shrinkage when restrained produces internal forces. But, because shrinkage develops gradually with time, shrinkage is always accompanied by creep and thus the internal forces due to shrinkage are well below the values that would develop if the shrinkage were to occur alone.

Similarly, the internal forces that develop due to gradual differential settlements of the supports in a continuous structure are greatly reduced by the effect of creep that occurs simultaneously with the settlement.

In the present chapter and in Chapter 5, we shall consider the analysis of the changes in internal forces in a statically indeterminate structure due to creep, shrinkage and differential settlement of supports. The well-known force or displacement method of structural analysis may be employed. In either method, two types of forces (or displacements) are to be considered: (a) external applied forces (or imposed displacements) introduced at their full values at instant t_0 and sustained without change in magnitude up to a later time t, and (b) forces (or displacements) developed gradually between zero value at time t_0 to their full values at time t. The first type of forces cause instantaneous displacement which is subsequently increased by the ratio φ, where $\varphi = \varphi(t, t_0)$, coefficient for creep at time t for age at loading t_0. The second type of forces produce at time t a total displacement, instantaneous plus creep, $(1 + \chi\varphi)$ times the instantaneous displacement that would occur if the full value of the force is introduced at t_0, where $\chi = \chi(t, t_0)$, the aging coefficient (see Section 1.7). This implies that the internal forces (or the displacements) develop with time at the same rate as relaxation of concrete (see Section 1.9).

Use of the coefficients φ or $\chi\varphi$ to calculate the increase in displacement due to creep – in the same way as done with strain – is strictly correct only when the structure considered is made of homogeneous material. In the preceding two chapters, we have seen that in a statically determinate structure, the presence of reinforcement reduces the axial strain and curvature caused by creep and hence reduces the associated displacements (see Section 3.3 and 3.4). The presence of reinforcement has a similar effect on the displacement in a statically indeterminate structure, but has a smaller effect on the statically indeterminate forces. Thus, the reinforcement is often ignored when the changes in the statically indeterminate forces due to creep or shrinkage are considered. The prestress loss due to creep, shrinkage and relaxation is predicted separately and is substituted by a set of external applied forces. However, the presence of prestressed or non-prestressed reinforcement should not be ignored when prediction of the displacement is the objective of the analysis or when more accuracy is desired. Also, the forces caused by the movements of the supports will be underestimated if the presence of the reinforcement is ignored; a correction to offset this error is suggested in Section 4.4.

Section 4.2 serves as a review of the general force method of structural analysis and introduces the symbols and terminology adopted. The analysis by the force method involves calculations of displacements due to known external forces applied on a statically determinate released structure. It also involves calculations of displacements of the released structure due to unit values of the statically indeterminate redundants. In Sections 4.3 and 4.4, the force method is applied to calculate the time-dependent changes in internal forces caused by creep, shrinkage of concrete, relaxation of steel and movement of supports in statically indeterminate structures. In these two sections, we shall ignore the presence of the reinforcement when calculating the displacements involved in the analysis by the force method. However, a correction is suggested in Section 4.4 to account for the reinforcement and avoid underestimation of the statically indeterminate forces caused by movements of supports.

An alternative solution which also employs the force method is presented in Section 4.5. The presence of all reinforcement is accounted for and the effect of prestress loss is automatically included. The general procedure of Section 2.5 is applied in a number of sections to calculate the axial strain and the curvature in a statically determinate released structure. The displacements involved in the analysis by the force method are calculated by numerical integration of the curvature and/or axial strain (see Section 3.8). Naturally, accounting for the reinforcement involves more computation (see Section 4.5).

4.2   The force method[1]

Consider, for example, the continuous beam shown in Fig. 4.1(a) subjected to vertical loads as shown. Here we shall consider the simple case when all the loads are applied at the same time and the beam is made of homogeneous material. The purpose of the analysis may be to find the reactions, the internal forces or the displacements; the term 'action' will be used here to mean any of these. The analysis by the force method involves five steps:

Step 1   Select a number of releases n by the removal of internal or external forces (redundants). Removal of the redundant forces $\{F\}$ leaves a statically determinate structure; for example, the continuous beam in Fig. 4.1(a) is released in Fig. 4.1(b). A system of n coordinates on the released structure indicates the chosen positive directions for the released forces and the corresponding displacements.

Step 2   With the given external loads applied on the released structure, calculate the displacements $\{D\}$ at the n coordinates. These represent inconsistencies to be eliminated by the redundant forces. The values $\{A_s\}$ of the actions are also determined at the desired positions of the released structure. In the example considered, $\{D\}$ represent the angular discontinuities at the intermediate supports (Fig. 4.1(c)).

Step 3   The released structures are subjected to a force $F_1 = 1$ and the displacements $f_{11}, f_{21}, \ldots, f_{n1}$ at the n coordinates are determined (see Fig. 4.1(d)). The process is repeated for unit values of the forces at each of the n coordinates, respectively. Thus a set of flexibility coefficients is generated, which forms the flexibility matrix $[f]_{n \times n}$; a general element f_{ij} is the displacement of the released structure at coordinate i due to a unit force $F_j = 1$. The values of the actions $[A_u]$ are also determined due to unit values of the redundants; any column j of the matrix $[A_u]$ is composed of the actions due to a force $F_j = 1$ on the released structure.

Step 4   The values of the redundant forces necessary to eliminate the inconsistencies in the displacements are determined by solving the compatibility equation:

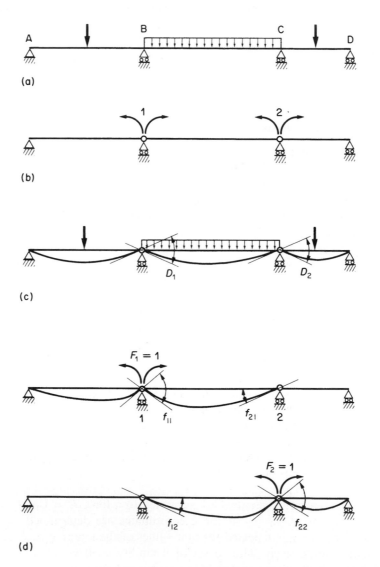

Figure 4.1 Analysis of a continuous beam by the force method: (a) statically indeterminate
structure; (b) released structure and coordinate system; (c) external forces
applied on released structure; (d) generation of flexibility matrix [*f*].

$$[f]\{F\} = -\{D\} \qquad\qquad (4.1)$$

The compatibility Equation (4.1) ensures that the forces $\{F\}$ are of such
a magnitude that the displacements of the released structure become
compatible with the actual structure.

Step 5   The values $\{A\}$ of the actions in the actual statically indeterminate structure are obtained by adding the values $\{A_s\}$ in the released structure, calculated in step 2, to the values caused by the redundants. This may be expressed by the following superposition equation:

$$\{A\} = \{A_s\} + [A_u] \{F\} \tag{4.2}$$

4.3   Analysis of time-dependent changes of internal forces by the force method

Forces applied at time t_0 on a structure made up of homogeneous material produce instantaneous strain which will increase due to creep. If the magnitude of the forces is maintained constant, strain at time t will be φ times the instantaneous strain; where $\varphi = \varphi(t, t_0)$ is the creep coefficient at time t when the age at loading is t_0. Because the material is homogeneous, the increase of strain by the ratio φ at all points, results in the same increase in the displacements. Thus, the creep coefficient φ used for strain can be applied directly to displacements.

In the example considered in Section 4.2 (Fig. 4.1), assume that the external loads are applied at time t_0 and the structure is made up of homogeneous material. At any time t greater than t_0, creep increases the values of $\{D\}$ and $[f]$ to $(1 + \varphi)$ times the values at t_0. This results in no change in the statically indeterminate forces and in the internal forces. The change in the actual statically indeterminate structure is only in the displacements which are magnified by the ratio $(1 + \varphi)$. The same conclusion can be reached by considering that the modulus of elasticity of the structure is $E_c(t_0)/(1 + \varphi)$ and performing a conventional elastic analysis, where $E_c(t_0)$ is the modulus of elasticity at age t_0.

Now let us consider a case in which creep affects the internal forces. Assume, for example, that the beam in Fig. 4.1(a) is made of three precast simple beams which are prestressed and placed in position at age t_0 and made continuous shortly after. The instantaneous deflections which occur at t_0, due to the self-weight of the beam, are those of simple beams with modulus of elasticity $E_c(t_0)$. Further deflection due to creep occurs after the beams have become continuous. The angular rotation of the beam ends at B and C must be compatible. This will result in the gradual development of the redundants $\{\Delta F\}$, which represent in this case the changes in the bending moments at coordinates 1 and 2 caused by creep.

To find the changes in the reactions, the internal forces or the displacements at any section occurring during a time interval t_0 to t, the analysis follows the five steps of the force method as outlined in Section 4.2, with the modifications discussed below. The time-dependent changes considered here may be caused by creep as in the above-mentioned example, or by shrinkage or support settlement or a combination of these.

In step 2 of the force method, calculate $\{\Delta D\}$, the changes in the displacement of the released structure at the coordinates that occur between t_0 and t. The displacement $\{\Delta D\}$ may be expressed as a sum of four terms:

$$\{\Delta D\} = \{\Delta D\}_{\text{loads}} + \{\Delta D\}_{\text{prestress loss}} + \{\Delta D\}_{\text{shrinkage}} + \{\Delta D\}_{\text{settlement}} \qquad (4.3)$$

$\{\Delta D\}_{\text{loads}}$ represents the displacements due to creep under the effect of prestress and other loads introduced at time t_0 and sustained at their full values up to time t, e.g. the structure self-weight. For calculation of the elements of this vector, multiply the instantaneous displacement at t_0 by the creep coefficient $\varphi(t, t_0)$. If the loads are applied at t_0 and the continuity is introduced at t_1 and we are concerned with the changes in the displacement between t_1 and a later time t_2, the creep coefficient would be $[\varphi(t_2, t_0) - \varphi(t_1, t_0)]$.

$\{\Delta D\}_{\text{prestress loss}}$ represents the displacements due to creep under the effect of prestress loss during the period t_0 to t. The loss of prestress should not be ignored in practice when the dead load and the load balanced by the prestress are of the same order of magnitude and of opposite signs. Thus the accuracy of the analysis may be sensitive to the accuracy in calculating and accounting for the effect of prestress loss. Prestress loss may be represented by a set of forces in the opposite direction to the prestress forces. The prestress loss develops gradually between time t_0 and t; thus, displacement due to the prestress loss is equal to $[1 + \chi\varphi(t, t_0)]$ times the instantaneous displacement due to the same forces if they were applied at time t_0.

$\{\Delta D\}_{\text{shrinkage}}$ and $\{\Delta D\}_{\text{settlement}}$ are displacements occurring in the released statically determinate structure; thus, in this step of analysis no forces are involved. These displacements are determined by geometry using the shrinkage or settlement values which would occur without restraint during the period $(t - t_0)$.

In the same step (2), also calculate $\{\Delta A_s\}$, the changes in the values of the required actions in the released structure occurring during the same period.

In step 3, generate an age-adjusted flexibility matrix $[\bar{f}]$ composed of the displacements of the released structure at the coordinates due to unit values of the redundants. These unit forces are assumed to be introduced gradually from zero at t_0 to unity at t. Any element $\bar{f}_{ij}$ represents the instantaneous plus creep displacements at coordinate i due to a unit force gradually introduced at coordinate j. $[\bar{f}]$ is generated in the same way as $[f]$, using for the calculation of the displacements the age-adjusted modulus of elasticity given by Equation (1.31), which is repeated here:

$$\bar{E}_c = \frac{E_c}{1 + \chi\varphi} \qquad (4.4)$$

where $\bar{E}_c = \bar{E}_c(t, t_0)$; $E_c = E_c(t_0)$ is the modulus of elasticity of concrete at age t_0.

The matrix $[\Delta A_\mathrm{u}]$, which is composed of the changes in the values of the actions due to unit change in the values of the redundant, is the same as $[A_\mathrm{u}]$ discussed in Section 4.2. Only when one of the actions is a displacement should the corresponding A_u value be magnified by the appropriate $(1 + \chi\varphi)$, as explained above for the flexibility coefficients.

In step 4 of the analysis, we find the changes in the redundants occurring between t_0 and t by solving the compatibility equations:

$$[\bar{f}]\,\{\Delta F\} = -\{\Delta D\} \tag{4.5}$$

In step 5, the changes in the actions caused by creep are determined by substitution in the equation:

$$\{\Delta A\} = \{\Delta A_\mathrm{s}\} + [\Delta A_\mathrm{u}]\,\{\Delta F\} \tag{4.6}$$

The value of the aging coefficient χ to be used in the above analysis may be taken from the graphs or the table in Appendix A. This implies that the prestress loss and the statically indeterminate forces develop with time at the same rate as the relaxation of concrete (see Section 1.8).

It is to be noted that the analysis discussed in the present section is concerned only with the changes $\{\Delta A\}$ in the values of the actions developing during a given period of time. Addition of $\{\Delta A\}$ to $\{A\}$, the values of the actions at the beginning of the period, gives the final values of the actions. Calculation of $\{A\}$ requires a separate analysis and may require use of the force method (Equation (4.2)). As an example, consider a statically indeterminate structure made up of parts of different creep properties and subjected at time t_0 to an external applied load or sudden settlement. To find the values of any actions at a later time, the analysis is to be performed in two stages and the force method may be used for each. In stage 1, determine $\{A\}$, the values of the actions at age t_0 immediately after application of the load or the settlement. The moduli of elasticity to be used in this analysis are the appropriate values for individual parts of the structure for instantaneous loading at time t_0. In the second stage, the time-dependent changes $\{\Delta A\}$ are determined, using the procedure described in the present section.

In the method of analysis suggested here, the presence of the reinforcement is to be consistently ignored in the calculation of the displacement vector $\{\Delta D\}$ and the age-adjusted flexibility matrix $[\bar{f}]$; this results in general in an overestimation of the elements of the two matrices. However, the consistency tends to reduce the error in calculation of the statically indeterminate forces $\{\Delta F\}$ by solution of Equation (4.5). For the same reason, the forces due to prestress loss – to be used in the calculation of $\{\Delta D\}_{\text{prestress loss}}$ – should be evaluated ignoring the presence of the non-prestressed steel. The error resulting from this approximation is generally acceptable in practice for calculation of the internal forces in statically indeterminate structures. The internal

forces calculated in this way may subsequently be employed to predict deflections, but the presence of reinforcement should not be ignored in the calculation of axial strains and curvature from which the displacements can be
determined, as discussed in Chapters 2 and 3.

An alternative procedure of analysis using Equation (4.5) is discussed in
Section 4.5, in which the presence of the reinforcement is accounted for in the
calculation of $[\bar{f}]$ and $\{\Delta D\}$.

Example 4.1 Shrinkage effect on a portal frame

Find the bending moment diagram in the concrete frame in Fig. 4.2(a)
due to shrinkage that gradually develops between a period t_0 to t. The
frame has a constant cross-section; the moment of inertia of the
concrete area is I_c. Ignore deformations due to axial forces.

The analysis for this problem is the same as that for a drop of temperature that produces a free strain equal to $\varepsilon_{cs}(t, t_0)$. The only difference

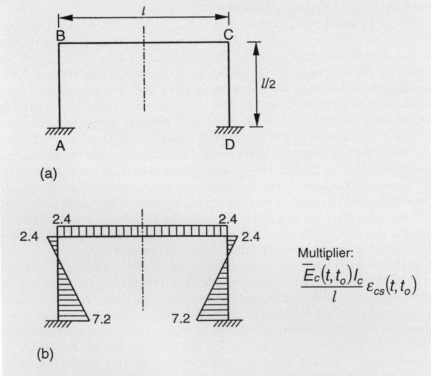

Multiplier:

$$\frac{\bar{E}_c(t, t_0) I_c}{l} \varepsilon_{cs}(t, t_0)$$

Figure 4.2 Analysis of internal forces caused by shrinkage in a plane frame: (a) frame
dimensions; (b) bending moment diagram.

is in the modulus of elasticity to be used in the analysis. With shrinkage, use the age-adjusted elasticity modulus.

$$\bar{E}_c(t, t_0) = \frac{E_c(t_0)}{1 + \chi\varphi(t, t_0)}$$

The bending moment diagram for this frame and the reactions are derived by a conventional elastic analysis, e.g. by use of the general force method or by moment distribution;[2] the results are given in Fig. 4.2(b). Note that the shrinkage ε_{cs} is a negative value; after application of the multiplier, the ordinates in Fig. 4.2(b) will have reversed signs.

To calculate the stress in concrete at any fibre, we should use the values of the internal forces as calculated by this analysis and the section properties, A_c and I_c of the concrete, excluding the reinforcement.

Example 4.2 Continuous beam constructed in two stages

The continuous prestressed beam ABC (Fig. 4.3(a)) is cast in two stages: AB is cast first and at age 7 days it is prestressed and its forms removed; span BC is cast in a second stage and its prestressing and removal of forms are performed when the ages of AB and BC are 60 and 7 days, respectively. Find the bending moment diagram at time infinity due to the self-weight of the beam only using the following creep and aging coefficients:

$$\varphi(\infty, 7) = 2.7 \qquad \chi(\infty, 7) = 0.74 \qquad \varphi(60, 7) = 1.1$$

$$\varphi(\infty, 60) = 2.3 \qquad \chi(\infty, 60) = 0.78.$$

Ratio of elasticity moduli for concrete at ages 60 and 7 days are:

$$E_c(60)/E_c(7) = 1.26.$$

Let t be the time measured from day of casting of AB. A statically determinate released structure and a system of one coordinate are shown in Fig. 4.3(b). At $t = 60$, uniform load q is applied on span BC of the continuous beam ABC, which has moduli of elasticity $E_c(60)$ for AB

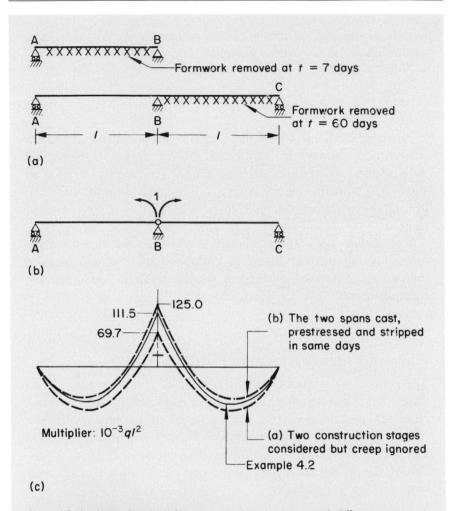

Figure 4.3 Analysis of internal forces in a continuous beam with different creep coefficients and different ages at loading of spans (Example 4.2): (a) continuous beam stripped in two stages; (b) statically determinate released structure; (c) bending moment diagram at $t = \infty$.

and $E_c(7)$ for BC. We use here the force method. Displacement of the released structure is:

$$D_1 = (D_1)_{AB} + (D_1)_{BC}$$

$$= 0 + \frac{ql^3}{24E_c(7)I_c}.$$

Flexibility coefficient is:

$$f_{11} = (f_{11})_{AB} + (f_{11})_{BC}$$

$$= \frac{l}{3E_c(60)I_c} + \frac{l}{3E_c(7)I_c}.$$

The statically indeterminate bending moment at B at $t = 60$ is:

$$F_1 = -D_1/f_{11}.$$

Substitution of $E_c(60) = 1.26E_c(7)$ in the above equations gives

$$F_1 = -0.0697\, ql^2.$$

The broken line (a) in Fig. 4.3(c) represents the bending moment diagram immediately after removal of the formwork of BC. If after this event the beam is released again, creep will produce, between $t = 60$, and ∞, the following change in displacement:

$$\Delta D_1 = (\Delta D_1)_{AB} + (\Delta D_1)_{BC}.$$

The first term on the right-hand side of this equation represents effects of creep on span AB due to load q introduced at $t = 7$ and the statically indeterminate force F_1 introduced at $t = 60$; thus,

$$(\Delta D_1)_{AB} = \frac{ql^3}{24E_c(7)I_c}[\varphi(\infty, 7) - \varphi(60, 7)] - \frac{0.0697\, ql^3}{3E_c(60)I_c}\varphi(\infty, 60).$$

On BC, the distributed load q and the force F are introduced at $t = 7$; creep produces a change in slope at B

$$(\Delta D_1)_{BC} = \frac{ql^3}{24E_c(7)I_c}\varphi(\infty, 7) - \frac{0.0697\, ql^3}{3E_c(7)I_c}\varphi(\infty, 7).$$

Substitution of the values of φ and $E_c(60) = 1.26\, E_c(7)$ in the above equations gives

$$\Delta D_1 = 0.0720\frac{ql^3}{E_c(7)I_c}.$$

The age-adjusted flexibility coefficient $\bar{f}_{11}$ is the sum of the rotations at the ends of the two spans due to a redundant force F_1 gradually introduced between $t = 60$ and ∞:

$$\bar{f}_{11} = (\bar{f}_{11})_{AB} + (\bar{f}_{11})_{BC}$$

$$(\bar{f}_{11})_{AB} = \frac{l}{3\bar{E}_c(\infty, 60)I_c}$$

$$(\bar{f}_{11})_{BC} = \frac{l}{3\bar{E}_c(\infty, 7)I_c}.$$

The age-adjusted moduli (Equation (1.31)) are:

$$\bar{E}_c(\infty, 60) = \frac{E_c(60)}{1 + \chi\varphi(\infty, 60)} = \left(\frac{1}{1 + 0.78 \times 2.3}\right) E_c(60) = 0.45E_c(7)$$

$$\bar{E}_c(\infty, 7) = \frac{E_c(7)}{1 + \chi\varphi(\infty, 7)} = \left(\frac{1}{1 + 0.74 \times 2.7}\right) E_c(7) = 0.34E_c(7).$$

Thus,

$$\bar{f}_{11} = \frac{l}{3 \times 0.45E_c(7)I_c} + \frac{l}{3 \times 0.34E_c(7)I_c} = 1.724\,\frac{l}{E_c(7)I_c}.$$

Solution of the compatibility Equation (4.5) gives the statically indeterminate moment at support B, developing gradually between $t = 60$ and ∞:

$$\Delta F_1 = -\bar{f}_{11}^{-1}\Delta D_1 = -0.0418\,ql^2.$$

The statically indeterminate bending moment at B at $t = \infty$ is

$$-0.0697\,ql^2 - 0.0418\,ql^2 = -0.1115\,ql^2.$$

The bending moment diagram is shown in Fig. 4.3(c). The two broken lines in the same figure indicate the bending moment diagram when: (a) the two construction stages are considered but creep is ignored, and (b) the beam is cast, prestressed and the forms removed in the two spans simultaneously.

*Example 4.3 Three-span continuous beam composed of precast
            elements*

Three precast prestressed simple beams are prestressed and made continuous at age t_0 by a reinforced concrete joint cast *in situ* (Fig. 4.4(a)).
It is required to find the bending moment diagram at a later age t. The
prestress tendon profile for each beam is as shown in Fig. 4.4(b). The
following data are given. The initial prestress at age t_0 creates a
uniformly distributed upward load of intensity $(2/3)q$; thus,

$$\frac{2}{3}q = \frac{8Pa}{l^2}$$

where P is the absolute value of the prestress force; a and l are defined in
Fig. 4.4(b); q is the weight per unit length of beam. Prestress loss is to be
assumed uniform and equal to 15 per cent of the initial prestress. Creep
coefficient $\varphi(t, t_0) = 2.5$; aging coefficient, $\chi(t, t_0) = 0.8$. Ignore cracking
at the joint.

Two statical systems need to be analysed. (a) Simple beams with
modulus of elasticity $E_c(t_0)$ subjected to the self-weight q/unit length
downwards plus a set of self-equilibrating forces representing the initial
prestress (Fig. 4.4(c)); the bending moment for this system is shown in
the same figure. (b) A continuous beam subjected to a set of self-
equilibrating forces representing the prestress loss and redundant connecting moments caused by creep; the modulus of elasticity to be used
with this loading is the age-adjusted modulus, $\bar{E}_c(t, t_0)$. The analysis
for the statically indeterminate bending moment due to loadings is
calculated below.

A statically determinate released structure is shown in Fig. 4.4(d).
Because of symmetry, the two coordinates representing the connecting
moments at B and C are given the same number 1.

Change in displacement in the released structure during the period t_0
to t (Equation (4.3)) are

$$\Delta D_1 = (\Delta D_1)_{\text{load}} + (\Delta D_1)_{\text{prestress loss}}$$

$$(\Delta D_1)_{\text{load}} = D_1(t_0)\varphi(t, t_0)$$

where $D_1(t_0)$ is the instantaneous displacement of the released structure
due to the loading in Fig. 4.4(c). Using Equation (C.6), Appendix C,

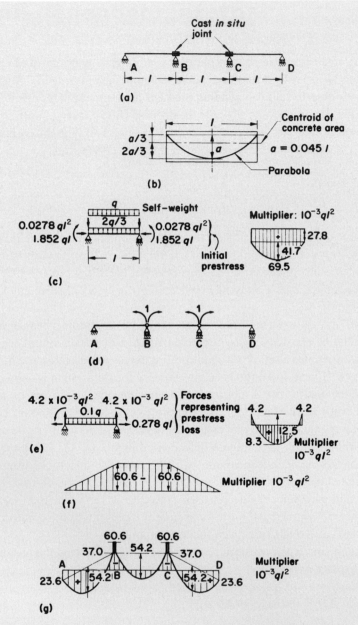

Figure 4.4 Bending moment developed by creep in precast simple beams made continuous by casting joints (Example 4.3): (a) three simple beams made continuous at age t_0 by a cast in situ joint; (b) typical prestress tendon profile for all beams; (c) loads and diagram of the bending moments introduced at age t_0; (d) statically determinate released structure and coordinate system; (e) forces and bending moment due to prestress loss in one span of released structure; (f) statically indeterminate bending moments; (g) bending moment diagram at time t.

$$D_1(t_0) = \frac{10^{-3}ql^2}{E_c(t_0)I_c} 2 \times \frac{l}{6} (2 \times 69.5 + 1 \times 27.8) = 55.6 \times 10^{-3} \frac{ql^3}{E_c(t_0)I_c}$$

$$(\Delta D_1)_{\text{load}} = 55.6 \times 10^{-3} \frac{ql^3}{E_c(t_0)I_c} 2.5 = 139.0 \times 10^{-3} \frac{ql^3}{E_c(t_0)I_c}.$$

The age-adjusted modulus of elasticity of concrete (Equation (1.31)) is:

$$\bar{E}_c(t, t_0) = \left(\frac{1}{1 + 0.8 \times 2.5}\right) E_c(t_0) = \frac{1}{3}E_c(t_0).$$

A set of self-equilibrating forces[3] representing the prestress loss and the corresponding bending moment diagram for a typical span of the released structure is shown in Fig. 4.4(c). The displacement due to these forces, using a modulus of elasticity $\bar{E}_c = E_c(t_0)/3$ (see Equation (C.6)) is:

$$(\Delta D_1)_{\text{presstress loss}} = \frac{10^{-3}ql^2}{[E_c(t_0)/3]I_c} 2 \times \frac{l}{6} (2 \times 8.3 - 1 \times 4.2)$$

$$= 12.5 \times 10^{-3} \frac{ql^3}{E_c(t_0)I_c}$$

$$(\Delta D_1) = (139.0 + 12.5)10^{-3} \frac{ql^3}{E_c(t_0)I_c}$$

$$= 151.5 \times 10^{-3} \frac{ql^3}{E_c(t_0)I_c}$$

Age-adjusted flexibility coefficient is

$$\bar{f}_{11} = \frac{l}{[E_c(t_0)/3]I_c} \left(\frac{1}{3} + \frac{1}{2}\right) = 2.5 \frac{l}{E_c(t_0)I_c}.$$

Substituting in Equation (4.5) and solving, gives

$$\Delta F_1 = -\frac{151.5}{2.5} 10^{-3}ql^2 = -60.6 \times 10^{-3}ql^2.$$

The statically indeterminate bending moment developed by creep and prestress loss is shown in Fig. 4.4(f). The diagrams of the bending moment at time t (Fig. 4.4(g)) are obtained by the superposition of the diagrams in Fig. 4.4(c), (e) and (f). Note that the negative bending moment at the joints B and C $(= -0.0606\,ql^2)$ is higher in absolute value than the bending moment on the adjacent sections: the higher value is plotted over a short length representing the length of the cast *in situ* joint.

Example 4.4 Post-tensioned continuous beam

A two-span continuous beam ABC (Fig. 4.5(a)) is built in two stages. Part AD is cast first and its scaffolding removed at time t_0 immediately after prestressing. Shortly after, part DC is cast and at time t_1 pre-stressed and its scaffolding removed. Find the bending moment diagram for the beam at a much later time t_2 due to prestressing plus the self-weight of the beam, q per unit length. The initial prestress creates an upward load of intensity of $0.75q$ and the prestress loss is 15 per cent of the initial value.

Assume that the time is measured from the day of casting of part AD and that the prestress for DC is applied at time t_1 when the age of DC is t_0. The following material properties are assumed to be known (data corresponds to $t_0 = 7$ days, $t_1 = 60$ days and $t_2 = \infty$):

$$\varphi(t_1, t_0) = 1.1 \qquad \chi(t_1, t_0) = 0.79 \qquad \varphi(t_2, t_0) = 2.7 \qquad \chi(t_2, t_0) = 0.74$$

$$\varphi(t_2, t_1) = 2.3 \qquad \chi(t_2, t_1) = 0.78 \qquad E_c(t_1)/E_c(t_0) = 1.26$$

The prestress loss starts to develop immediately after prestressing. However, for simplicity of presentation, we assume here that the loss is 15 per cent of the initial forces and the total amount of the loss occurs during the period t_1 to t_2.

Three statical systems need to be analysed:

(a) A simple beam with an overhang (Fig. 4.5(b)) subjected at t_0 to a downward load q and a system of self-equilibrating forces representing the initial prestress forces on AD. The bending moment diagram for this system is shown in Fig. 4.5(c).

(b) A continuous beam subjected at time t_1 to the self-weight of part DC and the forces due to the prestress of stage 2 (Fig. 4.5(d)). The moduli of elasticity to be used are $E_c(t_1)$ for AD and $E_c(t_0)$ for DC. The instantaneous bending moment diagram corresponding to this loading is shown in Fig. 4.5(e).

(c) A continuous beam subjected to a set of self-equilibrating forces representing the prestress loss and redundant forces caused by creep. With this system use the age-adjusted elasticity moduli

$$(\bar{E}_c)_{AD} = \bar{E}_c(t_2, t_1) = \frac{E_c(t_1)}{1 + 0.78 \times 2.3} = 0.36 E_c(t_1) = 0.45 \, E_c(t_0)$$

$$(\bar{E}_c)_{DC} = \bar{E}_c(t_2, t_0) = \frac{E_c(t_0)}{1 + 0.74 \times 2.7} = 0.34 E_c(t_0).$$

The released structure and the coordinate system shown in Fig. 4.5(f) will be used below to calculate the redundant force F_1 due to creep and prestress loss.

The term $(\Delta D_1)_{loads}$ is the displacement in the released structure caused by creep. Using virtual work (Equation (3.32)),

$$(\Delta D_1)_{loads} = \frac{1}{E_c(t_0)I_c} \left(\int_A^D M_c M_{u1} \, dl \right) [\varphi(t_2, t_0) - \varphi(t_1, t_0)]$$

$$+ \frac{1}{E_c(t_1)I_c} \left(\int_A^D M_e M_{u1} \, dl \right) \varphi(t_2, t_1)$$

$$+ \frac{1}{E_c(t_0)I_c} \left(\int_D^C M_e M_{u1} \, dl \right) \varphi(t_2, t_0)$$

where M_c and M_e are the bending moments shown in parts (c) and (e) of Fig. 4.5 and M_{u1} is the bending moment due to a unit value of the redundant at coordinate 1, Fig. 4.5(g). The values of the three integrals[4] are indicated separately in the following equation:

$$(\Delta D_1)_{loads} = 24.7 \times 10^{-3} \frac{ql^3}{E_c(t_0)I_c} - 21.5 \times 10^{-3}$$

$$\times \frac{ql^3}{E_c(t_1)I_c} + 20.1 \times 10^{-3} \frac{ql^3}{E_c(t_0)I_c}$$

$$(\Delta D_1)_{loads} = 27.7 \times 10^{-3} \frac{ql^3}{E_c(t_0)I_c}.$$

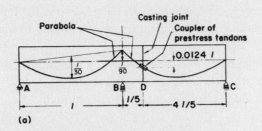

(a)

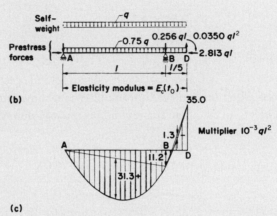

(b)

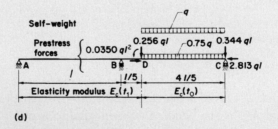

(c)

(d)

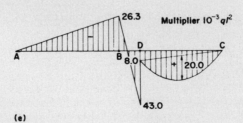

(e)

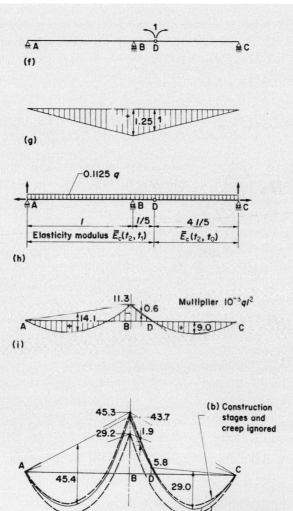

Figure 4.5 Analysis of the instantaneous and time-dependent bending moment in a continuous beam built and prestressed in two stages (Example 4.4): (a) a continuous beam cast and prestressed in two stages; (b) loads introduced at time t_0; (c) bending moment for the beam and loads in (b); (d) loads introduced at time t_1 on a continuous beam; (e) bending moment for the beam and loads in (d); (f) statically determinate released structure and coordinate system; (g) bending moment due to the unit value of the redundant F_1; (h) loads representing the prestress loss; (i) bending moment in the released structure due to prestress loss; (j) final bending moments at time t_2.

A system of forces representing the prestress loss is applied on the released structure in Fig. 4.5(h) and the corresponding bending moment is shown in Fig. 4.5(i). The displacement at coordinate 1 due to prestress loss is

$$(\Delta D_1)_{\text{prestress loss}} = \frac{1}{(\bar{E}_c)_{\text{AD}}I_c} \int_A^D M_i M_{\text{ul}} \, dl + \frac{1}{(\bar{E}_c)_{\text{DC}}I_c} \int_D^C M_i M_{\text{ul}} \, dl$$

where M_i is the bending moment shown in part (i) of Fig. 4.5.

The values of the two integrals in this equation are separately indicated in the following:

$$(\Delta D_1)_{\text{prestress loss}} = \frac{1}{0.45E_c(t_0)}\left(-0.06 \times 10^{-3}\frac{ql^3}{I_c}\right) + \frac{1}{0.34E_c(t_0)}$$

$$\left(2.4 \times 10^{-3}\frac{ql^3}{I_c}\right) = 7.0 \times 10^{-3}\frac{ql^3}{E_c(t_0)I_c}$$

$$\Delta D_1 = (27.7 + 7.0)10^{-3}\frac{ql^3}{E_c(t_0)I_c} = 34.7 \times 10^{-3}\frac{ql^3}{E_c(t_0)I_c}.$$

The age-adjusted flexibility coefficient

$$\bar{f}_{11} = \frac{1}{(\bar{E}_c)_{\text{AD}}I_c}\int_A^D M^2{}_{\text{ul}} \, dl + \frac{1}{(\bar{E}_c)_{\text{DC}}I_c}\int_D^C M_{\text{ul}}^2 \, dl = \frac{2.51l}{E_c(t_0)I_c}.$$

Substitution in Equation (4.5) and solving for the redundant value,

$$\Delta F_1 = -\frac{34.7}{2.51} \times 10^{-3}ql^2 = -13.8 \times 10^{-3}ql^2.$$

The bending moment diagram at time t_2 shown in Fig. 4.5(j) is obtained by superposition of ΔF_1 times M_{ul}, M_c, M_e and M_i. The two broken curves shown in Fig. 4.5(j) are approximate bending moment diagrams obtained as follows: (a) considering the construction stages but ignoring creep; (b) ignoring the construction stages and creep; thus applying the dead load and 0.85 the prestress forces directly on a continuous beam.

Figure 4.5(j) indicates that the bending moment diagram for a structure built in stages is gradually modified by creep to approach the bending moment which would occur if the structure were built in one stage.

It should be noted that at some sections the bending moment during the construction stages is higher than in the final stage.

4.4 Movement of supports of continuous structures

Sudden movement of a support in a statically indeterminate concrete structure produces instantaneous changes in the reactions and in the internal forces; subsequently these forces decrease gradually with time due to the effect of creep (i.e. relaxation occurs). In actual structures the movement of supports, such as the settlement due to soil consolidation, develops gradually over a period of time; creep also occurs during the same period and may continue to develop after the maximum settlement is reached. Thus, the changes in internal forces start from zero at the beginning of settlement, reaching maximum values at or near the end of the period of settlement and subsequent creep results in relaxation (reduction in values) of the induced forces.

This is illustrated by considering the reaction F at B caused by a downward settlement δ of the central support of the continuous beam in Fig. 4.6(a). When δ is sudden, a force of magnitude F_{sudden} is instantaneously induced at B. If subsequently δ is maintained constant, creep of concrete causes relaxation of the reaction as shown by curve A in Fig. 4.6(b). Curve B in the same figure represents the variation of the force F when the magnitude of the settlement is changed from zero to δ over a period of time. The force F increases from zero to a maximum value F_{max} – which is generally much smaller than F_{sudden} – and then decreases gradually.

Consider a continuous homogeneous structure subjected to support movements that develop gradually from 0 at age t_0 to final values $\{\delta\}$ at age t_1. The values at t_1 and at subsequent time t_2 of a reaction or internal force induced by the movement of supports may be calculated by the equations (for which the proof is given later in this section):

$$F(t_1) = F_{sudden} \frac{1}{1 + \chi \, \varphi(t_1, t_0)} \tag{4.7}$$

$$F(t_2) = F(t_1) \left(1 - \frac{E_c(t_1)}{E_c(t_e)} \frac{\varphi(t_2, t_e) - \varphi(t_1, t_e)}{1 + \chi\varphi(t_2, t_1)} \right) \tag{4.8}$$

where F_{sudden} is the value of the reaction or internal force when $\{\delta\}$ occurs

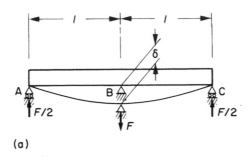

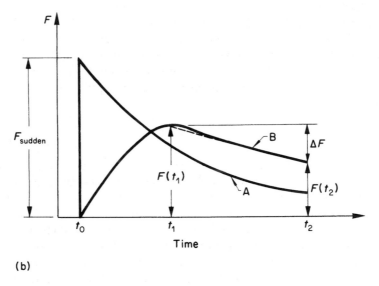

Figure 4.6 Time-dependent forces caused by support settlement in a continuous beam: (a) continuous beam; (b) reaction at central support versus time; A, sudden settlement; B, progressive settlement.

suddenly. The value F_{sudden} is obtained by conventional elastic analysis in which the value of the modulus of elasticity of concrete $E_c = E_c(t_0)$ and the cross-section properties of the members are those of *plain concrete* sections. φ and χ are creep and aging coefficients, which are functions of the time when creep is considered and the age at loading (see Sections 1.2 and 1.7). t_e is an age between t_0 and t_1. The value t_e can be determined, by trial from the graphs or equations of Appendix A, such that

$$\frac{1}{E_c(t_e)}[1 + \varphi(t_1, t_e)] = \frac{1}{E_c(t_0)}[1 + \chi\varphi(t_1, t_0)]. \tag{4.9}$$

In other words, a stress increment introduced at the effective time t_e and sustained without change in value to t_1 produces at t_1 a total strain of the same magnitude as would occur when the value of the stress increment is introduced gradually from zero at t_0 to full value at t_1.

If the movement of supports is introduced suddenly at age t_0, Equation (4.8) can be used to find the induced forces at any time t after t_0 by substitution of $t_1 = t_0 = t_e$ and $t_2 = t$; thus,

$$F(t) = F_{\text{sudden}} \left(1 - \frac{\varphi(t, t_0)}{1 + \chi\varphi(t, t_0)} \right) \tag{4.10}$$

The term between large parentheses in Equation (4.10) is equal to the relaxation function $r(t, t_0)$ divided by $E_c(t_0)$; see Equation (1.23).

The presence of the reinforcement may be accounted for as follows. In calculation of F_{sudden} use the cross-section properties of a transformed section composed of the area of concrete plus α times the area of steel; where $\alpha = E_s/E_c(t_0)$; also replace each creep coefficient $\varphi(t_i, t_j)$ in Equations (4.7) and (4.8) by $[\kappa\varphi(t_i, t_j)]$; where

$$\kappa = I_c/\bar{I} \tag{4.11}$$

κ is the curvature reduction factor (see Section 3.4); $\bar{I} = \bar{I}(t_i, t_j)$ is the moment of inertia of an age-adjusted transformed section for which $E_{\text{ref}} = \bar{E}_c(t_i, t_j)$ (see Equation (1.31) and Section 1.11.1); I_c is the moment of inertia of concrete. Both I_c and $\bar{I}$ are moments of inertia about an axis through the centroid of the age-adjusted transformed section. The above treatment is based approximately on Equation (3.27) which gives the change in curvature due to creep as the product $(\kappa\varphi)$ times the instantaneous curvature. No distinction is made between the effects of the reinforcement on axial strain and on curvature.

For proof of Equations (4.7) and (4.8), consider as an example the structure in Fig. 4.6(a). The instantaneous reaction at B due to a sudden settlement δ

$$F_{\text{sudden}} = \left(\frac{6}{l^3} E_c(t_0)I_c \right)\delta \tag{4.12}$$

where I_c is the moment of inertia of a concrete cross-section about an axis through its centroid. The term in the large parentheses represents the stiffness; that is the force when δ is unity.

Now consider that the settlement is introduced gradually from zero at t_0 up to δ at t_1; the reaction at B will also develop gradually from zero to a value $F(t_1)$ during the same period. The displacement δ may be expressed in terms of $F(t_1)$:

$$\delta = \left(\frac{l^3}{6\bar{E}_c(t_1, t_0)I_c}\right) F(t_1) \tag{4.13}$$

The term in the large parentheses is the age-adjusted flexibility, or the displacement due to a unit increment of force introduced gradually. $\bar{E}_c(t_1, t_0)$ is the age-adjusted modulus of elasticity of concrete (see Equation (1.31)),

$$\bar{E}_c(t_1, t_0) = \frac{E_c(t_0)}{1 + \chi\varphi(t_1, t_0)} \tag{4.14}$$

Equation (4.13) implies that the force F, and hence δ are developed with time at the same rate as relaxation of concrete (see Section 1.8).

Substitution of Equations (4.14) and (4.12) into (4.13) gives Equation (4.7).

Under the effect of the force $F(t_1)$, free creep would increase the deflection by the hypothetical increment:

$$\Delta\delta = \left(\frac{l^3}{6E_c(t_e)I_c}\right) F(t_1) \left[\varphi(t_2, t_e) - \varphi(t_1, t_e)\right] \tag{4.15}$$

In this equation, $F(t_1)$ is treated as if it were applied in its entire value at the effective time t_e.

Because the support settlement does not change during the period t_1 to t_2, an increment of force ΔF must develop such that

$$\Delta\delta + \left(\frac{l^3}{6\bar{E}_c(t_2, t_1)I_c}\right) \Delta F = 0 \tag{4.16}$$

where

$$\bar{E}_c(t_2, t_1) = \frac{E_c(t_1)}{1 + \chi\varphi(t_2, t_1)} \tag{4.17}$$

The force at B at time t_2 is

$$F(t_2) = F(t_1) + \Delta F \tag{4.18}$$

Solving for ΔF in Equation (4.16) and substitution of (4.15) and (4.17) into Equation (4.18) gives Equation (4.8).

The ascending part of curve B in Fig. 4.6(b) represents simultaneous gradual increase in force and in settlement, while the descending part represents the relaxation due to creep. Thus, one would expect curve B to be broken at t_1 as shown by the broken line. In practice, movement of supports, such as that caused by consolidation of clays, occurs over an infinite period of time. However, it is reasonable to consider for the analysis of forces that the full settlement occurs between ages t_0 and t_1, with the period (t_1, t_0) representing the time necessary for the major part (say 95 per cent) of the consolidation to occur. With settlement due to consolidation of soil, the transition between the ascending and descending parts of curve B, Fig. 4.6(b) would be smooth as shown by the continuous line.

Example 4.5 Two-span continuous beam: settlement of central support

The continuous concrete beam shown in Fig. 4.6(a) is subjected to a downwards settlement at B. Find the time variation of the force F and the reaction at the central support. Express F in terms of F_{sudden} the value of the instantaneous reaction when the settlement δ is suddenly introduced. Consider two cases:

(a) δ introduced suddenly at $t_0 = 14$ days and maintained constant to $t_2 = 10\ 000$ days.

(b) Settlement introduced gradually from zero at $t_0 = 14$ days to a value δ at $t_1 = 104$ days maintained constant thereafter up to $t_2 = 10\ 000$ days.

Use the following creep and aging coefficients

t_i	t_j	$\varphi(t_j, t_i)$	$\chi(t_j, t_i)$
14	104	1.14	0.79
14	500	1.79	0.76
14	2000	2.26	0.76
14	10000	2.57	0.76
23	104	1.01	
23	500	1.72	
23	2000	2.20	
23	10000	2.55	
104	500	1.17	0.81
104	2000	1.70	0.80
104	10000	1.96	0.79

The value $t_e = 23$ days is obtained by trial such that Equation (4.9) is satisfied. The ratio $E_c(t_1)/E_c(t_e) = 1.077$.

Use of Equation (4.10) with $t_0 = 14$ and $t = 104, 500, 2000,$ and $10\ 000$ gives the values of $F(t)$ which are plotted in Fig. 4.7, curve A.

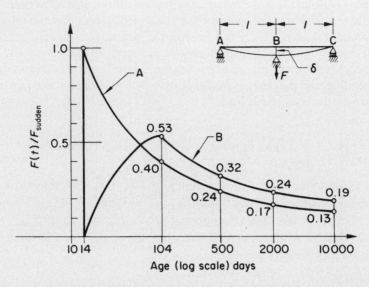

Figure 4.7 Values of the reaction at the central support versus time in a continuous
beam subjected to settlement of a support (Example 4.5): A, period of
settlement $(t_1 - t_0) = 0$; B, $(t_1 - t_0) = 90$ days.

When the settlement is gradually introduced, the starting value of F
is zero at $t_0 = 14$ days. Substitution in Equation (4.7) with $t_0 = 14$ and $t_1 =
104$ days gives the value of $F(t_1)$ at the end of the period in which the
settlement is introduced. Use of Equation (4.8), with $t_e = 23$, $t_1 = 104$
and $t_2 = 500$, 2000 and 10 000 gives the values of $F(t_2)$ plotted on curve
B, Fig. 4.7.

In practice, interest is in the maximum value of F; this is approximately equal to the value $F(t_1)$ with t_1 being the end of the period in
which the settlement occurs. Although Equations (4.7) and (4.8) give
the maximum value of F and its variation after the maximum is
reached, the two equations do not give the values of F between t_0 and t_1
(the ascending part of curve B, Fig. 4.7).

The above example is solved by a step-by-step procedure (see Section
4.6), assuming that the variation of settlement with time follows the
equation:

$$\frac{\delta(t)}{\delta(\infty)} = 1 - \exp\left(-\frac{3(t - t_0)}{t_{0.95} - t_0}\right) \tag{4.19}$$

where $\delta(t)$ and $\delta(\infty)$ are the settlement at any time t and the ultimate

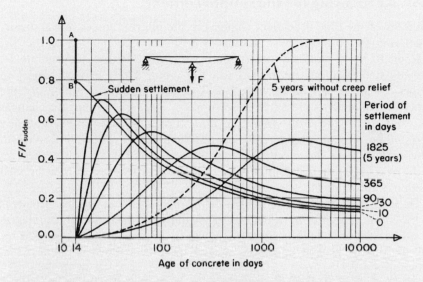

Figure 4.8 Time versus reaction by slow settlement of support occurring in a period of: 0, 10, 30, 365 days or 5 years (Example 4.5).

settlement at time infinity; t_0 is the time at which the settlement starts; $t_{0.95}$ is the time at which 95 per cent of the ultimate settlement occurs. Equation (4.19) closely approximates the standard-time consolidation curve for clays given by Terzaghi and Peck (in the form of a table)[5].

The results of the step-by-step analysis (employing Equation (4.31)) are shown in Fig. 4.8 in which the period $(t_{0.95} - t_0)$ – the time during which 95 per cent of settlement occurs – is considered equal to 0, 10, 30, 90, 365 days or 5 years. The graphs show the variation of F with time; the values of F are expressed in terms of F_{sudden} which is the instantaneous reaction at B if the full settlement occurs suddenly at $t_0 = 14$ days. The broken curve represents the case when $(t_{0.95} - t_0) = 5$ years, with creep ignored. The curves in Fig. 4.8 show clearly the pronounced effect of creep on the forces induced by slow settlement of a support.

When the settlement is sudden, the curve for F versus time has the same shape as the relaxation function $r(t, t_0)$, which represents the stress variation with time due to a strain imposed at age t_0 and sustained constant to age t (see Fig. A.3, Appendix A). The sudden drop, AB of force at age t_0 (Fig. 4.8), is caused by the creep which develops in the first few days but is considered as if it occurs at time t_0.

4.5   Accounting for the reinforcement

Analysis of the time-dependent changes in the internal forces in a statically indeterminate structure by Equation (4.5) involves calculation of the displacements of a statically determinate released structure to generate its age-adjusted flexibility matrix $[\bar{f}]$ and the vector $\{\Delta D\}$ of the changes in displacements occurring between two specified instants t_0 and t. By the procedure of analysis presented in Section 2.5, we can determine the changes $\Delta\varepsilon_O$ and $\Delta\psi$ in the axial strain and curvature in a section of a statically determinate structure, taking into account the presence of the reinforcement. The analysis gives the effects of creep, shrinkage and relaxation of steel on the stress and strain distribution, and thus the prestress loss in a prestressed section is automatically accounted for.

Once $\Delta\varepsilon_O$ and $\Delta\psi$ are determined, the changes $\{\Delta D\}$ in the displacements at the coordinates may be calculated by virtual work or by numerical integration (see Section 3.8). The equations given in Appendix C may be used for this purpose. This procedure of analysis described above is employed in Example 4.6.

Example 4.6 Three-span precast post-tensioned bridge

A three-span bridge (Fig. 4.9(a)) is made up of precast post-tensioned simple beams for which the cross-section at mid-spans is shown in Fig. 4.9(b). The beams are prestressed at age $\bar{t}$, placed in position and made continuous at age t_0 by casting concrete at the joints and by continuous prestress tendons as shown in Fig. 4.9(c). It is required to find the bending moment diagram at time t later than t_0. Assume no cracks are produced at the casting joint and that the joint results in perfect continuity. Also calculate the deflection at time t_0 at the centre of AB and the change in this value during the period t_0 to t.

To simplify the presentation, we shall assume that the difference between $\bar{t}$ and t_0 is small and consider that the prestressing, placing the beams in positions and casting of the joints, all occur at age t_0. We shall also ignore the area of the cast *in situ* concrete (hatched area in Fig. 4.9(b)). Other data are: area of concrete section for one beam, $A_c = 0.78\,\text{m}^2$ ($1200\,\text{in}^2$); moment of inertia about an axis through the centroid of the concrete area, $I_c = 0.159\,\text{m}^4$ ($382 \times 10^3\,\text{in}^4$); dead load of the precast and cast *in situ* concrete (assumed to come into effect at age t_0) = $9.1\,\text{kN/m}^2$ of area of deck; or the dead load per beam = $19.57\,\text{kN/m}$ ($1.344\,\text{kip/ft}$). A superimposed dead load of $5.0\,\text{kN/m}^2$ ($10.75\,\text{kN/m}$ per beam ($0.737\,\text{kip/ft}$)) is applied shortly after the structure is made con-

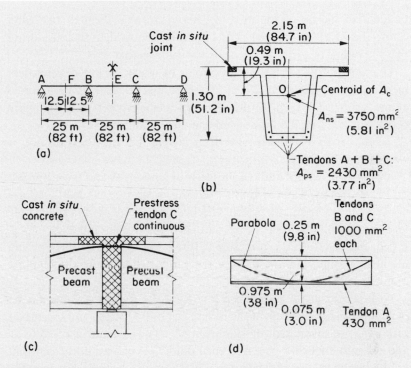

Figure 4.9 Continuous precast bridge of Example 4.6: (a) three-span bridge: (b) cross-section of one beam at mid-span; (c) joint of precast beams at supports B and C; (d) typical prestress tendon profiles in precast beams.

tinuous. Again, for the sake of simplicity, we shall consider that the superimposed load is applied at t_0 on the continuous structure.

The prestress in each beam is achieved by straight tendons A and parabolic tendons B and C. The prestressing of A and B is applied to simple beams, while C is inserted after placing the beams in position and the cable runs continuous over the whole length of the bridge. Further, we shall consider that cables B and C have identical profiles (Fig. 4.9(d)). The cross-section areas of prestress steel A_{ps} are 430, 1000 and 1000 mm^2 (0.67, 1.55, 1.55 in^2) for tendons A, B and C, respectively; the initial prestress forces are: 500, 1160 and 1160 kN (112, 260 and 260 kip). Consider that these forces exclude friction loss and that the prestress force is constant over the full length of a tendon.

Non-prestressed steel of total area, $A_{ns} = 3750$ mm^2 (5.81 in^2) is distributed over all surfaces of the cross-section; thus, we here assume that A_{ns} has the same centroid as A_c (point O in Fig. 4.9(b)) and that the moment of inertia of the area A_{ns} about an axis through the same centroid is $I_c(A_{ns}/A_c) = 0.764 \times 10^{-3}$ m^4; this is equivalent to considering that the radius of gyration for A_{ns} is the same as that of A_c.

The material properties are: modulus of elasticity for all reinforcement $E_{ps} = E_{ns} = 200$ GPa (29 000 ksi); modulus of elasticity of concrete at age t_0 $E_c(t_0) = 28$ GPa (4100 ksi); creep coefficient $\varphi(t, t_0) = 2.6$; aging coefficient $\chi(t, t_0) = 0.8$; free shrinkage during the period $(t - t_0) = \varepsilon_{cs}(t, t_0) = -240 \times 10^{-6}$; reduced relaxation during the same period, $\Delta\bar{\sigma}_{pr} = -90$ MPa (-13 ksi).

At t_0 the self-weight and the prestress of tendons A and B are applied on simple beams, while tendon C and the superimposed dead load are applied on a continuous beam. The bending moments for the simple and the continuous beams are calculated separately and then superposed; the result is shown in Fig. 4.10(a). Two values of the bending moment are indicated at B, with the larger value being the bending moment in the joint cast *in situ*.

With the axial force and bending moment known at time t_0, the instantaneous axial strain at the reference point O, $\varepsilon_O(t_0)$ and the curvature $\psi(t_0)$ are calculated (by Equation (2.32)) at a number of sections and given in Table 4.1. The reference point O is chosen at the centroid of the concrete and the reference modulus of elasticity used in the calculation of area properties is $E_{ref} = E_c(t_0)$.

The properties of the transformed section at age t_0 in Table 4.1 are calculated for a section composed of A_c plus $(\alpha(t_0)A_{ns})$. The area of prestress steel should have been accounted for in the calculation of the deformations due to the superimposed dead load, but this is ignored here.

The changes in axial strain and in curvature, $\Delta\varepsilon_o$ and $\Delta\psi$ during the period t_0 to t are calculated by Equation (2.40) and the results are given in Table 4.2. These calculations involve the properties of the age-adjusted transformed section which are included in Table 4.1, using as reference modulus $\bar{E}_{ref} = \bar{E}_c(t, t_0) = 9.09$ GPa (1320 ksi) (Equation (1.31)).

The released structure and the coordinate system are shown in Fig. 4.10(b). Because of symmetry, the change in displacement ΔD_1 needs to be calculated only at coordinate 1 and can be calculated from the curvature increments $\Delta\psi$ in Table 4.2. The increment in displacement ΔD_1 is equal to the sum of the changes in rotation at B of members BA and BC treated as simple beams. Employing Equations (C.6) and (C.7) gives

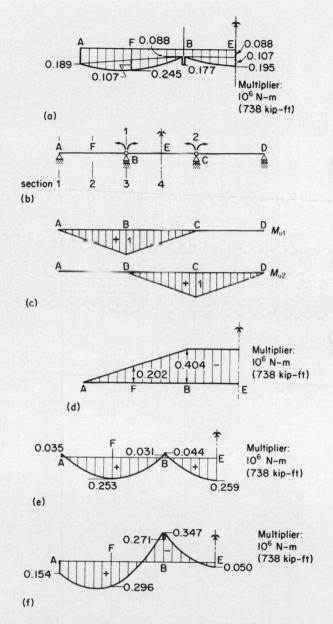

Figure 4.10 Analysis of the statically indeterminate forces and bending moment diagrams at t_0 and t for the continuous bridge of Fig. 4.9: (a) bending moment at time t_0; (b) released structure and coordinate system; (c) bending moment diagrams due to $F_1 = 1$ and $F_2 = 1$; (d) statically indeterminate bending moment developed during the period t_0 to t; (e) bending moment due to prestress loss; (f) final bending moment at time t.

Table 4.1 Cross-section properties[1] and calculation of instantaneous axial strain and curvature for a continuous bridge (Example 4.6)

Section number (see Fig. 4.10(b))	Concrete section properties			Transformed section properties at time t_0 $E_{ref} = E_c(t_0) = 28\,GPa$			Age-adjusted transformed section properties $E_{ref} = \bar{E}_c(t,t_0) = 9.09\,GPa$			Forces applied at time t_0: equivalents of prestress force and dead-load bending moment (Equation (2.31))		Instantaneous axial strain and curvature (Equation(2.32))	
	A_c	B_c	I_c	A	B	I	$\bar{A}$	$\bar{B}$	$\bar{I}$	N	M	$\varepsilon_O(t_0)$	$\psi(t_0)$
1	0.78	0	0.159	0.8068	0	0.1645	0.9160	-0.0036	0.1835	-2.82	0.189	-125	41.0
2	0.78	0	0.159	0.8068	0	0.1645	0.9160	0.0393	0.2047	-2.82	0.245	-125	53.2
3	0.78	0	0.159	0.8068	0	0.1645	0.9160	-0.0036	0.1835	-2.82	0.088	-125	19.1
4	0.78	0	0.159	0.8068	0	0.1645	0.9160	0.0393	0.2047	-2.82	0.195	-125	42.3
Multipliers	m^2	m^3	m^4	m^2	m^3	m^4	m^2	m^3	m^4	$10^6\,N$	$10^6\,N\text{-}m$	10^{-6}	$10^{-6}\,m^{-1}$

[1] Reference point O is chosen at the common centroid of A_c or A_{ns}.

Table 4.2 Changes in axial strain and in curvature of the released structure during the period t_0 to t in Example 4.6

| Section number (see Fig. 4.10(b)) | Calculation of restraining forces | | | | | | Total restraining forces (Equation (2.41)) | | Changes in axial strain and in curvature (Equation (2.40)) | |
| | Creep (Equation (2.42)) | | Shrinkage (Equation (2.43)) | | Relaxation (Equation (2.44)) | | | | | |
	ΔN	ΔM	ΔN	ΔM	ΔN	ΔM	ΔN	ΔM	$\Delta \varepsilon_O$	$\Delta \psi$
1	2.304	-0.1541	1.702	0	-0.219	0.0148	3.787	-0.1393	-455	74.6
2	2.304	-0.1999	1.702	0	-0.219	-0.1607	3.787	-0.3606	-467	283.4
3	2.304	-0.0718	1.702	0	-0.219	0.0148	3.787	-0.0570	-455	25.3
4	2.304	-0.1590	1.702	0	-0.219	-0.1607	3.787	-0.3197	-466	261.3
Multipliers	10^6 N	10^6 N-m	10^6 N	10^6 N-m	10^6 N	10^6 N-m	10^6 N	10^6 N-m	10^{-6}	10^{-6} m^{-1}

the change in displacement of the released structure during the time t_0 to t:

$$\Delta D_1 = \frac{25}{6} [0 \quad 2 \quad 1] \left\{ \begin{matrix} 74.6 \\ 283.4 \\ 25.3 \end{matrix} \right\} 10^{-6} + \frac{25}{6} [1 \quad 2 \quad 0] \left\{ \begin{matrix} 25.3 \\ 261.3 \\ 25.3 \end{matrix} \right\} 10^{-6}$$

$$= 4750 \times 10^{-6} \text{ radian.}$$

Use of Equation (C.8) and the curvature values $\psi(t_0)$ from Table 4.1 gives the instantaneous deflection at middle of span AB as

$$\frac{(25)^2}{96} [1 \quad 10 \quad 1] \left\{ \begin{matrix} 41.0 \\ 53.2 \\ 19.1 \end{matrix} \right\} 10^{-6} = 3.85 \times 10^{-3} \text{ m}$$

$$= 3.85 \text{ mm} \qquad (0.152 \text{ in}).$$

The change in deflection of the released structure during the period t_0 to t (using $\Delta\psi$ values from Table 4.2 and Equation (C.8)) is

$$\frac{(25)^2}{96} [1 \quad 10 \quad 1] \left\{ \begin{matrix} 74.6 \\ 283.4 \\ 25.3 \end{matrix} \right\} 10^{-6} = 19.10 \times 10^{-3} \text{ m}$$

$$= 19.1 \text{ mm} \qquad (0.752 \text{ in}).$$

For calculation of the age-adjusted flexibility coefficient, apply $F_1 = 1$ at coordinate 1; the diagram of the corresponding bending moment M_{u1} is shown in Fig. 4.10(c). Division of the ordinates of this diagram by $\bar{E}_{ref} \bar{I}_{centroid}$ at sections 1 to 4 gives the curvatures due to $F_1 = 1$. $\bar{I}_{centroid}$ is the moment of inertia of the age-adjusted transformed section about an axis through the centroid:

$$\bar{I}_{centroid} = \bar{I} - \frac{\bar{B}^2}{\bar{A}}$$

The values of the curvatures due to $F_1 = 1$, calculated in this fashion at the four sections considered, are:

$$\{\psi_{u1}\} = 10^{-9} \{0, 0.2710, 0.5995, 0.2710\} \text{ m}^{-1}/\text{N-m.}$$

The value $\bar{f}_{11}$ is the sum of the rotations just to the left and to the right of section 3, caused by $F_1 = 1$. These rotations can be calculated from the above curvatures, using Equations (C.6) and (C.7), giving

$$\bar{f}_{11} = \frac{25}{6}(2 \times 0.2710 + 1 \times 0.5995)2 \times 10^{-9} = 9.513 \times 10^{-9} \text{ (N-m)}^{-1}.$$

The age-adjusted flexibility coefficient $\bar{f}_{12}$ is the rotation at coordinate 1 due to $F_2 = 1$. Using a similar procedure as above gives

$$\bar{f}_{12} = \frac{25}{6}(2 \times 0.2710) \ 10^{-9} = 2.258 \times 10^{-9} \text{ (N m)}^{-1}.$$

The deflection at the centre of AB due to $F_1 = 1$ (by Equation (C.8))

$$\frac{(25)^2}{96}(10 \times 0.2710 + 0.5995) \ 10^{-9} = 21.55 \times 10^{-9} \text{ m/N-m}.$$

The force $F_2 = 1$ produces no deflection at the centre of AB.

Because of symmetry, the two redundants are equal and can be determined by solving one equation:

$$(\bar{f}_{11} + \bar{f}_{12})\Delta F_1 = -\Delta D_1$$

Thus,

$$\Delta F_1 = \Delta F_2 = \frac{-4750 \times 10^{-6}}{(9.513 + 2.258)10^{-9}} = -0.404 \times 10^6 \text{ N-m}.$$

The statically indeterminate bending moment diagram developed during the period t_0 to t is shown in Fig. 4.10(d).

When considering the bending moment due to prestressing, it is a common practice to consider the effect of the forces of the tendon on the concrete structure or on the concrete plus the non-prestressed steel when this steel is present. To determine the bending moment at time t, we calculate $\Delta\sigma_{ps}$ (the prestress loss) in each tendon by Equation (2.48). The summation $\Sigma(-A_{ps}\Delta\sigma_{ps}y_{ps})$ performed for all the tendons at any section gives the change in the bending moment of the released structure due to the prestress loss; where A_{ps} is the cross-section area of a tendon and y_{ps} is its distance below point O. This is calculated for various sections and plotted in Fig. 4.10(e). The final bending moment at time t is the superposition of the diagrams in Fig. 4.10(a), (d) and (e) and the result is given in Fig. 4.10(f).

The change in deflection of the actual structure can now be calculated by the superposition Equation (4.6) which is repeated here:

$$\{\Delta A\} = \{\Delta A_s\} + [\Delta A_u]\{\Delta F\}$$

where $\{\Delta A_s\}$ is the change in deflection of the released structure; $[\Delta A_u]$ are the changes in deflection due to $F_1 = 1$ and due to $F_2 = 1$; $\{\Delta F\}$ are the time-dependent redundant forces. Substitution of the values calculated above gives the change in deflection at the centre of AB during the period t_0 to t:

$$19.10 \times 10^{-3} + 10^{-9}[21.55 \quad 0]\begin{bmatrix} -0.404 \\ -0.404 \end{bmatrix} 10^6 = 10.39 \times 10^{-3}\,\text{m}$$

$$= 10.39\,\text{mm} \qquad (0.409\,\text{in}).$$

4.6   Step-by-step analysis by the force method

A step-by-step numerical procedure is presented in Section 1.10 for calculation of the strain of concrete caused by stress which is introduced gradually or step-wise in an arbitrary fashion. The procedure is also used to calculate the stress caused by imposed strain which is either constant with time (relaxation problem) or varying in arbitrary fashion.

In this and in Section 5.8, we shall use a similar procedure to calculate the internal forces in statically indeterminate structures caused by creep, shrinkage and settlement of supports. In the present section, the force method is employed for structures in which individual cross-sections are composed of homogeneous material (presence of reinforcement ignored). In Section 5.8 the step-by-step analysis is applied with the displacement method in concrete structures with composite cross-sections, taking into account the effect of the reinforcement.

The advantages of the step-by-step analysis are: (a) the time variation of forces or imposed displacement can be of any form (not necessarily affine to the time-relaxation curve as implied when the aging coefficient is used); (b) the method is applicable with any time functions chosen for creep, shrinkage or relaxation of steel or modulus of elasticity of concrete; (c) the changes in cross-section properties, e.g. due to cracking or modification of support conditions, can be accounted for in any time interval. The step-by-step analysis, however, involves a relatively large number of repetitive computations which makes it particularly suitable when a computer is used.

In the step-by-step analysis, the time is divided into intervals; the internal forces, the stresses or the displacements at the end of a time interval are calculated in terms of the forces or stresses applied in the first interval and the increments which have occurred in the preceding intervals. Increments of forces or stresses are introduced at the middle of the intervals (Fig. 4.11). Instantaneous applied loads, such as prestressing, are assumed, for the sake of consistency, to occur at the middle of an interval of length zero (e.g. intervals 1 and k in Fig. 4.11). Accurate results can be obtained with a small number of intervals (5 or 6); the length of the intervals should be relatively short in the early stages when the rates of change of modulus of elasticity, creep and shrinkage of concrete and often settlement of supports are greatest.

The general force method of structural analysis involves solution of the compatibility equation (see Section 4.2):

$$[f]\,\{F\} = \{-D\} \tag{4.20}$$

where $[f]$ is the flexibility matrix; $\{D\}$ are displacements of the released

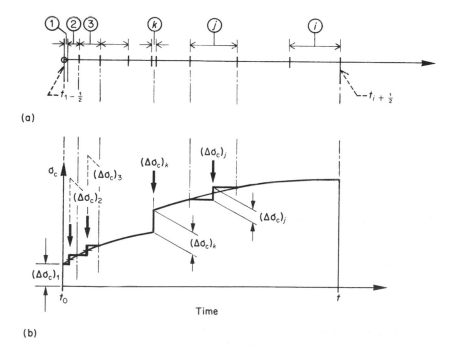

(a)

(b)

Figure 4.11 Division of: (a) time into intervals; (b) stresses into increments.

structure; $\{F\}$ are the redundant forces. The displacements $\{D\}$ represent inconsistencies in the released structure (with respect to the actual structure). The redundants $\{F\}$ must, therefore, be applied to eliminate the inconsistencies.

Any element of the flexibility matrix, f_{mn} is equal to the displacement at coordinate m due to unit load applied at coordinate n. Because of creep of concrete, the value of any element of the matrix $[f]$ depends upon the time for which the displacement is considered and the age of concrete at the time of the introduction of the unit load. Thus, we use here the symbol $[f(t_{i+\frac{1}{2}}, t_j)]$ to represent the matrix of flexibility at time $t_{i+\frac{1}{2}}$ when the age at loading is t_j. The subscripts $i - \frac{1}{2}$, i and $i + \frac{1}{2}$ respectively refer to the beginning, the middle and the end of interval i.

The forces $\{F\}_{i+\frac{1}{2}}$ and the displacement $\{D\}_{i+\frac{1}{2}}$ at the end of any interval i may be expressed as the sum of incremental forces $\{\Delta F\}_j$ and displacements, $\{\Delta D\}_j$ occurring at the middle of the intervals $j = 1, 2, \ldots, i$. Thus,

$$\{F\}_{i+\frac{1}{2}} = \sum_{j=1}^{i} \{\Delta F\}_j \tag{4.21}$$

$$\{D\}_{i+\frac{1}{2}} = \sum_{j=1}^{i} \{\Delta D\}_j \tag{4.22}$$

The compatibility Equation (4.20) applied at the end of the ith interval may be written in the form:

$$\sum_{j=1}^{i} \{[f(t_{i+\frac{1}{2}}, t_j)]\{\Delta F\}_j\} = -\sum_{j=1}^{i} \{\Delta D\}_j \tag{4.23}$$

The analysis for $\{\Delta F\}_i$ can be done in steps; in each step a new increment is calculated. In the ith step, the values $\{\Delta F\}_1$, $\{\Delta F\}_2$, $\ldots$, $\{\Delta F\}_{i-1}$ are known from the preceding steps and Equation (4.23) can be used to determine $\{\Delta F\}_i$. Equation (4.23) may be rewritten by separating the last term of the summation on the left-hand side and the substitution of Equation (4.22):

$$[f(t_{i+\frac{1}{2}}, t_i)] \{\Delta F\}_i = - \{D\}_{i+\frac{1}{2}} - \sum_{j=1}^{i-1} \{[f(t_{i+\frac{1}{2}}, t_j)]\{\Delta F\}_j\} \tag{4.24}$$

This recurrent equation can be solved successively with $i = 1, 2, \ldots$ to determine the values of the vector $\{\Delta F\}_1$, $\{\Delta F\}_2$, $\ldots$ and so on.

The flexibility matrices involved in the analysis differ only in the modulus of elasticity and the creep coefficient to be employed in the calculation.

The vector $\{D\}_{i+\frac{1}{2}}$ represents the total displacement of the released structure caused by applied loads, shrinkage or supports settlement. The displacement due to the applied load generally includes the instantaneous plus creep, but instantaneous displacements should be excluded if the loading is applied prior to the start of the period for which the changes of the internal forces are required.

The use of the recurrent Equation (4.24) is demonstrated below for a structure with one degree of indeterminacy.

Application   The two-span continuous concrete beam in Fig. 4.12(a) is subjected to a settlement of the central support, the magnitude of which, $\delta(t)$ varies with time in an arbitrary form. Equation (4.24) will be used to find the downward reaction F at the central support.

A statically determinate released structure with one coordinate is shown in Fig. 4.12(b). The instantaneous displacement at coordinate 1 due to a unit force at the same coordinate

$$f_{\text{instantaneous}} \quad -\frac{l^3}{6E_c I_c} \tag{4.25}$$

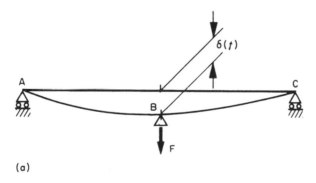

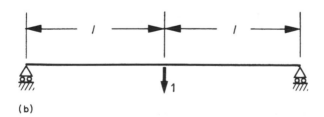

Figure 4.12 Reaction due to settlement of support of a continuous beam by a step-by-step procedure employing Equation (4.24): (a) continuous beam; (b) statically determinate released structure.

where I_c is the moment of inertia of the section; E_c is the modulus of elasticity of concrete at the time of application of the load. We have only one coordinate; thus, we use F to mean F_1 and f for f_{11}. If the unit load is applied at t_j, it will produce at time $t_{i+\frac{1}{2}}$ a displacement

$$f(t_{i+\frac{1}{2}},\, t_j) = C \frac{1}{E_c(t_j)} [1 + \varphi(t_{i+\frac{1}{2}},\, t_j)] \tag{4.26}$$

where C is a constant independent of time related to the geometry of the structure

$$C = \frac{l^3}{6I_c}. \tag{4.27}$$

At the end of any interval i

$$D_{i+\frac{1}{2}} = -\delta(t_{i+\frac{1}{2}}). \tag{4.28}$$

The minus sign is included in this equation because it represents a displacement caused by the redundant force F (rather than eliminated by it). Substitution of Equations (4.26) and (4.28) into Equation (4.24) gives

$$f(t_{i+\frac{1}{2}},\, t_i)(\Delta F)_i = \delta(t_{i+\frac{1}{2}}) - C \sum_{j=1}^{i-1} \left(\frac{1 + \varphi(t_{i+\frac{1}{2}},\, t_j)}{E_c(t_j)} (\Delta F)_j \right). \tag{4.29}$$

The magnitude of the reaction at the central support at the end of the ith interval is

$$F(t_{i+\frac{1}{2}}) = F(t_{i-\frac{1}{2}}) + (\Delta F)_i. \tag{4.30}$$

Solving Equation (4.29) for $(\Delta F)_i$ and substitution in Equation (4.30) gives

$$F(t_{i+\frac{1}{2}}) = F(t_{i-\frac{1}{2}})$$

$$+ [f(t_{i+\frac{1}{2}},\, t_i)]^{-1} \left[\delta(t_{i+\frac{1}{2}}) - C \sum_{j=1}^{i-1} \left(\frac{1 + \varphi(t_{i+\frac{1}{2}},\, t_j)}{E_c(t_j)} (\Delta F)_j \right) \right] \tag{4.31}$$

Equation (4.31) has been used to derive the graphs in Fig. 4.8 (see Example 4.5).

4.7   Example worked out in British units

Example 4.7 Two-span bridge: steel box and post-tensioned deck

The same bridge cross-section and method of construction of Example 2.7 are used for a continuous bridge of two equal spans, each $= l = 144\,\text{ft}$ (43.9 m); what will be the stress distribution at the section over the central support at time t? Again assume that at completion of installation of the precast elements, the structural steel section alone, acting as continuous beam, carries the weight of concrete and structural steel.

The bending moment over the central support at time t_0 immediately after installation of the precast deck $= -ql^2/8 = -5.4(144)^2/8 = -14\,000\,\text{kip-ft} = -168\,000\,\text{kip-in}$. The distribution of stress on the structural steel due to this bending moment is shown in Fig. 4.13(a); the same figure shows the stress distribution in the concrete deck cross-section due to the axial prestressing force introduced at time t_0.

The five steps of the force method (Sections 4.2 and 4.3) are followed to determine the time-dependent change in stresses in the section above the central support.

Step 1   A statically determinate released structure is shown in Fig. 4.13(b). The stress values required are:

$$\{\Delta A\} = \{\Delta\sigma_{c\ \text{top}}, \Delta\sigma_{c\ \text{bot}}, \Delta\sigma_{s\ \text{top}}, \Delta\sigma_{s\ \text{bot}}\}$$

These represent the stress changes in the period t_0 to t at top and bottom fibres of the concrete and the structural steel.

Step 2   In Example 2.7 we determined the time-dependent change in curvature at mid-span section as:

$$\Delta\psi(t, t_0) = 4.784 \times 10^{-6}\,\text{in}^{-1}.$$

The same change in curvature occurs at any other section of the released structure. Thus, the change in displacement in the released structure at coordinate 1 is:

$$\Delta D_1(t, t_0) = \Delta\psi l = 4.784 \times 10^{-6}\,(144 \times 12) = 8266 \times 10^{-6}.$$

$\{\Delta A_s\}$ in the present problem represents the stress changes in the

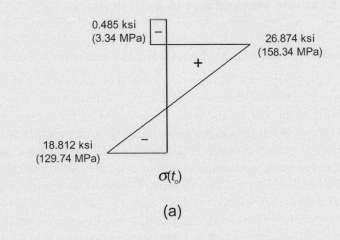

0.485 ksi
(3.34 MPa)

26.874 ksi
(158.34 MPa)

18.812 ksi
(129.74 MPa)

$\sigma(t_o)$

(a)

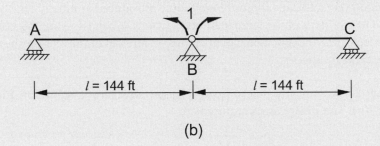

A

1

B

C

$l = 144$ ft

$l = 144$ ft

(b)

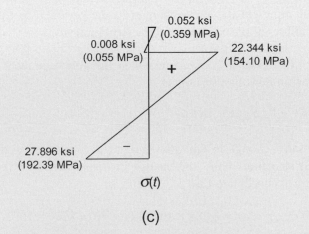

0.052 ksi
(0.359 MPa)

0.008 ksi
(0.055 MPa)

22.344 ksi
(154.10 MPa)

27.896 ksi
(192.39 MPa)

$\sigma(t)$

(c)

Figure 4.13 Analysis of stress distribution over the central support of a two-span continuous bridge (Example 4.7) (for bridge cross-section, see Fig. 2.16(a)): (a) stress at time t_0; (b) released structure and coordinate system; (c) stress distribution at time t.

released structure during the period t_0 to t. These are calculated in Example 2.7 and are constant over the span (the stress values in Fig. 2.16(c) minus the values in Fig. 2.16(b)):

$$\{\Delta A_s\} = \begin{Bmatrix} 0.177 \\ 0.296 \\ -7.785 \\ 1.324 \end{Bmatrix} \text{ksi}$$

Step 3 The age-adjusted elasticity modulus of concrete (Equation (4.4)), $\bar{E}_c = 1558\,\text{ksi}$.

Select the reference point O as shown in Fig. 2.16(a); properties of the age-adjusted section are:

$$\bar{A} = 10\,170\,\text{in}^2; \qquad \bar{B} = -242.1 \times 10^3 \,\text{in}^3; \qquad \bar{I} = 16.29 \times 10^6 \,\text{in}^4.$$

Due to $F_1 = 1\,\text{kip-in}$, the changes in strain at B are (Equation (2.19)):

$$(\Delta\varepsilon_{OB})_{\text{due to } F_1=1} = 1.451 \times 10^{-9} \,(\text{kip-in.})^{-1}$$

$$(\Delta\psi_B)_{\text{due to } F_1=1} = 60.97 \times 10^{-12} \,\text{in}^{-1}(\text{kip-in.})^{-1}$$

This change in curvature varies linearly between the above value at B and zero at the two ends. The corresponding change in displacement at coordinate 1 is:

$$\bar{f}_{11} = \Delta\psi_B \frac{2l}{3} = 60.97 \times 10^{-12} \left(\frac{2 \times 144 \times 12}{3} \right) = 70.24 \times 10^{-9} (\text{kip-in})^{-1}.$$

The stress changes at B in the four fibres considered due to $F_1 = 1\,\text{kip-in}$. are Equations (2.19) and (2.17):

$$[\Delta A_u] = 10^{-6} \begin{Bmatrix} -3.059 \\ -1.539 \\ -27.659 \\ 88.428 \end{Bmatrix} \frac{\text{ksi}}{\text{kip-in}}$$

Step 4 The time-dependent change in the statically indeterminate force (Equation (4.5)):

$$\Delta F_1(t, t_0) = (70.24 \times 10^{-9})^{-1}(-8266 \times 10^{-6}) = -117\,700\,\text{kip-in}.$$

Step 5   The stress changes in the period t_0 to t are (Equation (4.6)):

$$\{\Delta A\} = \begin{Bmatrix} 0.177 \\ 0.296 \\ -7.785 \\ 1.324 \end{Bmatrix} + 10^{-6} \begin{Bmatrix} -3.059 \\ -1.539 \\ -27.659 \\ 88.428 \end{Bmatrix} (-117\,700) = \begin{Bmatrix} 0.537 \\ 0.427 \\ -4.530 \\ -9.084 \end{Bmatrix} \text{ksi}.$$

Addition of these stress changes to the stress values at time t_0 gives the total stress distribution at time t at the section above the central support (Fig. 4.13(c)). It is interesting to compare the initial stress (−0.485 ksi) introduced by prestressing at time t_0 to the remaining compression at time t in the present example and in Example 2.7 (Fig. 2.16(c)). In the present example the remaining compression in concrete at time t dropped to almost zero.

4.8   General

The stresses in all reinforced or prestressed concrete structures, statically determinate or indeterminate, change with time due to the effects of creep, shrinkage of concrete and relaxation of prestress steel. In a statically determinate structure, the distribution of stresses over the area of concrete and reinforcement in any section varies with time, but the resultant of stresses in the two components combined remains unchanged. This is not the case with statically indeterminate structures, where statically indeterminate reactions are produced, causing gradual changes in the stress resultants in the sections.

The force method, employed in this chapter to analyse the time-dependent internal forces, is intended for computations by hand or using small desk calculators. It is, of course, possible to prepare computer programs to do parts of the computations or all the computations for a certain type of structure (for example, continuous beams). However, for a more general computer program, it is more convenient to use the displacement method, which is the subject of the following chapter.

Notes

1 For more detailed presentation and examples, see reference mentioned in note 3 of Chapter 3.
2 See reference mentioned in note 3 of Chapter 3.

3 See Section 5.11 of the reference mentioned in note 3 of Chapter 3.
4 A simple method for the evaluation of integrals for the calculation of displacements by virtual work can be found in Section 6.4 of the reference mentioned in note 3 of Chapter 3.
5 Terzaghi, K. and Peck, R.B. (1966), *Soil Mechanics in Engineering Practice*, Wiley, New York, page 240.

Chapter 5

Time-dependent internal forces in uncracked structures: analysis by the displacement method

Cast *in situ* segmental construction of 'Pont de la Fégire', near Lausanne, Switzerland.

5.1 Introduction

The force method is employed in Chapter 4 to calculate the time-dependent forces in a statically indeterminate structure caused by shrinkage and creep of concrete, relaxation of prestressed steel and movement of the supports. The general displacement method of structural analysis can be used for the same purpose. Computer programs for the elastic analysis of frames are now widely used by engineers and they are usually based on the displacement method. In Section 5.2, we shall review the general displacement method and, in Section 5.3, indicate how a conventional computer program for the analysis of an elastic framed structure can be used to determine the time-dependent changes in internal forces.

A step-by-step procedure suitable for computer use is presented in Section 5.8. It accounts for the effects of creep, shrinkage of concrete and relaxation of steel in statically determinate or indeterminate structures. The cross-section may be made up of one concrete type or composite and the structure may be composed of members of different ages and the presence of non-prestressed steel is accounted for in the analysis. The loading, prestressing forces or prescribed support displacements may be introduced gradually at an arbitrary rate or in stages and the boundary conditions may be changed in any stage. The method is suitable when precast segments are assembled and made continuous by prestressing or by cast *in situ* concrete or both. The '*segmental*' (or '*cantilever*') method of construction, mainly used for bridges, is an example of a case in which the step-by-step analysis is most fitting.

5.2   The displacement method

This section serves as a review of the general displacement method of analysis of framed structures, while the following two sections will indicate how this method can be used for the analysis of time-dependent changes in internal forces.

To explain the method consider, for example, the plane frame shown in Fig. 5.1(a) subjected to external applied loads (not shown in the figure). Assume that it is required to find m actions $\{A\}$ representing reaction components, internal forces or displacements at chosen sections. The analysis by the displacement method involves five steps.

Step 1   A coordinate system is established to identify the locations and the positive directions of the joint displacements (Fig. 5.1(b)). The number of coordinates n is equal to the number of possible independent joint displacements (degrees of freedom). There are generally two translations and a rotation at a free (unsupported) joint of a plane frame. The number of unknown displacements may be reduced by ignoring the axial deformations. For example, by considering that the length of the members of the frame in Fig. 5.1(b) remains unchanged, the degrees of freedom are reduced to coordinates 1, 3, 6 and 9.

Step 2   Restraining forces $\{F\}$ are introduced at the n coordinates to prevent the joint displacements. The forces $\{F\}$ are calculated by summing the fixed-end forces for the members meeting at the joints. Also determine $\{A_r\}$, values of the actions with the joints in the restrained position.

Step 3   The structure is now assumed to be deformed such that the displacement at coordinate j, $D_j = 1$, with the displacements prevented at all the other coordinates. The forces $S_{1j}, S_{2j}, \ldots, S_{nj}$ required to hold the frame in

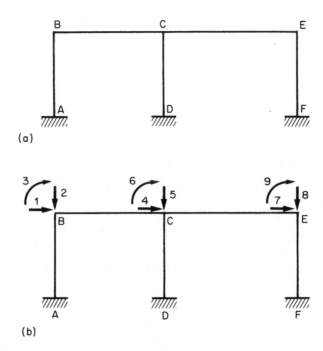

Figure 5.1 Example of a coordinate system (b) employed for the analysis of a plane frame (a) by the displacement method.

this configuration are determined at the n coordinates. This process is repeated for unit values of displacement at each of the coordinates, respectively. Thus a set of $n \times n$ stiffness coefficients is calculated, which forms the stiffness matrix $[S]_{n \times n}$ of the structure; a general element S_{ij} is the force required at coordinate i due to a unit displacement at coordinate j. The values of the actions $[A_u]$ are also determined due to unit values of the displacements; any column j of the matrix $[A_u]$ is composed of the values of the actions at the desired locations due to $D_j = 1$.

Step 4   The displacement $\{D\}$ in the actual (unrestrained) structure is obtained by solving the equilibrium equation:

$$[S]\{D\} = -\{F\} \tag{5.1}$$

The equilibrium Equation (5.1) indicates that the displacements $\{D\}$ must be of such a magnitude that the artificial restraining forces $\{F\}$ are eliminated.

Step 5   Finally, the required values $\{A\}$ of the actions in the actual structure are obtained by adding the values $\{A_r\}$ in the restrained structure (calculated in step 2) to the values caused by the joint displacements. This is expressed by the superposition equation:

$$\{A\}_{m \times 1} = \{A_r\}_{m \times 1} + [A_u]_{m \times n} \{D\}_{n \times 1} \qquad (5.2)$$

5.3   Time-dependent changes in fixed-end forces in a homogeneous member

In the analysis of statically indeterminate structures by the displacement method, the internal forces and the forces at the ends of individual members with fixed ends must be known in advance. In this section, we shall consider for a homogeneous beam with totally fixed ends, the changes in the fixed-end forces caused by creep, shrinkage and settlement of supports.

The totally fixed beam in Fig. 5.2(a) is made up of homogeneous material and subjected at age t_0 to a set of external applied loads such as gravity loads or prestress forces. Consider the changes in the forces at the ends of the beam, and hence the internal forces at any section that will occur during a later period t_1 to t_2, due to creep, gradual settlement of supports, shrinkage and prestress loss. We here assume that the amount of prestress loss is a known value, not affected by the internal forces resulting from the loss. We also consider the case when the support conditions change at time t_1.

Forces applied on the beam and sustained, without change in magnitude or alteration on the boundary conditions, produce no changes in the internal forces due to creep (see Section 4.3). However, under the same loads but with changes in support conditions, creep results in changes in the internal forces, as will be further discussed below.

The forces at the beam ends induced by shrinkage or gradual settlement of the support may be determined through conventional analysis by the force method, in which the modulus of elasticity is the age-adjusted modulus (see Example 4.1).

Prestress loss of a known magnitude may be represented as a set of self-equilibrating forces,[1] in the same way as the prestress itself, but generally with a reversed sign and smaller magnitude. The prestress loss is represented by a system of forces at the anchors and at the sections where the cable changes direction. The prestress loss develops gradually with time and so do the statically indeterminate forces it induces. Thus, the changes in the internal forces due to prestress loss are independent of the value of modulus of elasticity to be used in the analysis.

In Section 2.5, we have seen that creep, shrinkage and relaxation produce in a statically determinate structure changes in the stress distribution, but the stress resultants, N and M, remain unchanged. N and M are the resultants of normal stress on the entire section composed of its three components: concrete, non-prestressed reinforcement and prestressed steel. However, it is a

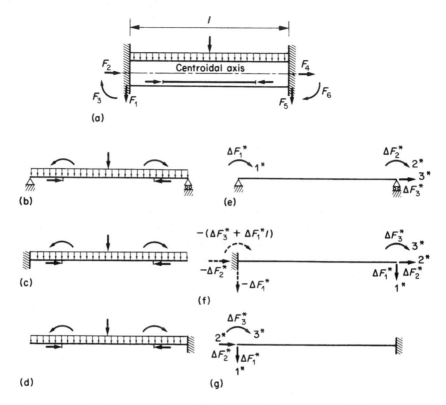

Figure 5.2 Analysis of the time-dependent changes in the end forces of a member caused by fixity introduced after loading: (a) totally fixed beam subjected at time t_0 to a system of forces; (b), (c), (d) statically determinate beams loaded at time t_0, statical system changed to totally fixed beam at time t_1; (e), (f), (g) coordinate systems.

common practice to calculate the internal forces due to prestressing by considering the forces exerted by the prestress tendons on the remainder of the structure, the concrete and non-prestressed reinforcement. This is the meaning adopted here where reference is made to the internal forces caused by prestress loss.

Now, consider that the beam in Fig. 5.2(a) is constructed in three different ways. At age t_0, we assume that the external loads are applied on one of the statically determinate systems in Fig. 5.2(b), (c) or (d). Subsequently, at age t_1 the beam is made totally fixed as shown in Fig. 5.2(a). Time-dependent changes in the forces at the end of the member will gradually develop; the equations derived below can be used to calculate the member-end forces at any time t_2 later than t_1.

A system of three coordinates, $1^* - 3^*$ is defined in each of Fig. 5.2(e), (f) and (g). If the statically determinate system in Fig. 5.2(b), (c) or (d) is left unchanged during the period t_1 to t_2, creep will change the displacement at the coordinates by the amount:

$$\{\Delta D^*\} = \{D^*(t_0)\}[\varphi(t_2, t_0) - \varphi(t_1, t_0)] \tag{5.3}$$

where $\{D^*(t_0)\}$ are the instantaneous displacement at t_0 due to the external loads on the statically determinate system; $\varphi(t_i, t_j)$ is the coefficient for creep at t_i when the age at loading is t_j.

The age-adjusted flexibility matrix is

$$[\bar{f}] = [f][1 + \chi\varphi(t_2, t_1)] \tag{5.4}$$

where $\chi = \chi(t_2, t_1)$ is the aging coefficient (see Section 1.7), $[f]$ is the flexibility matrix of a statically determinate beam (Fig. 5.2(b), (c) or (d)). The modulus of elasticity to be used in the calculation of the elements of $[f]$ is $E_c(t_1)$.

The compatibility Equation (4.5) can now be applied, which is repeated here:

$$[\bar{f}]\{\Delta F^*\} = \{-\Delta D^*\} \tag{5.5}$$

Substitution of Equations (5.3) and (5.4) in Equation (5.5) and solution gives the changes in the three end forces developed during the period t_1 to t_2:

$$\{\Delta F^*\} = \left(\frac{\varphi(t_2, t_0) - \varphi(t_1, t_0)}{1 + \chi\varphi(t_2, t_1)}\right)\{F^*\} \tag{5.6}$$

where

$$\{F^*\} = [f]^{-1}\{-D^*(t_0)\} \tag{5.7}$$

The three values $\{F^*\}$ in Equation (5.7) are equal to the three fixed-end forces at the same coordinates when calculated in a conventional way, i.e. for an elastic beam subjected to the external loads in Fig. 5.2(a), with no creep or change in support conditions.

Equation (5.6) gives the changes occurring during the period t_1 to t_2 in three of the six end forces. The changes in the other three are the statical equilibrants of the first three. As an example, see the three forces indicated by broken arrows at the left end of the beam in Fig. 5.2(f). It can be seen that the final fixed-end forces at time t_2 will not be the same in the three beams considered above.

Example 5.1 Cantilever: restraint of creep displacements

The cantilever in Fig. 5.3(a) is subjected at age t_0 to a uniformly distributed load q/unit length. At age t_1, end B is made totally fixed. Find the forces at the two ends at a later time t_2. Use the following creep and aging coefficients: $\varphi(t_1, t_0) = 0.9$; $\varphi(t_2, t_0) = 2.6$; $\chi(t_2, t_1) = 0.8$; $\varphi(t_2, t_1) = 2.45$.

If the beam were totally fixed at the two ends, with no creep or

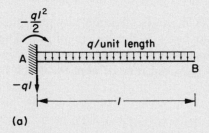

(a)

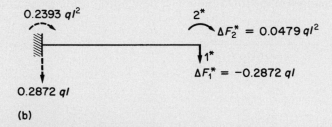

(b)

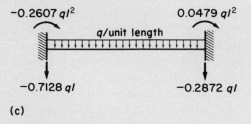

(c)

Figure 5.3 Analysis of time-dependent forces in a cantilever transformed into a totally fixed beam after loading (Example 5.1): (a) forces acting at time t_0; (b) changes in end forces between t_1 and t_2; (c) total forces at t_2.

change in support, the end forces at B caused by the load q would be:

$$\{F^*\} = \begin{Bmatrix} -ql/2 \\ ql^2/12 \end{Bmatrix}$$

Forces developed at end B of the cantilever during the period t_1 to t_2 (Equation (5.6)) are

$$\{\Delta F^*\} = \left(\frac{2.6 - 0.9}{1 + 0.8 \times 2.45}\right) \begin{Bmatrix} -ql/2 \\ ql^2/12 \end{Bmatrix} = \begin{Bmatrix} -0.2872ql \\ 0.0479ql^2 \end{Bmatrix}$$

These two forces and their equilibrants at end A are shown in Fig. 5.3(b). Superposition of the forces at the member ends in Fig. 5.3(a) and (b) gives the end forces at time t_2, shown in Fig. 5.3(c).

5.4 Analysis of time-dependent changes in internal forces in continuous structures

The method of analysis is explained, using as an example the plane frame shown in Fig. 5.4. This bridge structure is made up of three precast pre-stressed beams AB, CD and EF. At age t_0, prestress is applied in the factory, at which time each of the three members acted as a simple beam subjected to its self-weight and to the prestress. Precast elements in the form of a T are used for the inclined columns GH and IJ. Assume that the casting of the elements in the factory is done at the same time for all the elements. The precast elements are erected at age t_1 with provisional supports at B, C, D and E and, shortly after, the structure is made continuous by casting joints at B, C, D and E and post-tensioned cables inserted through ducts along the deck A to F. At the same time, the shores at B, C, D and E are removed. The

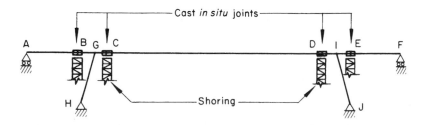

Figure 5.4 A frame composed of precast parts made continuous by cast *in situ* joints and post-tensioning.

analysis described below is concerned with the changes in the internal forces occurring between t_1 and a later time t_2.

We assume that a computer program is available for the analysis of elastic plane frames and will indicate here how such a program can be used for this problem. The axis of the frame is usually taken through the centroid of the cross-section and three degrees of freedom assumed at each joint. For the frame considered here, the joints are at the supports, the corners and at B, C, D and E.

The analysis is to be done in two stages, employing the same computer program in each to analyse a continuous frame. The presence of the reinforcement is ignored here; hence, the moment of inertia of any cross-section is that of a plain concrete section. In the first stage, calculate the displacements and the internal forces occurring instantaneously at t_1 after the continuity prestressing and removal of the shores. The modulus of elasticity to be used for the members is $E_c(t_1)$ and the loads to be applied are downward concentrated loads at B, C, D and E which are equal and opposite to the reaction on the shores due to the self-weight of the precast elements, before continuity. In addition, apply a set of self-equilibrating forces representing the effect of prestressing introduced at age t_1.

In the second stage, consider the effect of the forces developing gradually between t_1 and t_2. The modulus of elasticity to be used is the age-adjusted modulus $\bar{E}_c(t_2, t_1)$ (see Equation (1.31)). The forces to be applied form a system of self-equilibrating forces, $-\{\Delta F\}$; where $\{\Delta F\}$ are the changes in the fixed-end forces due to creep, shrinkage and prestress loss. Here each member is treated as a separate beam with fixed ends and the changes in the six forces at the member ends calculated according to the procedure of Section 5.3 (see Fig. 5.2). The six self-equilibrating forces calculated for each beam may be reversed and applied directly to the frame at the appropriate joints. Alternatively, the three forces to be applied at each joint are calculated by assemblage of forces at the ends of the members meeting at the joint.

The displacements and internal forces obtained in the analysis in the two stages mentioned above when superimposed on the corresponding values existing prior to t_1, give the final values existing at time t_2. Use of conventional linear computer programs to perform this analysis is discussed in detail with examples in Chapter 6.

5.5   Continuous composite structures

In this section we consider the time-dependent changes of internal forces in a statically indeterminate structure which has composite cross-sections. Consider the frame in Fig. 5.5(a) which has a composite cross-section for the part AD. The composite section is made up either of steel and concrete (Fig. 5.5(c)), or prestressed precast beam and cast *in situ* deck (Fig. 5.5(b)). Due to shrinkage, creep and prestress loss, internal forces develop and the changes

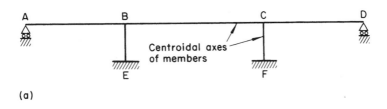

(a)

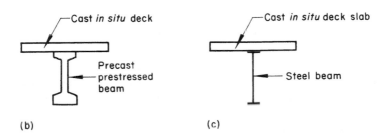

(b)   (c)

Figure 5.5 Example of a continuous composite structure: (a) statically indeterminate frame;
(b), (c) alternative composite cross-sections for part AD of the frame in (a).

for a specified period may be determined by application of the displacement method to the continuous frame in two stages as discussed in the preceding section. The first stage is concerned with the joint displacements and the member-end forces produced at time t_0 immediately after application of loads. The joints are artificially locked in this position, causing time-dependent fixed-end forces to develop gradually during a specified period t_0 to t. In the second stage of analysis, the artificial restraining forces are removed, producing changes in joint displacements and member-end forces, calculated by a second application of the displacement method. The following are additional remarks to be considered in the second stage of analysis when calculating the cross-section properties and the changes in fixed-end forces in composite members.

For any of the composite sections in Fig. 5.5, the cross-section to be used in the second stage of analysis should be the age-adjusted transformed section (see Section 1.11.1). The fixed-end forces to be used in the same stage are to be determined at the centroid of the age-adjusted transformed section.

The age-adjusted modulus of elasticity of concrete depends upon t_0 and t, the ages of concrete at the beginning and the end of the period considered. Thus, the centroid of the transformed section will be changing when analysing for different time periods or when considering the instantaneous effects of applied loads. This difficulty may be avoided by assuming that the axis of the frame passes through an arbitrary reference point in the cross-section, but this will result in coupling the effects of the axial forces and bending on the axial strain and curvature (see Section 2.3 and Equation (2.13)). Some computer programs allow the reference axis of the frame to be different from the centroidal axis, but in general this facility is not available. Hence, it may be necessary to determine the position of the centroid of the transformed section and calculate the fixed-end forces with respect to the centroid at the end sections of each member. Determining the correct position of the centroid is particularly important when considering the effect of the shrinkage of the deck slab.

Use of the displacement method for the analysis of a framed structure involves the assumption that the internal forces and the forces at the ends of a member with fixed ends are known *a priori*. Due to creep and shrinkage, the stress distribution in a composite statically determinate member changes with time (see Section 2.5), and if the member is statically indeterminate, the reactions and the stress resultants are also time-dependent. The statically indeterminate changes in internal forces in a composite member with fixed ends are discussed in the following section.

5.6   Time-dependent changes in the fixed-end forces in a composite member

Consider a member of a continuous structure subjected at time t_0 to external applied forces including prestressing. Assume that the axial force, N, and the bending moment, M, are known at all sections at time t_0 (determined by conventional analysis). Immediately after application of the loads, the joints are totally fixed as shown in Fig. 5.6(a) for a typical member which is assumed to have a composite cross-section. The time-dependent changes in the fixed-end forces due to creep and shrinkage of concrete and relaxation of prestressed steel are here analysed by the force method.

A system of three coordinates is chosen on a statically determinate released structure in Fig. 5.6(b). The analysis involves the solution of the following equation (see Equation (4.5)):

$$[\bar{f}] \{\Delta F\} = - \{\Delta D\} \tag{5.8}$$

where $[\bar{f}]$ is the age-adjusted flexibility matrix of the released structure corresponding to the three coordinates; $\{\Delta F\}$ are the changes in the redundants during the period t_0 to t; $\{\Delta D\}$ are the changes during the same period in the displacements of the released structure.

Coordinate 1 in Fig. 5.6(b) is assumed to be at the centroid of the age-adjusted transformed section (see Section 1.11.1).

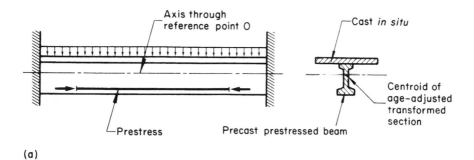

(a)

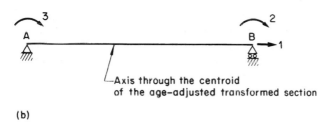

(b)

Figure 5.6 Analysis of changes of internal forces due to creep, shrinkage and relaxation in a composite member with fixed ends: (a) composite member, beam ends fixed at t_0 after application of external loads; (b) statically determinate released structure and coordinate system.

Solution of Equation (5.8) gives

$$\{\Delta F\} = [\bar{f}]^{-1} \{-\Delta D\} \qquad (5.9)$$

where $[\bar{f}]^{-1}$ is the age-adjusted stiffness corresponding to the coordinate system in Fig. 5.6(b). For a member with constant cross-section,[2]

$$[\bar{f}]^{-1} = \frac{\bar{E}_c}{l} \begin{bmatrix} \bar{A} & 0 & 0 \\ 0 & 4\bar{I} & 2\bar{I} \\ 0 & 2\bar{I} & 4\bar{I} \end{bmatrix} \qquad (5.10)$$

where l is the length of member; $\bar{A}$ and $\bar{I}$ are the area and moment of inertia about an axis through the centroid of the age-adjusted transformed section for which $E_{\text{ref}} = \bar{E}_c(t, t_0)$ is the age-adjusted elasticity modulus. Substitution of Equation (5.10) into (5.9) gives:

$$\{\Delta F\} = \frac{\bar{E}_c}{l} \begin{bmatrix} \bar{A} & 0 & 0 \\ 0 & 4\bar{I} & 2\bar{I} \\ 0 & 2\bar{I} & 4\bar{I} \end{bmatrix} \{-\Delta D\} \tag{5.11}$$

The changes $\{\Delta D\}$ in the displacements of the released structure may be determined by numerical integration or by virtual work using the equation (see Section 3.8):

$$\{\Delta D\} = \begin{cases} \int(\Delta \varepsilon_O) & N_{u1} & dl \\ \int(\Delta \psi) & M_{u2} & dl \\ \int(\Delta \psi) & M_{u3} & dl \end{cases} \tag{5.12}$$

where $\Delta \varepsilon_O$ and $\Delta \psi$ are the changes during the period considered in the strain at the reference point O and in the curvature in any section; N_{u1}, M_{u2} and M_{u3} are axial force and bending moments due to unit force at the three coordinates.

$\Delta \varepsilon_O$ and $\Delta \psi$ may be calculated by the method presented in Section 2.5 using Equation (2.40) which is rewritten here:

$$\begin{Bmatrix} \Delta \varepsilon_O \\ \Delta \psi \end{Bmatrix} = \frac{1}{\bar{E}_c(\bar{A}\bar{I} - \bar{B}^2)} \begin{bmatrix} \bar{I} & -\bar{B} \\ -\bar{B} & \bar{A} \end{bmatrix} \begin{Bmatrix} -\Delta N \\ -\Delta M \end{Bmatrix} \tag{5.13}$$

where $\{\Delta N, \Delta M\}$ are a normal force at O and a bending moment required to artificially prevent the change in strain in the section during the period t_0 to t. $\bar{B}$ is the first moment of area of the age-adjusted transformed section about an axis through the reference point O.

Because the reference point O is chosen at the centroid of $\bar{A}$, the value $\bar{B}$ is zero and Equation (5.13) is simplified to:

$$\begin{Bmatrix} \Delta \varepsilon_O \\ \Delta \psi \end{Bmatrix} = -\frac{1}{\bar{E}_c} \begin{Bmatrix} \Delta N/\bar{A} \\ \Delta M/\bar{I} \end{Bmatrix} \tag{5.14}$$

The value $\{\Delta N, \Delta M\}$ is obtained by summing up the forces required to prevent creep, shrinkage and relaxation (see Equations (2.41) to (2.44)).

In Examples 5.2 and 5.3, composite frames are analysed for the effects of creep and shrinkage, using the procedure discussed in Sections 5.5 and 5.6.

5.7   Artificial restraining forces

In Sections 5.5 and 5.6, a method is suggested for the analysis of the forces developed by creep, shrinkage of concrete and relaxation of prestress steel in a continuous structure. The procedure presented in Section 2.5 is employed, in which the strain due to creep, shrinkage and relaxation is first restrained by the introduction of the internal forces ΔN and ΔM (Equation (2.41)), which are subsequently released while the member ends are allowed to displace freely as in a simple beam. Then the member ends are restrained by the introduction of the fixed-end forces. This artificial restraint is also to be removed by the application of a set of equal and opposite forces at the joints on the continuous structure (see Example 5.2 to follow). An alternative procedure is to determine a set of external applied forces preventing the strain due to creep, shrinkage and relaxation at all sections and then remove this artificial restraint in one step by applying a set of equal and opposite forces on the continuous composite structure. The same method will be employed in Section 10.6 for the analysis of the effect of temperature on the continuous structure in which the cross-section and/or the temperature distribution is non-uniform.

The artificial restraining *internal* forces ΔN and ΔM can be introduced by the application of *external* forces at the ends of members as well as tangential and transverse forces as shown in Fig. 5.7. The intensities p and q of the tangential and transverse artificial restraining load are given by:

$$p = -\frac{d}{dx}(\Delta N) \tag{5.15}$$

$$q = -\frac{d^2}{dx^2}(\Delta M) \tag{5.16}$$

Two additional shear forces at the ends are necessary for equilibrium. The set of self-equilibrating forces shown in Fig. 5.7 is to be reversed and applied on the continuous composite structure.

When a computer is used, each member may be subdivided into parts for which the axial force ΔN may be considered constant while ΔM varies as a straight line. In this way, Equations (5.15) and (5.16) will give $p = 0$ and $q = 0$ and hence the restraining forces need to be applied only at the nodes.

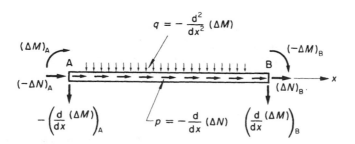

Figure 5.7 A set of self-equilibrating forces applied on a member to artificially prevent the strain due to creep, shrinkage and relaxation.

160   Concrete Structures

*Example 5.2 Steel bridge frame with concrete deck: effects of
shrinkage*

The bridge frame in Fig. 5.8(a) has a composite section for part AD
(Fig. 5.8(b)) and a steel section for the columns BE and CF. It is
required to find the changes in the reactions and in the stress distribution in the cross-section at G due to uniform shrinkage of deck slab
occurring during a period t_0 to t_1.

The cross-section properties of members are: for columns BE and
CF, area = 20000 mm² (31 in²) and moment of inertia about an axis
through centroid = 0.012 m⁴ (29000 in⁴); for part AD, the steel

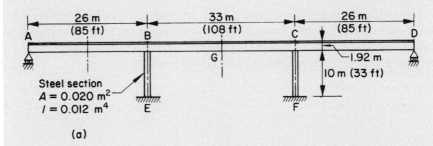

(a)

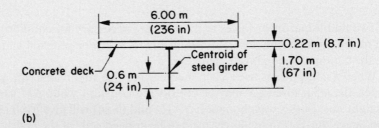

(b)

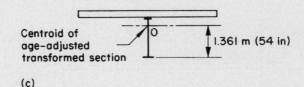

(c)

Figure 5.8 Analysis of statically indeterminate forces caused by creep and shrinkage
in a composite frame (Examples 5.2 and 5.3): (a) frame dimensions;
(b) cross-section properties for part AD; (c) location of centroid of
age-adjusted transformed section composed of area of concrete plus
$\bar{a}$ times area of steel.

cross-section area $= 39\,000\,\text{mm}^2$ ($60\,\text{in}^2$) and moment of inertia about its centroid $= 0.015\,\text{m}^4$ ($36\,000\,\text{in}^4$).

The material properties are:

$$E_c(t_0) = 30\,\text{GPa (4350 ksi)} \qquad E_s = 200\,\text{GPa (29\,000 ksi)}$$

$$\varphi(t_1, t_0) = 2.5 \qquad \chi(t_1, t_0) = 0.8 \qquad \varepsilon_{cs}(t_1, t_0) = -270 \times 10^{-6}$$

The following cross-section properties for part AD are needed in the analysis:

Age-adjusted transformed section

$$\bar{E}_c(t_1, t_0) = \frac{30 \times 10^9}{1 + 0.8 \times 2.5} = 10\,\text{GPa} \qquad \bar{a}(t_1, t_0) = \frac{200}{10} = 20.$$

The age-adjusted transformed section is composed of $A_c = 1.32\,\text{m}^2$ plus $\bar{a}A_s = 20 \times 0.039 = 0.780\,\text{m}^2$. A reference point O is chosen at the centroid of the age-adjusted transformed section at $1.361\,\text{m}$ above bottom fibre (Fig. 5.8(c)). Using $E_{ref} = \bar{E}_c = 10\,\text{GPa}$, the properties of the age-adjusted transformed section are:

$$\bar{A} = 2.10\,\text{m}^2 \qquad \bar{B} = 0 \qquad \bar{I} = 1.0232\,\text{m}^4.$$

Transformed section at t_0

$$E_c(t_0) = 30\,\text{GPa} \qquad a(t_0) = \frac{200}{30} = 6.667 \qquad E_{ref} = E_c(t_0)$$

Area and its first and second moment about an axis through the reference point O:

$$A = 1.58\,\text{m}^2 \qquad B = -0.3947\,\text{m}^3 \qquad I = 0.5221\,\text{m}^4.$$

The centroid of this transformed section is $1.611\,\text{m}$ above the bottom fibre and moment of inertia about an axis through the centroid is $0.4234\,\text{m}^4$.

Concrete deck slab Area, first and second moment of the concrete deck slab alone about an axis though the reference point O:

$$A_c = 1.32\,\text{m}^2 \qquad B_c = -0.5927\,\text{m}^3 \qquad I_c = 0.2714\,\text{m}^4.$$

The resultant of stresses if shrinkage were restrained at all sections of AD (Equation (2.43));

$$\begin{Bmatrix} \Delta N \\ \Delta M \end{Bmatrix} = -10 \times 10^9(-270 \times 10^{-6}) \begin{Bmatrix} 1.32 \\ -0.5927 \end{Bmatrix} = \begin{Bmatrix} 3.564 \times 10^6 \text{ N} \\ -1.600 \times 10^6 \text{ N-m} \end{Bmatrix}$$

Because ΔN and ΔM are constant in all sections of members AB, BC and CD, shrinkage can be prevented at all sections by the application of external forces only at the ends of the members as shown in Fig. 5.9(a). The stress distribution in the restrained condition is the same for all sections of AD and is shown in Fig. 5.9(b).

The restraining forces at the member ends are assembled at the joints and applied in a reversed direction on the continuous frame (Fig. 5.9(c)). The forces at the end of members at each of joints B and C cancel out, leaving only forces at A and D. Now the continuous frame in Fig. 5.9(c) is to be analysed in a conventional way, by computer or by hand, giving the internal forces shown in Fig. 5.9(d). The properties of the cross-sections of the members to be used in the analysis are the age-adjusted transformed section properties, using the same $E_{\text{ref}} = 10\,\text{GPa}$ for AD as well as for the columns. Line AD in Fig. 5.9(c) is at the level of the centroid of the age-adjusted transformed section (1.361 m above the soffit of the section in Fig. 5.8(c)). In the analysis of the continuous frame, the upper 1.361 m of each of members BE and CF in Fig. 5.9(c) is considered rigid.

The forces in Fig. 5.9(d) represent the internal forces which will eliminate the artificial restraint introduced in Fig. 5.9(a).

The statically indeterminate reactions caused by shrinkage are equal to the superposition of the reactions in Fig. 5.9(a) and (c). But since the forces in Fig. 5.9(a) produce no reactions, the reactions shown in Fig. 5.9(d) represent the total statically indeterminate values. The internal forces in Fig. 5.9(d) represent resultants of stresses in concrete and steel of the age-adjusted transformed sections caused by elimination of the artificial restraint.

To find the stress distribution at any section, we have to superpose the stress distribution in the restrained condition (Fig. 5.9(b)) to the stress distribution caused by the internal forces in Fig. 5.9(d), applied on the age-adjusted transformed section. The superposition is performed in Fig. 5.10 for the cross-section at G.

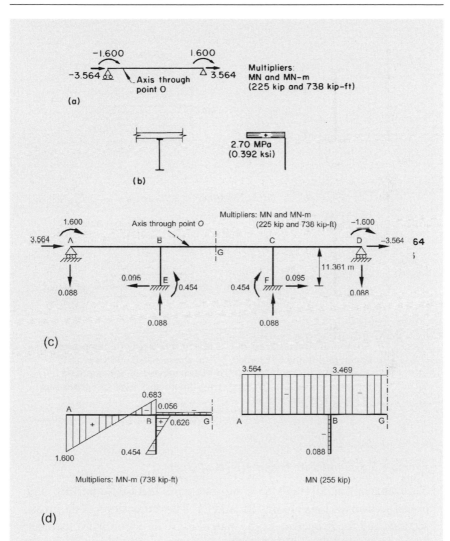

(a)

(b)

(c)

(d)

Figure 5.9 Analysis of internal forces caused by shrinkage in the composite continuous frame of Example 5.2: (a) resultants of stresses to restrain shrinkage of concrete at the ends of members AB, BC or CD; (b) stress distribution in any section of AD at time t_1 if shrinkage were fully restrained; (c) forces in (a) assembled and applied in a reversed direction on the continuous frame (the reactions corresponding to the applied forces are included in the figure); (d) bending moment and axial force diagrams for the frame in (c).

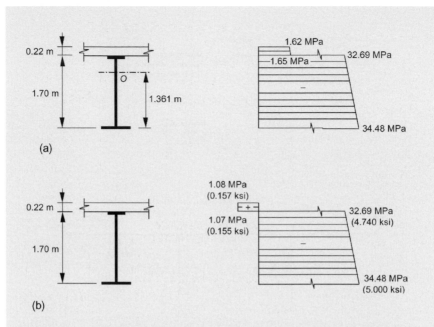

(a)

(b)

Figure 5.10 Analysis of stresses at section G due to shrinkage in a composite continuous frame of Example 5.2; (a) stress distribution due to $N = -3.469\,\text{MN}$ at O and $M = -0.056\,\text{MN-m}$ applied on age-adjusted transformed section; (b) total stress due to shrinkage (superposition of Figs 5.9(b) and 5.10(a)).

Example 5.3 Composite frame: effects of creep

The frame in Fig. 5.11(a) has a composite cross-section for part BC and a steel section for the columns BE and CF. The dimensions of the cross-sections and the properties of the materials are the same as for member BC in Example 5.2; see Fig. 5.8. The properties of the cross-sections of the columns BE and CF are given in Fig. 5.11(a). At time t_0, a uniformly distributed downward load of intensity $q = 40\,\text{kN/m}$ is applied on BC and sustained to a later time t_1. It is required to find the change in the bending moment due to creep during the period t_0 to t_1. Use the same creep and aging coefficients as in Example 5.2. Also find the stress distribution and the deflection at section G at time t_1.

The properties of the cross-section for member BC are the same as for part AD of the frame of Example 5.2, and thus this part of the calculation is not discussed here.

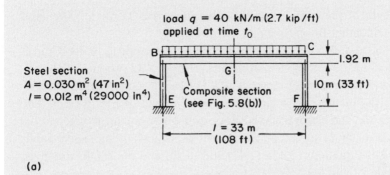

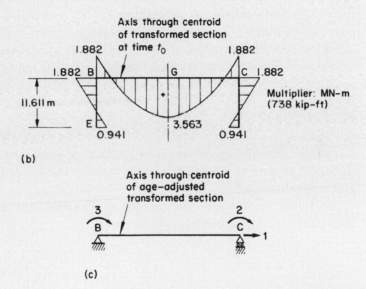

(c)

Figure 5.11 Composite frame of Example 5.3: (a) frame dimension (for cross-section of member BC, see Fig. 5.8(b)); (b) bending moment diagram at t_0; (c) released structure for analysis of changes of fixed-end forces in BC.

A conventional elastic analysis is performed for a continuous frame subjected to the load q, giving the bending moment at time t_0 shown in Fig. 5.11(b). The moments of inertia of the cross-sections used in the analysis are $0.4234 \, \text{m}^4$ for BC and $0.080 \, \text{m}^4$ for the columns. These are the centroidal moments of inertia of transformed sections, using $E_{\text{ref}} = E_c(t_0) = 30 \, \text{GPa}$ for all members. The centroid of the transformed section at age t_0 for member BC is $1.611 \, \text{m}$ above the bottom fibre; hence, the length of the columns used in the analysis is 11.611 m. The axial force in member BC at time t_0 is −0.2431 MN.

If, immediately after application of the load, at time t_0, joints B and C were locked preventing displacements, creep would produce change in the forces at the ends of member BC. For calculation of these changes, release the member as a simple beam as shown in Fig. 5.11(c). The changes $\Delta\varepsilon_O$ and $\Delta\psi$ in the axial strain and curvature due to creep in the released structure are calculated at various sections by successive applications of Equations (2.32), (2.42) and (2.40) and the results are given in Tables 5.1 and 5.2.

In the preparation of the two tables, the reference point O, at which the axial strain is calculated, is considered at the centroid of the age-adjusted transformed section. The values of the axial force and bending moments in member BC of the frame in Fig. 5.11(b) are transformed to their statical equivalents before listing in Table 5.1. (The centroidal axis is moved downwards 0.250 m; the value $0.250 \times 0.2431 = 0.062$ MN-m is added to the bending moment ordinates shown for part BC in Fig. 5.11(b).

The changes in the displacements $\{D\}$ at the three coordinates of Fig. 5.11(c) are calculated, assuming parabolic variation of $\Delta\varepsilon_O$ and $\Delta\psi$ over the length BC and employing Equations (C.5–7). The values obtained are listed in Table 5.2.

The forces necessary to prevent the displacements at the three coordinates are (Equation (5.11):

$$\{\Delta F\} = \frac{10 \times 10^9}{33} \begin{bmatrix} 2.10 & 0 & 0 \\ 0 & 4(1.0232) & 2(1.0232) \\ 0 & 2(1.0232) & 4(1.0232) \end{bmatrix} \begin{Bmatrix} 1691 \\ 807 \\ -807 \end{Bmatrix} 10^{-6}$$

$$= \begin{Bmatrix} 1.0761 \text{ MN} \\ 0.5004 \text{ MN-m} \\ -0.5004 \text{ MN-m} \end{Bmatrix}$$

The three forces $\{\Delta F\}$, together with their three equilibrants, are shown at the member ends in Fig. 5.12(a). This set of self-equilibrating forces is reversed in direction and applied on the frame in Fig. 5.12(b). Analysis of this frame by a conventional method gives the member-end forces shown in Fig. 5.12(c). The properties of the cross-section for member BC used in the analysis are those of the age-adjusted

Table 5.1 Instantaneous axial strain and curvature at t_0, immediately after application of the load q (Example 5.3, Fig. 5.11)

Member	Section	Properties of transformed[1] section at age t_0 ($E_{ref} = E_c(t_0) = 30\,GPa$)			Internal forces introduced at t_0		Axial strain and curvature at t_0 (Equation (2.32))		Properties of concrete area			Deflection at G (Equation (C.8))
		A	B	I	N	M	$\varepsilon_O(t_0)$	$\psi(t_0)$	A_c	B_c	I_c	$D(t_0)$
BC	B	1.58	−0.3947	0.5221	−0.2431	−1.821	−42.1	−148.1	1.32	−0.5927	0.2714	
	G	1.58	−0.3947	0.5221	−0.2431	3.624	64.9	280.5	1.32	−0.5927	0.2714	28.46
	C	1.58	−0.3947	0.5221	−0.2431	−1.821	−42.1	−148.1	1.32	−0.5927	0.2714	
Multiplier		m^2	m^3	m^4	$10^6\,N$	$10^6\,N\text{-}m$	10^{-6}	$10^{-6}\,m^{-1}$	m^2	m^3	m^4	$10^{-3}\,m$

[1] The reference point O is at the centroid of age-adjusted transformed section (Fig. 5.8(c))

Table 5.2 Changes in axial strain and in curvature and corresponding elongation and end rotations of the released structure in Fig. 5.11(c)

Member	Section	Internal forces to restrain creep (Equation (2.42))		Properties of age-adjusted transformed section ($E_{ref} = \bar{E}_c(t_1, t_0) = 10\,GPa$)			Changes in axial strain and in curvature (Equation (2.40))		Changes in displacements at the coordinates in Fig. 5.11(c) (Equations (C.5–7))			Change in deflection at G (Equation (C.8))
		ΔN	ΔM	$\bar{A}$	$\bar{B}$	$\bar{I}$	$\Delta\varepsilon_O$	$\Delta\psi$	ΔD_1	ΔD_2	ΔD_3	ΔD
BC	B	−0.8052	0.3810	2.10	0	1.0232	38.3	−37.2	−1691	−807	807	9.59
	G	2.015	−0.9415	2.10	0	1.0232	−96.0	92.0				
	C	−0.8052	0.3810	2.10	0	1.0232	38.3	−37.2				
Multipliers		10^6 N	10^6 N-m	m²	m³	m⁴	10^{-6}	10^{-6} m⁻¹	10^{-6} m	10^{-6} radian	10^{-6} radian	10^{-3} m

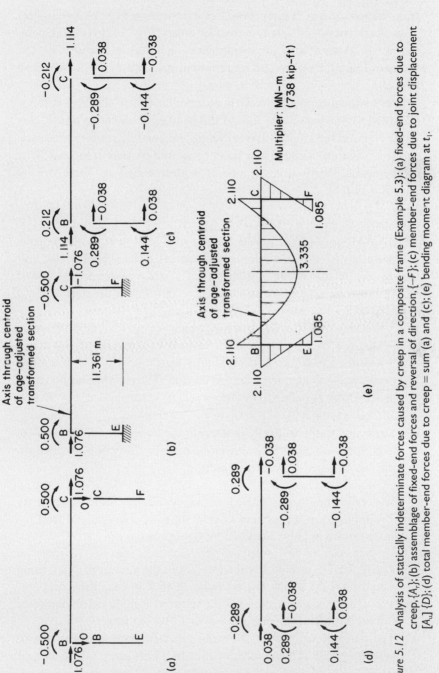

Figure 5.12 Analysis of statically indeterminate forces caused by creep in a composite frame (Example 5.3): (a) fixed-end forces due to creep, $\{A_r\}$; (b) assemblage of fixed-end forces and reversal of direction, $\{-F\}$; (c) member-end forces due to joint displacement $[A_u]\{D\}$; (d) total member-end forces due to creep = sum of (a) and (c); (e) bending moment diagram at t_1.

transformed section. Superposition of the forces in Fig. 5.12(a) and (c) gives the member-end forces caused by creep (Fig. 5.12(d)). Following the notations used with the displacement method in Section 5.2, the forces in Fig. 5.12(a), (b) and (c) represent respectively $\{A_r\}$, $\{-F\}$ and $[A_u]\{D\}$.

The bending moment at end B of member BC = $-1.821 - 0.289 = -2.110$ MN-m, which is the sum of the bending moment at time t_0 (see Table 5.1) and the change due to creep. The bending moments at various sections are calculated in a similar way and plotted in Fig. 5.12(e).

The stress distribution at time t_1 is determined in Table 5.3 by superposition of:

(a) Stress at time t_0, calculated for $N = -0.2431$ MN and $M = 3.624$ MN-m applied on the transformed section at t_0. The corresponding strain distribution is defined by: $\varepsilon_O(t_0) = 64.9 \times 10^{-6}$ and $\psi(t_0) = 280.5 \times 10^{-6}$ m^{-1} (Table 5.1). The stress values are calculated by multiplication of the strain by $E_s = 200$ GPa for the steel or $E_c(t_0) = 30$ GPa for concrete.

(b) Stress required to restrain creep, which is equal to the product of $[-\varphi(t_1, t_0)\bar{E}_c(t_1, t_0)/E_c(t_0)]$ and the stress in concrete calculated in (a).

(c) Stress due to $-\Delta N = -2.015$ MN and $-\Delta M = 0.9415$ MN-m applied on the age-adjusted transformed section. The corresponding strain distribution is defined by: $\Delta\varepsilon_O = -96.0 \times 10^{-6}$ and $\Delta\psi = 92.0 \times 10^{-6}$ m^{-1} (Table 5.2).

(d) Stress due to the statically indeterminate forces produced by creep: axial force = -0.038 MN and moment = -0.289 MN-m applied on the age-adjusted transformed section.

The stress values for the above four stages and their superposition are listed in Table 5.3 at the top and bottom fibres of concrete and steel.

The deflection at G is calculated by superposition of:

(a) The deflection at time t_0, calculated from the curvature $\psi(t_0)$, using Equation (C.8), which gives: $D(t_0) = 28.46 \times 10^{-3}$ m (Table 5.1).

(b) The deflection due to creep in the released system, calculated from the curvatures $\Delta\psi$ giving: $\Delta D = 9.59 \times 10^{-3}$ m (Table 5.2).

(c) The deflection due to a statically indeterminate moment due to creep = -0.289 MN-m constant over BC. This gives a deflection of -3.84×10^{-3} m. Hence, the total deflection at time t_1 is

Table 5.3 Stress distribution at section G (Example 5.3)

| | Stress in stages (MPa) | | Creep effect | | Creep effect = (b) − (c) + (d) | Stress at time t_1 = (a) + (b) + (c) + (d) | |
| | At time t_0 | | | | | | |
	(a)	(b)	(c)	(d)	MPa	MPa	ksi
Top of concrete	−2.756	2.297	−1.474	0.140	0.963	1.793	−0.260
Bottom of concrete	−0.906	0.755	−1.272	0.078	−0.439	−1.345	−0.195
Top of steel	−6.04	0	−25.44	1.55	−23.89	−29.93	−4.34
Bottom of steel	89.33	0	5.84	−8.04	−2.20	87.13	12.64

$$(28.46 + 9.59 - 3.84)10^{-3} = 34.21 \times 10^{-3} \, \text{m}$$

$$= 34.21 \, \text{mm} \, (1.347 \, \text{in}).$$

We can see by comparing the bending-moment diagrams in Figs 5.11(b) and 5.12(e) that creep increases the absolute values of the bending moment in the columns. Creep reduces the effective modulus of elasticity of concrete; thus, the flexural rigidity of BC is reduced while the rigidity of the steel column is unchanged. The change in relative rigidity is the reason for the increase in bending moment in the columns. It follows from this discussion that if the same composite cross-section is used in all members, creep will not result in any changes in internal forces or reactions. However, this is a hypothetical situation; in practice the shrinkage which occurs at the same time as creep will result in a change in the bending moments.

5.8   Step-by-step analysis by the displacement method

Modern concrete structures are often composed of precast or cast *in situ* elements assembled by prestressing. Bridges built by the segmental method are examples. The basis of a step-by-step numerical procedure, similar to that presented in Section 4.6, but using the general displacement method of analysis, will be presented here. The time is divided into intervals and the changes in stresses or internal forces are considered to occur at the middle of the intervals (Fig. 4.11).

Three different materials are generally involved: concrete, prestressed steel and non-prestressed reinforcement. In the three materials, the strains developed between t_0, the beginning of the first interval, and $t_{i + \frac{1}{2}}$, the end of the ith interval, are given by (see Equation (1.24)):

$$\varepsilon_c(t_{i + \frac{1}{2}}) = \sum_{j=1}^{i}\left(\frac{1 + \varphi(t_{i+\frac{1}{2}}, t_j)}{E_c(t_j)}(\Delta\sigma_c)_j\right) + \varepsilon_{cs}(t_{i + \frac{1}{2}}, t_0) \tag{5.17}$$

$$\varepsilon_{ps}(t_{i + \frac{1}{2}}) = \frac{1}{E_{ps}}\sum_{j=1}^{i}[(\Delta\sigma_{ps})_j - (\Delta\bar{\sigma}_{pr})_j] \tag{5.18}$$

$$\varepsilon_{ns}(t_{i + \frac{1}{2}}) = \frac{1}{E_{ns}}\sum_{j=1}^{i}(\Delta\sigma_{ns})_j \tag{5.19}$$

where σ and ε are the stress and strain with subscripts c, ps and ns referring to concrete, prestressed and non-prestressed steel, respectively; t is the age with subscript i (or j) indicating the middle of the ith (or jth) interval; t_0 is the age at the beginning of the period for which the analysis is considered; $\varepsilon_{cs}(t_{i+\frac{1}{2}}, t_0)$ is the shrinkage that would occur in concrete if it were free during the period t_0 to $t_{i+\frac{1}{2}}$; $\Delta\bar{\sigma}_{pr}$ is the reduced relaxation of prestressed steel (see Section 1.5); $(\Delta\sigma)_j$ is the change of stress at the middle of the jth interval.

The change in strain in the ith interval can be separated by taking the difference between the strain values calculated by each of the last three equations at the ends of the intervals $i-1$ and i:

$$(\Delta\varepsilon_c)_i = \frac{1 + \varphi(t_{i+\frac{1}{2}}, t_i)}{E_c(t_i)}(\Delta\sigma_c)_i$$

$$+ \sum_{j=1}^{i-1}\left(\frac{(\Delta\sigma_c)_j}{E_c(t_j)}[\varphi(t_{i+\frac{1}{2}}, t_j) - \varphi(t_{i-\frac{1}{2}}, t_j)]\right) + (\Delta\varepsilon_{cs})_i \tag{5.20}$$

$$(\Delta\varepsilon_{ps})_i = \frac{(\Delta\sigma_{ps})_i}{E_{ps}} - \frac{(\Delta\bar{\sigma}_{pr})_i}{E_{ps}} \tag{5.21}$$

$$(\Delta\varepsilon_{ns})_i = \frac{(\Delta\sigma_{ns})_i}{E_{ns}} \tag{5.22}$$

The last equation is a linear relationship between stress and strain in the non-prestressed steel. Equations (5.20) and (5.21) may be rewritten in pseudolinear forms:

$$(\Delta\varepsilon_c)_i = \frac{(\Delta\sigma_c)_i}{(E_{ce})_i} + (\Delta\bar{\varepsilon}_c)_i \tag{5.23}$$

$$(\Delta\varepsilon_{ps})_i = \frac{(\Delta\sigma_{ps})_i}{E_{ps}} + (\Delta\bar{\varepsilon}_{ps})_i \tag{5.24}$$

where $(E_{ce})_i$ is an effective modulus of elasticity of concrete to be used in an elastic analysis for the ith interval,

$$(E_{ce})_i = \frac{E_c(t_i)}{1 + \varphi(t_{i+\frac{1}{2}}, t_i)} \tag{5.25}$$

$(\Delta\bar{\varepsilon}_c)_i$ is equal to the sum of the second and third terms on the right-hand side of Equation (5.20). Similarly, $(\Delta\bar{\varepsilon}_{ps})_i$ is equal to the last term in Equation (5.21). The terms $(\Delta\bar{\varepsilon})_i$ in Equations (5.23) and (5.24) represent an 'initial' deformation independent of the stress increment in the ith interval. Thus,

$(\Delta\bar{\varepsilon})_i$ can be determined if the stress increments in the preceding increments are known.

In the step-by-step analysis, a complete analysis of the structure is performed for each time interval. Thus, when the analysis is done for the ith interval, the stress increments in the preceding intervals have been previously determined. In this way, the initial strains $(\Delta\bar{\varepsilon})_i$ are known values which can be treated as if they were produced by a change in temperature of known magnitude.

In the analysis of a plane frame by the displacement method, three nodal displacements are determined at each joint: translations in two orthogonal directions and a rotation. With the usual assumption that a plane cross-section remains plane during deformation, the strain and hence the stress at any fibre in a cross-section of a member can be calculated from the nodal displacements at its ends.

In the step-by-step procedure, a linear elastic analysis is executed for each time interval by the conventional displacement method. The cross-section properties to be used in this analysis are those of a transformed section composed of the area of concrete plus a_i times the area of steel; where a_i is a ratio varying with the interval and for the ith interval:

$$a_i = \frac{E_s}{(E_{ce})_i} \qquad (5.26)$$

where E_s is the respective modulus of elasticity of prestressed or non-prestressed steel.

In any interval i, the three materials are considered as if they were subjected to a change of temperature, producing the free strain $(\Delta\bar{\varepsilon})_i$ of known magnitude. The corresponding stress $(\Delta\sigma)_i$ in the three materials are unknowns to be determined by the analysis for the ith interval; the values $(\Delta\sigma)_i$ represent the stress due to external loading (if any) applied at the middle of the ith interval plus the stress due to the fictitious change in temperature mentioned above.

Analysis of stress due to arbitrary temperature distribution involves the following steps. First the strain due to temperature $((\Delta\bar{\varepsilon})_i$ in our case) is artificially restrained by internal forces ΔN and ΔM in each section (see Equations (2.25) and (2.26)). This is equivalent to the application of a set of self-equilibrating forces (see Fig. 5.7 and Equations (5.15) and (5.16)). The artificial restraint is then removed by application of a set of equal and opposite forces.

An example of analysis by this method and a listing of a computer program which performs the analysis can be found in the references mentioned in Note 3.

5.9 General

Chapters 2 to 5 are concerned with the analysis of stresses and deformations in uncracked reinforced or prestressed concrete structures accounting for the effects of the applied load including prestressing, creep and shrinkage of concrete and relaxation of prestressed steel. Creep is assumed to be proportional to stress and, thus, instantaneous strain and creep have a linear relationship with stress. Shrinkage and relaxation result in changes in concrete stress and must therefore also produce creep. In spite of this interdependence, the analysis is linear, which means that superposition of stresses, strains or displacements applies and the stresses or deformations due to applied loads or due to shrinkage or due to relaxation are proportional to the cause. Because of the linearity, conventional linear computer programs can be employed for the time-dependent analysis. This is demonstrated by examples in Chapter 6.

Creep, shrinkage and relaxation change stresses in concrete and steel. In statically determinate structures, the change is in the partitioning of the internal forces between concrete, prestressed and non-prestressed steel, but the resultants in the three components combined remain unchanged. In statically indeterminate structures, the reactions and the internal forces generally change with time.

Chapters 7, 8, 9 and 13 are concerned with the analysis of stresses and deformations when the tensile strength of concrete is exceeded at some sections of a structure producing cracking. The behavior is no longer linear.

Notes

1 See Section 14.5 of the reference mentioned in note 3 of Chapter 3.
2 See Appendix D of the reference mentioned in note 3 of Chapter 3.
3 A computer program in FORTRAN for analysis of the time-dependent displacements, internal forces and stresses reinforced and prestressed concrete structures, including the effects of cracking, is available. See Elbadry, M. and Ghali; A., *Manual of Computer Program CPF: Cracked Plane Frames in Prestressed Concrete, Research Report No. CE85-2*, revised 1993, Department of Civil Engineering, The University of Calgary, Calgary, Alberta, Canada.

Chapter 6

Analysis of time-dependent internal forces with conventional computer programs

The Confederation Bridge connecting Prince Edward Island and New Brunswick, Canada: Floating crane installing a 190m long segment on a pier

6.1   Introduction

Computers are routinely used in practice to analyse structures, particularly when linear stress–strain relationship of the material is acceptable and when displacements are small. These assumptions are commonly accepted in the analysis of structures in service. Thus, many of the available computer programs perform linear analysis, in which the strain is proportional to the stress and superposition of displacements, strains, stresses and internal forces is allowed. The present chapter demonstrates how conventional linear computer programs can be employed for approximate analysis of the time-dependent effects of creep and shrinkage of concrete and relaxation of prestressed steel. Only framed structures are considered here. These can be idealized as assemblages of beams (bars). Thus, the computer programs of concern are those for plane or space frames, plane or space trusses or plane grids.[1]

The procedure discussed in this chapter can be used to solve time-dependent problems of common occurrence in practice. As an example, consider the effects of shortening, due to creep and shrinkage, of a prestressed floor supported on columns constructed in an earlier stage. Analysis of the effect of differential shortening of columns in a high-rise building provides another example; the compressive stress and the change in length due to creep are commonly greater in interior than exterior columns. Bridge structures are frequently composed of members (segments), precast or cast-in-situ, made of concrete of different ages or of concrete and steel (e.g. cable stays). The precast members are erected with or without the use of temporary supports and made continuous with cast-in-situ joints or with post-tensioned tendons. In all these cases, the time-dependent analysis can be done by the application and the superposition of the results of conventional linear computer programs.

6.2   Assumptions and limitations

Immediate strain and creep of concrete are proportional to the stress (compressive or tensile) and the effect of cracking is ignored. Structures are idealized as prismatic bars (members) connected at nodes. The cross-sectional area properties of any bar are those of a homogeneous section. Thus, the presence of the reinforcing bars or the tendons in a cross-section is ignored in calculation of the cross-sectional area properties. Alternatively, a tendon or a reinforcing bar can be treated as a separate member connected to the nodes by rigid arms (Fig. 6.1(a)). The axes of members coincide with their centroidal axes. Because the cross-section of an individual member is considered homogeneous, no transformed cross-sectional area properties are required and the variation of the location of the centroids of transformed sections due to creep of concrete does not need to be considered. A composite member

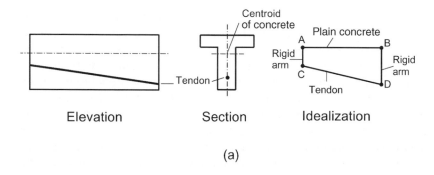

Elevation   Section   Idealization

(a)

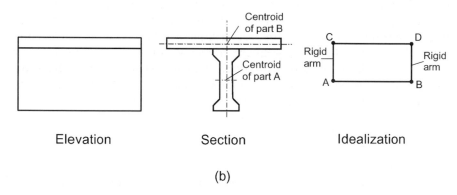

Elevation   Section   Idealization

(b)

Figure 6.1 Idealization of members: (a) prestressed member idealized as two bars; (b) composite member idealized as two bars of different material properties.

whose cross section consists of a precast part and a cast-in-situ part, or of concrete and steel, is treated as two homogeneous members connected by rigid arms joining the centroids of the two parts (Fig. 6.1(b)).

With the idealization using short rigid arms as shown in Figs. 6.1(a) and (b), the actual member should be divided into a number of short members (say 10; see Example 6.5). The internal forces obtained by analysis should be considered representatives of the actual structure only at mid-length of the short members. If the external loads are applied only at the nodes, the bending moment at mid-length of a member is the average of the two bending moment values at the two ends and the shearing force and the axial force are constants.

6.3   Problem statement

Consider a framed structure composed of members cast, prestressed or loaded in stages; each of these is treated as an event occurring at a specific instant. Introduction or removal of a support is considered an event. The subscript j is used to refer to the effects of the event occurring at the instant t_j. It is required to determine the changes in displacements, internal forces and reactions that occur between t_j and a later instant t_k, due to creep and shrinkage of concrete and relaxation of the prestressed reinforcement. When the changes in internal forces are known, the corresponding changes in strains and stresses can be determined by basic equations (e.g. Equations (2.19) and (2.20)). Section 6.4 describes two computer runs to solve this problem using a linear computer program.

As discussed in Chapter 5, in a statically determinate structure, the time-dependent phenomena affect only the displacements, while the reactions and the internal forces remain constants. The stress and stress resultants on a part of a composite cross-section can change with time, but when the structure is statically determinate the stress resultants in any cross section, as a whole, do not change with time. In other words, only the repartition of forces between the parts of a cross-section varies with time, without change in the resultants of stresses in all the parts combined.

6.4   Computer programs

This section describes the input and the output of typical linear computer programs for the analysis of framed structures, based on the displacement method (Section 5.2). Global axes must be defined by the user. The position of the nodes is specified by their coordinates (x, y) or (x, y, z) for plane or space structures, respectively. Figure 6.2 shows global axes, the nodal displacements (the degrees of freedom) and the order of numbering of the coordinates, representing displacements or forces, at a typical node of the five types of framed structures: plane truss, space truss, plane frame, space frame and grid. The analysis gives the nodal displacements, $\{D\}$ and the forces on the supported nodes in the global directions. It also gives a member end forces, $\{A\}$ for individual members in the directions of their local axes. Figure 6.3 shows local coordinates and their numbering and the positive directions of the member end forces for each of the five types of framed structures. An asterisk is used here in reference to local axes and local coordinates of members.

Input data description: The input data must give the nodal coordinates, the node numbers at the two ends of each member and its cross-sectional area. In addition, for a cross-section of a member of a plane frame the input must include the second moment of area, I about a centroidal principal axis

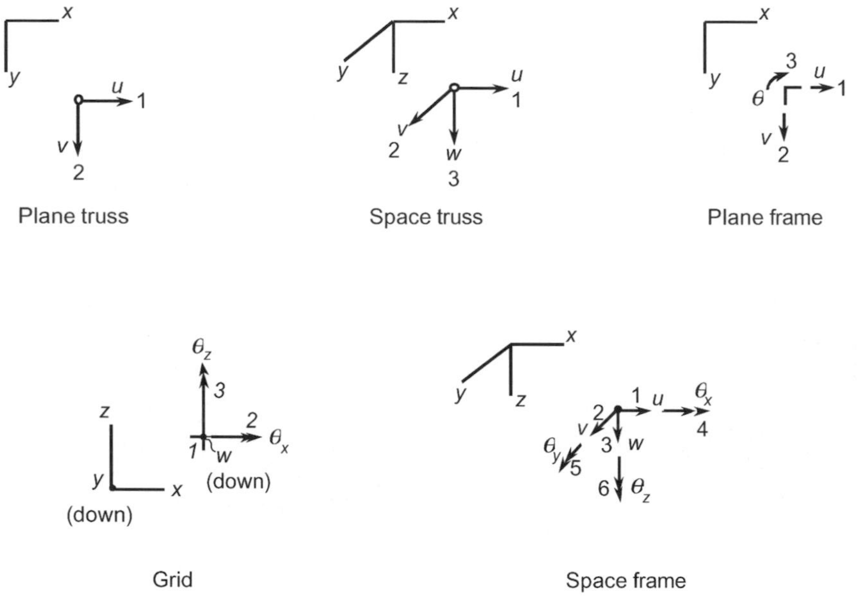

Plane truss   Space truss   Plane frame

Grid   Space frame

Figure 6.2 Global axes, degrees of freedom and the order of numbering of the coordinates at typical nodes.

perpendicular to the plane of the frame; for a space frame member the input must give I_{y*}, I_{z*} and J, the second moment of area about centroidal principal axes $y*$ and $z*$ and the torsion constant; for a member of a grid, the input must include the second moment of area, I_{z*} about centroidal principal axis $z*$ and the torsion constant J. All members are assumed to have constant cross-sections.

Images of an input data file are shown in each of Figs. 6.4 and 6.5. The first input file is for computer program SPACET[2] for the analysis of a space truss (Fig. 6.10), to be discussed in Example 6.4, Section 6.8. The second input file is for computer program PLANEF[3] for the analysis of a plane frame (Fig. 6.11), to be discussed in Example 6.5, Section 6.8. The integers and the real values given on the left-hand sides of Figs. 6.4 and 6.5 are the input data to be used by the computer; the words and the symbols on the right-hand side of the figures indicate to the user the contents of each data line.

Notation: The symbols employed in Figs. 6.4 and 6.5 are defined below:

NJ, NM, NSJ and *NLC*   number of joints, number of members, number of supported joints and number of load cases, respectively.

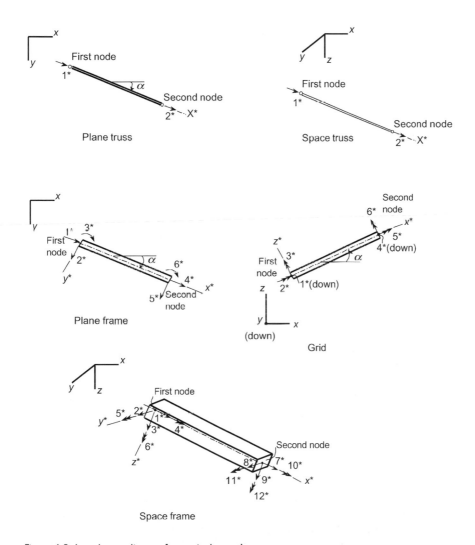

Figure 6.3 Local coordinates for typical members.

JS and *JE*	the joint numbers at the start and at the end of a member.
a, *I* and a_r	cross-sectional area, second moment of area and the reduced cross-sectional area for shear deformation (A large value is entered in Fig. 6.5 because shear deformation is ignored; this is also done in other examples of this chapter, where PLANEF is employed).

Space truss, Example 6.4, Computer run 1				
12 31 6 1				NJ, NM, NSJ, NLC
25.e9				Elasticity modulus
1	0.0	-1.5	0.0	Node no., x, y and z coordinates
2	0.0	1.5	0.0	
...				
12	18.0	0.0	3.0	
1	1	4 .4000		Member no., JS, JE and cross-sec. area
2	4	7 .4000		
...				
6	8	11 .4000		
7	1	2 .0096		
...				
10	10	11 .0048		
11	3	6 .0728		
...				
13	9	12 .0184		
14	1	3 .0184		
15	2	3 .0184		
...				
26	1	5 .0096		
...				
31	8	10 .0096		

1	1 1 0	0.0	0.0	0.0		Node no., restr. indics., prescr. displs.
2	1 1 0	0.0	0.0	0.0		
9	1 0 1	0.0	0.0	0.0		
10	0 1 1	0.0	0.0	0.0		
11	0 1 1	0.0	0.0	0.0		
12	0 0 0	0.0	0.0	0.0		
1	1	0.0	0.0	40.0e3		Load case no., node no., Fx, Fy and Fz
1	2	0.0	0.0	40.0e3		
1	4	0.0	0.0	80.0e3		
1	5	0.0	0.0	80.0e3		
1	7	0.0	0.0	80.0e3		
1	8	0.0	0.0	80.0e3		
1	10	0.0	0.0	40.0e3		
1	11	0.0	0.0	40.0e3		
10	0	0.0	0.0	0.0		Dummy, end of data of loaded nodes
10	0	0.0	0.0	0.0		Dummy, end of data of member loads

Figure 6.4 Image of input file (abbreviated) for computer program *SPACET* (Space Trusses); see note 1, p. 206). The input data are for the space truss of Example 6.4, Fig. 6.10.

F_x, F_y, F_z, M_z	forces at a joint, applied in directions of the nodal coordinates defined in Fig. 6.2.
$\{A_r\}$	fixed-end forces; these are the forces produced at fixed member ends due to external load, temperature variation, shrinkage, creep or relaxation.

Support conditions: A restraint indicator integer 1 or 0 in the input data signifies a free or a prescribed displacement in direction of one of the global

```
Plane frame, Example 6.5, Computer run 1
13   17   3   1                    NJ,NM,NSJ,NLC
25.e9   0.0                        Elasticity modulus, Poisson's ratio
 1     0.0   5.0                   Node no., x and y coordinates
 2     0.0   0.0
...
13   12.0   0.528
 1    1   2   .16    2.13e-3   1.e7   Member no., JS, JE, cross-sec. area, I, ar
 2    2   3   .936   55.86e-3  1.e7
 3    3   4   .936   55.86e-3  1.e7
 4    4   5   .936   55.86e-3  1.e7
 5    5   6   .936   55.86e-3  1.e7
 6    6   7   .936   55.86e-3  1.e7
 7    8   9   1.e-6  1.e-9     1.e7
...
12    2   8   1.e2   1.e2      1.e7
...
17    7  13   1.e2   1.e2      1.e7
 1    0   0   0  0.0   0.0   0.0   Node no., restr. indics., prsc. displs.
 7    0   1   0  0.0   0.0   0.0
13    0   1   0  0.0   0.0   0.0
 1    2  0.  31.2e3   0.          Ld. case no., node, Fx, Fy and Mz
 1    3  0.  62.4e3   0.
 1    4  0.  62.4e3   0.
 1    5  0.  62.4e3   0.
 1    6  0.  62.4e3   0.
 1    7  0.  31.2e3   0.
10    0  0.   0.0     0.          Dummy, end of data of loaded nodes
 1   7  -2640.e3  0.  0.  2640.e3  0.  0.      Case, member,{Ar}
 1   8  -2640.e3  0.  0.  2640.e3  0.  0.
 1   9  -2640.e3  0.  0.  2640.e3  0.  0.
 1  10  -2640.e3  0.  0.  2640.e3  0.  0.
 1  11  -2640.e3  0.  0.  2640.e3  0.  0.
10   0    0.      0.  0.    0.     0.  0.      Dummy, end of data of member loads
```

Figure 6.5 Image of input file (abbreviated) for computer program PLANEF, (Plane Frames); see note 1, p. 206. The input data are for the plane frame of Example 6.5, Fig. 6.11.

axes. The integer 0 denotes that the displacement has a prescribed real value included in the input; when a support prevents the displacement, the prescribed value should be 0.0. When the restraint indicator is 1, it signifies that the displacement is free; an arbitrary (dummy) real value should be entered in the space of prescribed displacement.

Load data: These are given in two sets of lines; each set is terminated by a 'dummy' line which starts by an integer >NLC. The first set is for forces applied at the nodes. The second set gives the fixed-end forces $\{A_r\}$ for individual members; two forces and six forces must be entered, respectively for a member of a space truss and a member of a plane frame. The fixed-end forces are included in the data only for members subjected to forces away from nodes or for members subjected to temperature variation. The values of the fixed-end forces are to be calculated by well-known equations given in many texts. Some computer programs calculate $\{A_r\}$ from input data describing the loads on the members; with such programs $\{A_r\}$ is not part of the input data.

Member end forces: In the displacement method of analysis, which is the basis of all computer programs, the member end forces for a member are determined by the superposition equation (see step 5 in Section 5.2):

$$\{A\} = \{A_r\} + [A_u] \{D^*\} \tag{6.1}$$

where $\{D^*\}$ is a vector of the displacements at the two ends of the member after they are transformed from the directions of the global axes to the directions of the local axes of the member (Fig. 6.3); $[A_u]$, which has the same meaning as the member's stiffness matrix, consists of the member end forces due to separate unit values of the displacements D_1^*, D_2^*, ... It is to be noted that for a concrete member $[A_u]$ is directly proportionate to the modulus of elasticity of concrete at the age considered. For the presentation that follows, define the symbol:

$$\{A_D\} = [A_u] \{D^*\} \tag{6.2}$$

$$\{A_D\} = \{A\} - \{A_r\} \tag{6.3}$$

$\{A_D\}$, which is equal to the second term on the right-hand side of Equation (6.1), is a vector of self-equilibrating forces that would be produced at the member ends by the introduction of the displacements $\{D^*\}$ at its two nodes.

6.5   Two computer runs

The problem stated in Section 6.3 can be solved by two computer runs using an appropriate linear computer program, such as the ones described in Section 6.4. For simplicity of presentation, we consider the case of the structure subjected to a single event at time t_j; that can be the application of external loads and/or prestressing or temperature change. The analysis is for the time-dependent effects of creep and shrinkage of concrete and relaxation of prestressed steel between t_j and a later instant t_k. Two computer runs are required:

Computer run 1: First the structure is analysed for the instantaneous forces introduced at t_j. The modulus of elasticity of concrete members is $E_c(t_j)$. The results give the instantaneous displacements, $\{D(t_j)\}$, the reactions and the member end forces $\{A(t_j)\}$.

The effect of prestressing introduced at t_j can be included in this run by treating the forces exerted by the tendons on the concrete as any other external force. Alternatively, when a tendon is idealized as a member (Fig. 6.1(a)), two axial restraining forces are to be entered for this member:

$$A_r(t_j)_{\text{prestress}} = \mp A_{\text{ps}}\sigma_p\,(t_j) \tag{6.4}$$

where A_{ps} and $\sigma_p\,(t_j)$ are the cross-sectional area of the tendon and its stress at time t_j, respectively. The minus and the plus sign are, respectively, for the force at the first and second ends of the member (Fig. 6.3).

Computer run 2: In this run the structure is idealized with the modulus of elasticity of concrete being the age-adjusted modulus, $\bar{E}_c\,(t_k,\ t_j)$ given by Equation (1.31), which is repeated here:

$$\bar{E}_c(t_k,\ t_j) = \frac{E_c(t_j)}{1 + \chi\varphi(t_k,\ t_j)} \tag{6.5}$$

where $\varphi\,(t_k,\ t_j)$ is creep coefficient at time t_k for loading at time t_j; $\chi\ (\equiv \chi\,(t_k,\ t_j)\,)$ is the aging coefficient; $E_c(t_j)$ is the modulus of elasticity of concrete at time t_j. The vector of fixed-end forces $\{A_r(t_k,\ t_j)\}$ is to be entered as loading data; where $\{A_r(t_k,\ t_j)\}$ is a vector of hypothetical forces that can be introduced gradually in the period t_j to t_k to prevent the changes in nodal displacements at member ends. The elements of the vector $\{A_r(t_k,\ t_j)\}$ for any member comprise a set of forces in equilibrium. Calculation of the elements of the vector $\{A_r(t_k,\ t_j)\}$ is discussed below, considering the separate effect of each of creep, shrinkage and relaxation.

Member fixed-end forces due to creep: The member end forces that restrain nodal displacements due to creep are:

$$\{A_r(t_k,\ t_j)\}_{\text{creep}} = -\frac{\bar{E}_c(t_k,\ t_j)}{E_c\,(t_j)}\ \varphi(t_k,\ t_j)\ \{A_D\,(t_j)\} \tag{6.6}$$

The vector $\{A_D(t_j)\}$ is given by Equation (6.3), using the results and the input data of Computer run 1. For the derivation of Equation (6.6), consider the hypothetical displacements change $[\varphi(t_k,\ t_j)\ \{D^*(t_k)\}]$ as if they were unrestrained. Premultiplication of this vector by $[-A_u]$ and substitution of Equation (6.2) give the values of the restraining forces for a member whose elasticity modulus is $E_c(t_j)$. Multiplication of the ratio $[\bar{E}_c\ (t_k,\ t_j)/E_c(t_j)]$, to account for the fact that the restraining forces are gradually introduced, gives Equation (6.6).

Member end forces due to shrinkage: The change of length of a concrete member subjected to shrinkage $\varepsilon_{\text{cs}}(t_k,\ t_j)$ can be prevented by the gradual introduction of axial member-end forces:

$$A_r(t_k, t_j)_{\text{axial, shrinkage}} = \pm \left[\overline{E}_c(t_k, t_j)A_c\right]\varepsilon_{cs} \qquad (6.7)$$

where A_c is the cross-sectional area of concrete member; the plus and the minus signs are respectively for the forces at the first and the second node of a member (see Fig. 6.3). Note that for shrinkage, ε_{cs} is a negative value.

Member end forces due to relaxation: When the effect of prestressing is represented in Computer run 1 by external forces exerted by the tendons on the concrete, it is only necessary in Computer run 2 to use an estimated prestress loss due to creep, shrinkage and relaxation combined to calculate external forces on the concrete, in the same way as for the prestress in Computer run 1 (with reversed signs and reduced magnitudes). When a tendon is idealized as an individual member, the relaxation effect can be represented in Computer run 2 by two axial restraining forces:

$$A_r(t_k, t_j)_{\text{axial, relaxation}} = \mp A_{ps}\Delta\overline{\sigma}_{pr}(t_k, t_j) \qquad (6.8)$$

where $\Delta\overline{\sigma}_{pr}(t_k, t_j)$ is the reduced relaxation; the negative and the positive signs in this equation are, respectively, for the force at the first and the second node of the member (Fig. 6.3). In verifying or in applying Equation (6.8), note that $\Delta\overline{\sigma}_{pr}(t_k, t_j)$ is commonly a negative value. When the tendons are idealized as separate members and Equation (6.8) is used, no estimated value of loss of prestress due to creep, shrinkage and relaxation is needed; the analysis will more accurately give the combined effect of creep, shrinkage and relaxation and the time-dependent changes in the internal forces.

6.6   Equivalent temperature parameters

In the preceding section two computer runs are proposed to analyse the time-dependent effects of creep, shrinkage and relaxation. In Computer run 2, the values of self-equilibrating forces $\{A_r(t_k, t_j)\}$ are entered as input data for individual members. It will be shown below that fictitious temperature parameters, to be calculated by Equations (6.9) and (6.10) can be employed as thermal data for computer programs that do not accept $\{A_r\}$ as input.

As example, consider a plane frame member AB having six end forces $\{A_r\}$ (Fig. 6.6(a)). The six forces represent a system in equilibrium. Figure 6.6(b) represents a conjugate beam of the same length and cross section as the beam in Fig. 6.6(a), but subdivided by a mid-length node. The conjugate beam is subjected to a rise of temperature T_0 for its two parts and temperature gradients T'_1 and T'_2 for parts AC and CB, respectively; where $T' = dT/dy$, with y being the coordinate of any fibre measured downward from the centroidal axis. It can be verified that the conjugate beam, with ends A and B fixed, has the same forces at the ends A and B as the actual member when:

(a)

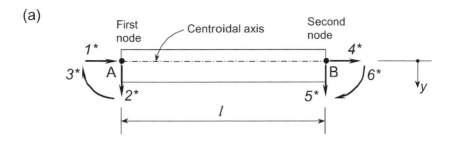

(b)

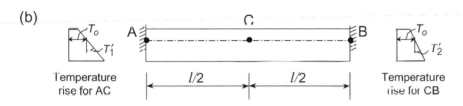

Figure 6.6 Equivalent temperature parameters: (a) actual member of a plane frame; (b) conjugate beam subjected to rise of temperature producing the same forces at ends A and B as in the actual beam.

$$
\begin{Bmatrix} T_{O} \\ T'_{1} \\ T'_{2} \end{Bmatrix} = \frac{1}{\bar{E}_{c}\alpha} \begin{bmatrix} 1/A & 0 & 0 \\ 0 & l/(6\mathrm{I}) & -1/\mathrm{I} \\ 0 & 5l/(6\mathrm{I}) & -1/\mathrm{I} \end{bmatrix} \begin{Bmatrix} A_{r1} \\ A_{r2} \\ A_{r3} \end{Bmatrix}
\tag{6.9}
$$

where A and I is the cross-sectional area and its second moment about centroidal axis; l is length of member; $\bar{E}_{c}$ $(\equiv\bar{E}_{c}(t_{k}, t_{j}))$ is the age-adjusted modulus and α is an arbitrary thermal expansion coefficient. The same values of $\bar{E}_{c}$ and α used in Equation (6.9) must be entered as input in Computer run 2. The fictitious temperature parameters T_{0}, T'_{1} and T'_{2} can be expressed (by an equation similar to Equation (6.9)) in terms of the fixed-end forces at end B instead of end A, to give the same result. The subdivision of members into two parts should not be done in Computer run 1. Also the subdivision is not necessary in Computer run 2, when the structure is a plane or a space truss. In this case, the input in Computer run 2 is a uniform rise of temperature, T_{0}; where

$$
T_{0} = \frac{A_{r}(t_{k}, t_{j})}{\alpha\bar{E}_{c}(t_{k}, t_{j})A}
\tag{6.10}
$$

where $A_{r}(t_{k}, t_{j})$ is an axial force at the first-end of the member (given by

Equation (6.4), (6.6) or (6.7). The first and the second nodes of members and the positive sign convention for member-end forces are defined in Fig. 6.3.

6.7   Multi-stage loading

The problem stated in Section 6.3 can be solved when the analysis for the time-dependent changes between time t_j and time t_k are required for the effect of events 1 to j, with the last event j occurring at t_j, with events 1 to $(j-1)$ occurring at earlier instants, $t_1, t_2, \ldots, t_{j-1}$. We recall, the term 'event' refers to the application of forces, the introduction of prestressing, the casting a new member or the removal or the introduction of a support. The two computer runs as discussed in Section 6.5 are to be applied, differing only in the calculation of the fixed-end forces $\{A_r(t_k, t_j)\}$ to be included in the input of Computer run 2. These forces are to be determined by a summation to replace Equation (6.6). The summation is to superimpose the effect of creep due to the forces introduced at $t_1, t_2, \ldots, t_j$, as well as due to the gradual changes in internal forces in the intervals $(t_2 - t_1), (t_3 - t_2), \ldots, (t_j - t_{j-1})$. As example, Equation (6.11) gives contribution to $\{A_r(t_k, t_j)\}_{\text{creep}}$ of the loads introduced at time t_i; where $t_i < t_j < t_k$:

$$\{A_r(t_k, t_j)\}_{\text{creep, load introduced at ti}}$$
$$= -\frac{\bar{E}_c(t_k, t_j)}{E_c(t_i)}[\varphi(t_k, t_i) - \varphi(t_j, t_i)]\{A_D(t_i)\} \tag{6.11}$$

The vector $\{A_D(t_i)\}$ is to be determined by Equation (6.3) using the results of a computer run having an input that includes the modulus of elasticity $E_c(t_i)$ and the loading introduced at t_i.

When the structure is subjected to more than one or two events, several computer runs are required. In this case it is more practical to apply the step-by-step procedure discussed in Section 5.8, employing a specialized computer program (see e.g. note 3, page 175).

6.8   Examples

The following are analysis examples of structures subjected to a single or two events and it is required to determine the change(s) in displacements and/or internal forces or stresses between time t_j and a later time t_k.

Example 6.1: Propped cantilever

The cantilever AB in Fig. 6.7(a) is subjected at time t_0 to a uniform load q. At time t_1 a simple support is introduced at B, thus preventing the

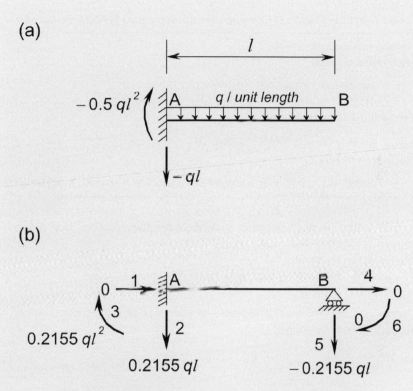

Figure 6.7 Propped cantilever, Example 6.1: (a) cantilever loaded at time t_0; (b) member end forces developed between time t_1 and t_2 due to the introduction of support B at t_1.

increase in deflection at B due to creep. Determine the change in the end forces at A and B between time t_1 and a later time t_2. Given data: $\varphi(t_1, t_0)$ = 0.9; $\varphi(t_2, t_0) = 2.6$; $\varphi(t_2, t_1) = 2.45$; $\chi(t_2, t_1) = 0.8$. Ignore the difference between $E_c(t_0)$ and $E_c(t_1)$.

The computer program PLANEF is used here, but the results can be checked by hand computation as discussed in Chapter 5. Table 6.1 shows the input data and the results of Computer run 1, analyzing the immediate effect of the load introduced at time t_0. Each of $E_c(t_0)$, q and l are considered equal to unity; the support conditions are those of a cantilever encastré at A and free at B (Fig. 6.7(a)); the end forces for a totally fixed beam subjected to uniform load are entered as the load input:

$$\{A_r(t_0)\} = \{0, -0.5\ ql, -0.0833\ ql^2, 0, -0.5\ ql, 0.0833\ ql^2\}$$

Table 6.1 Input and results of Computer run I with program PLANEF. Example 6.1

Analysis results; load case No. I

Nodal displacements

Node	u	v	θ
I	.00000E+00	.41668E–07	.10417E–06
2	.00000E+00	.12500E+00	.16667E+00

Forces at the supported nodes

Node	F_x	F_y	M_z
I	.00000E+00	–.10000E+01	–.50000E+00

Member end forces

Member	F_1^*	F_2^*	F_3^*	F_4^*	F_5^*	F_6^*
I	.00000E+00	–.10000E+01	–.50000E+00	.00000E+00	–.11102E–15	.55511E–15

The result of this computer run includes the member end forces immediately after load application:

$$\{A(t_0)\} = \{0, -ql, -0.5ql^2, 0, 0, 0\}$$

As expected, these are the forces at the ends of a cantilever. Apply Equation (6.3) to obtain:

$$\{A_D(t_0)\} = \{0, -0.5\ ql, -0.4167ql^2, 0, 0.5\ ql, -0.0833\ ql^2\}$$

These are the changes in end forces produced by varying the nodal displacements form null, when the nodal displacements are prevented, to the values $\{D^*\}$ included in the results of Computer run 1. Creep freely increases these displacements in the period t_0 to t_1. The hypothetical end forces that can prevent further increase in the period t_1 to t_2 are (Equation (6.11)):

$$\{A_r(t_2, t_1)\} = -\frac{\bar{E}_c(t_2, t_1)}{E_c(t_0)}\left[\varphi(t_2, t_0) - \varphi(t_1, t_0)\right]\{A_D(t_0)\}$$

The age-adjusted elasticity modulus is (Equation (6.5)):

$$\bar{E}_c(t_2, t_1) = \frac{E_c(t_1)}{1 + \chi\varphi(t_2, t_1)} = \frac{E_c(t_1)}{1 + 0.8(2.45)} = 0.3378\ E_c(t_0)$$

Substitution in Equation (6.11) gives a set of self-equilibrating end forces to be used as load input data in Computer run 2:

$$\{A_r(t_2, t_1)\} = -0.3378(2.6 - 0.9)\,\{A_D(t_0)\}$$

$$\{A_r(t_2, t_1)\} = \{0, 0.2872\,ql, 0.2393\,ql^2, 0, -0.2872\,ql, 0.0479\,ql^2\}$$

The same forces are obtained in Example 5.1 (Fig. 5.3(b)). Table 6.2 includes the input data and the analysis results of Computer run 2. We note that the age-adjusted elasticity modulus is used and the support conditions are those of end encastré at A and simply supported at B. The required changes in member end forces between time t_1 and t_2 are a set of self-equilibrating forces (Fig. 6.7(b)), which are copied here:

$$\{A(t_2, t_1)\} = \{0, 0.2155\,ql, 0.2154\,ql^2, 0, -0.2155\,ql, 0\}.$$

Table 6.2 Input (abbreviated) and results of Computer run 2 using program PLANEF. Example 6.1

Input data
Number of joints = 2;   Number of members = 1;   Number of load cases = 1
Number of joints with prescribed displacements = 2;   Elasticity modulus = 0.3378

Nodal coordinates and element information
Same as in Table 6.1

Support conditions

Node	Restraint indicators			Prescribed displacements		
	u	v	θ	u	v	θ
1	0	0	0	.00000E+00	00000E+00	00000E+00
2	1	0	1	.00000E+00	00000E+00	00000E+00

Forces applied at the nodes
Same as in Table 6.1

Member end forces with nodal displacement restrained

Ld. case	Member	A_{r1}	A_{r2}	A_{r3}	A_{r4}	A_{r5}	A_{r6}
1	1	.0000E+00	.2872E+00	.2393E+00	.0000E+00	−.2872E+00	.4786E−01

Analysis results; load case No. 1

Nodal displacements

Node	u	v	θ
1	.00000E+00	.17688E−07	.17688E−07
2	.00000E+00	−.17688E−07	−.35377E−01

Forces applied at the supported nodes

Node	F_x	F_y	M_z
1	.00000E+00	.21550E+00	.21540E+00
2	.00000E+00	−.21550E+00	.00000E+00

Member end forces

Member	F_1^*	F_2^*	F_3^*	F_4^*	F_5^*	F_6^*
1	.00000E+00	.21550E+00	.21540E+00	.00000E+00	−.21550E+00	.00000E+00

Example 6.2 Cantilever construction method

The girder ABC (Fig. 6.8(a)) is constructed as two separate cantilevers subjected at time t_0 to a uniform load q/unit length, representing the self weight. At time t_1 the two cantilevers are made continuous at B by a cast-in-situ joint. Determine the changes in member end forces for AB between t_1 and a later time t_2. Use the same creep and aging coefficients as in Example 1 and ignore the difference between $E_c(t_0)$ and $E_c(t_1)$.

Because of symmetry the computer analysis needs to be done for half the structure only (say part AB). Computer run 1 and calculation of $\{A_r(t_2, t_1)\}$ for use as load input in Computer run 2 are the same as in Example 1 (Table 6.1). In the current problem the support conditions at end B in Computer run 2 must be as indicated below with the remaining input data as in Table 6.2.

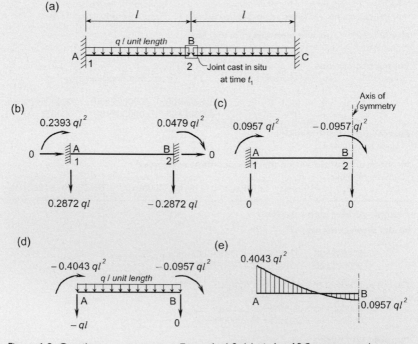

Figure 6.8 Cantilever construction, Example 6.2: (a) girder ABC constructed as two separate cantilevers subjected to uniform load at time t_0; (b) member end forces $A_r(t_2,t_1)$ calculated in Example 6.1; (c) changes in member end forces in Computer run 2; (d) superposition of member end forces in Fig. 6.7 (a) and Fig. 6.8 (c) to give $\{A(t_2)\}$; (e) bending moment diagram at time t_2.

Node	Restraint indicators			Restraint displacements		
	u	v	θ	u	v	θ
2	0	1	0	0.0	0.0	0.0

Parts of the input and the results of Computer run 2 are presented in Fig. 6.8, rather than in a table. Figure 6.8(b) shows $\{A_r(t_2, t_1)\}$; these are the forces that can artificially prevent the changes due to creep in the displacements at ends A and B of the cantilever AB. Figure 6.8(c) shows the results of Computer run 2; the computer applies the forces $\{A_r(t_2,t_1)\}$ in a reversed direction and determines the corresponding changes in member end forces and superposes them on $\{A_r(t_2, t_1)\}$. Figure 6.8(d) shows the sum of the forces in Fig. 6.8(c) and Fig. 6.7(a); this gives the forces on member AB at time t_2. The bending moment diagram at time t_2 is shown in Fig. 6.8(e).

Example 6.3 Cable-stayed shed

The line AB in Fig. 6.9 represents the centroidal axis of a concrete cantilever. At time t_1 the cantilever is subjected to its own weight $q = 25\,\text{kN-m}$ and a prestressing force, $P(t_1) = 200\,\text{kN}$, introduced by the steel cable AC. Calculate the changes in deflection at the tip of the cantilever and in the force in the cable in the period t_1 to a later time t_2, caused by

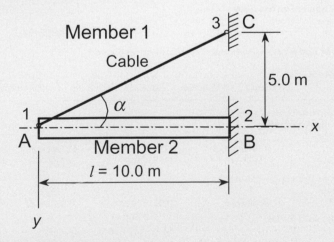

Figure 6.9 A cable-stayed shed, Example 6.3.

creep and shrinkage of concrete and relaxation of prestressed steel. Ignore cracking and presence of reinforcement in AB. Given data: $E_c(t_1)$ = 25 GPa; $\varphi(t_2, t_1) = 2$; $\chi = 0.8$; $\varepsilon_{cs}(t_2, t_1) = -300 \times 10^{-6}$; $\Delta\bar{\sigma}_{pr} = -50$ MPa. Cross-sectional area properties for AB: $A_c = 1.0 \, \text{m}^2$; $I = 0.1 \, \text{m}^4$. For the cable, $A_s = 250 \, \text{mm}^2$; $E_s = 200$ GPa.

Table 6.3 gives the input and the results of Computer run 1 using the program PLANEF. During the tensioning, the change in cable length can occur independently from the deformation of concrete; thus the translation at the tip of the cantilever is not compatible with the elongation of the cable. For this reason, the analysis in Table 6.3 is for a

Table 6.3 Input data and results of Computer run 1 using program PLANEF. Example 6.3, Fig. 6.9

Input data
Number of joints = 2;   Number of members = 1;   Number of load cases = 1
Number of joints with prescribed displacements = 1;   Elasticity modulus = 25.0E+09

Nodal coordinates

Node	x	y
1	0.0	0.0
2	10.0	0.0

Element information

Element	1st node	2nd node	a	I
1	1	2	.10000E+01	.10000E+00

Support conditions

Node	Restraint indicators			Prescribed displacements		
	u	v	θ	u	v	θ
2	0	0	0	.00000E+00	.00000E+00	.00000E+00

Forces applied at the nodes

Load case	Node	F_x	F_y	M_z
1	1	.17890E+06	−.89440E+06	.00000E+00

Member end forces with nodal displacement restrained

Ld. case	Member	A_{r1}	A_{r2}	A_{r3}	A_{r4}	A_{r5}	A_{r6}
1	1	.0000E+00	−.1250E+06	−.2083E+06	.0000E+00	−.1250E+06	.2083E+06

Analysis results; load case No. 1
Nodal displacements

Node	u	v	θ
1	.71560E−04	.57534E−03	.12200E−03
2	.71560E−10	.11853E−08	−.14730E−09

Forces at the supported nodes

Node	F_x	F_y	M_z
2	−.17890E+06	−.16056E+06	.35560E+06

Member end forces

Member	F_1^*	F_2^*	F_3^*	F_4^*	F_5^*	F_6^*
1	.17890E+06	−.89440E+05	−.29104E−10	−.17890E+06	−.16056E+06	.35560E+06

cantilever (without the cable) subjected to a uniform load $q = 25$ kN combined with $F_x = 178.9$ and $F_y = -89.4$ kN at node A; where F_x and F_y are the forces exerted by the cable on the concrete member at time t_1. The modulus of elasticity in Computer run 1 is equal to $E_c(t_1) = 25$ GPa. The results of Computer run 1 include the member end forces of AB at time t_1:

$$\{A(t_1)\}_{AB} = \{178.9\,\text{kN}, -89.4\,\text{kN}, 0, -178.9\,\text{kN}, -160.6\,\text{kN},$$
$$355.6\,\text{kN-m}\}$$

The changes in the forces at end of member AB from the fixed-end status are (Equation (6.3)).

$$\{A_D(t_1)\}_{AB} = \{178.9\,\text{kN}, 35.6\,\text{kN}, 208.3\,\text{kN-m}, -178.9\,\text{kN}, -35.6\,\text{kN},$$
$$147.3\,\text{kN-m}\}$$

In Computer run 2 (Table 6.4) the structure is composed of the two members AB and AC and the modulus of elasticity used is (Equation (6.5)):

$$E_c(t_2, t_1) = \frac{25\,\text{GPa}}{1 + 0.8(2.0)} = 9.615\,\text{GPa}$$

A transformed cross-sectional area equal to $A_s E_s / \overline{E}_c$ is used for the cable; a negligible value is entered for I. The end forces that can artificially prevent the time-dependent changes in displacements due to creep at the two nodes of member AB are (Equation (6.6)):

$$\{A_r(t_2, t_1)\}_{\text{creep } AB} = -\frac{9.615}{25.0}(2.0)\{A_D(t_1)\}_{AB}$$

$$\{A_r(t_2, t_1)\}_{\text{creep } AB} = \{-137.6\,\text{kN}, -27.4\,\text{kN}, -160.2\,\text{kN-m}, 137.6\,\text{kN},$$
$$27.4\,\text{kN}, -113.3\,\text{kN-m}\}$$

The axial force that can artificially prevent the change in length due to shrinkage of AB (Equation (6.7)):

$$A_r(t_2, t_1)_{\text{axial, shrinkage } AB} = \pm[9.615\,\text{GPa}\,(-300 \times 10^{-6})\,(1.0\,\text{m}^2) =$$
$$\mp 2884.5\,\text{kN}$$

The restraining forces for creep and shrinkage are entered on separate lines as load data (for member 1) in Computer run 2 (Table 6.4).

The relaxation in cable AC is a loss of tension presented in Computer run 2 by an axial compressive force in the member; thus the member end forces to be used in the input (Equation (6.8)):

Table 6.4 Input data and results of Computer run 2 using program PLANEF. Example 6.3, Fig. 6.9

Input data
Number of joints = 3;   Number of members = 2;   Number of load cases = 1
Number of joints with prescribed displacements = 2; Elasticity modulus = 9.615E+09

Nodal coordinates

Node	x	y
1	0.0	0.0
2	10.0	0.0
3	10.0	−5.0

Element information

Element	1st node	2nd node	a	I
1	1	2	.10000E+01	.10000E+00
2	1	3	.52000E−02	.10000E−06

Support conditions

Node	Restraint indicators			Prescribed displacements		
	u	v	θ	u	v	θ
2	0	0	0	.00000E+00	.00000E+00	.00000E+00
3	0	0	0	.00000E+00	.00000E+00	.00000E+00

Forces applied at the nodes

Load case	Node	F_x	F_y	M_z
1	1	.00000E+00	.00000E+00	.00000E+00

Member end forces with nodal displacement restrained

Ld. case	Member	A_{r1}	A_{r2}	A_{r3}	A_{r4}	A_{r5}	A_{r6}
1	1	−.1376E+06	−.2740E+05	−.1602E+06	.1376E+06	.2740E+05	−.1133E+05
1	1	−.2885E+07	.0000E+00	.0000E+00	.2885E+07	.0000E+00	.0000E+00
1	1	.1250E+05	.0000E+00	.0000E+00	−.1250E+05	.0000E+00	.0000E+00

Analysis results; load case No. 1

Nodal displacements

Node	u	v	θ
1	.31270E−02	.38513E−02	−.16116E−03
2	.31270E−08	.30455E−08	−.49711E−09
3	.12014E−08	−.24022E−08	−.56920E−09

Forces at the supported nodes

Node	F_x	F_y	M_z
2	−.15478E+05	−.77388E+04	.77890E+05
3	−.15478E+05	.77388E+04	.19580E+01

Member end forces

Member	F_1^*	F_2^*	F_3^*	F_4^*	F_5^*	F_6^*
1	−.15478E+05	.77388E+04	−.16808E+01	.15478E+05	−.77388E+04	.77890E+05
2	.17305E+05	.32547E+00	.16808E+01	−.17305E+05	−.32547E+00	.19580E+01

$$A_r(t_2, t_1)_{\text{axial, relaxation } AC} = \mp (250 \times 10^{-6}\,\text{m}^2)(-50\,\text{MPa}) = \pm\, 12.5\,\text{kN}$$

The forces $\{A_r\}$ due to relaxation are entered on a separate line (for member 2) in the input data in Table 6.4.

The results of Computer run 2 (Table 6.4) include the deflection increase at the tip of the cantilever, v = 3.9 mm and the changes in the end forces in member A_c, representing a drop of 17.3 kN in the tensile force in the cable.

Example 6.4 Composite space truss

Figure 6.10(a) depicts a cross-section of a concrete floor slab supported by structural steel members. The structure is idealized as a space truss shown in pictorial view, elevation and top views in Figs. 6.10(b), (c) and (d). The truss has a span of 36.0 m; but for symmetry half the span is analysed. Consider that the half truss is subjected at time t_1 to downward forces: P at each of nodes 1, 2, 10 and 11 and 2P at each of nodes 4, 5, 7 and 8; where P = 40 kN. Find the deflection at mid-span at time t_1 and the change in deflection, at the same location, occurring between time t_1 and a later time t_2 due to creep and shrinkage of concrete. Given data: for concrete, $E_c(t_1) = 25$ GPa; $\varphi(t_2, t_1) = 2.25$; $\chi(t_2, t_1) = 0.8$; $\varepsilon_{cs} = -400 \times 10^{-6}$; for structural steel $E_s = 200$ GPa. The material for members 1 to 6 is concrete; all other members are structural steel. The cross sectional areas of members are:

For each of members 1 to 6,   the cross-sectional area = 0.4 m²
For each of members 11 to 13,   the cross-sectional area = 9100 mm²
For each of member 14 of 25,   the cross-sectional area = 2300 mm²
For each remaining members,   the cross-sectional area = 1200 mm²

Light steel members running along lines 1–10 and 2–11 may be necessary during construction; these are here ignored.

The computer program SPACET (space trusses) is used in two runs. In Computer run 1, the modulus of elasticity is $E_c(t_1) = 25$ GPa; a transformed cross-sectional area $= A_s E_s / E_c(t_1)$ is entered for the steel members of the truss. An image of the input file (abbreviated) is shown in Fig. 6.4. Table 6.5 shows the results, which include the deflection at mid-span (nodes 10 or 11) at time $t_1 = 55.8$ mm.

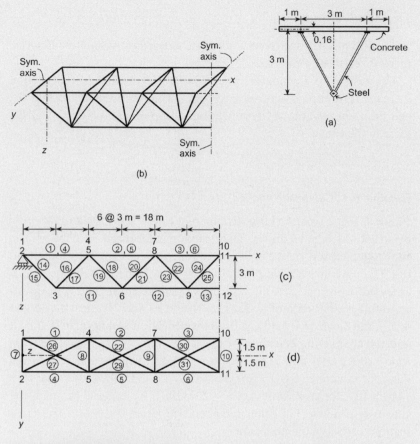

Figure 6.10 Concrete floor slab on structural steel members idealized as a space truss (Example 6.4): (a) cross-section; (b) pictorial view with the diagonal members in the x-y plane omitted for clarity; (c) elevation; (d) top view.

The age-adjusted elasticity modulus is (Equation (6.5)):

$$\bar{E}_c\,(t_2, t_1) = \frac{25\,\text{GPa}}{1 + 0.8(2.25)} = 8.929\,\text{GPa}$$

This modulus is used in Computer run 2 and a transformed cross-sectional area = $A_s E_s / \bar{E}_c$ is entered for the steel members. The load data are the two axial end forces $\{A_r(t_2,\,t_1)\}_{\text{creep}}$ calculated by Equation (6.6) for each of the concrete members (1 to 6):

Table 6.5 Abbreviated results of Computer run 1, Example 6.4. Space truss; immediate displacements and forces at time t_1

Nodal displacements

Node	u	v	w
1	.83225E–03	.60029E–03	.44022E–08
2	.83225E–03	–.60029E–03	.44022E–08
3	–.92308E–02	.10672E–24	.14765E–01
4	.71699E–03	–.27823E–03	.29254E–01
5	.71699E–03	.27823E–03	.29254E–01
6	–.65934E–02	.32702E–23	.39067E–01
7	.40567E–03	–.26566E–03	.48840E–01
8	.40567E–03	.26566E–03	.48840E–01
9	–.23736E–02	.99262E–23	.52367E–01
10	.41362E–09	–.27428E–03	.55758E–01
11	.41362E–09	.27428E–03	.55758E–01
12	–.23736E–08	.00000E+00	.00000E+00

Forces at the supported nodes

Node	F_x	F_y	M_z
1	.00000E+00	.00000E+00	–.24000E+06
2	.00000E+00	.00000E+00	–.24000E+06
9	.00000E+00	–.45097E–09	.00000E+00
10	–.72000E+06	.00000E+00	.00000E+00
11	–.72000E+06	.00000E+00	.00000E+00
12	.14400E+07	.00000E+00	.00000E+00

Member end forces

Member	F_1	F_2
1	.19209E+06	–.19209E+06
2	.51887E+06	–.51887E+06
3	.67612E+06	–.67612E+06
4	.19209E+06	–.19209E+06
5	.51887E+06	–.51887E+06
6	.67612E+06	–.67612E+06
7	.96046E+05	–.96046E+05
...		
31	.43425E+04	–.43425E+04

$$\{A_r(t_2, t_1)\}_{\text{creep}} = -\frac{\bar{E}_c\,(t_2, t_1)}{E_c\,(t_1)}\,\varphi(t_2,\, t_1)\,\{A_D(t_1)\}$$

The values of $\{A_D(t_1)\}$ are calculated by Equation (6.3) using the results of Computer run 1 and noting that $\{A_r(t_1)\} = \{0\}$ for all members. The artificial restraining forces are calculated below for member 1 as example:

$$\{A_D(t_1)\}_{\text{member 1}} = \{192.09, -192.09\}\ \text{kN}$$

Table 6.6 Abbreviated input and results of Computer run 2, Example 6.4. Space truss. Analysis of changes in displacements and internal forces between time t_1 and t_2

Elasticity modulus = .89286E+10
Member end forces with nodal displacement restrained

Ld. case	Member	A_{r1}	A_{r2}
1	1	−.1544E+06	.1544E+06
1	2	−.4170E+06	.4170E+06
1	3	−.5433E+06	.5433E+06
1	4	−.1544E+06	.1544E+06
1	5	−.4170E+06	.4170E+06
1	6	−.5433E+06	.5433E+06
1	1	−.1429E+07	.1429E+07
1	2	−.1429E+07	.1429E+07
1	3	−.1429E+07	.1429E+07
1	4	−.1429E+07	.1429E+07
1	5	−.1429E+07	.1429E+07
1	6	−.1429E+07	.1429E+07

Analysis results
Nodal displacements

Node	u	v	w
1	.87131E−02	−.20020E−03	.28363E−22
2	.87131E−02	.20020E−03	−.13222E−22
3	−.11291E−16	−.14889E−22	.86130E−02
4	.61614E−02	−.42660E−03	.14988E−01
5	.61614E−02	.42660E−03	.14988E−01
6	−.82952E−17	.49631E−23	.20936E−01
7	.31826E−02	−.46846E−03	.24353E−01
8	.31826E−02	.46846E−03	.24353E−01
9	−.28189E−17	−.12550E−45	.27301E−01
10	.29462E−08	−.48412E−03	.27543E−01
11	.29462E−08	.48412E−03	.27543E−01
12	−.28656E−23	.00000E+00	.00000E+00

Member end forces

Member	F_1*	F_2*
1	−.64064E+05	.64064E+05
2	−.72448E+05	.72448E+05
3	−.77460E+05	.77460E+05
4	−64064E+05	.64064E+05
5	−.72448E+05	.72448E+05
6	−.77460E+05	.77460E+05
7	−.32032E+05	.32032E+05
8	−.68256E+05	.68256E+05
9	−.74954E+05	.74954E+05
10	−.38730E+05	.38730E+05
. . .		
26	.71626E+05	−.71626E+05
27	.71626E+05	−.71626E+05
28	.80999E+05	−.80999E+05
29	.80999E+05	−.80999E+05
30	.86603E+05	−.86603E+05
31	.86603E+05	−.86603E+05

$$\{A_r(t_2, t_1)\}_{\text{creep, member 1}} = -\frac{8.929}{25.0}(2.25)\{192.09, -192.09\}$$

$$= \{-154.4, 154.4\}\,\text{kN}$$

The restraining forces for shrinkage are the same for any of the concrete members (1 to 6). Equation (6.7) gives:

$$\{A_r(t_2, t_1)\}_{\text{shrinkage members 1 to 6}} = 8.929\,\text{GPa}\,(-400 \times 10^{-6})\,0.4\,\{1, -1\}$$

$$= \{-1428.6, 1428.6\}\,\text{kN}$$

Table 6.6 gives abbreviated input and results of Computer run 2. Because this structure is statically determinate externally, creep and shrinkage do not affect the reactions (omitted in Table 6.6). The changes in displacements due to creep and shrinkage in the period t_1 to t_2 are given in Table 6.6, including the change of mid-span deflection of 27.5 mm (node 10 or 11). The changes in member end forces are given in Table 6.6 only for the members where the change is non zero.

Example 6.5 Prestressed portal frame

Figure 6.11(b) represents a portal frame idealization. Member BC has a post-tensioned T-section, shown in Fig. 6.11(a). Member AC is non-prestressed. The prestressing steel tendon, having parabolic profile, is idealized as straight steel members connected by rigid arms to nodes on the x-axis through the centroid of the gross concrete section of member BC. At time t_1 member BC is subjected to a uniform gravity load $q = 26\,\text{kN-m}$ (representing self weight and superimposed dead load), combined with a prestressing force, $P = 2640\,\text{kN}$, assumed constant over the length of the tendon. Find the changes in the force in the tendon and the deflection at mid-span due to creep and shrinkage of concrete and relaxation of prestressed steel occurring between t_1 and a later time t_2. Ignore presence of non-prestressed reinforcement and cracking. Given data: modulus of elasticity of concrete at time t_1, $E_c(t_1) = 25\,\text{GPa}$; creep and aging coefficients, $\varphi(t_2, t_1) = 2.0$ and $\chi(t_2, t_1) = 0.8$; free shrinkage, $\varepsilon_{cs}(t_2, t_1) = -300 \times 10^{-6}$; reduced relaxation (Section 1.5), $\Delta\bar{\sigma}_{pr} = -60\,\text{MPa}$; modulus of elasticity of prestressed steel $= 200\,\text{GPa}$. The cross-sections of the members have the following area properties:

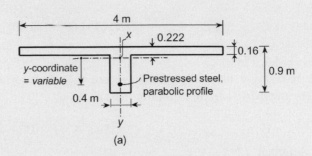

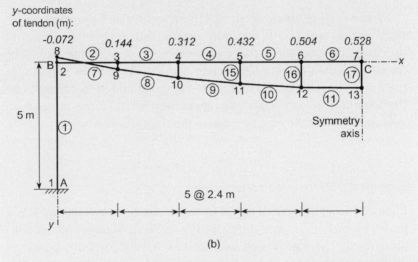

Figure 6.11 Prestressed portal frame of Example 6.5: (a) cross-section of member BC; (c) idealization of half the structure.

Member AB: cross-sectional area $= 0.16\,\mathrm{m^2}$; second moment of area $= 2.13 \times 10^{-3}\,\mathrm{m^4}$.

Member BC: cross-sectional area $= 0.936\,\mathrm{m^2}$; second moment of area $= 55.86 \times 10^{-3}\,\mathrm{m^4}$.

Tendon: cross-sectional area, $A_{ps} = 2200\,\mathrm{mm^2}$ and negligible second-moment of area.

The problem is solved by two computer runs using program PLANEF (Plane Frames, note 1, p. 206). The input file for Computer run 1 is shown in Fig. 6.5. The modulus of elasticity used is $E_c(t_1) = 25\,\mathrm{GPa}$. While the prestressing is being introduced, the tendon can elongate independently from the concrete. Thus, in Computer run 1, negligible cross-sectional areas is entered for members 7 to 11, which represent the tendon; in this way the tendon does not contribute to the stiffness of the frame. Two axial forces are entered as $\{A_r\}$, to represent initial tension = 2640 kN in each of members 7 to 11. Large values are entered for the cross-sectional area properties to represent rigid members 12 to 17 connecting the nodes of the tendon to nodes on the centroid of BC. Table 6.7 gives abbreviated results of Computer run 1.

The age-adjusted elasticity modulus of concrete is used in Computer run 2 (Equation (6.5)):

$$\bar{E}_c(t_2,\ t_1) = \frac{25\,\text{GPa}}{1 + 0.8(2.0)} = 9.615\,\text{GPa}$$

Table 6.7 Abbreviated results of Computer run 1 for the portal frame of Example 6.5, using program PLANEF

Nodal displacements

Node	u	v	w
1	−.48997E−08	.39000E−09	−.84349E−09
2	.13645E−02	.39000E−03	.25057E−02
3	.10922E−02	.61245E−02	.22427E−02
4	.81954E−03	.11027E−01	.18195E−02
5	.54654E−03	.14764E−01	.12780E−02
6	.27332E−03	.17100E−01	.65836E−03
7	.13086E−13	.17894E−01	.20542E−13
8	.15449E−02	.39001E−03	.25057E−02
9	.76927E−03	.61245E−02	.22427E−02
10	.25185E−03	.11027E−01	.18195E−02
11	−.55401E−05	.14764E−01	.12780E−02
12	−.58490E−04	.17100E−01	.65836E−03
13	.42656E−20	.17894E−01	.27272E−19

Forces at the supported nodes

Node	F_x	F_y	M_z
1	.25047E+05	−.31200E+06	.35933E+05
7	−.26649E+07	.00000E+00	−.38885E+06
13	.26399E+07	.00000E+00	−.51617E+00

Member end forces

Member	$F_1{}^*$	$F_2{}^*$	$F_3{}^*$	$F_4{}^*$	$F_5{}^*$	$F_6{}^*$
1	.31200E+06	.25047E+05	.35933E+05	−.31200E+06	−.25047E+05	.89303E+05
2	.26544E+07	−.44157E+05	.10001E+06	−.26544E+07	.44157E+05	−.20599E+06
3	.26586E+07	−.34051E+05	.20538E+06	−.26586E+07	.34051E+05	−.28711E+06
4	.26618E+07	−.24165E+05	.28612E+06	−.26618E+07	.24165E+05	−.34412E+06
5	.26639E+07	−.14436E+05	.34321E+06	−.26639E+07	.14436E+05	−.37785E+06
6	.26649E+07	−.48013E+04	.37732E+06	−.26649E+07	.48013E+04	−.38885E+06
7	−.26400E+07	−.12826E−02	.11827E−02	.26400E+07	.12826E−02	−.42733E−02
8	−.26400E+07	−.86862E−03	.33526E−02	.26400E+07	.86862E−03	−.54424E−02
9	−.26400E+07	−.50763E−03	.50243E−02	.26400E+07	.50763E−03	−.62442E−02
10	−.26400E+07	−.24704E−03	.61547E−02	.26400E+07	.24704E−03	−.67479E−02
11	−.26400E+07	−.71642E−04	.67716E−02	.26400E+07	.71642E−04	−.69436E−02

Table 6.8 Abbreviated input data and results of Computer run 2 for the portal frame of Example 6.5, using program PLANEF

Member end forces with nodal displacement restrained

Ld. case	Member	A_{r1}	A_{r2}	A_{r3}	A_{r4}	A_{r5}	A_{r6}
I	I	−.2400E+06	−.1927E+05	−.2764E+05	.2400E+06	.1927E+05	−.6870E+05
I	2	−.2042E+07	.3397E+05	−.7693E+05	.2042E+07	−.3397E+05	.1585E+06
I	3	−.2045E+07	.2619E+05	−.1580E+06	.2045E+07	−.2619E+05	.2209E+06
I	4	−.2048E+07	.1859E+05	−.2201E+06	.2048E+07	−.1859E+05	.2647E+06
I	5	−.2049E+07	.1111E+05	−.2640E+06	.2049E+07	−.1111E+05	.2907E+06
I	6	−.2050E+07	.3693E+04	−.2903E+06	.2050E+07	−.3693E+04	.2991E+06
I	I	−.4615E+06	.0000E+00	.0000E+00	.4615E+06	.0000E+00	.0000E+00
I	2	−.2700E+07	.0000E+00	.0000E+00	.2700E+07	.0000E+00	.0000E+00
I	3	−2700E+07	.0000E+00	.0000E+00	.2700E+07	.0000E+00	.0000E+00
I	4	−.2700E+07	.0000E+00	.0000E+00	.2700E+07	.0000E+00	.0000E+00
I	5	−.2700E+07	.0000E+00	.0000E+00	.2700E+07	.0000E+00	.0000E+00
I	6	−.2700E+07	.0000E+00	.0000E+00	.2700E+07	.0000E+00	.0000E+00
I	7	.1320E+06	.0000E+00	.0000E+00	−.1320E+06	.0000E+00	.0000E+00
I	8	.1320E+06	.0000E+00	.0000E+00	−.1320E+06	.0000E+00	.0000E+00
I	9	.1320E+06	.0000E+00	.0000E+00	−.1320E+06	.0000E+00	.0000E+00
I	10	.1320E+06	.0000E+00	.0000E+00	−.1320E+06	.0000E+00	.0000E+00
I	11	.1320E+06	.0000E+00	.0000E+00	−.1320E+06	.0000E+00	.0000E+00

Analysis results; load case No. I

Nodal displacements

Node	u	v	θ
1	−.10241E−07	.22799E−08	−.14497E−08
2	.59852E−02	.22799E−02	.64905E−02
3	.48107E−02	.17343E−01	.59470E−02
4	.36251E−02	.30379E−01	.48319E−02
5	.24258E−02	.40286E−01	.33686E−02
6	.12157E−02	.46433E−01	.17224E−02
7	.58205E−13	.48512E−01	.53870E−13
8	.64525E−02	.22799E−02	.64905E−02
9	.39544E−02	.17343E−01	.59470E−02
10	.21176E−02	.30379E−01	.48319E−02
11	.97055E−03	.40286E−01	.33686E−02
12	.34761E−03	.46433E−01	.17224E−02
13	.76495E−15	.48512E−01	.13720E−18

Forces at the supported nodes

Node	F_x	F_y	M_z
1	.86860E+03	.26096E−04	−.38882E+04
7	.19104E+06	.00000E+00	−.93086E+05
13	−.19190E+06	.00000E+00	−.99877E+00

Member end forces

Member	$F_1{}^*$	$F_2{}^*$	$F_3{}^*$	$F_4{}^*$	$F_5{}^*$	$F_6{}^*$
1	−.26096E−04	.86860E+03	−.38882E+04	.26096E−04	−.86860E+03	.82302E+04
2	−.33753E+06	−.30456E+05	−.32595E+05	.33753E+05	.30456E+05	−.40501E+05
3	−.29902E+06	−.20993E+05	.34956E+05	.29902E+06	.20993E+05	−.85331E+05
4	−.24999E+06	−.12543E+05	.70033E+05	.24999E+06	.12543E+05	−.10013E+06
5	−.21136E+06	−.63670E+04	.83440E+05	.21136E+06	.63670E+04	−.98733E+05
6	−.19103E+06	−.19190E+04	.88486E+05	.19103E+06	.19190E+04	−.93086E+05
7	.33977E+06	−.19886E−02	−.22704E−03	−.33977E+06	.19886E−02	−.45648E−02
8	.30063E+06	−.13754E−02	.28021E−02	−.30063E+06	.13754E−02	−.61112E−02
9	.25117E+06	−.82658E−03	.48621E−02	−.25117E+06	.82658E−03	−.68483E−02
10	.21233E+06	−.42217E−03	.60855E−02	−.21233E+06	.42217E−03	−.70992E−02
11	.19191E+06	−.12775E−03	.67470E−02	−.19191E+06	.12775E−03	−.70536E−02

Table 6.8 gives abbreviated input and results of Computer run 2. A transformed cross-sectional area $= A_{ps}E_s/\bar{E}_c$ $(t_2, t_1) = 0.0022(200/9.615) = 0.04576\,\text{m}^2$ is entered for each of members 7 to 11. The forces $\{A_r\}$ that can restrain the nodal displacements due to creep and shrinkage of members 1 to 6 are entered separately for each of the two causes. Also the forces $\{A_r\}_{\text{relaxation}}$ are entered for members 7 to 11. These forces are calculated using Equations (6.6), (6.7) and (6.8). As example, we show below the calculation for member 2:

$$A_r(t_2, t_1)_{\text{creep member 2}} = -\frac{\bar{E}_c\,(t_2, t_1)}{E_c\,(t_1)}\,\varphi(t_2, t_1)\,\{A_D(t_1)\}_{\text{member 2}}$$

$$= -\frac{9.615}{25}\,(2.0)\,\{2654.4\,\text{kN}, -44.2\,\text{kN}, 100.0\,\text{kN.m}, -2654.4\,\text{kN},$$

$$44.2\,\text{kN}, -206.0\,\text{kN.m}\}$$

$$= \{-2042.0\,\text{kN}, 33.97\,\text{kN}, -76.9\,\text{kN.m}, 2042.0\,\text{kN}, -33.97\,\text{kN},$$

$$158.5\,\text{kN.m}\}$$

$$A_r(t_2, t_1)_{\text{axial, shrinkage member 2}} = \pm\,[\bar{E}_c(t_2, t_1)\,\varepsilon_{cs}\,(t_2, t_1)\,A_c\text{ member 2}]$$

$$= \pm[9.615 \times 10^9\,(-300 \times 10^{-6})\,(0.936)]$$

$$= \mp\,2700\,\text{kN}$$

$$A_r(t_2, t_1)_{\text{axial, relaxation member 2}} = \mp\,A_{ps}\text{ member 2}\,\Delta\bar{\sigma}_{pr}\,(t_2, t_1)$$

$$= \mp\,(2200 \times 10^{-6})\,(-60 \times 10^6) = \pm\,132.0\,\text{kN}$$

The results in Table 6.8 include the increase in deflection at mid-span in the period t_1 to t_2 (node 7) $= 48.5\,\text{mm}$. They also include the change in tension in the tendon member 11 $= -191.9\,\text{kN}$; this represents a drop in tension at a section halfway between nodes 6 and 7. The change in force in the tendon is the combined effect of creep, shrinkage and relaxation and the accompanying variation of internal forces.

6.9   General

Conventional linear computer programs for framed structures are employed in this chapter to calculate approximately the time-dependent effects of creep and shrinkage of concrete and relaxation of prestressed steel in various structures. A number of computer runs (at least two), depending on the number of load stages, is required for the analysis. The approach can be useful in the absence of specialized computer programs that can perform the analysis more accurately in a single computer run for structures constructed and/or loaded in stages. Cracking requires non-linear analysis that cannot be considered in the procedure presented in this chapter. The non-linear analysis that considers cracking is discussed in Chapter 13.

Notes

1 See, for example, the computer programs described in an appendix of Ghali, A. and Neville, A. M., *Structural Analysis: A Unified and Classical Approach*, 4th edn., E & FN Spon, London 1997, 831 pp. This set of programs is available from Liliane Ghali; 3911 Vincent Drive N.W., Calgary, Canada T3A 0G9. The set includes the programs PLANEF (Plane Frames) and SPACET (Space Trusses) used to solve examples in this chapter.
2 See Note 1, above.
3 See Note 1, above.

Chapter 7

Stress and strain of cracked sections

Western Canadian Place, Calgary. Partial prestressing used in all floors. (Courtesy Cohos Evamy & Partners, Calgary.)

7.1   Introduction

Cracks occur in reinforced and partially prestressed members when the stresses exceed the tensile strength of concrete. After cracking, the stresses in concrete normal to the plane of the crack cannot be tensile. Thus, the internal forces in a section at the crack location must be resisted by the reinforcement and the uncracked part of the concrete cross-section. The part of the concrete cross-section area which continues to be effective in resisting the internal forces is subjected mainly to compression and some tension not exceeding the tensile strength of concrete. At sections away from cracks, concrete in tension also contributes in resisting the internal forces and hence to the stiffness of the member.

Two extreme states are to be considered in the calculation of displacements in a cracked member, as will be further discussed in Chapter 8. In state 1, the full area of the concrete cross-section is considered effective and the strains in the concrete and the reinforcement are assumed to be compatible. In state 2, concrete in tension is ignored; thus, the cross-section is assumed to be composed of the reinforcement and concrete in compression. The cross-section in state 2 is said to be *fully cracked*.

The actual elongation or curvature of a cracked member can be calculated by interpolation between the two extreme states 1 and 2.

In Chapters 2 and 3 we analysed the stresses, axial strain and curvature in an uncracked section, including the effects of creep, shrinkage and relaxation of prestressed steel. The section was assumed to be subjected to an axial force and/or a bending moment. The values and the time of application of these forces were assumed to be known. With prestressing, the initial prestress force was assumed to be known, but the changes in the stresses in the prestressed and non-prestressed steel due to creep, shrinkage and relaxation were determined by the analysis. The full concrete cross-section area was considered to be effective, whether the stresses were tensile or compressive.

In the present chapter, fully cracked reinforced concrete sections without prestressing are analysed. The section is assumed to be subjected to an axial force, N, and a bending moment, M, of known magnitudes. With the concrete in tension ignored, these forces are resisted by the concrete in compression and by the reinforcement. The analysis will give axial strain, curvature and corresponding stresses immediately after application of N and M and after a period of time in which creep and shrinkage occur.

Analysis of a partially prestressed section is also included in this chapter. The section is assumed to remain in state 1 (uncracked) under the effect of prestress and loads of long duration, such as the dead load. After a given period of time, during which creep, shrinkage and relaxation have occurred, live load is assumed to be applied, producing cracking. With this assumption, the equations of Chapter 2 can be used to determine the stress and strain in concrete and the reinforcement at the time of prestressing and after a specified period during which creep, shrinkage and relaxation have occurred. The additional internal forces produced by the live load are assumed to produce instantaneous changes in stress and strain and also cause cracking which reduces the effective area of the section. The instantaneous changes in stress and strain are calculated but no time-dependent effects are considered. It is believed that these assumptions are not too restrictive and they represent most practical situations. Other assumptions adopted in the analysis are stated in the following section.

If the load which produces cracking is sustained, the effects of creep and shrinkage which occur after cracking are the same as for a reinforced concrete section without prestressing.

7.2 Basic assumptions

Concrete in the tension zone is assumed to be ineffective in resisting internal forces acting on a cracked cross-section. The effective area of the cross-section is composed of the area of the compressive zone and the area of reinforcement.

Plane cross-sections are assumed to remain plane after the deformation and strains in concrete and steel are assumed to be compatible. These two assumptions are satisfied by using in the analysis the area properties of a *transformed fully cracked section* composed of: A_c, the area of the compression zone and αA_s where $\alpha = E_s/E_c$; E_s is the modulus of elasticity of the reinforcement. E_c is the modulus of elasticity of concrete at the time of application of the load when the analysis is concerned with instantaneous stress and strain. When creep and shrinkage are considered, E_c is the age-adjusted modulus (see Section 1.11).

Due to creep and shrinkage, the depth of the compression zone changes; thus, A_c is time-dependent. In the analysis of stress and strain changes due to creep and shrinkage during a time interval, A_c is considered a constant equal to the area of the compression zone at the beginning of the time interval. This assumption greatly simplifies the analysis, but involves negligible error.

7.3 Sign convention

A positive bending moment M, produces compression at the top fibre (Fig. 7.1(a)). The axial force, N, is positive when tensile. N acts at an arbitrarily chosen reference point O. The eccentricity $e = M/N$ and the coordinate y of any fibre are measured downward from O. Tensile stress and the corresponding strain are positive. Positive M produces positive curvature ψ.

The above is a review of some of the conventions adopted throughout this book (see Section 2.2).

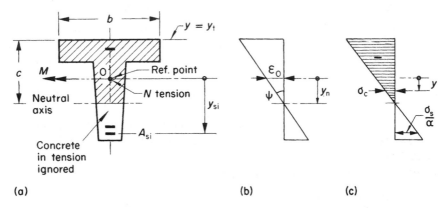

(a)   (b)   (c)

Figure 7.1 Stress (c) and strain (b) distributions in a fully cracked reinforced concrete
section (a) (state 2) subjected to M and N. Convention for positive M, N, y, y_n
and y_s.

7.4   Instantaneous stress and strain

Consider a concrete section reinforced by a number of layers of steel and
subjected to a bending moment M and a normal force N at an arbitrarily
chosen reference point O (Fig. 7.1(a)). The values of M and N are such
that the top fibre is in compression and the bottom is in tension, producing
cracking at the bottom face.

The equations, graphs and tables presented in this section and subsections
7.4.1 and 7.4.2 are based on the assumption that the top fibre is in compression and the bottom part of the section is cracked due to tension. When the
bottom part of the section is in compression and the tension zone and cracking are at the top, the equations apply if the direction of the y-axis is reversed
and all reference to the top fibre will be considered to mean the bottom fibre.
In this case, the flange of a T section will be at the tension zone; the graphs
and tables for a rectangular section (of width equal to the width of the web)
will apply as long as the compression zone is a rectangle.

Not included here are the situations when the stresses over all the section
are of the same sign. When all the stresses are compressive, the equations for
uncracked sections presented in Chapter 2 apply. When all the stresses are
tensile, the concrete is assumed to be ineffective in the fully cracked state 2
and the internal forces are resisted only by the steel. In this case, creep and
shrinkage have no effect on the stress and strain distribution over the section.

The stress and strain distributions shown in Fig. 7.1(b) and (c) are assumed
to be produced by the combined effect of M and N as shown in Fig. 7.1(a).
The resultant of M and N is located at eccentricity e given by

$$e = M/N \tag{7.1}$$

Positive e means that the resultant is situated below the reference point O. The location of the neutral axis depends on the value of e, not on the separate values of M and N. This is true in an uncracked or a fully cracked state 1 and 2, but the depth of the compression zone is, of course, not the same in the two states. In the analysis presented below, the area considered effective in resisting the internal forces is composed of the area of the compression zone plus the area of the reinforcement. The equations given below enable determination of the depth c of the compressive zone without time effect, and when c is known, the properties of the transformed area can be determined and the stress and strain calculated in the same way as for an uncracked section.

The strain at any fibre (Fig. 7.1(b)) is

$$\varepsilon = \varepsilon_O + y\psi \tag{7.2}$$

The y-coordinate of the neutral axis is:

$$y_n = -\varepsilon_O / \psi \tag{7.3}$$

The stress in concrete at any fibre is

$$\sigma_c = \begin{cases} E_c \left(1 - \dfrac{y}{y_n}\right)\varepsilon_O & y < y_n \\[2mm] 0 & y \geq y_n \end{cases} \tag{7.4} \tag{7.5}$$

It may be noted that in Fig. 7.1(b), ε_O is a negative quantity since O is chosen in the compression zone. The stress in any steel layer at coordinate y_s is:

$$\sigma_s = E_s \left(1 - \frac{y_s}{y_n}\right)\varepsilon_O \tag{7.6}$$

Integrating the stresses over the area and taking moment about an axis through O gives:

$$\varepsilon_O \left\{ E_c \int_{y_t}^{y_n} \left(1 - \frac{y}{y_n}\right) dA + E_s \Sigma \left[A_s \left(1 - \frac{y_s}{y_n}\right) \right] \right\} = N \tag{7.7}$$

$$\varepsilon_O \left\{ E_c \int_{y_t}^{y_n} y \left(1 - \frac{y}{y_n}\right) dA + E_s \Sigma \left[A_s y_s \left(1 - \frac{y_s}{y_n}\right) \right] \right\} = M \tag{7.8}$$

where

dA = an elemental area of concrete in compression
A_s and y_s = the area of steel in one layer of reinforcement and its coordinate, measured downwards from the reference point O
y_t = the y-coordinate at the top fibre
y_n = the y-coordinate of the neutral axis
E_s and E_c = the moduli of elasticity of steel and concrete.

The summations in Equations (7.7) and (7.8) are for all steel layers.

When the section is subjected to bending moment only, N can be set equal to zero in Equation (7.7), giving the following equation which can be solved for the coordinate y_n, defining the position of the neutral axis:

$$\int_{y_t}^{y_n} (y_n - y)\mathrm{d}A + a \sum [A_s(y_n - y_s)] = 0 \tag{7.9}$$

where $a = E_s/E_c$.

Equation (7.9) indicates that when N = 0, the first moment of the transformed area of the fully cracked cross-section about the neutral axis is zero. Thus, the neutral axis is at the centroid of the transformed fully cracked section (the area of concrete in compression plus a times the area of reinforcement).

When N ≠ 0, the neutral axis does not coincide with the centroid of the transformed area. The equation to be solved for y_n is obtained by division of Equation (7.8) by (7.7):

$$\frac{\int_{y_t}^{y_n} y(y_n - y)\mathrm{d}A + a \sum [A_s y_s(y_n - y_s)]}{\int_{y_t}^{y_n} (y_n - y)\mathrm{d}A + a \sum [A_s(y_n - y_s)]} - e = 0 \tag{7.10}$$

For an arbitrary cross-section, the value y_n that satisfies Equation (7.9) or (7.10) may be determined by trial. In subsection 7.4.1, Equations (7.9) and (7.10) are applied for a cross-section in the form of a T or a rectangle.

Once the position of the neutral axis is determined, the properties of the transformed fully cracked section are determined in the conventional way, giving A the area, B the first moment and I the moment of inertia about an axis through the reference point O. Now the general equations of Section 2.3 may be applied to determine ε_O, ψ and the stress at any fibre.

7.4.1   Remarks on determination of neutral axis position

Equations (7.9) or (7.10) can be used to determine the position of the neutral axis, and thus the depth c of compression zone, for any section having a vertical axis of symmetry. Equation (7.9) applies when the section is subjected to a moment, M without a normal force. Equation (7.10) applies when M is combined with a normal force, N.

For a section of arbitrary shape, a trial value of the coordinate y_n of the neutral axis is assumed, the integral in Equation (7.9) or the two integrals in Equation (7.10) are evaluated, ignoring concrete in tension. By iteration a value y_n, between y_t and y_b, is determined to satisfy one or the other of the two equations; where y_t and y_b are the y coordinates of the top and bottom fibres, respectively.

Both Equations (7.9) and (7.10) are based on the assumption that the extreme top and bottom fibres are in compression and in tension, respectively. Thus the equations apply when:

$$\sigma_{t1} \leqslant 0 \quad \text{while} \quad \sigma_{b1} \geqslant 0 \qquad (7.11)$$

where σ is stress at concrete fibre; the subscripts t and b refer to top and bottom fibres and the subscript 1 refers to state 1 in which cracking is ignored. When the extreme top and bottom fibres are in tension and compression, respectively, Equation (7.9) or (7.10) applies when the direction of the y-axis is reversed to point upwards and the symbol y_t in the equations is treated as coordinate of bottom fibre. It is here assumed that at least one of σ_{t1} and σ_{b1} exceeds the tensile strength of concrete, causing cracking.

When a section is subjected to a moment, without a normal force, solution of Equation (7.9) gives the position of the neutral axis at the centroid of the transformed section, with concrete in tension ignored. In this case, the equation has a solution y_n between y_t and the y-coordinate of the extreme tension reinforcement. However, when a section is subjected to a normal force, N combined with a moment, M, the neutral axis can be not within the height of the section; in which case Equation (7.10) has no solution for y_n that is between y_t and y_b. The following are limitations on the use of Equation (7.10), depending upon the values of M and N. It is here assumed that the compression zone is at top fibre:

(1) When N is compressive, both σ_{t1} and σ_{b1} are compressive when:

$$\frac{I_1 - y_t B_1}{B_1 - y_t A_1} \geqslant \frac{M}{N} \geqslant \frac{I_1 - y_b B_1}{B_1 - y_b A_1} \qquad (7.12)$$

where A_1, B_1 and I_1 are area of transformed uncracked section (state 1) and its first moment and second moment about an axis through the reference point O (Fig. 7.1). In this case the section is uncracked and use of Equation (7.10) is not needed.

(2) When the section is made of plain concrete, without reinforcement, Equation (7.10) applies only when resultant force is compressive and situated within the height of the section; that is when,

$$y_t \leqslant \frac{M}{N} \leqslant y_b \tag{7.13}$$

(3) When the section has two or more reinforcement layers and the normal force N is tensile, Equation (7.10) applies only when:

$$\frac{\Sigma I_s - y_t \Sigma B_s}{\Sigma B_s - y_t \Sigma A_s} \leqslant \frac{M}{N} \leqslant \frac{\Sigma I_s - y_b \Sigma B_s}{\Sigma B_s - y_b \Sigma A_s} \tag{7.14}$$

where ΣA_s, ΣB_s and ΣI_s are sum of cross-sectional areas of reinforcement layers and their first and second moments about an axis through the reference point, O (Fig. 7.1). This inequality gives lower and upper limits of a range of (M/N) within which Equation (7.10) does not apply. The lower and the upper limits of the range are respectively equal to the third and the first terms in Equation (7.14). When (M/N) is equal to the lower limit or to the upper limit, the neutral axis coincides with the bottom or top fibres, respectively. In other words, when (M/N) is within this range, Equation (7.10) has no solution for y_n that lies between y_t and y_b. In this case the resultant tensile force is resisted entirely by the reinforcement; the strain and stress in any reinforcement layer can be determined by Equations (2.19) and (2.20), substituting ΣA_s, ΣB_s and ΣI_s for A, B and I, respectively.

(4) When the section has only one reinforcement layer and the normal force N is tensile, the compression zone is at top fibre and Equation (7.10) applies when $(M/N) \geqslant y_s$; where y_s is the y-coordinate of the reinforcement layer. But when $(M/N) < y_s$ the compression one is at the bottom; the direction of the y axis must be reversed; the coordinate of the reinforcement layer becomes $(-y_s)$ before Equation (7.10) can be applied.

7.4.2   Neutral axis position in a T or rectangular fully cracked section

The equations of the preceding section are applied below for a T section reinforced by steel layers A_{ns} and A'_{ns} near the bottom and top fibres (Fig. 7.2). The section is also assumed to have one layer of prestress steel A_{ps} situated anywhere in the tension zone. Presence of A_{ps} simply adds an area

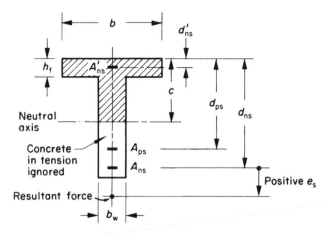

Figure 7.2 Definition of symbols employed in Section 7.4.2.

$a_{ps}A_{ps}$ to the transformed effective area; where $a_{ps} = E_{ps}/E_c$, with E_{ps} and E_c being the moduli of elasticity of prestressed steel and concrete. The equations presented below are applicable for a rectangular section by setting $b_w = h$.

Consider the case when the section in Fig. 7.2 is subjected to a positive bending moment without an axial force. Application of Equation (7.9) gives the following quadratic equation from which the depth c of the compression zone can be determined:

$$\tfrac{1}{2}b_w c^2 + [h_f(b - b_w) + a_{ns}A_{ns} + a_{ps}A_{ps} + (a_{ns} - 1)\,A'_{ns}]c$$
$$- [\tfrac{1}{2}(b - b_w)h_f^2 + a_{ns}A_{ns}d_{ns} + a_{ps}A_{ps}d_{ps}$$
$$+ (a_{ns} - 1)\,A'_{ns}d'_{ns}] = 0 \quad \text{when } c \geqslant h_f \tag{7.15}$$

where d_{ns}, d_{ps} and d'_{ns} are distances from the extreme compression fibre to the reinforcements A_{ns}, A_{ps} and A'_{ns} respectively. b and b_w are widths of the flange and of the web, respectively, and h_f is the thickness of the flange. $a_{ns} = E_{ns}/E_c$, with E_{ns} being the modulus of elasticity of non-prestressed steel.

Solution of the quadratic Equation (7.15) gives the depth of the compression zone in a T section subjected to bending moment:

$$c = \frac{-a_2 + \sqrt{(a_2^2 - 4a_1a_3)}}{2a_1} \tag{7.16}$$

where

$$a_1 = b_w/2 \tag{7.17}$$

$$a_2 = h_f(b - b_w) + a_{ns}A_{ns} + a_{ps}A_{ps} + (a_{ns} - 1)A'_{ns} \tag{7.18}$$

$$a_3 = -\tfrac{1}{2}h_f^2(b - b_w) - a_{ns}A_{ns}d_{ns} - a_{ps}A_{ps}d_{ps} - (a_{ns} - 1)A'_{ns}d'_{ns} \tag{7.19}$$

When the section is subjected to a bending moment M and a normal force N in any position, the two actions may be replaced by a resultant normal force N at the appropriate eccentricity. Let e_s be the eccentricity of the resultant measured downwards from the bottom reinforcement. Thus, e_s is a negative quantity when the resultant normal force is situated above A_{ns} (Fig. 7.2). The depth of the compression zone c can be determined by solving the following cubic equation, which is derived from Equation (7.10):

$$b_w(\tfrac{1}{2}c^2)(d_{ns} - \tfrac{1}{3}c)$$

$$+ (b - b_w)h_f[c(d_{ns} - \tfrac{1}{2}h_f) - \tfrac{1}{2}h_f(d_{ns} - \tfrac{2}{3}h_f)]$$

$$+ (a_{ns} - 1)A'_{ns}(c - d'_{ns})(d_{ns} - d'_{ns}) - a_{ps}A_{ps}(d_{ps} - c)(d_{ns} - d_{ps})$$

$$+ e_s[b_w(\tfrac{1}{2}c^2) + (b - b_w)h_f(c - \tfrac{1}{2}h_f) + (a_{ns} - 1)A'_{ns}(c - d'_{ns})$$

$$- a_{ps}A_{ps}(d_{ps} - c) - a_{ns}A_{ns}(d_{ns} - c)] = 0 \qquad \text{when } c \geqslant h_f \tag{7.20}$$

Equation (7.20) may be conveniently solved by trial, employing a programmable calculator. A direct solution is also possible (see Appendix D).

In the derivations of Equations (7.15) and (7.20), the height c of the compression zone is assumed to be greater or equal to h_f (Fig. 7.2). If $c < h_f$, the area for the fully cracked T section in Fig. 7.2 will be the same as that for a rectangular section of width b. Equation (7.15) or (7.20) applies for a rectangular section, simply by setting $b_w = b$.

It should be noted that Equation (7.20) applies when the top fibre of the T section is in compression while the bottom fibre is in tension. This occurs only when the normal force is tensile, situated below the centroid of the tensile reinforcement (A_{ns} plus A_{ps}) or when the normal force is compressive, situated above approximately 0.7 the depth of the section.

7.4.3   Graphs and tables for the properties of transformed fully cracked rectangular and T sections

Figure 7.3 shows a T section subjected to a bending moment or to a bending moment combined with an axial force that produces cracking. The section is provided with only one layer of reinforcement A_s in the tension zone. The graphs and tables presented below give the depth c of the compression zone, the distance $\bar{y}$ between the extreme compression fibre and the centroid of the transformed fully cracked section and its moment of inertia I about an axis through the centroid. Each of c, $\bar{y}$ and I depends on the dimensions of the

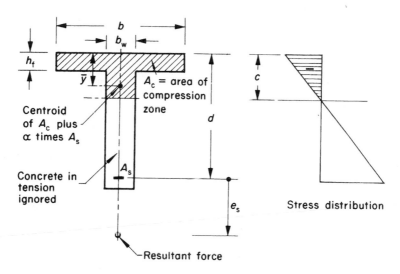

Figure 7.3 Definition of symbols used in the graphs on Figs 7.4 to 7.6 and Tables 7.1 to 7.4.

section and the product αA_s, where $\alpha = E_s/F_c$, the ratio of elasticity moduli of steel and concrete. The computer programs described in Appendix G can be used in lieu of the graphs and the tables.

For the use of the graphs or the tables with a section having more than one layer of steel in the tension zone with different elasticity moduli (as, for example, in the section in Fig. 7.2), the value αA_s to be used in the graphs or tables is:

$$\alpha A_s = \Sigma \alpha_i A_{si} \tag{7.21}$$

and the area αA_s is to be considered situated at distance d from the top edge given by:

$$d = \frac{\Sigma \alpha_i A_{si} d_i}{\Sigma \alpha_i A_{si}} \tag{7.22}$$

When the section is subjected to bending without axial force, the height c of the compression zone depends on αA_s and the dimensions d, h_f, b and b_w (Fig. 7.3). When the section is subjected to a moment M and a normal force N, the height c is a function of the same parameters plus e_s, where e_s is the eccentricity of the resultant of M and N measured downwards from the tension reinforcement (Fig. 7.3).

The graphs in Fig. 7.4 give the value of c for a fully cracked rectangular section subjected to a moment and a normal force. This pair of forces must be

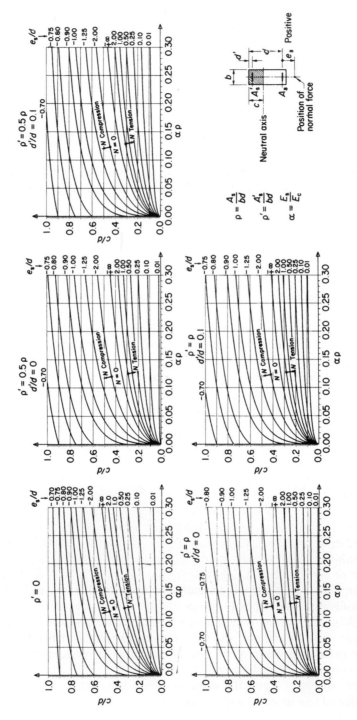

Figure 7.4 Depth of the compression zone in a fully cracked rectangular section subjected to eccentric normal force.

replaced by statical equivalents M and N, with N located at the same level as A_s. The resultant of the pair is thus a force N situated at a distance

$$e_s = M/N \qquad (7.23)$$

where e_s is an eccentricity of the resultant measured downwards from A_s.

The use of the graphs in Fig. 7.4 is limited to a rectangular cracked section with the compression zone at the top part of the section. This occurs only when N is tension and e_s/d has a value greater than zero, or when N is compression and e_s/d has a value smaller than -0.7. The limiting values 0 and -0.7 are approximate quantities which depend upon the reinforcement ratios ρ and ρ' where

$$\rho = A_s/bd \qquad (7.24)$$

and

$$\rho' = A'_s/bd \qquad (7.25)$$

where b is the breadth of the section and d is the distance between the bottom reinforcement A_s and the extreme compression fibre. A'_s is the area of an additional layer near the top, situated at a distance d' from the extreme compression fibre.

The case of a section subjected to a positive moment with no axial force, is the same as for $e_s = \infty$ with N a small tensile force or $e_s = -\infty$ with N a small compressive force. In each of the graphs in Fig. 7.4, the curve labelled $e_s = \pm\infty$ is to be used when the section is subjected to a moment without axial force. Other curves are usable when N is tension or compression.

Figs 7.5 and 7.6 give the position of the centroid and the moment of inertia of a transformed fully cracked rectangular section for which the depth c of the compression zone is predetermined.

Tables 7.1 and 7.2 can be used for the same purpose as Fig. 7.4 when the section is in the form of a T. In order to reduce the number of variables, the tables are limited to T sections without steel in the compression zone or when this reinforcement is ignored. The two tables naturally give the identical results as Fig. 7.4 in the special case when $\rho' = 0$ and $b_w/b = 1$; where b and b_w are widths of flange and web, respectively.

Once the depth c of the compression zone of a fully cracked T section is determined, Tables 7.3 and 7.4 can be used to determine the centroid and the moment of inertia about an axis through the centroid of the transformed section.

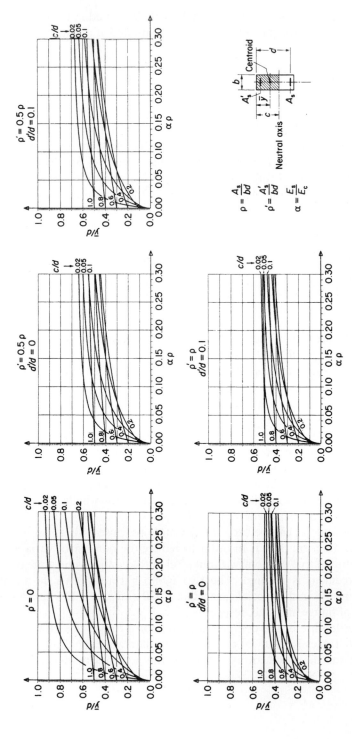

Figure 7.5 Distance from the top fibre to the centroid of the transformed area of a fully cracked rectangular section.

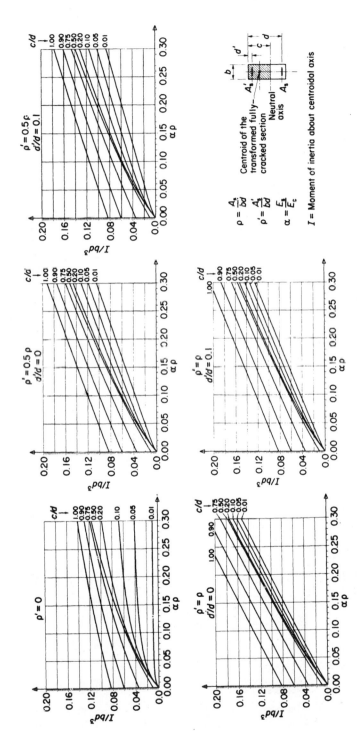

Figure 7.6 Moment of inertia about an axis through the centroid of the transformed area of a fully cracked rectangular section.

Table 7.1 Depth of compression zone in a fully cracked T section subjected to eccentric compressive force; $c = $ (coefficient from table $\times 10^{-3})d$

$b_w/b = 0.05$

$a\rho$	$e_s/d=-0.9$ h_f/d					$e_s/d=-1.0$ h_f/d					$e_s/d=-1.5$ h_f/d					$e_s/d=-2.0$ h_f/d					$e_s/d=-5.0$ h_f/d					$e_s/d=-\infty$ h_f/d				
	.05	.10	.15	.20	.30	.05	.10	.15	.20	.30	.05	.10	.15	.20	.30	.05	.10	.15	.20	.30	.05	.10	.15	.20	.30	.05	.10	.15	.20	.30
.0005	820	909	829	586	320	246	150	137	137	137	52	52	52	52	52	43	43	43	43	43	34	34	34	34	34	31	31	31	31	31
.0010	828	912	836	613	339	333	214	173	170	170	77	72	72	72	72	61	60	60	60	60	48	48	48	48	48	43	43	43	43	43
.0030	853	923	859	683	403	490	372	283	245	239	162	123	120	120	120	124	102	102	102	102	91	82	82	82	82	79	74	74	74	74
.0050	870	931	875	725	453	567	461	361	304	278	227	164	151	151	151	177	133	129	129	129	129	105	105	105	105	111	95	95	95	95
.0075	886	939	891	762	504	629	534	433	363	314	292	209	184	181	181	232	168	155	155	155	171	130	127	127	127	147	116	115	115	115
.0100	898	946	902	788	544	671	585	488	412	345	345	249	214	204	204	278	200	178	176	176	208	153	145	145	145	180	136	131	131	131
.0125	908	951	911	809	578	703	625	531	452	373	389	284	241	226	224	317	229	200	194	194	241	175	161	160	160	209	154	146	146	146
.0150	916	955	919	825	606	729	656	566	487	398	426	316	266	246	242	352	256	220	210	210	271	196	176	174	174	236	172	159	158	158
.0200	928	961	931	850	652	767	703	621	543	443	486	371	312	283	271	410	304	258	240	237	323	234	205	198	198	284	205	183	180	180

$b_w/b = 0.1$

$a\rho$	$e_s/d=-0.9$ h_f/d					$e_s/d=-1.0$ h_f/d					$e_s/d=-1.5$ h_f/d					$e_s/d=-2.0$ h_f/d					$e_s/d=-5.0$ h_f/d					$e_s/d=-\infty$ h_f/d				
	.05	.10	.15	.20	.30	.05	.10	.15	.20	.30	.05	.10	.15	.20	.30	.05	.10	.15	.20	.30	.05	.10	.15	.20	.30	.05	.10	.15	.20	.30
.0010	647	700	653	524	338	292	208	173	170	170	77	72	72	72	72	61	60	60	60	60	48	48	48	48	48	43	43	43	43	43
.0020	670	717	676	562	371	373	290	231	212	211	120	100	100	100	100	93	84	84	84	84	70	68	68	68	68	62	61	61	61	61
.0060	731	766	734	651	467	519	452	378	325	294	242	181	165	164	164	192	147	140	140	140	143	115	114	114	114	124	103	103	103	103
.0100	768	797	770	701	531	591	534	464	403	344	322	245	213	204	204	263	198	178	176	176	200	153	145	145	145	174	135	131	131	131
.0150	799	824	801	742	588	649	600	535	473	397	394	308	264	246	242	329	251	220	210	210	257	194	176	174	174	226	171	159	158	158
.0200	822	843	824	772	630	689	645	586	525	440	448	359	308	282	271	382	297	256	240	237	304	231	204	198	198	269	203	183	180	180
.0250	840	859	841	794	663	720	679	624	566	477	491	401	346	315	296	424	335	289	268	260	343	263	230	219	218	306	232	206	199	199
.0300	854	871	855	812	690	744	707	655	600	509	527	437	379	344	318	460	369	319	293	280	377	292	254	239	236	338	259	227	217	216
.0400	875	890	876	840	731	780	748	702	651	560	583	496	435	395	358	518	426	370	338	315	434	343	297	275	267	392	306	266	249	245

Table 7.1 continued

$b_w/b = 0.2$

$a\rho$	$e_s/d = -0.9$ h_f/d					$e_s/d = -1.0$ h_f/d					$e_s/d = -1.5$ h_f/d					$e_s'/d = -2.0$ h_f/d					$e_s/d = -5.0$ h_f/d					$e_s/d = -\infty$ h_f/d				
	.05	.10	.15	.20	.30	.05	.10	.15	.20	.30	.05	.10	.15	.20	.30	.05	.10	.15	.20	.30	.05	.10	.15	.20	.30	.05	.10	.15	.20	.30
.0020	542	569	546	485	368	318	269	227	211	211	116	100	100	100	100	92	84	84	84	84	70	68	68	68	68	62	61	61	61	61
.0040	582	604	584	533	417	395	351	304	272	260	177	142	137	137	137	141	117	116	116	116	107	94	94	94	94	93	85	85	85	85
.0120	669	684	670	634	534	534	502	459	417	365	318	261	232	221	221	266	215	195	191	191	209	168	158	157	157	184	149	143	143	143
.0200	717	730	718	688	599	604	577	539	499	436	400	340	302	281	271	343	285	253	240	237	277	225	203	198	198	247	199	183	180	180
.0300	757	768	758	732	654	659	636	603	566	500	470	410	368	340	318	412	350	312	291	280	341	282	251	239	236	307	251	225	217	216
.0400	785	794	786	764	693	698	678	648	614	548	521	463	419	388	357	464	402	360	334	315	390	328	292	274	267	354	294	263	248	245
.0500	807	815	808	787	723	728	709	682	650	586	561	505	461	428	391	505	443	400	371	345	430	367	328	306	293	394	331	295	277	270
.0600	824	831	825	806	746	751	734	709	679	618	593	540	496	462	421	538	478	434	404	372	464	400	359	334	316	427	363	324	303	291
.0800	850	856	850	834	782	787	772	750	723	666	644	594	552	518	472	592	534	490	458	419	519	455	412	383	357	481	416	374	349	328

$b_w/b = 0.5$

$a\rho$	$e_s/d = -0.9$ h_f/d					$e_s/d = -1.0$ h_f/d					$e_s/d = -1.5$ h_f/d					$e_s'/d = -2.0$ h_f/d					$e_s/d = -5.0$ h_f/d					$e_s/d = -\infty$ h_f/d				
	.05	.10	.15	.20	.30	.05	.10	.15	.20	.30	.05	.10	.15	.20	.30	.05	.10	.15	.20	.30	.05	.10	.15	.20	.30	.05	.10	.15	.20	.30
.0050	472	480	473	458	418	335	321	303	288	278	174	157	151	151	151	144	131	129	129	129	114	105	105	105	105	101	95	95	95	95
.0100	525	531	526	513	477	408	396	381	364	342	242	221	209	204	204	205	185	177	176	176	165	149	145	145	145	147	133	131	131	131
.0300	632	636	632	623	595	543	535	523	508	480	383	361	343	330	318	337	314	297	286	280	282	259	245	238	236	256	234	221	217	216
.0500	687	690	688	680	656	611	604	594	581	555	460	440	422	407	388	413	390	372	358	344	353	329	311	300	293	324	299	283	274	270
.0750	732	735	733	726	705	666	660	651	640	615	526	506	489	474	452	479	456	438	423	404	416	392	373	360	346	385	360	342	330	319
.1000	764	766	764	758	740	704	699	691	681	658	573	555	538	523	500	527	506	488	472	451	464	440	421	406	389	431	407	388	374	360
.1250	788	790	788	783	766	733	728	721	712	691	610	593	577	562	539	565	545	527	512	490	502	479	460	445	426	469	445	426	411	394
.1500	807	808	807	802	786	756	752	745	737	717	639	623	608	594	571	596	577	560	545	522	534	511	492	477	457	501	477	458	443	424
.2000	835	836	835	831	817	791	787	781	774	756	685	671	657	644	622	644	527	561	597	574	584	563	545	530	508	552	529	510	495	474

Table 7.1 continued

$b_w/b = 1.0$

e_s/d						
$a\rho$	-0.9	-1.0	-1.5	-2.0	-5.0	$-\infty$
.0100	448	340	204	176	145	131
.0200	507	412	271	237	198	180
.0600	620	546	408	367	316	291
.1000	678	614	483	441	385	358
.1500	724	668	546	504	447	417
.2000	757	706	592	550	493	463
.2500	781	735	627	587	530	500
.3000	801	758	656	617	560	530
.4000	830	792	700	663	609	579

Position of normal force

Positive e_s

$$\alpha = E_s/E_c$$
$$\rho = A_s/(bd)$$

Table 7.2 Depth of the compression zone in a fully cracked T section subjected to eccentric tensile force; c = (coefficient from table × 10^{-3})d

$b_w/b = 0.05$

αρ	$e_s/d = .1$					$e_s/d = .5$					$e_s/d = 1.0$					$e_s/d = 2.0$					$e_s/d = 5.0$					$e_s/d = \infty$				
	h_f/d					h_f/d					h_f/d					h_f/d					h_f/d					h_f/d				
	.05	.10	.15	.20	.30	.05	.10	.15	.20	.30	.05	.10	.15	.20	.30	.05	.10	.15	.20	.30	.05	.10	.15	.20	.30	.05	.10	.15	.20	.30
.0005	9	9	9	9	9	18	18	18	18	18	22	22	22	22	22	25	25	25	25	25	28	28	28	28	28	31	31	31	31	31
.0010	13	13	13	13	13	25	25	25	25	25	31	31	31	31	31	35	35	35	35	35	40	40	40	40	40	43	43	43	43	43
.0030	23	23	23	23	23	43	43	43	43	43	53	53	53	53	53	62	61	61	61	61	71	68	68	68	68	79	74	74	74	74
.0050	29	29	29	29	29	56	56	56	56	56	71	68	68	68	68	85	78	78	78	78	99	87	87	87	87	111	95	95	95	95
.0075	36	36	36	36	36	71	68	68	68	68	92	83	83	83	83	112	95	95	95	95	130	106	106	106	106	147	116	115	115	115
.0100	42	42	42	42	42	86	79	79	79	79	112	95	95	95	95	137	110	109	109	109	159	123	121	121	121	180	136	131	131	131
.0125	46	46	46	46	46	99	88	88	88	88	131	106	106	106	106	160	123	121	121	121	186	139	134	134	134	209	154	146	146	146
.0150	51	51	51	51	51	113	96	96	96	96	149	117	116	116	116	181	137	132	132	132	210	155	146	146	146	236	172	159	158	158
.0200	59	59	59	59	59	138	110	110	110	110	182	137	133	133	133	220	162	151	151	151	254	185	168	167	167	284	205	183	180	180

$b_w/b = 0.1$

αρ	$e_s/d = .1$					$e_s/d = .5$					$e_s/d = 1.0$					$e_s/d = 2.0$					$e_s/d = 5.0$					$e_s/d = \infty$				
	h_f/d					h_f/d					h_f/d					h_f/d					h_f/d					h_f/d				
	.05	.10	.15	.20	.30	.05	.10	.15	.20	.30	.05	.10	.15	.20	.30	.05	.10	.15	.20	.30	.05	.10	.15	.20	.30	.05	.10	.15	.20	.30
.0010	13	13	13	13	13	25	25	25	25	25	31	31	31	31	31	35	35	35	35	35	40	40	40	40	40	43	43	43	43	43
.0020	18	18	18	18	18	35	35	35	35	35	43	43	43	43	43	50	50	50	50	50	56	56	56	56	56	62	61	61	61	61
.0060	32	32	32	32	32	62	61	61	61	61	79	74	74	74	74	95	85	85	85	85	110	95	95	95	95	124	103	103	103	103
.0100	42	42	42	42	42	85	79	79	79	79	111	95	95	95	95	134	110	109	109	109	155	123	121	121	121	174	135	131	131	131
.0150	51	51	51	51	51	111	96	96	96	96	145	117	116	116	116	175	136	132	132	132	202	154	146	146	146	226	171	159	158	158
.0200	59	59	59	59	59	135	110	110	110	110	176	137	133	133	133	211	161	151	151	151	242	183	168	167	167	269	203	183	180	180
.0250	67	65	65	65	65	156	122	122	122	122	203	156	147	147	147	243	184	168	168	168	276	209	188	185	185	306	232	206	199	199
.0300	75	71	71	71	71	176	133	133	133	133	228	174	161	160	160	271	205	185	182	182	307	234	207	201	201	338	259	227	217	216
.0400	90	82	82	82	82	213	162	152	152	152	272	206	186	183	183	320	244	215	208	207	359	277	242	229	228	392	306	266	249	245

Table 7.2 continued

$b_w/b = 0.2$

	$e_s/d = .1$					$e_s/d = .5$					$e_s/d = 1.0$					$e_s/d = 2.0$					$e_s/d = 5.0$					$e_s/d = \infty$				
	\multicolumn{5}{c}{h_f/d}					\multicolumn{5}{c}{h_f/d}					\multicolumn{5}{c}{h_f/d}					\multicolumn{5}{c}{h_f/d}					\multicolumn{5}{c}{h_f/d}					\multicolumn{5}{c}{h_f/d}
$a\rho$	.05	.10	.15	.20	.30	.05	.10	.15	.20	.30	.05	.10	.15	.20	.30	.05	.10	.15	.20	.30	.05	.10	.15	.20	.30	.05	.10	.15	.20	.30
.0020	18	18	18	18	18	35	35	35	35	35	43	43	43	43	43	50	50	50	50	50	56	56	56	56	56	62	61	61	61	61
.0040	26	26	26	26	26	50	50	50	50	50	62	61	61	61	61	73	70	70	70	70	83	78	78	78	78	93	85	85	85	85
.0120	45	45	45	45	45	94	86	86	86	86	121	104	104	104	104	145	120	119	119	119	165	135	132	132	132	184	149	143	143	143
.0200	59	59	59	59	59	130	110	110	110	110	167	136	133	133	133	197	160	151	151	151	224	181	168	167	167	247	199	183	180	180
.0300	75	71	71	71	71	167	137	133	133	133	212	171	161	160	160	249	201	185	182	182	281	228	206	201	201	307	251	225	217	216
.0400	89	82	82	82	82	199	161	152	152	152	251	202	185	183	183	292	237	214	208	207	326	268	240	229	228	354	294	263	248	245
.0500	102	92	92	92	92	227	183	170	169	169	283	230	208	203	203	328	269	241	230	229	364	302	270	255	251	394	331	295	277	270
.0600	115	100	100	100	100	252	203	186	184	184	312	255	229	221	220	359	297	266	252	248	396	333	297	279	271	427	363	324	303	291
.0800	138	116	115	115	115	295	240	217	210	210	361	299	268	253	250	410	346	309	290	281	449	384	344	322	306	481	416	374	349	328

$b_w/b = 0.5$

	$e_s/d = .1$					$e_s/d = .5$					$e_s/d = 1.0$					$e_s/d = 2.0$					$e_s/d = 5.0$					$e_s/d = \infty$				
	\multicolumn{5}{c}{h_f/d}					\multicolumn{5}{c}{h_f/d}					\multicolumn{5}{c}{h_f/d}					\multicolumn{5}{c}{h_f/d}					\multicolumn{5}{c}{h_f/d}					\multicolumn{5}{c}{h_f/d}
$a\rho$	.05	.10	.15	.20	.30	.05	.10	.15	.20	.30	.05	.10	.15	.20	.30	.05	.10	.15	.20	.30	.05	.10	.15	.20	.30	.05	.10	.15	.20	.30
.0050	29	29	29	29	29	56	56	56	56	56	70	68	68	68	68	81	78	78	78	78	92	87	87	87	87	101	95	95	95	95
.0100	42	42	42	42	42	82	79	79	79	79	102	95	95	95	95	119	109	109	109	109	134	122	121	121	121	147	133	131	131	131
.0300	73	71	71	71	71	149	135	133	133	133	184	166	161	160	160	213	192	184	182	182	236	215	204	201	201	256	234	221	217	216
.0500	97	92	92	92	92	196	177	170	169	169	239	217	206	203	203	273	249	236	230	229	300	276	261	253	251	324	299	283	274	270
.0750	122	112	111	111	111	240	218	207	204	204	290	266	251	244	243	328	303	287	278	273	359	334	317	306	298	385	360	342	330	319
.1000	143	129	128	128	128	276	252	239	233	232	331	305	289	280	275	372	346	329	317	308	405	379	361	348	336	431	407	388	374	360
.1250	161	145	142	142	142	306	282	267	259	256	365	339	321	310	302	408	382	364	351	339	442	417	398	384	369	469	445	426	411	394
.1500	177	160	155	155	155	333	308	291	282	277	394	368	349	337	327	439	413	394	380	366	473	448	429	415	398	501	477	458	443	424
.2000	206	186	178	177	177	378	352	334	322	313	442	416	397	383	369	489	464	444	429	412	524	500	481	466	446	552	529	510	495	474

Table 7.2 continued

$b_w/b = 1.0$

$\alpha\rho$	\.1	\.5	e_s/d 1.0	2.0	5.0	∞
.0100	42	79	95	109	121	131
.0200	59	110	133	151	167	180
.0600	100	184	220	248	271	291
.1000	128	232	275	308	335	358
.1500	155	277	326	364	393	417
.2000	177	313	367	406	438	463
.2500	197	344	400	442	474	500
.3000	214	370	429	472	504	530
.4000	243	414	476	520	553	579

Position of normal force

$\alpha = E_s/E_c$

$\rho = A_s/(bd)$

Table 7.3 Distance from top fibre to centroid of transformed[1] fully cracked T section; $\bar{y}$ = (coefficient from table × 10^{-3})d

$b_w/b = 0.05$

$\alpha\rho$	c/d = .10					c/d = .20					c/d = .30					c/d = .50					c/d = .75					c/d = 1.00				
	\multicolumn{5}{c}{h_t/d}																									
	.05	.10	.15	.20	.30	.05	.10	.15	.20	.30	.05	.10	.15	.20	.30	.05	.10	.15	.20	.30	.05	.10	.15	.20	.30	.05	.10	.15	.20	.30
.0005	36	54	54	54	54	46	59	79	102	102	62	67	85	105	151	108	95	103	119	159	184	145	139	147	177	272	207	187	185	203
.0010	45	59	59	59	59	54	63	82	104	104	69	72	87	108	152	114	99	106	121	160	188	148	142	149	178	276	210	189	186	204
.0030	79	77	77	77	77	85	81	94	113	113	98	88	99	116	158	138	113	116	129	166	207	160	151	156	183	290	221	197	193	209
.0050	111	95	95	95	95	114	97	105	121	121	125	104	110	125	163	160	127	127	137	171	224	173	160	163	188	304	231	206	199	213
.0075	148	116	116	116	116	149	117	119	132	132	156	123	123	135	170	186	145	139	147	177	245	187	171	172	194	320	244	215	208	219
.0100	182	136	136	136	136	180	136	133	142	142	185	141	136	145	177	211	161	151	156	184	265	202	182	181	200	336	256	225	215	225
.0125	214	155	155	155	155	209	155	146	152	152	212	159	149	155	184	234	177	163	165	190	284	215	193	189	206	351	268	235	223	230
.0150	243	173	173	173	173	237	172	159	162	162	237	176	161	164	190	256	192	174	174	196	302	229	203	198	212	366	279	244	231	236
.0200	295	208	208	208	208	286	205	183	181	181	284	207	185	183	203	296	221	197	192	209	335	254	223	214	224	393	301	262	246	247

$b_w/b = 0.1$

$\alpha\rho$	c/d = .10					c/d = .20					c/d = .30					c/d = .50					c/d = .75					c/d = 1.00				
	\multicolumn{5}{c}{h_t/d}																									
	.05	.10	.15	.20	.30	.05	.10	.15	.20	.30	.05	.10	.15	.20	.30	.05	.10	.15	.20	.30	.05	.10	.15	.20	.30	.05	.10	.15	.20	.30
.0010	46	59	59	59	59	62	67	84	104	104	87	82	94	111	152	152	127	127	136	168	250	202	186	184	201	357	290	259	245	246
.0020	63	68	68	68	68	76	75	89	108	108	99	90	99	115	155	161	133	131	140	170	256	207	189	187	203	361	294	262	248	248
.0060	125	103	103	103	103	128	107	112	126	126	143	119	120	131	166	194	157	149	154	180	279	225	204	199	212	378	308	274	258	256
.0100	178	136	136	136	136	174	137	134	142	142	183	146	140	147	177	224	179	167	168	190	301	243	219	211	221	394	322	286	268	264
.0150	237	173	173	173	173	226	171	159	162	162	229	177	164	166	190	260	206	188	185	202	327	264	236	226	232	413	339	300	281	274
.0200	288	208	208	208	208	272	203	183	181	181	269	207	187	184	203	292	231	207	201	214	351	284	253	240	242	431	354	314	293	283
.0250	332	240	240	240	240	312	233	206	199	199	306	234	208	202	215	321	254	226	217	226	374	303	269	254	253	447	369	327	304	292
.0300	372	269	269	269	269	348	260	227	217	217	339	259	228	218	227	348	276	244	232	237	394	321	284	267	262	463	384	340	316	301
.0400	438	321	321	321	321	410	309	267	250	250	396	306	266	250	250	397	316	278	261	258	432	354	312	291	282	492	410	364	337	318

[1] The same table may be used for age-adjusted transformed section replacing α by $\bar{\alpha} = E_s[1 + \chi\varphi(t, t_0)]/E_c(t_0)$.

Table 7.3 continued

$b_w/b = 0.2$

$a\rho$	c/d = .10					c/d = .20					c/d = .30					c/d = .50					c/d = .75					c/d = 1.00				
	\multicolumn{5}{c}{h_t/d}																											
	.05	.10	.15	.20	.30	.05	.10	.15	.20	.30	.05	.10	.15	.20	.30	.05	.10	.15	.20	.30	.05	.10	.15	.20	.30	.05	.10	.15	.20	.30
.0020	64	68	68	68	68	85	81	92	108	108	117	105	109	121	155	197	170	162	164	184	308	268	247	237	240	425	375	344	325	312
.0040	93	86	86	86	86	107	96	103	117	117	134	118	119	129	161	208	179	169	170	188	315	274	252	242	244	430	380	348	329	315
.0120	194	151	151	151	151	184	151	145	150	150	196	164	156	159	182	250	213	198	194	207	342	298	273	261	259	448	397	364	344	327
.0200	274	208	208	208	208	250	200	183	181	181	250	206	189	187	203	287	244	224	217	224	367	320	293	279	273	465	413	379	357	339
.0300	355	269	269	269	269	318	253	226	217	217	307	252	228	219	227	329	280	255	244	245	396	347	317	300	291	485	432	397	374	353
.0400	419	321	321	321	321	375	299	264	250	250	357	294	263	250	250	366	313	284	269	265	422	371	339	320	307	503	449	413	389	366
.0500	472	366	366	366	366	423	341	299	279	279	399	331	295	277	271	399	343	311	293	284	446	393	360	339	323	520	466	429	404	379
.0600	516	406	406	406	406	464	377	331	307	307	437	365	324	303	291	429	370	335	315	302	468	414	379	357	338	536	482	444	419	391
.0800	585	472	472	472	472	531	439	387	357	357	499	422	376	349	328	481	419	380	355	335	508	452	414	390	366	565	511	472	445	415

$b_w/b = 0.5$

$a\rho$	c/d = .10					c/d = .20					c/d = .30					c/d = .50					c/d = .75					c/d = 1.00				
	\multicolumn{5}{c}{h_t/d}																											
	.05	.10	.15	.20	.30	.05	.10	.15	.20	.30	.05	.10	.15	.20	.30	.05	.10	.15	.20	.30	.05	.10	.15	.20	.30	.05	.10	.15	.20	.30
.0050	101	95	95	95	95	120	112	114	121	121	156	146	144	147	163	243	229	221	218	222	361	344	332	324	317	482	463	449	438	423
.0100	154	136	136	136	136	152	140	138	142	142	179	166	162	163	177	256	241	233	229	231	368	352	339	331	323	487	468	454	442	428
.0300	315	269	269	269	269	262	236	222	217	217	259	239	227	223	227	305	287	276	269	267	398	380	367	357	347	505	487	472	460	444
.0500	424	366	366	366	366	346	312	291	279	279	324	299	284	274	271	348	328	314	306	299	424	406	392	382	370	522	504	489	476	460
.0750	520	457	457	457	457	428	388	362	345	345	392	363	343	330	320	394	373	357	347	336	455	436	421	410	396	542	524	508	496	479
.1000	589	524	524	524	524	491	449	420	400	400	447	416	394	378	362	434	412	395	383	369	482	463	447	435	420	560	542	526	514	496
.1250	640	577	557	577	577	542	500	468	446	446	493	461	437	419	399	470	447	429	415	399	507	487	471	459	443	577	559	543	531	512
.1500	680	620	620	620	620	584	541	509	485	485	532	500	474	456	433	501	477	459	444	427	529	509	493	480	463	593	574	559	546	528
.2000	738	683	683	683	683	648	607	574	549	549	594	562	536	516	490	553	529	510	495	474	568	548	532	519	500	621	603	587	574	555

Table 7.3 continued

h_f b d c $\bar{y}$ A_s b_w Centroid

$$\alpha = E_s/E_c$$
$$\rho = A_s/(bd)$$

| | $b_w/b = 1.0$ | | | | | |
| | | | c/d | | | |
$\alpha\rho$	.1	.2	.3	.5	.75	1.0
.0100	136	142	177	264	383	504
.0200	208	181	203	278	391	509
.0600	406	307	291	330	421	528
.1000	524	400	362	375	448	545
.1500	619	485	433	423	479	565
.2000	683	549	489	464	506	583
.2500	728	599	536	500	531	599
.3000	762	640	574	531	553	615
.4000	809	699	635	583	592	642

Table 7.4 Moment of inertia of a transformed fully cracked T section about centroidal axis $I = $ (coefficient from table $\times\, 10^{-4})bd^3$

$b_w/b = 0.05$

$a\rho$	c/d = .10 h_t/d				c/d = .20 h_t/d					c/d = .30 h_t/d					c/d = .50 h_t/d					c/d = .75 h_t/d					c/d = 1.00 h_t/d				
	.10	.15	.20	.30	.05	.10	.15	.20	.30	.05	.10	.15	.20	.30	.05	.10	.15	.20	.30	.05	.10	.15	.20	.30	.05	.10	.15	.20	.30
.0005	5	5	5	5	5	5	7	10	10	7	7	8	11	26	17	18	18	20	32	46	50	50	51	59	99	111	114	114	118
.0010	9	9	9	9	9	10	11	14	14	11	11	12	15	29	21	22	22	24	35	50	54	54	54	62	102	115	117	117	121
.0030	27	27	27	27	27	27	28	30	30	28	28	29	31	43	36	38	38	39	49	62	68	68	69	75	112	127	130	131	134
.0050	43	43	43	43	43	43	44	46	46	44	45	45	47	58	51	53	53	54	63	75	82	83	83	89	122	139	143	143	146
.0075	63	63	63	63	62	63	64	65	65	62	64	64	65	75	68	72	72	72	80	89	99	100	100	105	133	153	159	159	161
.0100	82	82	82	82	79	82	82	83	83	79	83	83	84	92	84	90	90	90	97	103	115	117	117	121	145	167	174	175	176
.0125	101	101	101	101	95	101	101	101	101	96	101	102	102	109	99	107	108	108	114	116	130	133	134	137	155	181	189	190	191
.0150	118	118	118	118	110	118	118	119	119	111	118	119	120	125	113	123	125	125	130	129	145	149	150	153	166	194	203	205	206
.0200	151	151	151	151	137	151	151	153	153	138	151	154	154	157	139	155	158	159	162	152	174	180	181	183	185	219	231	234	235

$b_w/b = 0.1$

$a\rho$	c/d = .10 h_t/d				c/d = .20 h_t/d					c/d = .30 h_t/d					c/d = .50 h_t/d					c/d = .75 h_t/d					c/d = 1.00 h_t/d				
	.10	.15	.20	.30	.05	.10	.15	.20	.30	.05	.10	.15	.20	.30	.05	.10	.15	.20	.30	.05	.10	.15	.20	.30	.05	.10	.15	.20	.30
.0010	9	9	9	9	10	10	11	14	14	13	13	14	16	29	29	31	31	32	41	75	85	87	87	91	157	185	195	197	198
.0020	18	18	18	18	19	19	20	22	22	21	22	22	24	36	36	39	39	40	48	80	91	94	94	97	161	190	200	203	204
.0060	51	51	51	51	51	52	52	53	53	52	54	54	55	64	63	68	68	69	75	102	116	120	120	122	177	209	222	225	226
.0100	82	82	82	82	80	82	83	83	83	80	84	84	84	92	88	96	97	97	102	122	139	144	145	147	192	228	242	247	248
.0150	118	118	118	118	111	118	119	119	119	112	119	120	120	125	117	128	131	131	134	145	167	174	174	177	210	250	267	273	275
.0200	151	151	151	151	140	151	153	153	153	140	151	154	154	157	143	159	163	163	165	167	193	203	205	206	227	272	291	299	301
.0250	181	181	181	181	165	182	186	186	186	165	182	186	186	189	157	187	193	194	196	188	218	230	233	234	242	292	314	323	326
.0300	209	209	209	209	187	210	216	217	217	188	210	217	218	219	189	214	222	224	225	206	242	256	261	262	257	311	336	347	351
.0400	258	258	258	258	225	261	273	276	276	228	261	273	276	277	229	264	277	281	282	241	286	305	313	315	284	348	378	392	398

Table 7.4 continued

$b_w/b = 0.2$

$a\rho$	c/d = .10					c/d = .20					c/d = .30					c/d = .50					c/d = .75					c/d = 1.00				
			h_t/d					h_t/d					h_t/d					h_t/d					h_t/d					h_t/d		
	.05	.10	.15	.20	.30	.05	.10	.15	.20	.30	.05	.10	.15	.20	.30	.05	.10	.15	.20	.30	.05	.10	.15	.20	.30	.05	.10	.15	.20	.30
.0020	18	18	18	18	18	19	19	20	22	22	24	24	25	26	36	48	53	53	54	59	118	136	143	145	146	248	290	313	323	327
.0040	35	35	35	35	35	36	36	36	38	38	39	40	40	41	51	61	66	67	67	72	128	147	155	157	158	255	298	321	332	337
.0120	93	97	97	97	97	94	97	97	98	98	95	99	100	100	105	108	118	121	121	123	164	188	198	202	202	280	328	354	367	374
.0200	140	151	151	151	151	143	151	153	153	153	143	152	154	154	157	151	165	170	171	173	197	226	239	244	245	303	356	386	401	409
.0300	187	209	209	209	209	194	211	217	217	217	195	211	217	218	219	199	220	228	230	231	235	270	287	295	297	331	390	423	441	452
.0400	224	258	258	258	258	236	263	273	276	276	239	264	274	276	277	241	269	281	285	286	270	311	332	342	346	356	421	458	479	493
.0500	255	301	301	301	301	272	310	325	330	330	278	311	325	330	332	279	314	330	337	339	302	349	375	387	393	380	450	492	515	532
.0600	280	339	339	339	339	303	351	372	380	380	312	354	373	381	383	313	355	376	385	389	331	385	414	429	438	402	478	524	550	570
.0800	320	401	401	401	401	354	420	453	469	469	368	427	457	471	478	372	429	459	474	481	383	449	487	508	521	443	528	582	614	641

$b_w/b = 0.5$

$a\rho$	c/d = .10					c/d = .20					c/d = .30					c/d = .50					c/d = .75					c/d = 1.00				
			h_t/d					h_t/d					h_t/d					h_t/d					h_t/d					h_t/d		
	.05	.10	.15	.20	.30	.05	.10	.15	.20	.30	.05	.10	.15	.20	.30	.05	.10	.15	.20	.30	.05	.10	.15	.20	.30	.05	.10	.15	.20	.30
.0050	43	43	43	43	43	44	45	45	46	46	51	52	52	53	58	92	99	101	102	103	225	244	255	261	264	483	523	551	569	585
.0100	81	82	82	82	82	82	83	83	83	83	85	88	88	88	92	120	128	131	132	133	245	265	278	284	287	497	537	566	584	602
.0300	197	209	209	209	209	207	214	217	217	217	207	215	218	218	219	224	236	242	245	245	321	346	361	370	376	547	592	623	645	665
.0500	276	301	301	301	301	303	319	327	330	330	307	321	328	331	332	314	332	341	346	348	390	419	438	449	458	595	643	677	701	725
.0750	344	387	387	387	387	396	424	440	448	448	410	433	446	452	455	413	437	451	459	464	468	503	526	540	553	649	702	740	767	796
.1000	394	452	452	452	452	469	508	532	546	546	493	525	545	556	564	498	529	548	560	568	539	578	606	624	640	699	756	798	828	861
.1250	431	502	502	502	502	527	577	609	629	629	563	604	630	646	659	573	610	635	650	663	603	647	679	700	720	746	807	852	885	922
.1500	459	542	542	542	542	575	634	674	700	700	622	671	704	725	744	639	682	712	731	749	661	710	746	770	795	789	853	902	938	980
.2000	501	602	602	602	602	648	724	779	816	816	717	781	826	857	889	750	805	844	871	899	762	821	864	895	929	866	938	993	1034	1085

Table 7.4 continued

$$\alpha = E_s / E_c$$
$$\rho = A_s / (bd)$$

	b_w/b = 1.0					
			c/d			
αρ	.1	.2	.3	.5	.75	1.0
.0100	82	83	92	159	390	858
.0200	151	153	157	212	427	882
.0600	339	380	383	405	568	974
.1000	452	546	564	572	696	1060
.1500	542	700	744	753	839	1159
.2000	602	816	889	907	968	1249
.2500	645	906	1007	1041	1083	1333
.3000	677	978	1106	1158	1188	1410
.4000	722	1086	1261	1354	1370	1547

Example 7.1 Cracked T section subjected to bending

The T section shown in Fig. 7.7(a) is subjected to a bending moment of 1000 kN-m (8850 kip-in). It is required to find the stress and strain distributions ignoring the concrete in tension. Effects of creep and shrinkage are not considered in this example. The cross-section dimensions are indicated in Fig. 7.7(a); $E_c = 30$ GPa (4350 ksi); $E_s = 200$ GPa (29 000 ksi).

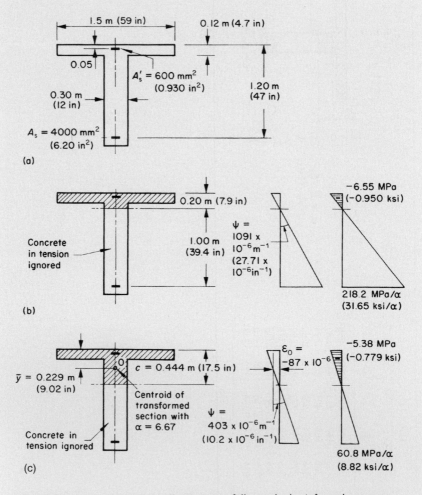

Figure 7.7 Strain and stress distributions in a fully cracked reinforced concrete section (Examples 7.1 and 7.2): (a) cross-section dimensions; (b) effective area, strain and stress due to bending (Example 7.1); (c) effective area, strain and stress due to bending and normal force (Example 7.2).

$$a = \frac{E_s}{E_c} = \frac{200}{30} = 6.667.$$

In the absence of prestress steel, $A_{ps} = 0$ and the symbols A_{ns} and A'_{ns} have the same meaning as A_s and A'_s.

Substitution in Equations (7.17–19) gives:

$$a_1 = 0.15\,\text{m} \qquad a_2 = 174.07 \times 10^{-3}\,\text{m}^2 \qquad a_3 = -40.812 \times 10^{-3}\,\text{m}^3$$

Equation (7.16) gives the depth of the compression zone

$$c = \frac{-174.07 \times 10^{-3} + \sqrt{[(174.07 \times 10^{-3})^2 + 4(0.15)(40.812 \times 10^{-3})]}}{2 \times 0.15}$$

$$= 0.200\,\text{m} \quad (7.9\,\text{in}).$$

The moment of inertia of the transformed section about the centroidal axis (which is the same as the neutral axis):

$$I = 0.3 \cdot \frac{0.200^3}{3} + (1.5 - 0.3)0.12\left(\frac{0.12^2}{12} + 0.14^2\right)$$

$$+ 6.667(0.004)1.000^2 + 5.667(0.0006)0.15^2$$

$$= 30.54 \times 10^{-3}\,\text{m}^4 \quad (3.53\,\text{ft}^4).$$

Alternatively, if A'_s is ignored, Tables 7.1 and 7.4 can be used giving $c = \bar{y} = 0.202\,\text{m}$ and $I = 30.46 \times 10^{-3}\,\text{m}^4$. The curvature

$$\psi = \frac{1000 \times 10^3}{30 \times 10^9 \times 30.54 \times 10^{-3}} = 1091 \times 10^{-6}\,\text{m}^{-1}\,(28 \times 10^{-6}\,\text{in}^{-1}).$$

Stress at the top fibre $= 30 \times 10^9 \times 1091 \times 10^{-6}(-0.200)$

$$= -6.55\,\text{MPa} \quad (-0.950\,\text{ksi}).$$

Stress in steel $= 200 \times 10^9 \times 1091 \times 10^{-6}(1.000)$

$$= 218.2\,\text{MPa} \quad (31.65\,\text{ksi}).$$

Strain and stress distributions are shown in Fig. 7.7(b).

Example 7.2 Cracked T section subjected to M and N

Solve Example 7.1, assuming that the section is subjected to a bending moment of 1000 kN-m (8850 kip-in) and a normal force of −800 kN (−180 kip) at a point 1.0 m (40 in) below the top edge of the section. The cross-section dimensions and moduli of elasticity of steel and concrete are the same as in Example 7.1 (Fig. 7.7(a)).

The resultant force on the section is a normal force of −800 kN at a distance 0.25 m above the top edge. Thus, $e_s = -(0.25 + 1.20) = -1.45$ m. Substituting in Equation (7.20) and solving for c, the height of the compression zone, gives:

$$c = 0.444\,\text{m} \qquad (17.5\,\text{in}).$$

The effective area is shown in Fig. 7.7(c). The transformed section is composed of the area of concrete in compression plus $\alpha(A_s + A'_s)$ with $\alpha = 200/30 = 6.667$. The distance between point O, the centroid of the transformed section, and the top edge is calculated to be $\bar{y} = 0.229$ m (Fig. 7.7(c)). The area and moment of inertia of the transformed section about an axis through its centroid

$$A = 0.3073\,\text{m}^2 \quad I = 31.73 \times 10^{-3}\,\text{m}^4.$$

If A'_s is ignored, Tables 7.1, 7.3 and 7.4 may be used, giving:

$$c = 0.46\,\text{m} \quad \bar{y} = 0.24\,\text{m} \quad I = 30 \times 10^{-3}\,\text{m}^4.$$

Transform the given bending moment and normal force into an equivalent system of a normal force N at the centroid of the transformed section combined with a bending moment M.

$$N = -800\,\text{kN}$$

$$M = 1000 \times 10^3 - 800 \times 10^3(1.000 - 0.229)$$

$$= 383.2\,\text{kN m} \quad (3400\,\text{kip in}).$$

The strain at O and the curvature (Equation (2.16))

$$\varepsilon_O = \frac{1}{30 \times 10^9} \frac{-800 \times 10^3}{0.3073} = -87 \times 10^{-6}$$

$$\psi = \frac{1}{30 \times 10^9} \frac{383.2 \times 10^3}{31.73 \times 10^{-3}}$$

$$= 403 \times 10^{-6}\,\mathrm{m}^{-1} \quad (10.2 \times 10^{-6}\,\mathrm{in}^{-1}).$$

Stress at the top fibre $= 30 \times 10^9[-87 + 403(-0.229)]10^{-6} = -5.38$ MPa.
Stress in bottom steel $= 200 \times 10^9[-87 + 403 \times 0.971]10^{-6} = 60.8$ MPa.
The strain and stress distributions are shown in Fig. 7.7(c).

7.5   Effects of creep and shrinkage on a reinforced concrete section without prestress

Consider a cross-section cracked due to the application of a positive bending moment M and an axial tensile (or compressive) force N at an arbitrarily chosen point O (Fig. 7.1(a)). The internal forces M and N are assumed to have been introduced at age t_0. The instantaneous strain and stress distribu-tions immediately after application of M and N are assumed to be available (see Section 7.4). It is required to find the changes in strain and in stress due to creep and shrinkage occurring between t_0 and t, where $t > t_0$.

In a fully cracked section, only the part of the concrete area subjected to compression is considered effective in resisting the internal forces. Creep and shrinkage generally result in a shift of the neutral axis towards the bottom of the section. Thus, to be strictly consistent, the effective area of the cross-section must be modified according to the new position of the neutral axis. However, this would hamper the validity of the superposition involved in the analysis. To avoid this difficulty, the effective area of the cracked section is assumed to be unchanged by creep or shrinkage. The error resulting from this assumption can be assessed at the end of the analysis and corrected by iter-ation procedure. But, because the error is usually small, the iteration is hardly justified.

With the above simplification, the analysis for the changes in axial strain and in curvature and the corresponding stresses can be done by the procedure given in Section 2.5.2. The resulting equations are given in Section 3.4 and repeated here.

A reference point O is chosen at the *centroid of the age-adjusted trans-formed section*, composed of the area of the compression zone plus $\bar{a}(t, t_0)$ times the area of steel (Figs. 7.8 and 7.9(a)); where $\bar{a}(t, t_0) = E_s/\bar{E}_c(t, t_0)$, with $\bar{E}_c(t, t_0)$ the age-adjusted modulus of elasticity of concrete (see Equation (1.31)). Creep and shrinkage produce the following changes in axial strain at O, in curvature and in stresses:

$$\Delta \varepsilon_O = \eta[\varphi(t, t_0)(\varepsilon_O + \psi y_c) + \varepsilon_{cs}(t, t_0)] \tag{7.26}$$

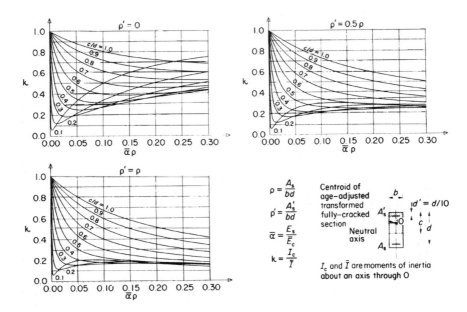

Figure 7.8 Curvature reduction κ for a fully cracked rectangular section.

$$\Delta\psi = \kappa\left[\varphi(t,\,t_0)\left(\psi + \varepsilon_0\frac{y_c}{r_c^2}\right) + \varepsilon_{cs}(t,\,t_0)\frac{y_c}{r_c^2}\right] \tag{7.27}$$

$$\Delta\sigma_c = \bar{E}_c(t,\,t_0)[-\varphi(t,\,t_0)(\varepsilon_0 + \psi y) - \varepsilon_{cs}(t,\,t_0) + \Delta\varepsilon_0 + \Delta\psi y] \tag{7.28}$$

$$\Delta\sigma_s = E_s(\Delta\varepsilon_0 + \Delta\psi y_s) \tag{7.29}$$

where

    $\varepsilon_0,\ \psi$ = the axial strain at O and the curvature at time t_0 immediately after application of M and N (Fig. 7.9(b))

    $\varphi(t,\,t_0)$ = coefficient for creep at time t for age at loading t_0

    $\varepsilon_{cs}(t,\,t_0)$ = the shrinkage that would occur in concrete if it were free, during the period $(t - t_0)$

    y_c = the y-coordinate of the centroid of the concrete area in compression (based on the stress distribution at age t_0). y_c is measured downwards from O

$$r_c^2 = I_c/A_c \tag{7.30}$$

with A_c and I_c being the area of the compression zone and its moment of inertia about an axis through O.

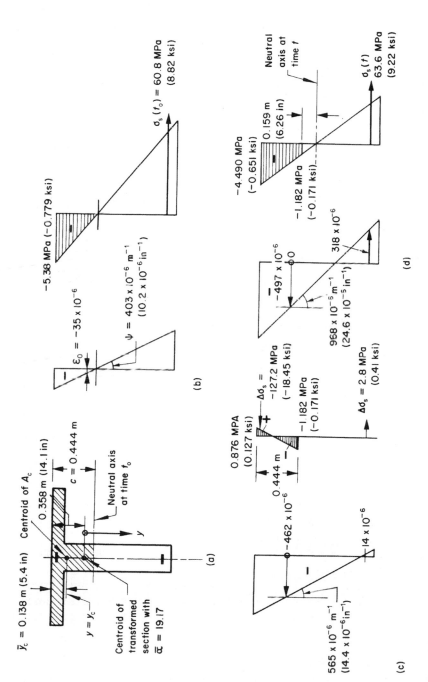

Figure 7.9 Analysis of changes in strain and stress due to creep and shrinkage of a fully cracked reinforced concrete section (Example 7.3): (a) effective area; (b) strain and stress distributions at t_0; (c) changes in strain and stress due to creep and shrinkage; (d) strain and stress at t.

Table 7.5 Curvature reduction factor κ for a fully cracked T section (for use in Equation (6.23)) κ = coefficient from table × 10^{-3}

$b_w/b = 0.05$

$\bar{a}\rho$	c/d = .10					c/d = .20					c/d = .30					c/d = .50					c/d = .75					c/d = 1.00				
	.05	.10	.15	.20	.30	.05	.10	.15	.20	.30	.05	.10	.15	.20	.30	.05	.10	.15	.20	.30	.05	.10	.15	.20	.30	.05	.10	.15	.20	.30
.0005	43	160	160	160	160	170	236	420	623	623	408	426	522	661	862	774	773	781	809	891	928	927	927	929	942	973	971	971	971	973
.0010	36	94	94	94	94	105	140	270	455	455	266	276	356	496	758	636	632	643	681	804	868	865	864	868	892	948	945	944	943	947
.0030	59	58	58	58	58	80	75	126	229	229	146	135	171	258	516	394	381	387	425	582	700	691	686	691	736	865	857	852	851	860
.0050	90	65	65	65	65	98	74	98	165	165	136	111	128	185	397	312	289	289	318	461	600	584	576	580	630	801	788	780	777	789
.0075	127	81	81	81	81	128	86	92	134	134	149	110	112	149	315	274	240	233	252	372	524	500	488	489	539	741	721	710	705	717
.0100	161	100	100	100	100	157	101	96	123	123	170	118	110	134	267	263	219	206	217	318	479	447	431	429	474	697	670	655	649	660
.0125	193	118	118	118	118	186	118	103	120	120	192	130	114	129	237	263	211	193	198	282	451	412	392	388	427	663	631	613	604	614
.0150	223	136	136	136	136	213	134	112	121	121	215	143	121	128	218	270	210	187	187	257	435	389	366	358	392	637	600	579	568	576
.0200	276	171	171	171	171	262	167	133	130	130	258	171	138	135	196	292	219	187	180	228	420	363	333	321	344	602	556	529	515	518

$b_w/b = 0.1$

$\bar{a}\rho$	c/d = .10					c/d = .20					c/d = .30					c/d = .50					c/d = .75					c/d = 1.00				
	.05	.10	.15	.20	.30	.05	.10	.15	.20	.30	.05	.10	.15	.20	.30	.05	.10	.15	.20	.30	.05	.10	.15	.20	.30	.05	.10	.15	.20	.30
.0010	41	94	94	94	94	160	179	286	455	455	387	387	430	530	758	758	759	760	771	834	925	925	924	924	930	973	972	971	971	971
.0020	47	63	63	63	63	107	111	175	300	300	256	250	281	366	613	618	617	616	631	717	863	863	860	860	870	949	947	945	944	944
.0060	102	71	71	71	71	111	84	98	149	149	164	139	144	183	358	390	377	370	380	469	696	691	684	681	697	869	863	857	854	853
.0100	156	100	100	100	100	149	103	97	123	123	174	134	125	145	267	324	300	286	290	360	602	590	579	573	589	809	798	790	784	782
.0150	215	136	136	136	136	198	133	113	121	121	204	150	130	135	218	301	265	245	241	292	535	516	499	490	501	754	738	725	717	712
.0200	267	171	171	171	171	243	164	133	130	130	239	172	143	139	196	303	257	230	221	256	499	471	451	438	444	714	693	677	666	659
.0250	313	203	203	203	203	284	193	154	142	142	273	196	160	148	186	314	260	228	214	237	479	445	420	405	405	686	660	640	626	616
.0300	353	233	233	233	233	321	221	174	157	157	305	219	177	160	184	330	268	232	214	226	469	429	401	383	378	664	634	612	596	583
.0400	421	288	288	288	288	385	271	214	186	186	362	265	213	186	189	365	292	249	224	220	466	417	382	359	345	638	601	573	553	534

Table 7.5 continued

$b_w/b = 0.2$

$\bar{a}\rho$	$c/d = .10$					$c/d = .20$					$c/d = .30$					$c/d = .50$					$c/d = .75$					$c/d = 1.00$				
	\(h_f/d \) .05	.10	.15	.20	.30	.05	.10	.15	.20	.30	.05	.10	.15	.20	.30	.05	.10	.15	.20	.30	.05	.10	.15	.20	.30	.05	.10	.15	.20	.30
.0020	49	63	63	63	63	150	148	194	300	300	356	355	366	417	613	733	741	739	741	775	919	921	921	920	921	973	973	972	972	971
.0040	71	61	61	61	61	114	103	124	189	189	241	233	238	274	448	590	596	593	594	637	853	857	856	854	855	949	948	947	945	944
.0120	169	114	114	114	114	153	115	105	120	120	185	158	146	154	242	378	373	363	359	391	684	686	681	676	675	869	867	863	859	854
.0200	251	171	171	171	171	214	157	132	130	130	215	174	152	147	196	329	313	296	287	305	595	592	583	575	571	811	807	800	794	786
.0300	334	233	233	233	233	282	210	172	157	157	263	209	178	164	184	322	295	273	258	263	536	527	514	502	493	760	752	742	733	722
.0400	400	288	288	288	288	340	257	211	186	186	310	247	208	187	189	336	301	273	254	243	507	492	475	461	446	723	712	700	689	674
.0500	455	335	335	335	335	390	300	246	216	216	353	283	238	212	200	355	315	283	260	246	493	474	454	437	417	698	684	669	656	638
.0600	500	376	376	376	376	433	338	280	244	244	391	316	268	236	215	378	332	297	271	250	489	466	443	423	400	680	663	647	632	611
.0800	571	445	445	445	445	503	403	338	295	295	457	376	321	283	247	424	371	330	299	267	496	466	438	415	384	659	638	617	599	573

$b_w/b = 0.5$

$\bar{a}\rho$	$c/d = .10$					$c/d = .20$					$c/d = .30$					$c/d = .50$					$c/d = .75$					$c/d = 1.00$				
	\(h_f/d \) .05	.10	.15	.20	.30	.05	.10	.15	.20	.30	.05	.10	.15	.20	.30	.05	.10	.15	.20	.30	.05	.10	.15	.20	.30	.05	.10	.15	.20	.30
.0050	74	65	65	65	65	135	130	135	165	165	305	309	307	316	397	691	701	702	701	707	909	912	912	911	972	972	972	972	972	971
.0100	123	100	100	100	100	124	114	110	123	123	215	213	209	212	267	542	552	553	551	556	837	842	843	842	841	947	947	947	946	945
.0300	287	233	233	233	233	211	184	166	157	157	206	192	180	172	184	354	355	352	347	344	661	667	667	665	660	866	866	866	865	864
.0500	401	335	335	335	335	296	260	234	216	216	259	238	220	207	200	324	340	313	305	296	576	580	579	575	567	808	808	807	804	799
.0750	500	429	429	429	429	382	340	307	283	283	325	298	276	258	239	335	326	315	304	289	525	526	523	517	506	758	758	755	752	744
.1000	572	500	500	500	500	449	405	369	341	341	382	352	327	306	279	360	347	334	321	302	503	502	496	490	476	724	723	719	715	705
.1250	625	556	556	556	556	504	458	421	391	391	431	400	372	349	318	388	374	358	344	321	496	493	486	478	462	701	699	694	689	678
.1500	667	600	600	600	600	549	503	465	434	434	474	441	413	388	353	417	400	384	368	343	497	492	484	475	458	685	682	677	671	659
.2000	727	667	667	667	667	618	574	536	504	504	542	509	480	454	415	470	451	433	415	387	512	504	494	484	463	669	664	658	650	636

Table 7.5 continued

$b_w/b = 1.0$

$\bar{\alpha}\rho$	.1	.2	.3	.5	.75	1.0
			c/d			
.0100	100	123	267	660	902	971
.0200	171	130	196	510	826	945
.0600	376	244	215	336	646	863
.1000	500	341	279	318	563	805
.1500	600	434	353	337	515	755
.2000	667	504	415	367	497	722
.2500	714	558	466	399	493	699
.3000	750	602	510	431	496	685
.4000	800	668	579	487	515	670

$\bar{\alpha} = E_s/\bar{E}_c$

$\zeta = A_s/(bd)$

$\bar{E}_c = \dfrac{E_c(t_0)}{1 + \chi\varphi(t, t_0)}$

η and κ are axial strain and curvature reduction factors given by:

$$\eta = A_c/\bar{A} \tag{7.31}$$

$$\kappa = I_c/\bar{I} \tag{7.32}$$

where $\bar{A}$ and $\bar{I}$ are the area and moment of inertia about an axis through O of an age-adjusted transformed section composed of A_c plus $\bar{a}(t, t_0)A_s$. The coefficient η and κ represent the restraining effect of the reinforcement on the axial strain and curvature due to creep and shrinkage.

Figure 7.8 and Table 7.5 give the values of κ for fully cracked rectangular and T sections, respectively. Use of Fig. 7.8 and Table 7.5 must be preceded by determination of c (from Fig. 7.4 and Table 7.1 or 7.2). Location of O, the centroid of the age-adjusted transformed fully cracked section may be determined by the graphs of Fig. 7.5 or Table 7.3 replacing a by $\bar{a}$.

Provided that the depth c is known, calculation of the axial strain reduction factor η by Equation (7.27) involves simple calculation; thus no tables or graphs are provided here for η.

7.5.1 Approximate equation for the change in curvature due to creep in a reinforced concrete section subjected to bending

An approximation Equation (3.27) is suggested in Section 3.5 for the curvature due to creep in a reinforced concrete section subjected to bending without axial force. Extension of use of this approximation for a cracked section would result in a relatively larger margin of error. This is so because the term $\varepsilon_{O}y_c/r_c^2$ for the cracked section is not negligible enough compared to ψ to justify ignoring the first of these two quantities when using Equation (7.27).

Example 7.3 Cracked T section: creep and shrinkage effects

Find the changes in strain and stress distributions due to creep and shrinkage in the cross-section of Example 7.2 (Fig. 7.7(a)). Consider that the result of Example 7.2 represents the stress and strain at age t_0 and use the following data:

$$\varphi(t, t_0) = 2.5 \quad \chi(t, t_0) = 0.75 \quad \varepsilon_{cs}(t, t_0) = -300 \times 10^{-6}.$$

The effective area of the section is considered unchangeable with time. Thus, using the result of Example 7.2, the depth of the effective part of the section $c = 0.444$ and the stress distribution at time t_0 is as shown in Fig. 7.7(c).

The area of the effective part of concrete, $A_c = 0.2766\,\mathrm{m}^2$. The distance of the centroid of A_c from top, $\bar{y}_c = 0.138\,\mathrm{m}$ (Fig. 7.9(a)).

The age-adjusted modulus of elasticity of concrete (Equation (1.31))

$$\bar{E}_c(t, t_0) = \frac{30 \times 10^9}{1 + 0.75 \times 2.5} = 10.43 \, \text{GPa} \quad (1500 \, \text{ksi})$$

$$\bar{a}(t, t_0) = \frac{200}{10.43} = 19.17.$$

The area of a transformed section composed of A_c plus $\bar{a}(A_s + A'_s)$ is

$$\bar{A} = 0.3648 \, \text{m}^2 \quad (560 \, \text{in}^2).$$

For use of Equations (7.26–31), a reference point O must be chosen at the centroid of the transformed effective area. This centroid is calculated and is found to be at $\bar{y} = 0.358 \, \text{m}$ below the top edge.

The moment of inertia of A_c about an axis through O is

$$I_c = 17.56 \times 10^{-3} \, \text{m}^4 \quad r_c^2 = I_c/A_c = 0.0635 \, \text{m}^2.$$

The moment of inertia of the transformed section is

$$\bar{I} = 73.01 \times 10^{-3} \, \text{m}^4.$$

The axial strain and curvature reduction factors (Equations (7.30) and (7.31)) are

$$\eta = \frac{0.2766}{0.3648} = 0.7582 \quad \kappa = \frac{17.56}{73.01} = 0.2404.$$

If the area A'_s is ignored, Tables 7.3 to 7.5 can be used to calculate $\bar{y}$, $\bar{I}$ and κ.

The y-coordinate of the centroid of A_c (see Fig. 7.9(a)) is

$$y_c = -(0.358 - 0.138) = -0.220 \, \text{m}.$$

The strain and stress distributions at time t_0 are shown in Fig. 7.9(b) (copied from the result of Example 7.2, Fig. 7.7(c)):

$$\varepsilon_O = -35 \times 10^{-6} \quad \psi = 403 \times 10^{-6} \, \text{m}^{-1}.$$

(Note that the reference point O is lower in Fig. 7.9(b) compared to Fig. 7.7(c).)

Changes in strain at O and in curvature due to creep and shrinkage (Equations (7.26) and (7.27)) are

$$\Delta\varepsilon_O = 0.7582\{2.5[-35 + 403(-0.22)]10^{-6} - 300 \times 10^{-6}\} = -462 \times 10^{-6}$$

$$\Delta\psi = 0.2404\left[2.5\left(403 - 35\frac{(-0.22)}{0.0635}\right)10^{-6} - 300 \times 10^{-6}\frac{(-0.22)}{0.0635}\right]$$

$$= 565 \times 10^{-6}\text{m}^{-1}$$

Changes in concrete stresses due to creep and shrinkage (Equation (7.28)) are at the top edge;

$$(\Delta\sigma_c)_{top} = 10.43 \times 10^9 \{-2.5[-35 + 403(-0.358)]$$

$$+ 300 - 462 + 565(-0.358)\} 10^{-6}$$

$$= 0.876\,\text{MPa}\quad(0.127\,\text{ksi})$$

at the lower edge of the effective area;

$$(\Delta\sigma_c)_{\text{at } 0.444\text{m below top edge}} = 10.43 \times 10^9(300 - 462 + 565 \times 0.086)10^{-6}$$

$$= -1.182\,\text{MPa}\quad(-0.171\,\text{ksi}).$$

Changes in stress in steel due to creep and shrinkage (Equation (7.29)) are:

$$(\Delta\sigma_s)_{bot} = 200 \times 10^9 (-462 + 565 \times 0.842)10^{-6} = 2.8\,\text{MPa}\,(0.41\,\text{ksi})$$

$$(\Delta\sigma_s)_{top} = 200 \times 10^9 (-462 - 565 \times 0.308)10^{-6}$$

$$= -127.2\,\text{MPa}\quad(-18.45\,\text{ksi}).$$

The changes in strain and stress distributions due to creep and shrinkage are shown in Fig. 7.9(c). The final strain and stress distributions at time t are obtained by summing up the values in Figs. 7.9(b) and (c); the results are shown in Fig. 7.9(d).

From the stress distribution in Fig. 7.9(d), it is seen that the neutral axis has moved downwards due to the effects of creep and shrinkage. Thus, according to the above solution, a part of the area of the web, of height 0.159 m above the neutral axis at time t, is ignored although it would have been subjected to compressive stress. Because the ignored area is close to the neutral axis, the error involved is small.

7.6   Partial prestressed sections

Consider a prestressed concrete section which is also reinforced by non-prestressed steel. The prestress is applied at age t_0 at which time a part of the dead load is also introduced and, shortly after, a superimposed dead load is applied. At a much later date t, the live load comes into effect and produces cracking. What is the procedure of analysis to determine the strain and stress distributions at age t after cracking? The term *partial prestressing* is used throughout this book to refer to the case when the prestressing forces are not sufficient to prevent cracking at all load stages.

We shall assume here that all the time-dependent changes due to creep and shrinkage of concrete and relaxation of prestressed steel take place prior to age t and that no cracking occurs up to this date. Thus, the method of analysis presented in Section 2.5 for uncracked sections can be applied to determine the strain and stress distributions at age t just before application of the live load. The problem that needs to be discussed in the present section may be stated as follows. Given the stress distribution in an uncracked section reinforced by prestressed and non-prestressed reinforcement, what are the instantaneous changes in stress and strain caused by the application of an additional bending moment and axial force causing cracking?

Figure 7.10(a) shows a cross-section with several layers of prestressed and non-prestressed reinforcement. At time t, the distribution of stress on the section is assumed to be known and $\sigma_c(t)$, the concrete stress, is assumed to vary linearly over the depth without producing cracking. This stress distribution may be completely defined by the stress value $\sigma_O(t)$ at an arbitrary reference point O and stress diagram slope, $\gamma(t) = d\sigma/dy$. The additional bending moment M and axial force N at O are applied, producing cracking of the section. It is required to find the changes in strain and in stress due to M and N.

Partition each of M and N in two parts, such that (see Fig. 7.10(c) and (e)):

$$M = M_1 + M_2 \tag{7.33}$$

$$N = N_1 + N_2 \tag{7.34}$$

M_1 and N_1 represent the part of the internal forces that will bring the stresses in the concrete to zero and M_2 and N_2 represent the remainder of the internal forces. With M_1 and N_1, the section is in state 1 (uncracked). Cracking is

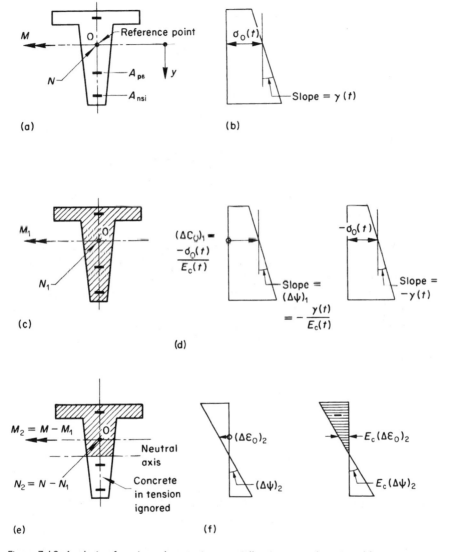

Figure 7.10 Analysis of strain and stress in a partially prestressed section: (a) cross-section
dimensions; (b) concrete stress $\sigma_c(t)$ immediately before application of M and
N; (c) decompression forces M_1 and N_1 on uncracked section; (d) strain and
stress changes due to application of M_1 and N_1; (e) M_2 and N_2 on a fully cracked
section; (f) strain and stress changes due to application of M_2 and N_2.

produced only by the part M_2 combined with N_2. Thus, for the analysis, two loading stages need to be considered:

(1) M_1 and N_1 applied on uncracked section;
(2) M_2 and N_2 applied on a fully cracked section.

The strain changes in the two stages are given by Fig. 7.10(d) and (f):

$$(\Delta\varepsilon)_1 = (\Delta\varepsilon_O)_1 + (\Delta\psi)_1 y \tag{7.35}$$

$$(\Delta\varepsilon)_2 = (\Delta\varepsilon_O)_2 + (\Delta\psi)_2 y \tag{7.36}$$

The total instantaneous change in strain due to M and N is

$$\Delta\varepsilon = (\Delta\varepsilon)_1 + (\Delta\varepsilon)_2 \tag{7.37}$$

The stress produced in stage 1 is simply equal to the stress in Fig. 7.10(b) reversed in sign, as shown in Fig. 7.10(d). The corresponding strain in stage 1 is obtained by division of stress values by $E_c(t)$; the strain distribution in stage 1 is also shown in Fig. 7.10(d). Thus, the stress in concrete is zero after application of M_1 and N_1. The final stress in concrete is given by the analysis of the effects of M_2 and N_2 only (Fig. 7.10(f)). It should, however, be noted that M_1 and N_1 bring to zero the stress in concrete but not in steel.

The values of M_1 and N_1 are equal and opposite to the resultants of stresses $\sigma_c(t)$ on the concrete and a times this stress on steel, with $\sigma_c(t)$ being the stress existing before application of M and N (Fig. 7.10(b)). M_1 and N_1 are sometimes referred[1] to as decompression forces, because $\sigma_c(t)$ is generally compressive. (In all the stress and strain diagrams in Fig. 7.10, the variables ε_O, ψ, σ_O and γ are plotted as positive quantities.) The decompression forces are given by:

$$N_1 = -\int \sigma dA \tag{7.38}$$

$$M_1 = -\int \sigma y dA \tag{7.39}$$

When the stress varies over the full height of the section as *one* straight line, the integrals in Equations (7.38) and (7.39) may be eliminated (see Equations (2.2–8)):

$$N_1 = -(A\sigma_O + B\gamma) \tag{7.40}$$

$$M_1 = -(B\sigma_O + I\gamma) \tag{7.41}$$

where A is the area of a transformed section composed of the full concrete area plus a times the area of steel, prestressed and non-prestressed;

$a = E_s/E_c(t)$ with E_s and $E_c(t)$ being the moduli of elasticity of steel and of concrete at the time of application of M and N. B and I are the first and second moments of the same transformed area about an axis through the reference point O. $\sigma_O = \sigma_O(t)$ is the stress in concrete at the reference point O at time t immediately before application of the live load; $\gamma = \gamma(t)$ is the slope of the stress diagram

$$\gamma = \gamma(t) = \frac{d}{dy}\,\sigma(t) \qquad\qquad (7.42)$$

If O is chosen at the centroid of the above-mentioned transformed area, $B = 0$ and Equations (7.40) and (7.41) become:

$$N_1 = -A\sigma_O \qquad\qquad (7.43)$$

$$M_1 = -I\gamma \qquad\qquad (7.44)$$

The changes in axial strain and curvature due to M_1 and N_1 simply are:

$$(\Delta\varepsilon_O)_1 = -\frac{1}{E_c}\,\sigma_O \qquad\qquad (7.45)$$

$$(\Delta\psi)_1 = -\frac{1}{E_c}\,\gamma \qquad\qquad (7.46)$$

The strain and stress distributions due to M_2 and N_2 require more elaborate calculation, following the procedure for a cracked section presented in Section 7.4.

In a composite section, made of more than one type of concrete, the distribution of stress $\sigma(t)$ is generally represented by one straight line for each part of the section and thus Equations (7.40) and (7.41) must be adjusted. If we assume that cracking occurs only in one part, say part i of the cross-section, the values σ_O and γ in Equations (7.40) and (7.41) are to be substituted by σ_i and γ_i which define one straight line of distribution of stress $\sigma(t)$ over part i; other non-cracked parts are to be treated in the same way as the non-prestressed steel, but using appropriate moduli of elasticity $E_c(t)$. The forces N_1 and M_1 calculated in this way represent the decompression forces which will bring the stress in concrete to zero in part i of the section; the stress in other parts will change but will not necessarily become zero.

7.7 Flow chart

The steps of analysis of the strain and stress presented in Chapter 2 and the present chapter apply to the whole range from reinforced concrete without prestressing to fully prestressed concrete where no cracking is allowed.

The flow chart in Fig. 2.14 shows how the procedures discussed in the two chapters can be applied in a general case to determine the instantaneous and time-dependent changes in strain and stress due to the application at time t_0 of a normal force N and a bending moment M on a section for which the initial strain and stress are known.

Example 7.4 Pre-tensioned tie before and after cracking

Fig. 7.11 shows a square cross-section of a precast pretensioned tie. Immediately before transfer, the force in the tendon is 1100 kN (247 kip), the age of concrete t_0 and no dead load is simultaneously applied with the prestress. At a much older age t, a normal tensile force 1200 kN (270 kip) is applied at the centre of the section. It is required to find the axial strain and stress in the concrete and steel immediately after prestressing, and just before and after application of the 1200 kN force. The following data are given: the moduli of elasticity of concrete and steel, $E_c(t_0) = 24\,\text{GPa}$ (3480 ksi); $E_c(t) = 35\,\text{GPa}$ (5076 ksi); $E_s = 200\,\text{GPa}$ (29 000 ksi) (for prestressed and non-prestressed reinforcements); creep coefficient $\varphi(t, t_0) = 2.4$; aging coefficient $\chi(t, t_0) = 0.80$; during the period $(t - t_0)$, the reduced relaxation $\Delta\bar{\sigma}_{pr} = -90\,\text{MPa}$ (−13 ksi) and the free shrinkage $\varepsilon_{cs}(t, t_0) = -270 \times 10^{-6}$.

(a) Strain and stress immediately after transfer
The area of the transformed section is composed of $A_c + \alpha(A_{ps} + A_{ns})$, where $\alpha = E_s/E_c(t_0)$.

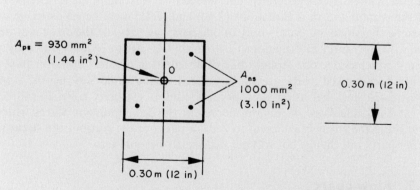

Figure 7.11 Cross-section of a partially prestressed tie analysed for strain and stress in Example 7.4.

$$A_c = 0.30 \times 0.30 - (930 + 1000)10^{-6} = 0.0881\,\text{m}^2$$

$$a = 200/24 = 8.33$$

$$A = 0.0881 + 8.33(930 + 1000)10^{-6} = 0.1042\,\text{m}^2.$$

The axial strain at transfer (Equation (2.33)) is

$$\varepsilon(t_0) = -\frac{1100 \times 10^3}{24 \times 10^9 \times 0.1042} = -440 \times 10^{-6}.$$

The stress in concrete (Equation (2.35)) is

$$\sigma(t_0) = 24 \times 10^9(-440 \times 10^{-6}) = -10.559\,\text{MPa} \quad (-1.532\,\text{ksi}).$$

The stress in non-prestressed and in prestressed steel is

$$\sigma_{ns} = 200 \times 10^9(-440 \times 10^{-6}) = -88.0\,\text{MPa} \quad (-12.8\,\text{ksi})$$

$$\sigma_{ps} = \frac{1100 \times 10^3}{930 \times 10^{-6}} + 200 \times 10^9(-440 \times 10^{-6})$$

$$= 1094.8\,\text{MPa} \quad (158.8\,\text{ksi}).$$

(b) Changes in strain and in stress due to creep, shrinkage and relaxation
The transformed section to be used here is composed of $A_c + \bar{a}(A_{ps} + A_{ns})$; where $\bar{a} = E_s/\bar{E}_c(t, t_0)$
   Using Equation (1.31)

$$\bar{E}_c = \frac{24 \times 10^9}{1 + 2.4 \times 0.8} = 8.215\,\text{GPa} \quad (1192\,\text{ksi})$$

$$\bar{a} = \frac{200}{8.215} = 24.33.$$

The transformed area

$$\bar{A} = 0.0881 + 24.33(930 + 1000)10^{-6} = 0.1351\,\text{m}^2.$$

   The artificial force that would be necessary to prevent strain due to creep, shrinkage and relaxation (Equations (2.41–44)) is

$$\Delta N = -8.215 \times 10^9 \times 2.4 \times 0.0881 \quad (-440 \times 10^{-6})$$

$$-8.215 \times 10^9(-270 \times 10^{-6})0.0881 + 930 \times 10^{-6}(-90 \times 10^6)$$

$$= 0.8759 \times 10^6 \, \mathrm{N} \quad (196.9 \, \mathrm{kip}).$$

The change in axial strain in concrete when the restraint is removed (Equation (2.40)) is

$$\Delta \varepsilon = -\frac{0.8758 \times 10^6}{8.215 \times 10^9 \times 0.1351} = -789 \times 10^{-6}.$$

The change in concrete stress (Equations (2.45) and (2.46)) is

$$\Delta \sigma = -8.215 \times 10^9[2.4(-440 \times 10^{-6}) - 270 \times 10^{-6}]$$

$$+ 8.215 \times 10^9(-789 \times 10^{-6})$$

$$= 4.407 \, \mathrm{MPa} \quad (0.6392 \, \mathrm{ksi}).$$

Changes in stress in non-prestressed and prestressed steels (Equations (2.47) and (2.48)) are

$$\Delta \sigma_{\mathrm{ns}} = 200 \times 10^9(-789 \times 10^{-6}) = -157.9 \, \mathrm{MPa} \quad (-22.90 \, \mathrm{ksi})$$

$$\Delta \sigma_{\mathrm{ps}} = -90 \times 10^6 + 200 \times 10^9(-789 \times 10^{-6})$$

$$= -247.9 \, \mathrm{MPa} \quad (-35.95 \, \mathrm{ksi}).$$

The stress in concrete after creep, shrinkage and relaxation is

$$\sigma(t) = -10.559 + 4.407 = -6.152 \, \mathrm{MPa} \quad (-0.8923 \, \mathrm{ksi}).$$

(c) Changes in strain and stress in the decompression stage
The transformed area to be used here is composed of $A_{\mathrm{c}} + \alpha(A_{\mathrm{ps}} + A_{\mathrm{ns}})$; where $\alpha = E_s/E_c(t)$

$$\alpha = 200/35 = 5.71.$$

The transformed area is

$$A = 0.0881 + 5.71(930 + 1000)10^{-6} = 0.0991 \, \mathrm{m}^2.$$

The decompression force (Equation (7.43)) is

$$N_1 = -0.0991(-6.152 \times 10^6) = 609.8 \, \text{kN} \quad (137.1 \, \text{kip}).$$

The change in strain due to N_1 (Equation (7.45)) is

$$(\Delta\varepsilon)_1 = \frac{6.152 \times 10^6}{35 \times 10^9} = 176 \times 10^{-6}.$$

The change in stress in the two types of reinforcement is

$$(\Delta\sigma_{\text{ns}})_1 = (\Delta\sigma_{\text{ps}})_1 = 200 \times 10^9 \times 176 \times 10^{-6} = 35.2 \, \text{MPa} \quad (5.11 \, \text{ksi}).$$

(d) Changes in strain and stress in the cracking stage
All the concrete area will be in tension; thus, the transformed area is composed of $\alpha(A_{\text{ps}} + A_{\text{ns}})$, with α the same as in (c) above.
Transformed area is

$$A = 5.71(930 + 1000)10^{-6} = 0.0110 \, \text{m}^2.$$

Force producing cracking (Equation (7.34)) is

$$N_2 = 1200 - 609.8 = 590.2 \, \text{kN} \quad (113 \, \text{kip}).$$

The change in strain due to N_2 (Equation (2.16)) is

$$(\Delta\varepsilon)_2 = \frac{590.2 \times 10^3}{35 \times 10^9 \times 0.0110} = 1530 \times 10^{-6}.$$

The change in stress in any of the two types of reinforcement is

$$(\Delta\sigma_{\text{ns}})_2 = (\Delta\sigma_{\text{ps}})_2 = 200 \times 10^9 \times 1530 \times 10^{-6}.$$

$$= 306.0 \, \text{MPa} \quad (44.4 \, \text{ksi}).$$

In this example, the entire concrete area is ineffective in stage 2 and N_2 is resisted only by the steel with cross-section $A_{\text{ps}} + A_{\text{ns}} = 1930 \times 10^{-6} \, \text{m}^2$. With $E_{\text{s}} = 200 \, \text{GPa}$ for the two types of steel, the strain increment $(\Delta\varepsilon)_2$ can also be calculated as follows:

$$(\Delta\varepsilon)_2 = \frac{590.2 \times 10^3}{200 \times 10^9 \times 1930 \times 10^{-6}} = 1530 \times 10^{-6}.$$

The stress in non-prestressed steel is

$$-88.0 - 157.9 + 35.2 + 306.0 = 95.3 \, \text{MPa} \quad (13.8 \, \text{ksi}).$$

The stress in prestressed steel is

$$1094.8 - 247.9 + 35.2 + 306.0 = 1188.1 \, \text{MPa} \quad (172.3 \, \text{ksi}).$$

The strain in the non-prestressed steel immediately before cracking is the sum of strain values calculated in steps (a), (b) and (c) = $(-440 - 789 + 176)10^{-6} = -1053 \times 10^{-6}$. At cracking, the change in strain in the pre-stressed or non-prestressed steel is $(\Delta\varepsilon)_2 = 1530 \times 10^{-6}$. Thus, the strain in the non-prestressed steel after cracking is 477×10^{-6}. At this stage the concrete is not participating in resisting any force. The strains in concrete and steel are no more compatible and slip must occur in the vicinity of cracks. This will be discussed further in Chapter 8.

Example 7.5 Pre-tensioned section in flexure: live-load cracking

Fig. 7.12(a) shows the cross-section of a pre-tensioned partially pre-stressed beam. A 700 kN-m (6200 kip-in) bending moment due to a dead load is introduced at age t_0 at the same time as the prestress transfer. This bending moment includes the effect of the superimposed dead load introduced shortly after transfer, but is considered here as if it were applied simultaneously with the prestress transfer. At time t, long after t_0, a live load is applied, producing a bending moment of 400 kN-m (3540 kip-in). Find the strain and stress distributions immediately after application of the live load bending moment, given the following data.

Tension in prestressed tendon before transfer = 1250 kN (281 kip); moduli of elasticity of concrete at ages t_0 and t, $E_c(t_0)$ = 24 GPa (3480 ksi) and $E_c(t)$ = 30 GPa (4350 ksi); E_s = 200 GPa (29 000 ksi) for all reinforcements; $\varphi(t, t_0)$ = 2.0; $\chi(t, t_0)$ = 0.8; reduced relaxation for the period $(t - t_0) = -90$ MPa (−13 ksi); shrinkage during the same period, $\varepsilon_{cs}(t, t_0) = -300 \times 10^{-6}$.

As in Example 7.4, the analysis may be done in five parts:

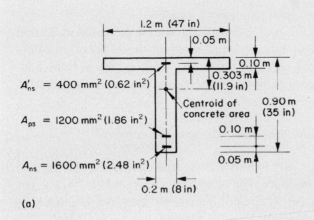

$A'_{ns} = 400 \text{ mm}^2 (0.62 \text{ in}^2)$

$A_{p3} = 1200 \text{ mm}^2 (1.86 \text{ in}^2)$

$A_{ns} = 1600 \text{ mm}^2 (2.48 \text{ in}^2)$

1.2 m (47 in)

0.05 m

0.10 m

0.303 m
(11.9 in)

Centroid of concrete area

0.90 m
(35 in)

0.10 m

0.05 m

0.2 m (8 in)

(a)

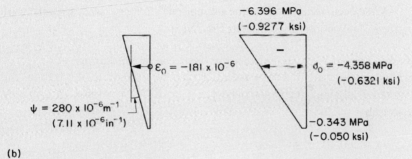

−6.396 MPa
(−0.9277 ksi)

$\varepsilon_0 = -181 \times 10^{-6}$

$\sigma_0 = -4.358 \text{ MPa}$
(−0.6321 ksi)

$\psi = 280 \times 10^{-6} \text{m}^{-1}$
$(7.11 \times 10^{-6} \text{in}^{-1})$

−0.343 MPa
(−0.050 ksi)

(b)

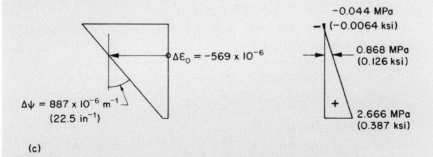

−0.044 MPa
(−0.0064 ksi)

$\Delta\varepsilon_0 = -569 \times 10^{-6}$

0.868 MPa
(0.126 ksi)

$\Delta\psi = 887 \times 10^{-6} \text{ m}^{-1}$
(22.5 in^{-1})

2.666 MPa
(0.387 ksi)

(c)

Figure 7.12 Analysis of instantaneous and time-dependent strain and stress in a pre-tensioned cross-section before cracking (Example 7.5); (a) cross-section dimensions; (b) strain and stress immediately after prestress transfer; (c) changes in stress and strain due to creep, shrinkage and relaxation.

(a) Strain and stress immediately after transfer
The calculations in this part follow the procedure presented in Section
2.3 and applied in Example 2.2. Thus, here only the results of the calculations are presented (Fig. 7.12(b)). The stress in the bottom non-
prestressed reinforcement, $\sigma_{ns} = -5.6\,\text{MPa}$ and in the prestressed steel,
$\sigma_{ps} = 1030.5\,\text{MPa}$.

*(b) Changes in strain and in stress due to creep, shrinkage
     and relaxation*
The analysis for this part follows the method discussed in Section 2.5
and applied in Example 2.2. The results are shown in Fig. 7.12(c). The
changes in stress in the bottom non-prestressed steel, $\Delta\sigma_{ns} = -16.8\,\text{MPa}$
and in the prestress steel, $\Delta\sigma_{ps} = -124.5\,\text{MPa}$.

   After occurrence of the time-dependent changes, the distribution of
stress $\sigma(t)$ becomes as shown in Fig. 7.13(b).

(c) Changes in strain and stress in the decompression stage
The transformed area to be used here is composed of A_c plus α times the
area of all reinforcements; where $\alpha = E_s/E_c(t)$; A_c = area of concrete
section = $0.2768\,\text{m}^2$:

$$\alpha = 200/30 = 6.667.$$

   Choose reference point O at the centroid of A_c, at 0.303 m below the
top edge (Fig. 7.13(a)). The moment of inertia of A_c about an axis
through O = $21.78 \times 10^{-3}\,\text{m}^4$; $A_c = 0.2768\,\text{m}^2$.
   The area of the transformed section, its first and second moments
about an axis through O are:

$$A = 0.2768 + 6.667(1600 + 1200 + 400)10^{-6} = 0.2981\,\text{m}^2$$

$$B = 6.667(1600 \times 0.547 + 1200 \times 0.447 - 400 \times 0.253)10^{-6}$$

$$= 8.734 \times 10^{-3}\,\text{m}^3$$

$$I = 21.78 \times 10^{-3} + 6.667(1600 \times 0.547^2$$

$$+ 1200 \times 0.447^2 + 400 \times 0.253^2)10^{-6}$$

$$= 26.74 \times 10^{-3}\,\text{m}^4.$$

The stress distribution in Fig. 7.13(b) may be defined by the value of
stress at O and the slope of diagram:

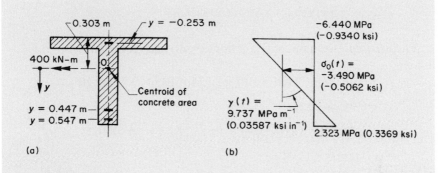

(a)   (b)

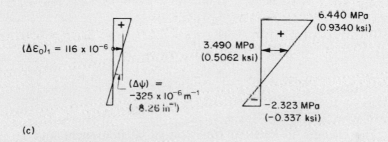

(c)

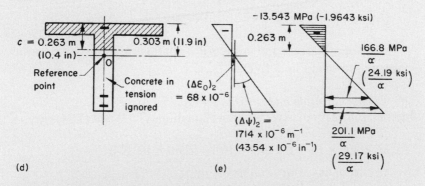

(d)   (e)

Figure 7.13 Changes in strain and stress in the cross-section of Fig. 7.12 due to a bending moment producing cracking (Example 7.5): (a) effective cross-section area before cracking; (b) stress at time *t* immediately before application of bending moment due to live load (Fig. 7.12(b), (c)); (c) strain and stress changes in the decompression stage; (d) effective cross-section area after cracking; (e) changes in strain and stress at cracking.

$\sigma_O(t) = -3.490\,\text{MPa} \quad \gamma(t) = 9.737\,\text{MPa/m}.$

The decompression forces (Equations (7.40) and (7.41)) are

$N_1 = - [0.2981(-3.490 \times 10^6) + 8.734 \times 10^{-3} \times 9.737 \times 10^6]$

$\quad = 0.955 \times 10^6 \text{N}$

$M_1 = - [8.734 \times 10^{-3}(-3.490 \times 10^6) + 26.74 \times 10^{-3} \times 9.737 \times 10^6]$

$\quad = -229.9 \times 10^3 \text{N-m}.$

The changes in strain at O and in curvature (Equations (7.45) and (7.46)) are

$(\Delta\varepsilon_O)_1 = \dfrac{3.490 \times 10^6}{30 \times 10^9} = 116 \times 10^{-6}$

$(\Delta\psi)_1 = -\dfrac{9.737 \times 10^6}{30 \times 10^9} = -325 \times 10^{-6}\,\text{m}^{-1}.$

The changes in stress in the bottom reinforcement and in the prestressed steel are:

$(\Delta\sigma_{ns})_1 = 200 \times 10^9 (116 - 325 \times 0.547)10^{-6} = -12.3\,\text{MPa}$

$(\Delta\sigma_{ps})_1 = 200 \times 10^9 (116 - 325 \times 0.447)10^{-6} = -5.8\,\text{MPa}.$

The changes in strain and in stress distributions in the decompression stage are shown in Fig. 7.13(c).

(d) Changes in strain and stress in the cracking stage
Internal forces producing cracking (Equations (7.33) and (7.34)) are

$M_2 = 400 \times 10^3 - (-229.9 \times 10^3) = 629.9 \times 10^3 \text{N-m}$

$N_2 = 0 - 0.955 \times 10^6 = -0.955 \times 10^6 \text{N}.$

Eccentricity of the resultant of M_2 and N_2 measured from the bottom reinforcement

$e_s = \dfrac{629.9 \times 10^3}{-0.955 \times 10^6} - 0.547 = -1.206\,\text{m}.$

Substitution in Equation (7.20) and solution by trial or use of Table 7.1 gives the depth of the compression zone (Fig. 7.13(d)):

$$c = 0.263 \, \text{m}.$$

The transformed section to be used here is composed of the area of concrete in compression plus a times the area of all reinforcements; where $a = E_s/E_c(t) = 200/30 = 6.667$.

The transformed area, its first and second moments about an axis through the reference point O are (Tables 7.3 and 4 may be used for this purpose):

$$A = 0.1736 \, \text{m}^2 \quad B = -25.484 \times 10^{-3} \, \text{m}^3 \quad I = 13.270 \times 10^{-3} \, \text{m}^4.$$

Changes in axial strain and in curvature produced by M_2 and N_2 (Equation (2.15) with $E_{\text{ref}} = 30 \, \text{GPa}$) are

$$(\Delta \varepsilon_O)_2 = 68 \times 10^{-6} \quad (\Delta \psi)_2 = 1714 \times 10^{-6} \, \text{m}^{-1}.$$

The distributions of strain and stress changes are shown in Fig. 7.13(e).

The changes in stress in the bottom reinforcement and in the prestress steel are:

$$(\Delta \sigma_{\text{ns}})_2 = 200 \times 10^9 (68 + 1714 \times 0.547) 10^{-6} = 201.1 \, \text{MPa} \quad (29.17 \, \text{ksi})$$

$$(\Delta \sigma_{\text{ps}})_2 = 200 \times 10^9 (68 + 1714 \times 0.447) 10^{-6} = 166.8 \, \text{MPa} \quad (24.19 \, \text{ksi})$$

(e) Strain and stress immediately after cracking
The stress diagram in Fig. 7.13(e), obtained by multiplying the strain diagram in the same figure by the value $E_c(t) = 30 \, \text{GPa}$, represents the final stress in concrete after cracking. The final stress in the reinforcement may be obtained by summing up the stress values calculated above in steps (a) to (d). Thus, the stress in the bottom non-prestressed steel is

$$-5.6 - 16.8 - 12.3 + 201.1 = 166.4 \, \text{MPa} \quad (24.13 \, \text{ksi}).$$

The stress in the prestressed steel is

$$1030.5 - 124.5 - 5.8 + 166.8 = 1067.0 \, \text{MPa} \quad (155 \, \text{ksi}).$$

Similarly, summing up the strains (Fig. 7.12(b) and (c) and Figs 7.13(c) and (e)) gives the strain at the reference point O:

$$(-181 - 569 + 116 + 68)10^{-6} = -566 \times 10^{-6}$$

and curvature

$$(280 + 887 - 325 + 1714)10^{-6} = 2556 \times 10^{-6} \, m^{-1} \, (64.92 \times 10^{-6} \, in^{-1}).$$

7.8   Example worked out in British units

Example 7.6 The section of Example 2.6: live-load cracking

The cross-section of Example 2.6 (Fig. 2.15) is subjected at time t to an additional bending moment = 9600 kip-in (1080 kN-m), representing the effect of live load. Determine the stress and strain distributions immediately after application of the live load moment. Consider $E_c(t) =$ 4000 ksi (28 GPa); other data are the same as in Example 2.6. Assume that cracking occurs due to the live load moment.

The strain and stress distributions existing at time t, before application of the live load have been determined in Example 2.6 (Fig. 2.15(c)). The stress parameters are:

$$\sigma_o(t) = -0.506 \, ksi; \qquad \gamma(t) = 0.0130 \, ksi/in.$$

(a) The decompression stage
Properties of the transformed section at time t are (assuming the reference point O at top fibre):

$$A = 1145 \, in^2; \qquad B = 19.43 \times 10^3 \, in^3; \qquad I = 533.6 \times 10^3 \, in^4$$

The decompression forces are (Equations (7.40) and (7.41)):

$$N_1 = -[1145(-0.506) + 19.43 \times 10^3(0.0130)] = 327 \, kip$$

$$M_1 = -[19.43 \times 10^3(-0.506) + 533.6 \times 10^3(0.0130)] = 2908 \, kip\text{-}in.$$

The strain changes by decompression are (Equations (7.45) and (7.46)):

$$(\Delta\varepsilon_O)_1 = -\frac{1}{4000}(-0.506) = 127 \times 10^{-6}$$

$$(\Delta\psi)_1 = -\frac{1}{4000}(0.0130) = -3.24 \times 10^{-6} \text{ in}^{-1}.$$

(b) The cracking state

The forces on the section due to live load are: $N = 0$; $M = 9600$ kip-in.
The forces to be applied on a fully cracked section are (Equations (7.33) and (7.34)):

$$N_2 = 0 - 327 = -327 \text{ kip}; \qquad M_2 = 9600 - 2908 = 6692 \text{ kip-in}.$$

This combination of forces is equivalent to a compressive force of -327 kip at 20.5 in above the top edge ($e_s = -57.5$ in; see Fig. 7.14(a))

The depth of the compressive zone $c = 13.9$ in (by solving Equation (7.20)). The properties of the fully cracked section are:

$$A = 635 \text{ in}^2; \qquad B = 5824 \text{ in}^3; \qquad I = 141.8 \times 10^3 \text{ in}^4.$$

The changes in strain and stress due to N_2 and M_2 are (Equations (2.19) and (2.20)):

$$(\Delta\varepsilon_O)_2 = -380 \times 10^{-6}; \qquad (\Delta\psi)_2 = 27.41 \times 10^{-6} \text{ in}^{-1}$$

$$(\Delta\sigma_O)_2 = -1.520 \text{ ksi}; \qquad (\Delta\gamma)_2 = 0.1096 \text{ ksi/in}.$$

The corresponding change in stress in the bottom non-prestressed steel layer is (Equation (2.17)):

$$(\Delta\sigma_{ns})_2 = 29\,000 \, [-380 \times 10^{-6} + 27.41 \times 10^{-6}(37)] = 18.39 \text{ ksi}.$$

Summing up the strain changes calculated above to the strain existing before the live load application (Fig. 2.15(c)) gives the total strain after cracking (Fig. 7.14(b)):

$$\varepsilon_O = -870 \times 10^{-6} + 127 \times 10^{-6} - 380 \times 10^{-6} = -1123 \times 10^{-6}$$

$$\psi = 12.59 \times 10^{-6} - 3.24 \times 10^{-6} + 27.41 \times 10^{-6} = 36.76 \times 10^{-6} \text{in}^{-1}$$

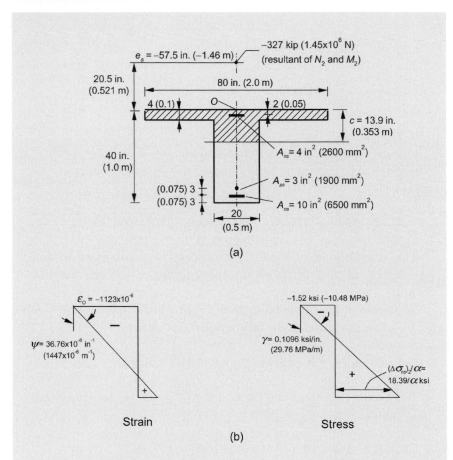

Figure 7.14 Analysis of strain and stress distributions in a partially prestressed cracked section (Example 7.6).

The total stress in concrete after the live load application is simply equal to the stress due to N_2 and M_2 (Fig. 7.14(b)).

7.9   General

The methods of analysis presented in this chapter give the axial strain and curvature in a fully cracked section, referred to as state 2. In this state the concrete in tension is fully ignored. It is well established that concrete, although weak in tension, participates in the rigidity of cracked members as will be discussed in Chapter 8. The axial strain and curvature calculated for a fully cracked section do not represent actual values from which displacements can be calculated. They only represent upper bounds of strain and curvature.

Lower bounds of strain and curvature can be obtained by assuming that there is no cracking and using the full area of the concrete section, regardless of stress value or its sign. The values of axial strain and curvature to be used in calculation of displacements are derived by interpolation between these two bounds. How this is done, is the subject of the next chapter.

The procedures of analysis presented in this chapter can be easily programmed for modern desk calculators or microcomputers and indeed this should be the route to follow if such computations are to be done repeatedly. The analysis can also be done by the computer programs SCS and TDA, described in Appendix G. The programs can be executed on microcomputers using the software provided on the Internet. The program SCS, Stresses in Cracked Sections, determines the neutral axis position and calculates the stress and the strain distributions, ignoring concrete in tension. The program TDA, Time Dependent Analysis, accounts for the effects of creep and shrinkage of concrete and relaxation of prestressed reinforcement.

Note

1 See Tadros, M K (1982), Expedient serviceability analysis of cracked prestressed concrete beams. *Prestressed Concrete Inst. J.*, **27**, No. 6, 67–86.

Chapter 8

Displacements of cracked members

Launching girder for construction of a cast *in situ* 2.043 km long bridge at Lake Gruyère, Switzerland.

8.1   Introduction

Cracks are generally expected to occur in reinforced concrete structures without or with partial prestress, when the tensile stresses exceed the strength of concrete in tension. Reduction in stiffness of members due to cracking must be considered in the calculation of displacements in reinforced concrete structures. This chapter presents a method to predict the elongation and curvature of a reinforced concrete cracked member subjected to axial force and/or bending moment.

The weakest section in a cracked member is obviously at the location of a crack. Away from a crack, the concrete in the tension zone is capable of resisting some tensile stress and thus contributing to the stiffness (the rigidity) of the member. Thus, the stiffness of a cracked member varies from a minimum value at the location of the crack to a maximum value midway between cracks. For calculation of displacements, a mean value of the flexibility of the member is employed.

Two extreme states are considered: the uncracked condition in which concrete and steel are assumed to behave elastically and exhibit compatible deformations, and the fully cracked condition with the concrete in tension ignored. The elongations or the curvatures are calculated with these two assumptions and the actual deformations in a cracked member are predicted by interpolation between these two extreme conditions. A dimensionless coefficient, ζ, is used for the interpolation; it represents the extent of cracking. At the start of cracking, $\zeta = 0$ and its value approaches unity with the increase of the values of the applied axial force and/or bending moment.

The same coefficient ζ can be used to predict the width of a crack. The spacings between the cracks depend on several factors other than the magnitude of the applied loads. There exist several empirical equations for the prediction of the spacing between cracks and no doubt more equations will evolve from research. A chosen procedure for the prediction of crack spacing is presented in Appendix E. In the numerical examples of the present chapter, the spacing between the cracks will simply be assumed when the width of cracks is calculated.

The interpolation procedure described above gives mean values of axial strain and curvature at various sections of a structure, which can be subsequently employed for calculation of displacements (see Section 3.8). The internal forces are assumed to be known and supposed to be of a magnitude well below the ultimate strength of the sections (service conditions). The change in internal forces due to cracking in a statically indeterminate structure is briefly discussed in Section 8.11.

The reduction in stiffness due to cracking associated with shear stress is more difficult to evaluate. The truss model often employed in the calculation of the ultimate strength of members subjected to shearing force or twisting moment is sometimes used to assess an upper limit of the displacements after cracking. This is briefly discussed at the end of this chapter.

8.2   Basic assumptions

Consider a reinforced concrete member subjected to an axial force or a bending moment. When the stress in concrete has never exceeded its tensile strength, the member is free from cracks. The reinforcement and concrete undergo compatible strains. This condition is referred to as *state 1*.

When the tensile strength in concrete is exceeded, cracks occur. At the location of a crack, the tensile stress is assumed to be resisted completely by the reinforcement. The tensile zone is assumed to be fully cracked and this condition is referred to as *state 2*.

In both conditions, states 1 and 2, Bernoulli's assumption is adopted; namely, plane cross-sections remain plane after deformation or cracking. Analysis of strain and stress in states 1 and 2 in accordance with these assumptions is covered in Chapters 2, 3 and 7.

In a section situated between two cracks, bond between the concrete and the reinforcing bars restrains the elongation of the steel and thus a part of the tensile force in the reinforcement at a crack is transmitted to the concrete situated between the cracks. The stress and strain in the section will be in an intermediate condition between states 1 and 2. Thus, the strain in a reinforcing bar varies from a maximum value at the cracks to a minimum value midway between the cracks. The rigidity varies between consecutive cracks in a similar way. Therefore an effective or mean value of the member stiffness must be considered in the calculation of the elongation or curvature of the member.[1] The contribution of the concrete in the tension zone to the rigidity of the member is sometimes referred to as *tension stiffening*. Ignoring the effect of tension stiffening generally results in overestimation of deflection or crack width. To account for tension-stiffening effects, additional assumptions are required, which will be discussed in the following sections.

8.3   Strain due to axial tension

A reinforced concrete member subjected to axial tension N (Fig. 8.1(a)) will be free from cracks when the value of N is lower than

$$N_r = f_{ct}(A_c + aA_s) = f_{ct}A_1 \tag{8.1}$$

where f_{ct} is the strength of concrete in tension; N_r is the value of the axial force that produces first cracking; A_c and A_s are the cross-section areas of concrete and steel and $a = E_s/E_c$, with E_s being the modulus of elasticity of steel and E_c the secant modulus of elasticity of concrete for a loading of short

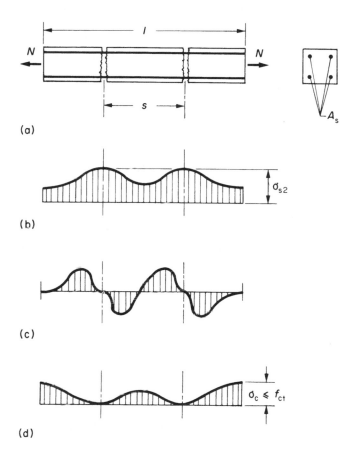

Figure 8.1 Stresses in a reinforced concrete member cracked due to axial force; (a) cracking of a tie; (b) stress in reinforcement; (c) bond stress; (d) stress in concrete ($\sigma_c \leqslant f_{ct}$).

duration. (The effect of creep is not considered in this section.) A_1 is the area of a transformed section in state 1, composed of A_c plus aA_s.

Just before cracking, the section is in state 1; the stress in concrete is f_{ct} and the stress in steel is af_{ct}. Immediately after cracking, the section at a crack is in state 2, the stress in steel

$$\sigma_{sr} = N_r/A_s \qquad (8.2)$$

When N_r is reached, the first crack occurs. At the crack, the tensile stress in concrete drops to zero and the total tension is resisted by the steel reinforcement (state 2). The sudden increase in stress in steel produces strain in steel that is incompatible with the strain in the adjacent concrete and results in widening of the crack.

Away from the crack, concrete, bonded to the reinforcement, tends to restrain its elongation and the bond stress τ transmits a part of the tensile force from the bar to the surrounding concrete. At a certain distance s from the first crack, strain compatibility is recovered (state 1) and the tensile strength in concrete is again reached, causing a second crack (Fig. 8.1(a)).

Figure 8.1(b), (c) and (d) shows the variation of steel stress, bond stress and concrete stress over the length of a cracked member subjected to an axial force $N > N_r$.

At a crack, the section is in state 2, the concrete stress is zero and the steel stress and strain when $N > N_r$

$$\sigma_{s2} = N/A_s \tag{8.3}$$

$$\varepsilon_{s2} = N/E_s A_s \tag{8.4}$$

Midway between consecutive cracks, the tensile stress in concrete has some unknown value smaller than f_{ct} and the steel stress has value smaller than σ_{s2}. Thus, the strain in the reinforcement varies along the length of the member; a mean value of the steel strain is

$$\varepsilon_{sm} = \Delta l/l \tag{8.5}$$

where l is the original length of the member and Δl is the member extension. The symbol ε_{sm} represents an overall mean strain value for the cracked member. Obviously, ε_{sm} is smaller than ε_{s2} which is the steel strain at the cracked section. Let

$$\varepsilon_{sm} = \varepsilon_{s2} - \Delta\varepsilon_s \tag{8.6}$$

where $\Delta\varepsilon_s$ is a reduction in steel strain caused by the participation of concrete in carrying the tensile stress between the cracks. Fig. 8.2 shows the variation of the mean strain ε_{sm} with the applied load N; it follows a curve situated between the two straight lines representing ε_{s1} and ε_{s2}. Here, ε_{s1} is a hypothetical strain in the reinforcement, assuming that state 1 continues to apply when $N > N_r$. Thus,

$$\varepsilon_{s1} = \varepsilon_{c1} = \frac{N}{E_c(A_c + \alpha A_s)} = \frac{N}{E_c A_1} \tag{8.7}$$

where A_1 is the area of the transformed section in state 1.

The value $\Delta\varepsilon_s$ represents the difference between the mean steel strain ε_{sm} and the steel strain in a fully cracked section. This difference has a maximum

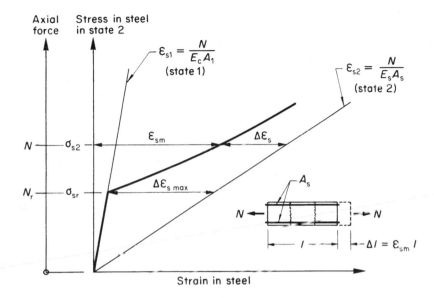

Figure 8.2 Axial force versus mean strain in a member subjected to axial tension.

value, $\Delta\varepsilon_{s\ max}$, at the start of cracking, when $N = N_r$. Based on experimental evidence, it is assumed that $\Delta\varepsilon_s$ has hyperbolic variation with σ_{s2} as follows:

$$\Delta\varepsilon_s = \Delta\varepsilon_{s\ max}\ \frac{\sigma_{sr}}{\sigma_{s2}} \tag{8.8}$$

From the geometry of the graph in Fig. 8.2

$$\Delta\varepsilon_{s\ max} = (\varepsilon_{s2} - \varepsilon_{s1})\frac{\sigma_{sr}}{\sigma_{s2}} \tag{8.9}$$

Substitution of Equations (8.8) and (8.9) into Equation (8.6) gives for a cracked member an overall strain value, which is also the mean strain in steel:

$$\varepsilon_{sm} = (1 - \zeta)\varepsilon_{s1} + \zeta\varepsilon_{s2} \tag{8.10}$$

where ζ is a dimensionless coefficient, between 0 and 1, representing the extent of cracking. $\zeta = 0$ for an uncracked section ($N < N_r$), and $0 < \zeta < 1$ for a cracked section. The value of ζ is given by:

$$\zeta = 1 - \left(\frac{\sigma_{sr}}{\sigma_{s2}}\right)^2 \quad (\text{with } \sigma_{s2} > \sigma_{sr}) \tag{8.11}$$

or

$$\zeta = 1 - \left(\frac{N_r}{N}\right)^2 \quad \text{(with } N > N_r\text{)} \tag{8.12}$$

In Equation (8.10), the mean strain in steel is determined by interpolation between the steel strains ε_{s1} and ε_{s2} in states 1 and 2. The interpolation co-efficient ζ depends upon the ratio of the steel stresses σ_{sr} and σ_{s2} in a fully cracked section when the applied forces are N_r and N, respectively. The use of this equation will be extended in the following sections to be applied for members subjected to bending.

In order to take into account the bond properties of the reinforcing bars and the influence of duration of the application or the repetition of loading, the Eurocode 2–1991[2] (EC2–91) introduces the coefficients β_1 and β_2 into Equation (8.11) as follows:

$$\zeta = 1 - \beta_1\beta_2\left(\frac{\sigma_{sr}}{\sigma_{s2}}\right)^2 \quad \text{(with } \sigma_{s2} \geq \sigma_{sr}\text{)} \tag{8.13}$$

where $\beta_1 = 1$ and 0.5 for high bond bars and for plain bars, respectively. $\beta_2 = 1$ and 0.5, respectively for first loading and for loads applied in a sustained manner or for a large number of load cycles.

With this modification, the graph of ε_{sm} (Fig. 8.2) will have a horizontal plateau at cracking level as shown in Fig. 8.3 (line AC).

The second term in Equation (8.10) ($\zeta\varepsilon_{s2}$) represents the supplementary strain of steel compared with the strain of concrete.[3] Thus, the average width of a crack is

$$w_m = s_{rm}\zeta\varepsilon_{s2} \tag{8.14}$$

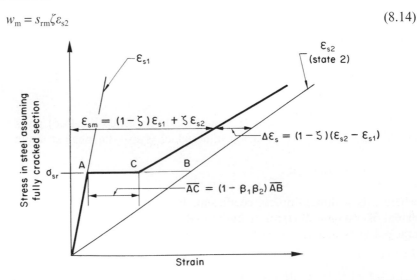

Figure 8.3 Mean strain in the reinforcement of a cracked member (according to EC2–91).

where s_{rm} is the average spacing between cracks; it depends upon factors including the bond properties, the amount of cover of the reinforcement and the shape of distribution of tensile stress over the section. Empirical equations based on experiments are generally used to predict value of s_{rm}. This is further discussed in Appendix E.

Example 8.1   Mean axial strain in a tie

Find the mean strain, excluding the effect of creep in a reinforced concrete member (Fig. 8.1(a)) having a square cross-section $0.20 \times 0.20\,\mathrm{m}^2$ ($62\,\mathrm{in}^2$) subjected to an axial tensile force $N = 200\,\mathrm{kN}$ ($45\,\mathrm{kip}$), given the following data: $A_s = 804\,\mathrm{mm}^2$ ($1.25\,\mathrm{in}^2$); $E_s = 200\,\mathrm{GPa}$ ($29\,000\,\mathrm{ksi}$); $E_c = 30\,\mathrm{GPa}$ ($4350\,\mathrm{ksi}$); $f_{ct} = 2.0\,\mathrm{MPa}$ ($290\,\mathrm{psi}$); $\beta_1 = 1$ and $\beta_2 = 0.5$. What is the width of a crack assuming $s_{rm} = 200\,\mathrm{mm}$ ($8\,\mathrm{in}$)?

Equation (8.1) gives $N_r = 89.1\,\mathrm{kN}$ ($20.0\,\mathrm{kip}$). The stresses in steel, assuming state 2 prevails (Equations (8.2) and (8.3)):

$$\sigma_{sr} = \frac{N_r}{A_s} = 111\,\mathrm{MPa} \qquad \sigma_{s2} = 249\,\mathrm{MPa}.$$

Substitution in Equation (8.13) gives $\zeta = 0.90$. The strains in steel due to N, calculated with the assumption that the section is in states 1 and 2, are (Equations (8.7) and (8.4)):

$$\varepsilon_{s1} = 150 \times 10^{-6} \qquad \varepsilon_{s2} = 1244 \times 10^{-6}.$$

The mean strain for the member (Equation (8.10)) is

$$\varepsilon_{sm} = 150 \times 10^{-6}(1 - 0.90) + 1244 \times 10^{-6} \times 0.90 = 1134 \times 10^{-6}.$$

The width of a crack (Equation (8.14)) is

$$w_m = 200 \times 0.90 \times 1244 \times 10^{-6} = 0.22\,\mathrm{mm}(8.8 \times 10^{-3}\,\mathrm{in}).$$

8.4   Curvature due to bending

A reinforced concrete member subjected to a bending moment (Fig. 8.4) will be free from cracks when the bending moment is less than

$$M_r = W_1 f_{ct} \tag{8.15}$$

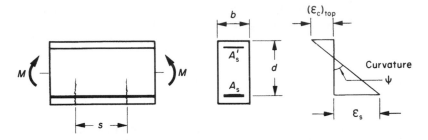

Figure 8.4 A reinforced concrete member in flexure.

where M_r is the value of the bending moment that produces first cracking; W_1 is the section modulus in state 1. Thus, W_1 is calculated for the cross-section area of concrete plus a times the cross-section area of steel. f_{ct} is the tensile strength of concrete in flexure (modulus of rupture).

For a bending moment $M > M_r$, cracking occurs and the steel stress along the reinforcement varies from a maximum value at the crack location to a minimum value at the middle of the spacing between the cracks. Assuming that the concrete between the cracks has the same effect on the mean strain in steel as in the case of axial force, Equation (8.10) can be adopted. Thus,

$$\varepsilon_{sm} = (1 - \zeta)\varepsilon_{s1} + \zeta\varepsilon_{s2} \tag{8.16}$$

where

$$\zeta = 1 - \beta_1\beta_2 \left(\frac{\sigma_{sr}}{\sigma_{s2}}\right)^2 = 1 - \beta_1\beta_2 \left(\frac{M_r}{M}\right)^2 \tag{8.17}$$

Here σ_{sr} and σ_{s2} are the steel stresses calculated for M_r and M, with assumption that the section is fully cracked.

For spacing between cracks s_{rm}, the width of one crack can be calculated by Equation (8.14), which is repeated here:

$$w_m = s_{rm}\zeta\varepsilon_{s2} \tag{8.18}$$

The curvature at an uncracked or a cracked section can be expressed in terms of the bending moment and flexural rigidity or in terms of strains as follows:

$$\psi = \frac{M}{EI} \tag{8.19}$$

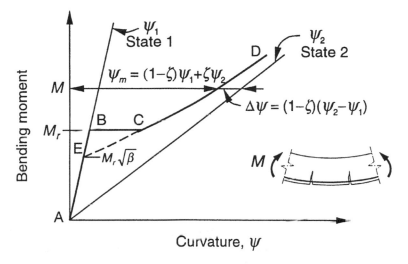

Figure 8.5 Moment versus curvature in a reinforced concrete member in flexure.

or

$$\psi = \frac{\varepsilon_s - (\varepsilon_c)_{top}}{d} \tag{8.20}$$

where ψ is the curvature; E is the modulus of elasticity; I is the moment of inertia of the section; ε_s is the strain in steel reinforcement and $(\varepsilon_c)_{top}$ is the strain at the extreme fibre of the compression zone and d is the distance between steel in tension and the extreme compression fibre (Fig. 8.4). Assume that cracking has an effect on curvature similar to its effect on the strain in axial tension. Thus, the mean curvature is expressed in this form:

$$\psi_m = (1 - \zeta)\psi_1 + \zeta\psi_2 \tag{8.21}$$

where ψ_1 and ψ_2 are the curvatures corresponding to a bending moment M, with the assumptions that the section is in states 1 and 2, respectively.

Thus, the coefficient ζ is employed to interpolate between the curvatures in states 1 and 2 to obtain the mean curvature. This is illustrated in the moment–curvature graph in Fig. 8.5. The cracked member has a mean flexural rigidity given by:

$$(EI)_m = \frac{M}{\psi_m} \tag{8.22}$$

The curvatures ψ_1 and ψ_2 are given by:

$$\psi_1 = \frac{M}{E_c I_1} \tag{8.23}$$

$$\psi_2 = \frac{M}{E_c I_2} \tag{8.24}$$

where I_1 and I_2 are the moments of inertia of a transformed uncracked and fully cracked section about an axis through their respective centroids. $E_c = E_{ref}$ is the modulus of elasticity of concrete, the value used as a reference elasticity modulus in the calculation of I_1 and I_2. The use of Equation (8.21) is demonstrated in Example 8.2.

8.4.1   Provisions of codes

The interpolation between states 1 and 2 to calculate the mean curvature as done in Equation (8.21) is adopted in MC-90 and EC2–91.[4] The EC2–91 allows use of the same coefficient ζ to calculate, by the same equation, mean values for deformation parameters such as curvature, strain, rotation or deflection.

The MC-90 considers that the M-ψ relation shown by the lines ABCD in Fig. 8.5 is most representative of actual practice with the exception of the part EBC. This part is replaced by the dashed line, which is an extension of the curve CD (Equation (8.21)) until it intersects AB at point E. Thus, for practical application, the M–ψ relation follows the straight line AE when $0 \leqslant M \leqslant (M_r \sqrt{\beta})$; where $(M_r \sqrt{\beta})$ represents a reduced value of the cracking moment; $\beta = \beta_1 \beta_2$.

When $(M_r \sqrt{\beta}) \leqslant M \leqslant M_y$, the M–ψ relation is the non-linear part ED, following hyperbolic Equation (8.21); where M_y is the moment which produces yielding of the reinforcement. If the concrete is in a virgin state and the loading is of short-term character, the M–ψ relation is more closely presented by the lines ABCD. Replacement of the part EBC by EC takes into consideration the behaviour of a member which has been cracked due to loads, shrinkage and temperature variations during construction.

The MC-90 also differs in the value of the coefficient β_2, which is considered equal to 0.8 (instead of 1.0) for first loading.

The deflection of members can be calculated most accurately by numerical integration of the curvatures at various sections (see Appendix C). The EC2–91 allows, for simplicity, to calculate the deflection twice, assuming the whole member to be in uncracked and fully cracked condition in turn (states 1 and 2), and then to employ Equation (8.21), substituting the deflection values for the curvatures.

ACI318-01[5] also allows a similar interpolation between the moment of inertia of a gross concrete section neglecting the reinforcement and the moment of inertia of a transformed fully cracked section to calculate an 'effective moment of inertia', I_e to be used in the deflection calculation. This is based on an empirical equation by Branson, discussed further in Section 9.5.

Example 8.2 Rectangular section subjected to bending moment

Calculate the mean curvature in a reinforced concrete member of a rectangular cross-section (Fig. 8.4) due to a bending moment $M = 250$ kN-m (221 kip-ft), excluding creep effect and employing the following data: $b = 400$ mm (16 in); $h = 800$ mm (32 in); $d = 750$ mm (30 in); $d' = 50$ mm (2 in); $A_s = 2120$ mm² (3.29 in²); $A'_s = 760$ mm² (1.18 in²); $E_s = 200$ GPa (29 000 ksi); $E_c = 30$ GPa (4350 ksi); $f_{ct} = 2.5$ MPa (360 psi); $\beta_1 = 1$ and $\beta_2 = 0.5$.

Assuming the spacing between cracks $s_{rm} = 300$ mm (12 in), find the width of a crack.

The moment of inertia and the section modulus of transformed uncracked section are (graphs of Fig. 3.5 may be employed):

$$I_1 = 0.0191\,\text{m}^4 \quad W'_1 = 0.0488\,\text{m}^3.$$

Equation (8.15) gives $M_r = 122$ kN-m (90.0 kip-ft). Substitution in Equation (8.17) gives $\zeta = 0.88$.

Depth of compression zone in state 2 (by Equation (7.16) or the graphs of Fig. 8.4):

$$c = 0.191\,\text{m} \quad (7.52\,\text{in}).$$

The centroid of the transformed fully cracked section coincides with the neutral axis. The moment of inertia (calculated from first principles or by use of graphs of Fig. 7.6) is

$$I_2 = 0.00543\,\text{m}^4.$$

The curvatures due to $M = 250$ kN-m, assuming the section to be in states 1 and 2 (Equations (8.23) and (8.24)) are:

$$\psi_1 = 437 \times 10^{-6}\,\text{m}^{-1} \quad \psi_2 = 1530 \times 10^{-6}\,\text{m}^{-1}.$$

The mean curvature (Equation (8.21)) is

$$\psi_m = [(1 - 0.88)437 + 0.88 \times 1530]10^{-6} = 1400 \times 10^{-6}\,\text{m}^{-1}.$$

The strain in steel in state 2 is

$$\varepsilon_{s2} = \psi_2 y_s = 1530 \times 10^{-6}(0.75 - 0.191) = 856 \times 10^{-6}.$$

The width of a crack (Equation (8.14)) is

$$w_m = 300 \times 0.88 \times 856 \times 10^{-6} = 0.23\,\text{mm} \ (0.0091\,\text{in}).$$

8.5   Curvature due to a bending moment combined with an axial force

Fig. 8.6 shows a reinforced concrete member subjected to a bending moment M and an axial force N at the centroid of the transformed uncracked section. The values of M and N are assumed to be large enough to produce cracking at the bottom fibre.

The use of the equations of Section 8.4 will be extended to calculate the mean steel strain and the mean curvature in a cracked member subjected to N and M.

The eccentricity of the axial force is:

$$e = M/N \tag{8.25}$$

Our sign convention is as follows: N is positive when tensile and M is positive when it produces tension at the bottom fibre. It thus follows that e is positive when the resultant of M and N is situated below the centroid of the transformed uncracked section (Fig. 8.6).

Without change in eccentricity, we can find the values of N_r and the corresponding M_r that produce at the bottom fibre a tensile stress f_{ct}, the strength of concrete in tension:

$$N_r = f_{ct} \left(\frac{1}{A} + \frac{e}{W_{bot}} \right)_1^{-1} \tag{8.26}$$

$$M_r = eN_r \tag{8.27}$$

where A_1 and W_1 are the area and section modulus with respect to the bottom fibre of the transformed uncracked section.

Equation (8.26), of course, does not apply when the bottom fibre is in compression. This occurs when the resultant normal force on the section is tensile, acting at a point above the top edge of the core of the transformed uncracked section ($e \leqslant -(W_{bot}/A)_1$). This occurs also when the resultant normal force is compressive acting within the core ($(W_{top}/A)_1 \geqslant e \geqslant -(W_{bot}/A)$)

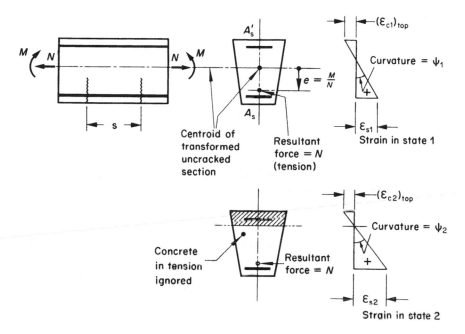

Figure 8.6 Curvature due to an eccentric normal force on a reinforced concrete section in states I and 2.

$A)_1$), where W_{bot} and W_{top} are section moduli with respect to the bottom and the top fibres; the subscript 1 refers to the transformed uncracked section.

When $N > N_r$ and $M > M_r$, cracking occurs and the mean strain in the reinforcement can be calculated by:

$$\varepsilon_m = (1 - \zeta)\varepsilon_{s1} + \zeta\varepsilon_{s2} \tag{8.28}$$

where

$$\zeta = 1 - \beta_1\beta_2\left(\frac{\sigma_{sr}}{\sigma_{s2}}\right)^2 \tag{8.29}$$

or

$$\zeta = 1 - \beta_1\beta_2\left(\frac{M_r}{M}\right)^2 \tag{8.30}$$

or

$$\zeta = 1 - \beta_1\beta_2\left(\frac{N_r}{N}\right)^2 \tag{8.31}$$

The symbols in Equations (8.28) and (8.29) are defined below:

ε_{s1} and ε_{s2} = strain in the bottom steel due to M combined with N, on a section
in states 1 and 2, respectively
σ_{s2} = stress in the bottom steel due to M and N on a section in state 2
σ_{sr} = stress in the bottom steel due to M_r and N_r on a section in state 2

It is to be noted that in a fully cracked section, the position of the neutral axis depends on the eccentricity $e = M/N$, not on the separate values of M and N. Because e is assumed to be unchanged, $(M/N) = (M_r/N_r)$ and

$$\frac{\sigma_{sr}}{\sigma_{s2}} = \frac{M_r}{M} = \frac{N_r}{N} \tag{8.32}$$

Assuming that the cracks are spaced at a distance s_{rm}, the width of a crack

$$w_m = s_{rm}\zeta\varepsilon_{s2} \tag{8.33}$$

The mean curvature in the cracked member

$$\psi_m = (1 - \zeta)\psi_1 + \zeta\psi_2 \tag{8.34}$$

where ψ_1 and ψ_2 are the curvatures corresponding to a bending moment M and an axial force N, with the assumptions that the section is in states 1 and 2, respectively.

Example 8.3 Rectangular section subjected to M and N

Calculate the mean curvature for the reinforced concrete section of Example 8.2 subjected to $M = 250\,\text{kN-m}$ (184 kip-ft) combined with an axial force $N = -200\,\text{kN}$ (–45 kip) at mid-height. All other data are the same as in Example 8.2. Assuming spacing between cracks, $s_{rm} = 300\,\text{mm}$, find the width of a crack.

The area of the transformed section in state 1

$$A_1 = 0.336\,\text{m}^2.$$

The centroid of A_1 is very close to mid-height; the eccentricity is considered to be measured from mid-height:

$$e = \frac{250}{-200} = -1.25 \, \text{m}.$$

Substitution in Equations (8.26) and (8.27) (with $f_{ct} = 2.5 \, \text{MPa}$, $W_1 = 0.0488 \, \text{m}^3$; see Example 8.2) gives:

$M_r = 138 \, \text{kN-m}.$

Substitution in Equation (8.30) gives

$\zeta = 0.85.$

The presence of N does not change the curvature in state 1 from what is calculated in Example 8.2. Thus,

$\psi_1 = 437 \times 10^{-6} \, \text{m}^{-1}.$

Solution of Equation (7.20) or use of graphs in Fig. 7.4 gives the depth of the compression zone:

$c = 0.241 \, \text{m} \, (9.49 \, \text{in}).$

Distance from the top fibre to the centroid of the transformed section in state 2 (Fig. 7.5) is

$\bar{y} = 0.195 \, \text{m} \, (7.68 \, \text{in}).$

The area and the moment of inertia of the transformed section in state 2 about an axis through its centroid (Fig. 7.6) are

$A_2 = 0.115 \, \text{m}^2, \quad I_2 = 0.00544 \, \text{m}^4.$

The applied forces $N = -200 \, \text{kN}$ at mid-height combined with $M = 250 \, \text{kN-m}$ may be replaced by an equivalent system of $N' = -200 \, \text{kN}$ at the centroid of the transformed section in state 2 combined with $M' = 209 \, \text{kN-m}$.

The curvature in state 2 is

$$\psi_2 = \frac{209 \times 10^3}{30 \times 10^9 \times 0.00544} = 1280 \times 10^{-6} \, \text{m}^{-1}.$$

The mean curvature (Equation (8.34)) is

$$\psi_m = [(1 - 0.85)437 + 0.85 \times 1280]10^{-6}$$
$$= 1150 \times 10^{-6} \text{m}^{-1} \ (29.2 \times 10^{-6} \text{in}^{-1}).$$

The axial strain at the centroid of the fully cracked section is

$$\varepsilon_{O2} = -\frac{200 \times 10^3}{30 \times 10^9 \times 0.115} = -58.0 \times 10^{-6}.$$

The strain in the bottom steel in state 2 is

$$\varepsilon_{s2} = 10^{-6}[-58.0 + 1280(0.75 - 0.195)] = 652 \times 10^{-6}.$$

Crack width (Equation (8.33)) is

$$w_m = 300 \times 0.85 \times 652 \times 10^{-6} = 0.17 \text{mm} \ (0.0067 \text{in}).$$

8.5.1   Effect of load history

Calculation of N_r and M_r by Equations (8.26) and (8.27) implies that M and N are increased simultaneously from zero until cracking occurs, without change in the eccentricity $e = M/N$. This represents the case when M and N are caused by an external applied load of a gradually increasing magnitude.

If N is introduced first and maintained at a constant value and subsequently M is gradually increased, cracking will occur when

$$M_r = \left(f_{ct} - \frac{N}{A_1}\right)W_1 \tag{8.35}$$

This means that the values of M_r and the coefficient ζ, representing the extent of cracking, depend upon the history of loading.

An important case in practice is when the axial force N is a compressive force due to partial prestressing. The axial force N is generally introduced with its full value before the cracking bending moment M_r. Thus, use of Equation (8.35) is more appropriate, and the first cracking occurs due to the combination N and M_r.

In a fully cracked section, the position of the neutral axis depends upon the eccentricity $e = M/N$. Thus, the combination of M_r and N has a different neutral axis from the combination of M and N. With the two combinations, the ratio σ_{s1}/σ_{s2} is not equal to M_r/M. It is, therefore, necessary to calculate σ_{s1} and σ_{s2} separately for a fully cracked section, once due to M_r and N and another time with M and N. The ratio $(\sigma_{s1}/\sigma_{s2})$ can then be used to determine ζ by Equation (8.29), rather than the ratio (M_r/M) with Equation (8.30). This would result in slightly different values for the mean curvature and crack width. In Example 8.3, this modification would give: $M_r = 151\,\text{kN-m}$; $\zeta = 0.88$; $\psi_m = 1170 \times 10^{-6}\,\text{m}^{-1}$ and $w_m = 0.17\,\text{mm}$, compared with $\psi_m = 1150 \times 10^{-6}\,\text{m}^{-1}$ and $w_m = 0.17\,\text{mm}$ previously calculated. Because the difference is small, it is suggested that Equations (8.26), (8.27) and (8.30) (or (8.31)) be employed in all cases, regardless of loading history.

8.6 Summary and idealized model for calculation of deformations of cracked members subjected to N and/or M

In the preceding sections, equations were presented for calculation of an interpolation coefficient ζ for calculation of the mean strain in a reinforced concrete member subjected to axial tension (Equation (8.12)) and the curvature due to a bending moment without or combined with an axial force (Equations (8.17) and (8.30) respectively). These equations are repeated here and the symbols are defined again for easy reference.

Axial tension (Fig. 8.1)
The mean axial strain

$$\varepsilon_{Om} = (1 - \zeta)\varepsilon_{O1} + \zeta\varepsilon_{O2} \tag{8.36}$$

where

$$\zeta = 1 - \beta_1\beta_2\left(\frac{N_r}{N}\right)^2 \tag{8.37}$$

ε_{O1} and $\varepsilon_{O2} =$ axial strain values due to N, calculated with the assumptions that the section is uncracked and fully cracked (states 1 and 2), respectively

$\zeta =$ dimensionless coefficient employed for interpolation between the steel strain values in states 1 and 2

$N_r =$ value of the axial force that produces tension in the concrete equal to its strength f_{ct}. The value N_r is given by:

$$N_r = f_{ct}(A_c + aA_s) \tag{8.38}$$

$$a = E_s/E_c \tag{8.39}$$

$\beta_1 = 1$ or 0.5 for high bond or for plain bars, respectively
$\beta_2 = 1$ or 0.5. The value 1 is to be used for first loading and 0.5 is for the case when the load is applied in a sustained manner or with a large number of load cycles.

Bending moment (Fig. 8.4)
The mean curvature

$$\psi_{\mathrm{m}} = (1 - \zeta)\psi_1 + \zeta\psi_2 \tag{8.40}$$

where

$$\zeta = 1 - \beta_1\beta_2\left(\frac{M_{\mathrm{r}}}{M}\right)^2 \tag{8.41}$$

ψ_1 and ψ_2 = values of the curvature due to M, calculated with the assumptions that the section is respectively uncracked and fully cracked (states 1 and 2).

$\quad M_{\mathrm{r}}$ = the value of the bending moment that produces tensile stress f_{ct} at the extreme fibre:

$$M_{\mathrm{r}} = f_{\mathrm{ct}}W_1 \tag{8.42}$$

where W_1 is the modulus of the transformed section. Other symbols are the same as defined earlier in this section.

Bending moment combined with axial force (Fig. 8.6)
The mean axial strain and curvature

$$\varepsilon_{\mathrm{Om}} = (1 - \zeta)\varepsilon_{\mathrm{O1}} + \zeta\varepsilon_{\mathrm{O2}} \tag{8.43}$$

$$\psi_{\mathrm{m}} = (1 - \zeta)\psi_1 + \zeta\psi_2 \tag{8.44}$$

where

$$\zeta = 1 - \beta_1\beta_2\left(\frac{M_{\mathrm{r}}}{M}\right)^2 = 1 - \beta_1\beta_2\left(\frac{N_{\mathrm{r}}}{N}\right)^2 \tag{8.45}$$

The pairs $\varepsilon_{\mathrm{O1}}$ with ψ_1 and $\varepsilon_{\mathrm{O2}}$ with ψ_2 are values of axial strain at a reference point O and the curvatures calculated with the assumptions that the section is respectively uncracked and fully cracked (states 1 and 2). M_{r} and N_{r} are the values of the bending moment and the corresponding axial force that produces tensile stress f_{ct} at the extreme fibre. The eccentricity e is assumed to be unchanged; thus,

$$e = \frac{M}{N} = \frac{M_r}{N_r} \tag{8.46}$$

The value of M_r is given by:

$$M_r = ef_{ct}\left(\frac{1}{A_1} + \frac{e}{W_1}\right)^{-1} \tag{8.47}$$

Other symbols have the same meaning as defined earlier in this section.

The mean crack width due to any of the above internal forces is given by:

$$w_m = s_{rm}\zeta\varepsilon_{s2} \tag{8.48}$$

where s_{rm} is the mean crack spacing; ε_{s2} is the steel stress due to N and/or M on a fully cracked section.

When the section is subjected to N or M, or N and M combined, the interpolation coefficient ζ may be expressed in terms of concrete stresses:

$$\zeta = 1 - \beta_1\beta_2\left(\frac{f_{ct}}{\sigma_{1\,max}}\right)^2 \tag{8.49}$$

where $\sigma_{1\,max}$ is the value of the tensile stress at the extreme fibre which would occur due to the applied N and/or M, with the assumption of no cracking (state 1). f_{ct} is the concrete strength in tension. If the stress is caused mainly by flexure (see Section 8.4), f_{ct} will represent the tensile strength in flexure, which is sometimes called the *modulus of rupture* and considered somewhat larger than the value for axial tension.

Equation (8.49) gives the same result as Equation (8.37), because the same linear relationship between f_{ct} and N_r applies between $\sigma_{1\,max}$ and N. Similarly, Equation (8.49) gives the same result as Equation (8.41) or (8.45).

Fig. 8.7 shows a physical model which idealizes the behaviour of a cracked member in accordance with the equations of this section. An element of unit length is considered to be composed of two parts: a part of length $(1 - \zeta)$ in state 1 (uncracked) and a part of length ζ in state 2 (fully cracked).

The axial deformation of this idealized member and the angular rotation per unit length (the curvature) are the same as in the actual cracked member.

Equations (8.43) to (8.49) are applicable for partially prestressed sections, but it must be noted in this case that M and N represent the part of the bending moment and of the normal force after deduction of the decompression forces (i.e. use the values of M_2 and N_2 obtained by Equations (7.33) and (7.34). This is further explained by Example 8.5.

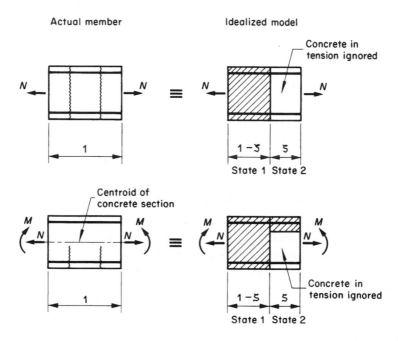

Figure 8.7 Representation of an element of unit length of a cracked member by a model composed of uncracked and fully cracked parts such that the extension or curvature is the same as in the actual member.

8.6.1   Note on crack width calculation

The value of ε_{s2}, to be used in the crack width calculation by Equation (8.48), is the steel strain due to N and M on a fully cracked section, ignoring the concrete in tension. Here it is assumed that the stress on concrete is zero prior to the application of N and M. If the section is subjected to initial stress due, for example, to the effect of shrinkage occurring prior to the application of N and M, the forces N and M should be replaced by $N_2 = N - N_1$ and $M_2 = M - M_1$; where N_1 and M_1 are two forces just sufficient to eliminate the initial stress. The values of N_1 and M_1 may be calculated by Equations (7.40) and (7.41), which are used to calculate the decompression forces in partially prestressed sections.

8.7   Time-dependent deformations of cracked members

Partially prestressed members are often designed in such a way that cracking does not occur under the effect of the dead load. Thus, cracking due to the live load is of a transient nature; hence the effects of creep, shrinkage and relaxation of prestress steel need be considered only for uncracked sections.

If this is not the case, or in the case of a reinforced concrete member where cracking occurs for a load of long duration, the time-dependent effects may be accounted for in the calculation of the axial strain and curvature in states 1 and 2 as covered in Chapters 2, 3 and 7. Interpolation between the two states may be done to find the deformations in the cracked member accounting for the tension stiffening. The equations presented in Section 8.6 for the interpolation coefficient ζ are applicable (noting that with a loading of long duration, $\beta_2 = 0.5$).

Example 8.4 Non-prestressed simple beam: variation of curvature over span

The reinforced concrete simple beam of the constant cross-section shown in Fig. 8.8(a) has bottom and top steel area ratios, $\rho = 0.6$ per cent and $\rho' = 0.15$ per cent. At time t_0, uniform load $q = 17.0$ kN/m (1.17 kip/ft) is applied. It is required to find the curvatures at t_0 and at a later time t and to draw sketches of the variations of the curvature over the span. The following data are given:

$E_s = 200$ GPa (29 000 ksi); $E_c(t_0) = 30.0$ GPa (4350 ksi); $f_{ct} = 2.5$ MPa (0.36 ksi); $\beta_1 = 1.0$; $\beta_2 = 1.0$ for calculation of instantaneous curvature and 0.5 for long-term curvature; creep coefficient $\varphi(t, t_0) = 2.5$; aging coefficient, $\chi(t, t_0) = 0.8$; free shrinkage, $\varepsilon_{cs}(t, t_0) = -250 \times 10^{-6}$.

What is the deflection at mid-span at time t?

(a) Curvature at time t_0
The following sections' properties will be used in the analysis of curvatures at t_0:

Transformed uncracked section (state 1) Area, $A_1 = 0.2027$ m^2; centroid O_1 is at 0.331 m below top edge; moment of inertia about an axis through O_1, $I_1 = 7.436 \times 10^{-3}$ m^4; section modulus $W_1 = 23.33 \times 10^{-3}$ m^3.

Transformed cracked section (state 2) Depth of compression zone (Equation (7.16)), $c = 0.145$ m; centroid O_2 lies on neutral axis; moment of inertia about an axis through O_2, $I_2 = 1.809 \times 10^{-3}$ m^4.

The bending moment at mid-span $= 17 \times 8^2/8 = 136$ kN-m. The bending moment which produces cracking (Equation (8.15))

$$M_r = 23.33 \times 10^{-3} \times 2.5 \times 10^6 = 58.3 \text{ kN-m.}$$

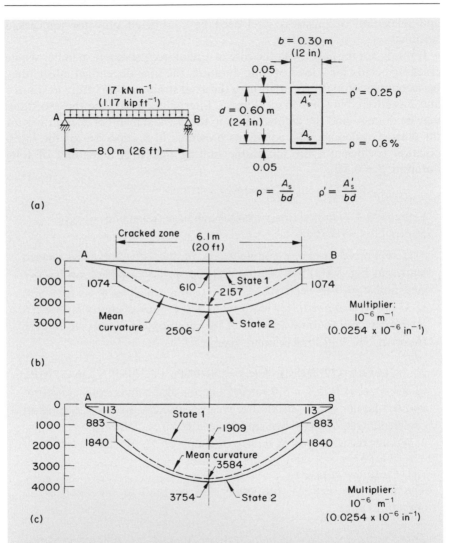

Figure 8.8 Curvature in a reinforced concrete beam (Example 8.4): (a) span, load and cross-section dimensions; (b) curvature at time t_0; (c) curvature at time t.

The interpolation coefficient for instantaneous curvature (Equation (8.41)) is

$$\zeta = 1 - 1.0 \times 1.0' \left(\frac{58.3}{136}\right)^2 = 0.82.$$

The interpolation coefficient for long-term curvature (Equation (8.41)) is

$$\zeta = 1 - 1.0 \times 0.5 \left(\frac{58.3}{136}\right)^2 = 0.91.$$

The curvature at t_0, assuming states 1 and 2 (Equations (8.23) and (8.24)):

State 1

$$\psi_1(t_0) = \frac{136 \times 10^3}{30 \times 10^9 \times 7.436 \times 10^{-3}} = 610 \times 10^{-6}\,\text{m}^{-1}$$

State 2

$$\psi_2(t_0) = \frac{136 \times 10^3}{30 \times 10^9 \times 1.809 \times 10^{-3}} = 2506 \times 10^{-6}\,\text{m}^{-1}$$

Interpolation
Mean curvature at time t_0 (Equation (8.40))

$$\psi(t_0) = (1 - 0.82)610 \times 10^{-6} + 0.82 \times 2506 \times 10^{-6} = 2157 \times 10^{-6}\,\text{m}^{-1}.$$

With parabolic variation of the bending moment over the span, the value $M_r = 58.3$ kN-m is reached at distance 0.98 m from the support. Thus, cracking occurs over the central 6.05 m (19.8 ft) of the span.

Fig. 8.8(b) shows the variation of the curvatures at time t_0, with the assumptions of states 1 and 2; the mean curvature is also shown with the broken curve.

(b) Curvatures at time t
The age-adjusted modulus of elasticity of concrete (Equation (1.31))

$$\bar{E}_c(t, t_0) = \frac{30 \times 10^9}{1 + 0.8 \times 2.5} = 10\,\text{GPa}.$$

$$\bar{a} = \frac{E_s}{\bar{E}_c(t, t_0)} = \frac{200}{10} = 20.$$

The following sections' properties are required for the age-adjusted transformed sections in states 1 and 2.

Age-adjusted transformed section in state 1 $\bar{A}_1 = 0.2207\,\text{m}^2$; centroid $\bar{O}_1$ is at 0.344 m below top edge. Moment of inertia about an axis through $\bar{O}_1$, $\bar{I}_1 = 8.724 \times 10^{-3}\,\text{m}^4$; y = coordinate of the centroid of the concrete area (measured downwards from $\bar{O}_1$); $y_c = -0.020\,\text{m}$; area of concrete, $A_c = 0.1937\,\text{m}^2$; moment of inertia of A_c about an axis through $\bar{O}_1$, $I_c = 6.937 \times 10^{-3}\,\text{m}^4$; $r_c^2 = I_c/A_c = 35.34 \times 10^{-3}\,\text{m}^2$.

The curvature reduction factor (Equation (3.18)) is

$$\kappa_1 = \frac{6.937 \times 10^{-3}}{8.724 \times 10^{-3}} = 0.795.$$

Age-adjusted transformed section in state 2 $\bar{A}_2 = 70.1 \times 10^{-3}\,\text{m}^2$; centroid $\bar{O}_2$ is at 0.233 m below top edge; moment of inertia about an axis through $\bar{O}_2$, $\bar{I}_2 = 4.277 \times 10^{-3}\,\text{m}^4$; y-coordinate of centroid of concrete area in compression (measured downwards from $\bar{O}_2$); $y_c = -0.161\,\text{m}$; area of the compression zone; $A_c = 0.0431\,\text{m}^2$; moment of inertia of A_c about an axis through $\bar{O}_2$, $I_c = 1.190 \times 10^{-3}\,\text{m}^4$; $r_c^2 = I_c/A_c = 27.62 \times 10^{-3}\,\text{m}^2$.

The curvature reduction factor (Equation (7.31)) is

$$\kappa_2 = \frac{1.190}{4.277} = 0.278.$$

Changes in curvature due to creep and shrinkage

State 1
The curvature at $t_0 = 610 \times 10^{-6}\,\text{m}^{-1}$; the corresponding axial strain at $\bar{O}_1$ $= 610 \times 10^{-6}\,(0.344 - 0.331) = 8 \times 10^{-6}$.

The change in curvature during the period t_0 to t (Equation (3.16)),

$$\Delta\psi = 0.795\left[2.5\left(610 \times 10^{-6} + 8 \times 10^{-6}\frac{-0.020}{35.34 \times 10^{-3}}\right)\right.$$

$$+ (-250 \times 10^{-6})\frac{-0.020}{35.34 \times 10^{-3}}\right]$$

$$= 1299 \times 10^{-6}\,\text{m}^{-1}.$$

The curvature at time t (state 1)

$$\psi_1(t) = (610 + 1299)10^{-6} = 1909 \times 10^{-6}\,\text{m}^{-1}.$$

State 2

The curvature at $t_0 = 2506 \times 10^{-6} \ \text{m}^{-1}$; the corresponding axial strain at $\bar{O}_2 = 2506 \times 10^{-6} \ (0.233 - 0.145) = 222 \times 10^{-6}$.

The change in curvature during the period t_0 to t (Equation (7.27))

$$\Delta \psi = 0.278 \Bigg[2.5 \bigg(2506 \times 10^{-6} + 222 \times 10^{-6} \frac{-0.161}{27.62 \times 10^{-3}} \bigg)$$

$$+ (-250 \times 10^{-6}) \frac{-0.161}{27.62 \times 10^{-3}} \Bigg]$$

$$= 1248 \times 10^{-6} \ \text{m}^{-1}.$$

The curvature at time t (state 2)

$$\psi_2(t) = (2506 + 1248)10^{-6} = 3754 \times 10^{-6} \ \text{m}^{-1}.$$

Interpolation

Mean curvature at time t (Equation (8.40))

$$\psi(t) = (1 - 0.91)1909 \times 10^{-6} + 0.91 \times 3754 \times 10^{-6}$$

$$= 3584 \times 10^{-6} \ \text{m}^{-1}$$

$$= 91.13 \times 10^{-6} \ \text{in}^{-1}.$$

The curvature at the end section is caused only by shrinkage and may be calculated by Equation (3.16). However, if we ignore this value and calculate the deflection by assuming parabolic variation of curvature, with zero at ends and maximum at the centre, we obtain (Equation (C.8)):

$$\text{Deflection at centre} = 3584 \times 10^{-6} \frac{8^2}{96}$$

$$= 0.0239 \, \text{m}$$

$$= 23.9 \, \text{mm} \ (0.948 \, \text{in}).$$

By numerical integration, a more accurate value of the deflection at the centre is 23.5 mm (0.925 in).

It can be seen in Fig. 8.8(b) and (c)[6] that once M_r is exceeded, the line

representing the mean curvature starts to deviate from the curve for
state 1 and quickly becomes closer to the curve for state 2. Thus,
prediction of deflection in design may start by considering state 2 which
gives the upper bound for deflection and the designer may find this
computation sufficient when the upper bound is not excessive.

Solution of this example is done, using no graphs in order to demon-
strate the computation for a general case with any cross-section. However, with a rectangular section, the graphs in Figs 3.5 and 7.4–7.6 can
be used to determine the section properties involved in the calculation.
Additional graphs are presented in Chapter 9 which further simplify
the prediction of deflection when the cross-section is a rectangle.

Example 8.5 Pre-tensioned simple beam: variation of curvature over span

Find the mean curvature at a section at mid-span of a partially
prestressed beam shown in Fig. 8.9(a), after application of a live

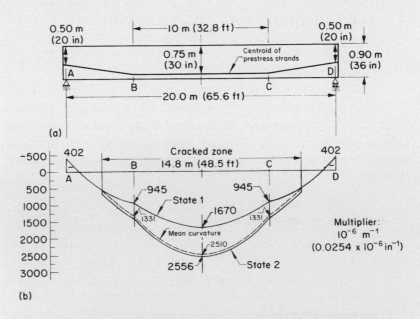

(a)

(b)

Figure 8.9 Curvature of a partially prestressed beam (Example 8.5): (a) tendon
profile; (b) curvature after creep and shrinkage and application of live
load. For beam cross-section see Fig. 7.12(a).

load producing cracking. Also sketch the corresponding variation of curvature over the span and calculate the deflection at the centre.

Fig. 7.12(a) shows the cross-section at mid-span. The section is constant over the span, with the exception of the location of the prestressed steel. The beam is pretensioned with a tendon depressed at points B and C, resulting in the profile shown in Fig. 8.9(a). The beam carries uniform dead and live loads of intensities 14.0 and 8.0 kN/m, respectively (0.96 and 0.55 kip/ft), resulting in bending moments at mid-span of 700 and 400 kN-m (6200 and 3540 kip-in). Assume a high-bond quality of reinforcement and tensile strength of concrete $f_{ct} = 2.5$ MPa. Other data are the same as in Example 7.5.

The stress and strain in the section at mid-span have been analysed in Example 7.5. The curvature in state 2 is obtained by summing up the values of curvatures shown in Fig. 7.12(b) and (c) and 7.13(c) and (e). This gives the following value of curvature in state 2:

$$\psi_2 = 2556 \times 10^{-6} \text{ m}^{-1}.$$

Cracking is produced at time t only after application of a live load. Immediately before application of the live load, after occurrence of prestress loss, the curvature at mid-span is 1167×10^{-6} m^{-1} (sum of curvature values indicated in Fig. 7.12(b) and (c)). Assuming no cracking (state 1), the live load would produce additional curvature of 499×10^{-6} m^{-1}. This is calculated by dividing the live-load moment by $[E_c(t)I_1(t)]$, where $E_c(t) = 30$ GPa is the modulus of elasticity of concrete at time t and $I_1(t) = 26.74 \times 10^{-3}$ m^4 is the centroidal moment of inertia of transformed uncracked section at time t. Thus, after live-load application, the total curvature in state 1 is

$$\psi_1 = (1167 + 499)10^{-6} = 1670 \times 10^{-6} \text{ m}^{-1}.$$

The stress at the bottom fibre due to the live-load moment on the uncracked section is 8.580 MPa. Addition of this value to the stress of 2.323 MPa existing before application of the live load (Fig. 7.13(b)) gives the stress at the bottom fibre after the live-load application with the assumptions of state 1

$$\sigma_{1 \text{ max}} = 2.323 + 8.580 = 10.903 \text{ MPa}.$$

The interpolation coefficient between states 1 and 2 (Equation (8.49)) is

$$\zeta = 1 - \beta_1\beta_2\left(\frac{f_{ct}}{\sigma_{1\,max}}\right)^2$$

$$= 1 - 1.0 \times 1.0\left(\frac{2.5}{10.903}\right)^2 = 0.95.$$

$\beta_1 = 1.0$ because of the high-bond quality of the reinforcement and $\beta_2 = 1.0$, assuming that the deflection is calculated for non-repetitive loading.

The mean curvature at mid-span (Equation (8.44)) is

$$\psi_m = (1 - 0.95)1670 \times 10^{-6} + 0.95 \times 2556 \times 10^{-6}$$

$$= 2510 \times 10^{-6}\,m^{-1}(63.8 \times 10^{-6}\,in^{-1}).$$

The curvature variation over the span is shown in Fig. 8.9(b).[7] The length of the zone where cracking occurs is 14.8 m. Over this zone, three lines are plotted for curvatures in states 1 and 2 and mean curvature.

If we assume parabolic variation and use the values of the mean curvature at the ends and the centre, we obtain by Equation (C.8):

$$\text{Deflection at the centre} = \frac{20^2}{96}\,[2(-402) + 10 \times 2510]10^{-6}$$

$$= 101.2\,mm\ (3.99\,in).$$

Using five sections instead of three and employing Equation (C.16) gives a more accurate value for the central deflection after application of live load of 86.2 mm (3.39 in). The dead-load deflection, including effects of creep, shrinkage and relaxation is 38.4 mm (1.51 in).

In the design of a partially prestressed cross-section, the amount of non-prestressed steel may be decreased and the prestressed steel increased such that the ultimate strength in flexure is unchanged. The amount of deflection is one criterion for the decision on the amounts of prestressed steel and non-prestressed reinforcement. The calculated deflection in this example may be considered excessive. Assuming that the yield stresses of the non-prestressed reinforcement and the pre-stressed steel are 400 and 1600 MPa (58 and 230 ksi), the area of the bottom non-prestressed reinforcement may be reduced from 1600 to 400 mm², with the addition of prestressed steel of area 300 mm² at the same level without substantial change in the flexural strength of the section. If the stress before transfer is the same in all prestressed steel as in the original design, the tension in the added prestressed steel before transfer is 312.5 kN.

With the second design, the curvatures in states 1 and 2 at mid-span, after application of the live load, will respectively be 1109×10^{-6} and 1976×10^{-6} m^{-1} and the corresponding mean curvature will be 1897×10^{-6} m^{-1}. The deflection just before and after the application of the live load will respectively be 6.0 and 43.1 mm (0.24 and 1.70 in) and the length of the cracked zone after the live-load application will be 12.5 m (40.8 ft).

8.8   Shear deformations

Reinforced concrete members are often designed in such a way that inclined cracks due to shear are expected to occur even at service load. After the development of such cracks shear deformations can be large. To predict the ultimate shear strength, the behaviour of a beam cracked by shear is often idealized as that of a truss model in which compression is resisted by concrete and tension by stirrups and flexural reinforcement. The same model has been employed for evaluation of the deflection in a member cracked by shear. However, the computation involves several assumptions and relies on empirical rules. The mean shear deformations are somewhere between those given by an uncracked member and those given by a truss model.[8]

8.9   Angle of twist due to torsion

Cracks due to twisting moments in reinforced concrete members result in reduction of the torsional rigidity. The reduction in rigidity due to cracking by torsion is much more important than the corresponding reduction in case of flexure. In the following the angles of twist per unit length θ_1 and θ_2 are derived for uncracked or fully cracked conditions (states 1 and 2). This gives lower and upper bounds of the angle of twist. When the value of the twisting moment exceeds the value T_r that produces first cracking, the angle of twist per unit length θ_m will be some value between θ_1 and θ_2, but it is difficult to find an expression that can reliably predict the value of θ_m. For this reason, only expressions for θ_1 and θ_2 will be derived below.

In many structures, for example grids or curved beams, the drastic reduction in stiffness due to cracking by torsion results in a redistribution of stresses and the twisting moment drops at the expense of an increase of the bending moment in another section without excessive deformation of the structure. When such redistribution cannot occur, excessive deformations due to torsion must be avoided, for example, by the introduction of appropriate prestressing.

8.9.1 Twisting of an uncracked member

According to the theory of elasticity, the angle of twist per unit length is

$$\theta_1 = \frac{T}{G_c J_1} \tag{8.50}$$

where T is the twisting moment, G_c is the shear modulus of concrete and J_1 is the torsion constant. For a rectangular section,

$$J_1 = cb^3\left[\frac{1}{3} - 0.21\frac{b}{c}\left(1 - \frac{b^4}{12c^4}\right)\right] \tag{8.51}$$

where c and b are the two sides of the rectangle with $b \leq c$. The maximum shear stress is at the middle of the longer side c and its value

$$\tau_{max} = \frac{T}{\mu b c^2} \tag{8.52}$$

where μ is a dimensionless coefficient which varies with the aspect ratio c/b as follows:[9]

c/b	1.0	1.5	1.75	2.0	2.5	3.0	4.0	6.0	8.0	10.0	∞
μ	0.208	0.231	0.239	0.246	0.258	0.267	0.282	0.299	0.307	0.313	0.333

For a closed hollow section

$$J_1 = 4A_0^2\left[\int(ds/t)\right]^{-1} \tag{8.53}$$

where t is the wall thickness; A_0 is the area enclosed by a line through the centre of the thickness and the integral is carried out over the circumference.

The shear flow (the shearing force per unit length of the circumference) is given by:

$$\tau t = T/2A_0 \tag{8.54}$$

where τ is the shear stress.

8.9.2   Twisting of a fully cracked member

The discussion here is limited to a hollow box section (Fig. 8.10(a)). The truss model usually adopted in the calculation of strength of reinforced concrete members in shear or torsion is used here for the calculation of the angle of twist in state 2. After cracking, the sides of the hollow section are assumed to act as a truss in which the compression is resisted by concrete inclined members and the tension is resisted by the stirrups and by the longitudinal

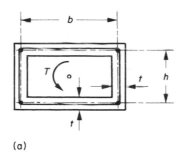

(a)

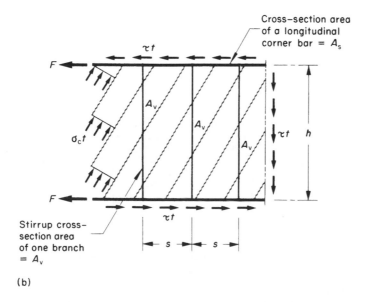

(b)

Figure 8.10 Torsion in a box girder: (a) cross-section; (b) free body diagram of a wall of a cracked box girder.

reinforcement, assumed to be lumped at the four corners. Fig. 8.10(b) is a free-body diagram showing the forces acting on a part of the wall of the box. Fig. 8.11(a) is a truss idealization of a cracked box girder. The members in the hidden faces of the box are not shown for clarity. The external applied twisting moment is replaced by the forces $\tau t h$ and $\tau t b$ as shown, where τt is the shear flow (Equation (8.54)):

$$\tau t = T/2hb \qquad (8.55)$$

where h and b are height and breadth of the truss model (see Fig. 8.11(a)).

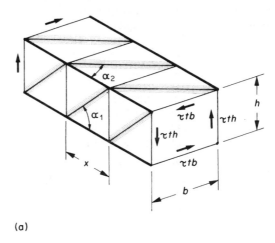

(a)

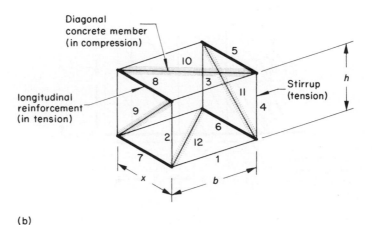

(b)

Figure 8.11 Truss idealization of a box girder cracked by twisting (see Fig. 8.10): (a) space truss model; (b) a typical panel of truss model.

A typical panel of the space truss is shown in Fig. 8.11(b). By statics, the forces in the twelve members of the panel due to a unit twisting moment are:

forces resisted by stirrups

$$F_1 = F_3 = \frac{1}{2h} \qquad F_2 = F_4 = \frac{1}{2b}; \tag{8.56}$$

forces in the longitudinal bars

$$F_5 = F_6 = F_7 = F_8 = \frac{x}{2hb}; \tag{8.57}$$

forces in the diagonal members

$$F_9 = F_{11} = -\frac{1}{2h \sin a_1} \qquad F_{10} = F_{12} = -\frac{1}{2h \sin a_2}; \tag{8.58}$$

where a_1 and a_2 are angles defined in Fig. 8.11(a). It is suggested that the distance x in Fig. 8.11(b) be selected such that the angles a_1 and a_2 are close to 45 degrees.

The angle of twist per unit length of the cracked member is considered the same as the relative rotation of the two cross-sections of the panel in Fig. 8.11(b) divided by the distance x between them. Considering virtual work, the angle of twist per unit length is:

$$\theta = \frac{T}{x} \sum_{i=1}^{12} \left(\frac{F^2 l}{AE} \right)_i \tag{8.59}$$

where F_i is the force in the ith member due to a unit twisting moment; l_i is its length; $E_i = E_s$ for the members in tension (the stirrups and the longitudinal bars) and $E_i = E_c$ for the diagonal members in compression; A_i is the cross-section area. For members representing the stirrups ($i = 1$ to 4)

$$A_i = A_v \frac{x}{s} \tag{8.60}$$

where A_v is the cross-sectional area of a stirrup and s is the spacing between stirrups. For longitudinal bars A_i is the area of the longitudinal reinforcement lumped at one corner. The area of the diagonal compression member is usually considered equal to

$$A_9 = A_{11} = th \cos a_1 \tag{8.61}$$

and

$$A_{10} = A_{12} = tb \cos a_2 \tag{8.62}$$

8.10   Examples worked out in British units

Example 8.6 Live-load deflection of a cracked pre-tensioned beam

Consider that the analyses conducted in Examples 2.6 and 7.6 are for the cross-section at the centre of a simply supported beam of span 80 ft (24 m). What is the deflection at mid-span after application of the live load? Assume $f_{ct} = 0.50$ ksi (3.4 MPa); $\beta_1 = 1.0$; $\beta_2 = 0.5$. Other data are the same as in Examples 2.6 and 7.6. Assume parabolic variation of curvature over the span and ignore the curvature at the two ends.

The curvature at mid-span after application of the live load calculated at a cracked section in Example 7.6 is:

$$\psi_2 = 36.76 \times 10^{-6} \text{ in}^{-1}.$$

Properties of the transformed uncracked section at time t are ($E_c(t) = 4000$ ksi and reference point O at top fibre):

$$A = 1145 \text{ in}^2; \qquad B = 19.43 \times 10^3 \text{ in}^3; \qquad I = 533.6 \times 10^3 \text{ in}^4.$$

The curvature change due to $M = 9600$ kip-in, applied on an uncracked section is (Equation (2.19)):

$$(\Delta\varepsilon_{O1})_{\text{live load}} = -200 \times 10^{-6}; \qquad (\Delta\psi_1)_{\text{live load}} = 11.77 \times 10^{-6} \text{in}^{-1}.$$

The corresponding stress change at bottom fibre is:

$$(\Delta\sigma_{bot})_{\text{live load}} = 1.083 \text{ ksi}.$$

Add the change in stress to the stress value existing before application of the live load to obtain the total stress ignoring cracking:

$$(\sigma_{bot})_{\text{non-cracked}} = 0.013 + 1.083 = 1.096 \text{ ksi}.$$

Similarly, the total curvature ignoring cracking:

$\psi_1 = 12.59 \times 10^{-6} + 11.77 \times 10^{-6} = 24.36 \times 10^{-6}\ \text{in}^{-1}$.

The interpolation coefficient (Equation (8.49)):

$$\zeta = 1 - 1.0(0.5)\left(\frac{0.500}{1.096}\right)^2 = 0.896.$$

The mean curvature (Equation (8.44)):

$$\psi_m = (1 - 0.896)24.36 \times 10^{-6} + 0.896(36.76 \times 10^{-6}) = 35.47 \times 10^{-6}\ \text{in}^{-1}.$$

The deflection at mid-span after application of the live load (Equation (C.8)):

$$D_{\text{mid-span}} = \frac{(80 \times 12)^2}{96}[0 + 10(35.47 \times 10^{-6}) + 0] = 3.41\ \text{in}.$$

Example 8.7 Parametric study

At time t after occurrence of creep, shrinkage and relaxation, the structure of Example 3.6 (Fig. 3.9(a)) is subjected to a uniform live load $p = 1.00\ \text{kip/ft}$. (14.6 kN-m). The intensity p is sufficient to produce cracking at mid-span. The tensile strength of concrete at time t is $f_{ct} = 0.360\ \text{ksi}$ (2.50 MPa). The modulus of elasticity of concrete $E_c(t) = 4350\ \text{ksi}$ (30.0 GPa). Other data are the same as in Example 3.6. The objective of the analysis is to determine the stresses, crack width and mid-span deflection immediately after application of p and to study the influence of varying the non-prestressed steel ratio $\rho_{ns} = A_{ns}/bh$ on the results; where A_{ns} is the non-prestressed bottom steel; the same amount of non-prestressed steel is also provided at the top. The effects of varying the creep and shrinkage parameters will also be discussed.

The live-load bending moment at mid-span is 3750 kip-in. Table 8.1, which gives the results of the analysis, includes the load intensity p_{cr} and the corresponding mid-span bending moment M_{cr} when cracking first occurs. The table also gives the stress changes $\Delta\sigma_{ps}$ and $\Delta\sigma_{ns}$ in the stress in the prestressed and non-prestressed steels due to $p = 1\ \text{kip/ft}$. The last column of the table gives the results for the case $\rho_{ns} = 0.4$ per cent and

Table 8.1 Stresses, mid-span deflection and crack width after live-load application of the structure of Example 8.7 (Fig. 3.9(a))

Non-prestressed steel ratio ρ_{ns} (per cent)		0	0.2	0.4	0.6	0.8	1.0	0.4 with reduced Φ & ε_{cs}
Live-load bending moment at which cracking occurs (kip-in)		2600	2400	2300	2200	2100	2000	2700
Ratio of uniform load intensity p_{cr} at which cracking occurs to p ($p = 1$ kip/ft)		0.69	0.65	0.61	0.58	0.56	0.54	0.73
Deflection after application of p (10^{-3} in)		1250	1229	1182	1128	1074	1022	976
Steel stresses after application of p (ksi)	σ_{ns}(bot)	−7	−6	−5	−5	−4	−4	−2
	σ_{ps}	180	181	182	183	183	183	187
Stress changes in steel caused by p (ksi)	$\Delta\sigma_{ns}$(bot)	29	27	25	23	21	20	18
	$\Delta\sigma_{ps}$	22	20	19	17	16	15	14
Ratio of crack width to crack spacing (10^{-3})		0.72	0.68	0.65	0.62	0.58	0.55	0.39

Conversion factors: 1 kip/ft = 14.6 kN/m; 1 ksi = 6.9 MPa.

reduced creep and shrinkage parameters $\varphi(t, t_0) = 1.5$ and $\varepsilon_{cs} = -150 \times 10^{-6}$ (from $\varphi(t, t_0) = 3.0$ and $\varepsilon_{cs} = -300 \times 10^{-6}$).

Based on the results in Table 8.1, the following remarks can be made:

(a) Presence of the non-prestressed steel reduces the deflection; in other words, the deflection is overestimated if the presence of the non-prestressed steel is ignored.

(b) The level of loading at which cracking occurs drops because of the presence of the non-prestressed steel; thus for certain load intensity, ignoring the non-prestressed steel may indicate that cracking does not occur contrary to reality.

(c) The steel stress increments $\Delta\sigma_{ps}$ and $\Delta\sigma_{ns}$ decrease with the increase in ρ_{ns}. Thus, presence of the non-prestressed steel increases safety against fatigue.

(d) The width of cracks can be controlled by the appropriate choice of ρ_{ns}. The ratio of the mean crack width w_m to the mean crack spacing s_{rm} is given in the table, rather than the value of w_m. This is so because w_m is proportional to s_{rm} (see Equation (8.48)) and the value s_{rm} depends on ρ_{ns} and on how the non-prestressed steel is arranged in the section. In general S_{rm} decreases with the increase in ρ_{ns}. Thus, w_m decreases faster than the ratio w_m/s_{rm} as ρ_{ns} is increased.

(e) It is interesting to note that the stress in the bottom non-prestressed steel is compressive in spite of cracking.

8.11   General

Strain in cracked sections is determined by two analyses: ignoring cracking (state 1) and assuming that the concrete cannot carry any stress in tension (state 2). Values of the axial strain and curvature are determined in the two states and the values in the actual condition are obtained by interpolation between the two analyses, using an empirical coefficient, ζ. In this way, account is made of the additional stiffness which concrete in tension provides to a section in state 2.

Branson[10] accounts for the tension stiffening by interpolation between moments of inertia of the cross-section in states 1 and 2 about axes through their respective centroids, using an empirical interpolation Equation (9.26), to calculate an 'effective' moment of inertia to be used in calculation of deflection. More important than the type of empirical procedure to be used for the interpolation is the correct analysis of the two limiting states 1 and 2.

It should be noted that when the section changes from state 1 to state 2, the centroid is shifted towards the compression zone. Thus, in the case of a section subjected to an eccentricity normal force, e.g. prestressing, a substantial change in eccentricity is associated with cracking. The moment about a centroidal axis changes, and so does the moment of inertia of the section. This is automatically accounted for when the equations used to calculate the axial strain and curvature employ cross-section properties (A, B and I) with respect to a reference point O used for both states 1 and 2.

Cracking changes cross-section properties and thus is associated with alteration in the reactions and internal forces when the structure is statically indeterminate. Analysis of these statically indeterminate forces has to be made by iterative methods, which are treated in references on structural analysis (see also Chapter 13). The equations presented in this chapter, which give the axial strain and curvature due to specified values of M and N, can be incorporated in an iterative analysis to determine the statically indeterminate forces in cracked reinforced or prestressed concrete structures.[11]

Notes

1 Favre, R., Beeby, A.W., Falkner, H., Koprna, M. and Schiessl, P. (1985), *Cracking and Deformations*, CEB Manual. Printed and distributed by the Swiss Federal Institute of Technology, Lausanne, Switzerland.
2 See the reference mentioned in Note 5, page 19.
3 See reference mentioned in Note 1 above.
4 See references mentioned in Notes 2 and 5 on page 19, respectively.
5 ACI Committee 318, *Building Code Requirement for Structural Concrete*, 2001, American Concrete Institute, Farmington Hills, Michigan 48333-9094.
6 The graphs in Fig. 8.8(b) and (c) are prepared using the computer program RPM, 'Reinforced and Prestressed Members', Elbadry, M. and Ghali, A., American Concrete Institute, P.O. Box 9094, Farmington Hills, MI 48333-9094, USA. RPM analyses strain, stress, change in length, end rotation and deflection of a reinforced member with or without prestressing. The member can have variable depth and can be a simple beam, a cantilever or can be part of a continuous beam or a frame. Cracking, tension stiffening, creep and shrinkage of concrete and relaxation of prestressing reinforcement are accounted for.
7 This figure was prepared using the computer program RPM; see Note 6, above.
8 See reference in Note 1, above.
9 Timoshenko, S. and Young, D. (1962), *Elements of Strength of Materials*, 4th edn, Van Nostrand, Princeton, New Jersey, pp. 91–2.
10 See reference mentioned in Note 2, page 348. See also Branson, D.E. and Trost, H. (1982), Application of the I-effective method in calculation deflections of partially prestressed members. *Prestressed Concrete Institute Journal*, Chicago, Illinois, K27, No. 5, Sept.–Oct., pp. 62–77.
11 A computer program in FORTRAN for analysis of the time-dependent internal forces, stresses and displacements in cracked reinforced and prestressed concrete structures is available. See Elbadry, M. and Ghali, A., *Manual of Computer Program CPF: Cracked Plane Frames in Prestressed Concrete, Research Report No. CE85-2*, revised 1993, Department of Civil Engineering, University of Calgary, Calgary, Alberta, Canada.

Chapter 9

Simplified prediction
of deflections

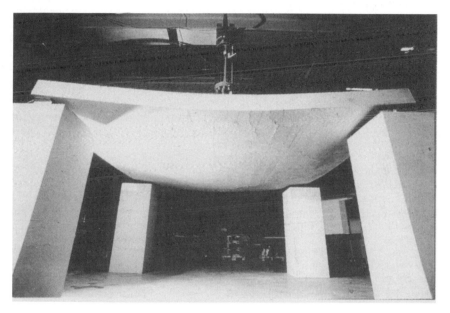

Long-term deflection and cracking of a reinforced concrete slab. A test series conducted at the Swiss Federal Institute of Technology, Lausanne.

9.1 Introduction

In many practical situations, designers are interested in prediction of probable maximum deflections of reinforced concrete members. Accuracy in prediction is often of little or no concern. For this purpose, two methods are presented in this chapter for prediction of maximum deflections of reinforced concrete members, accounting for long-term effects of creep and shrinkage. We are here concerned only with the transverse deflection associated with curvature ψ in simple or continuous members subjected to bending moments with or without axial forces. Prestressed beams are treated as reinforced concrete members for which the magnitude of the axial force and the bending moment are known; thus, prestress losses must be determined by a separate estimate.

In earlier chapters, we discussed how to obtain the axial strain ε_o and curvature ψ at sections of reinforced concrete frames in uncracked and cracked conditions (states 1 and 2). Ignoring concrete in tension in state 2 underestimates the rigidity of the sections. To account for the stiffening effect of concrete in the tension zone, mean values of axial strain and curvature are calculated by empirical interpolation between values of ε_o and ψ in states 1 and 2. The mean values can be employed to calculate displacements at any section by conventional methods, which generally require knowledge of the variation of the mean values over the entire length of all members (see Section 3.8 or Appendix C).

The interpolation mentioned above is done by using a coefficient ζ, which depends upon the value of the internal forces. Thus, any member, even with a constant cross-section, behaves in general as a beam of variable rigidity. The simplified methods presented in this chapter avoid this difficulty by calculation of limiting deflection values for states 1 and 2, considering the member to have a constant section in each state; thus, well-known expressions for deflection of members of constant rigidity can be applied. The interpolation is then done for the two limiting deflection values rather than for axial strain or curvature.

Through conventional linear analysis, a 'basic' deflection value is calculated, assuming that the member is made of homogeneous elastic material without cracking. The basic value is then multiplied by coefficients which account for the stiffening effect of reinforcement, cracking and creep. The deflection due to shrinkage is determined by a simple expression which also includes a coefficient depending on the amount of reinforcement and its position in the section. The coefficients needed in these calculations when the section is rectangular are presented in graphs in this chapter and in Appendix F. This appendix also includes expressions for the coefficients for cross-sections of any shape.

Section 9.9 is concerned with the deflection of flat slabs by simplified procedures similar to the methods suggested for beams.

9.2   Curvature coefficients, κ

Consider a reinforced concrete cross-section without prestressing (Fig. 9.1), subjected to a bending moment M introduced at time t_0. The following expressions give the instantaneous curvature and the changes in curvature caused by creep and shrinkage between t_0 and a later time t:

$$\psi(t_0) = \kappa_s \psi_c \tag{9.1}$$

$$(\Delta\psi)_\varphi = \psi(t_0)\varphi\kappa_\varphi \tag{9.2}$$

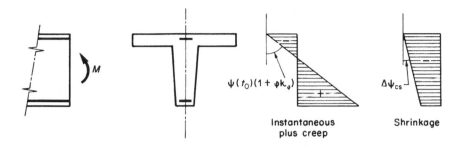

Figure 9.1 Curvature at a reinforced concrete cross-section subjected to bending moment.

$$(\Delta\psi)_{cs} = -\frac{\varepsilon_{cs}}{d}\kappa_{cs} \tag{9.3}$$

where ψ_c is the instantaneous curvature at a hypothetical uncracked concrete section without reinforcement:

$$\psi_c = \frac{M}{E_c(t_0)I_g} \tag{9.4}$$

I_g = moment of inertia of gross concrete section about an axis through its centroid

$E_c(t_0)$ = modulus of elasticity of concrete at time t_0

$\psi(t_0)$ = instantaneous curvature

$(\Delta\psi)_\varphi$ and $(\Delta\psi)_{cs}$ = curvature changes caused by creep and by shrinkage

$\varepsilon_{cs} = \varepsilon_{cs}(t, t_0)$ = value of free shrinkage during the period considered.

Following the sign convention adopted through this book, positive strain represents elongation; hence, shrinkage of concrete is a negative quantity.

A positive bending moment produces positive curvature (Fig. 9.1). In a cross-section with top and bottom reinforcements, shrinkage is restrained by the reinforcement and the result is smaller shrinkage at the face of the section with heavier reinforcement. In a simple beam subjected to gravity load, the heavier reinforcement is generally at the bottom. Thus, the curvature due to shrinkage is of the same sign as the curvature due to the positive bending moment due to load. For the same reason, in a cantilever with heavier reinforcement at the top, curvatures due to shrinkage and due to gravity load are cumulative.

κ_s, κ_ϕ and κ_{cs} are dimensionless coefficients depending on the geometrical properties of the cross-section, the ratio $a(t_0) = E_s/E_c(t_0)$ and the product

$\chi\varphi(t, t_0)$; where E_s is the modulus of elasticity of steel; $E_c(t_0)$ is the modulus of elasticity of concrete at time t_0; φ and χ are creep and aging coefficients, functions of the ages t_0 and t (see Section 1.7).

Equations (9.1)–(9.4) are applicable to uncracked sections in state 1 or fully cracked sections in state 2, employing coefficients κ_{s1}, $\kappa_{\varphi1}$ and κ_{cs1} for state 1 and κ_{s2}, $\kappa_{\varphi2}$ and κ_{cs2} for state 2.

For state 2, cracking is assumed to occur at t_0 due to the bending moment M. The concrete in tension is ignored; thus, the cross-section in state 2 is composed of the area of concrete in compression plus the area of the reinforcement. For T or rectangular cross-sections, the depth of the compression zone may be determined by Equation (7.16). The geometrical properties of the cracked section are assumed to undergo no further changes during the period of creep and shrinkage.

The graphs in Figs F.1 to F.10 of Appendix F give the values of the κ-coefficients in the two states for rectangular cross-sections. For easy reference, the variables in these graphs are listed in Table 9.1. Expressions for the coefficients for a general cross-section are also given in Appendix F. These are derived from Equations (2.16) and (3.16).

9.3   Deflection prediction by interpolation between uncracked and cracked states

In a simplified procedure suggested in Section 9.4, the probable maximum deflection in reinforced concrete members, including the effects of creep and shrinkage, is predicted by empirical interpolation between lower and upper bounds, D_1 and D_2. The values of D_1 and D_2 are determined, assuming the member to have a constant cross-section in states 1 and 2, respectively. An empirical coefficient ζ is employed to determine the probable deflection between the two limits D_1 and D_2. The difference between this simplified procedure and the method discussed in Chapter 8 is that the interpolation is performed on the deflection at one section to be defined below, rather than on the curvature at various sections of the member.

The interpolation coefficient used in Chapter 8 depends on the value of the bending moment and the cracking moment at the section considered (see Equation (8.41)). Here the interpolation coefficient for deflection is based on the bending moment at one section which is referred to as the '*determinant*' section. Similarly, the properties of the cross-section in states 1 and 2 will be based on the reinforcement at the determinant section.

If we apply any of Equations (C.4), (C.8), (C.12) or (C.16) to calculate the deflection in a simple beam in terms of curvature at various sections, it becomes evident that the maximum deflection is largely dependent upon the curvature at mid-span. This is so because the largest curvature is at this section and this value is multiplied by the largest coefficient in each equation. Thus, for a simple beam, the determinant section is considered at mid-span.

Table 9.1 Graphs for curvature coefficients K_s, K_φ and K_{cs} for rectangular reinforced concrete sections (Figs. F.1 to F.10)

Coefficient	d/h	d'/h	χφ	Figure number in Appendix F
		Parameters		
K_{s1}	1.0 0.9 0.8	0 to 0.2	–	F.1
K_{s2}	1.0 0.9 0.8	0 to 0.2	–	F.2
$K_{\varphi1}$	1.0	0 to 0.2	1.0 2.0 3.0 4.0	F.3
	0.9	0 to 0.2	1.0 2.0 3.0 4.0	F.4
	0.8	0 to 0.2	1.0 2.0 3.0 4.0	F.5
$K_{\varphi2}$	0.8 to 1.0	0	1.0 2.0 3.0 4.0	F.6
		0.1	1.0 2.0 3.0 4.0	F.7
		0.2	1.0 2.0 3.0 4.0	F.8
K_{cs1}	1.0 0.8	0 to 0.2	any value	F.9
K_{cs2}	0.8 to 1.0	0 0.2	2.0	F.10

Note: The value of some parameters is indicated by a range for which the graph may be employed. For preparation of the graphs, the value at the middle of the range is used in the calculations.

Similarly, for a cantilever, the determinant section is at or near its fixed end (see Equations (C.17), (C.19) and (C.21)).

Equations (C.4), (C.8), (C.12) and (C.16) are also applicable to members of continuous structures. Because the coefficient of curvature is largest for the value at mid-span, this section may again be considered determinant. But in this case, the curvatures at the end sections are generally not small and may have a comparatively larger influence on the calculated deflection. It should be recognized here that we are dealing with an approximation; the choice of the determinant section is a matter of judgement.

9.3.1 Instantaneous and creep deflections

Consider a simple beam (Fig. 9.2) subjected at age t_0 to a uniform load of intensity q. Variation of the instantaneous deflection $D(t_0)$ at mid-span with the load intensity, is as shown. For any load intensity q, the deflection $D(t_0)$ is some value between lower and upper bounds $D_1(t_0)$ and $D_2(t_0)$ where:

$D_1(t_0)$ = instantaneous deflection in state 1: all sections are assumed to be uncracked.

$D_2(t_0)$ = instantaneous deflection in state 2: all sections are assumed to be fully cracked. Contribution of concrete in tension to the stiffness of the member is ignored.

A basic deflection value, D_c, is calculated by conventional analysis, assuming the load q applied on a member of linear homogeneous material of modulus of elasticity $E_c(t_0)$ and having a constant cross-section with a moment of inertia equal to that of the gross concrete section, without considering reinforcement.

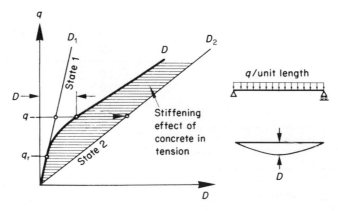

Figure 9.2 Instantaneous deflection at mid-span of a reinforced concrete simple beam versus intensity of load.

The instantaneous deflections at mid-span in states 1 and 2 are determined by the equations:

$$D_1(t_0) = \kappa_{s1} D_c \tag{9.5}$$

$$D_2(t_0) = \kappa_{s2} D_c \tag{9.6}$$

κ_{s1} and κ_{s2} are curvature coefficients calculated at a determinant section, which in this case is the section at mid-span. Equations (9.5) and (9.6) follow from Equation (9.1), if we consider that the reinforcement effect on the flexural rigidity is the same in all sections as in the determinant section.

Similarly, the changes in deflection due to creep in states 1 and 2 (see Equation (9.2)) are

$$(\Delta D_1)_\varphi = D_1(t_0)\varphi\kappa_{\varphi 1} \tag{9.7}$$

$$(\Delta D_2)_\varphi = D_2(t_0)\varphi\kappa_{\varphi 2} \tag{9.8}$$

$\kappa_{\varphi 1}$ and $\kappa_{\varphi 2}$ are curvature coefficients related to creep, to be calculated for the determinant section.

9.3.2   Deflection of beams due to uniform shrinkage

Consider a non-prestressed reinforced concrete simple beam without cracks (Fig. 9.3(a)). Assume that the cross-section is constant with area of the bottom reinforcement A_s larger than the top reinforcement A'_s. Uniform shrinkage of concrete occurring during a specified period produces at all sections a curvature of magnitude given by:

$$(\Delta\psi)_{cs} = \varepsilon_{cs}\left(\frac{A_c y_c}{\bar{I}}\right) \tag{9.9}$$

or

$$(\Delta\psi)_{cs} = -\varepsilon_{cs}\frac{\kappa_{cs}}{d} \tag{9.10}$$

where ε_{cs} is the value of the free shrinkage (generally a negative quantity); A_c is the cross-section area of concrete; $\bar{I}$ is the moment of inertia of the age-adjusted transformed section about its centroid; y_c is the y-coordinate of the centroid of A_c; y is measured downwards from point O, the centroid of the age-adjusted transformed section. Note that in the cross-section considered in Fig. 9.3(a), y_c is a negative value. The age-adjusted transformed section is composed of A_c plus $\bar{a}$ times area of the reinforcement; where $\bar{a} = E_s/\bar{E}_c$; E_s is

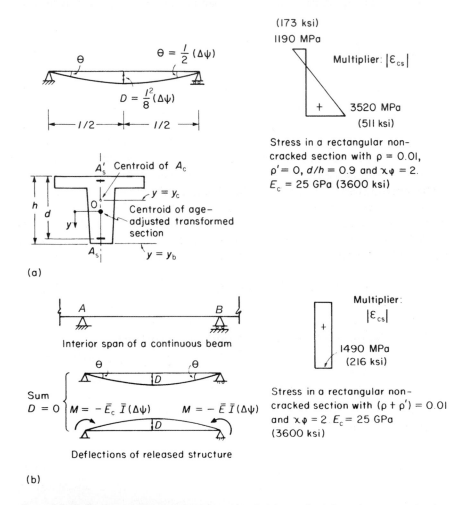

Figure 9.3 Deflection and stresses produced by shrinkage of reinforced concrete simple (a) and continuous (b) beams.

the modulus of elasticity of steel; $\bar{E}_c = E_c/(1 + \chi\varphi)$ is the age-adjusted modulus of elasticity of concrete; φ and χ are creep and aging coefficients and E_c is the modulus of elasticity of concrete at the start of the period considered; κ_{cs} is the curvature coefficient defined in Section 9.2 and given in graphs in Figs F.9 and F.10. Equations (9.9) and (9.10) may be derived from Equation (3.16).

For a simple beam of span l, the deflection at mid-span due to shrinkage is

$$(\Delta D)_{cs} = (\Delta\psi)_{cs}\frac{l^2}{8} \tag{9.11}$$

The stress at any fibre due to shrinkage (see Equations (3.15) and (3.19)) is

$$(\Delta\sigma)_{cs} = \bar{E}_c\left[(\Delta\psi)_{cs}\, y - \varepsilon_{cs}\left(1 - \frac{A_c}{\bar{A}}\right)\right] \tag{9.12}$$

where $\bar{A}$ is the area of the age-adjusted section; y is the coordinate of the fibre considered. At the bottom fibre, $y = y_b$, the stress is the largest tensile.

$$(\Delta\sigma)_{cs\,bot} = \bar{E}_c\left[(\Delta\psi)_{cs}\, y_b - \varepsilon_{cs}\left(1 - \frac{A_c}{\bar{A}}\right)\right] \tag{9.13}$$

The stress distribution shown in Fig. 9.3(a) is calculated for a rectangular cross-section with $\rho = 1$ per cent; $d/h = 0.9$; $E_c = 25$ GPa (3600 ksi) and $\chi\varphi = 2$. For a free shrinkage $\varepsilon_{cs} = -300 \times 10^{-6}$, the tensile stress at the bottom fibre is 1.06 MPa (0.154 ksi). Presence of this tensile stress allows cracking to occur at smaller external applied loads (smaller value of N_r or M_r; see Sections 8.4 and 8.5).

Thus, it can be concluded that uniform shrinkage can affect the deflection in direct and indirect ways. First, it produces curvature which increases the deflection in a simple beam. Second, it produces tension at the bottom fibre, enhancing cracking and causing further increase in deflection.

In a cantilever, heavier reinforcement is commonly at the top and the curvature due to shrinkage will be given by Equation (9.9).

The corresponding downward deflection at the free end is

$$(\Delta D)_{cs} = -(\Delta\psi)_{cs}\frac{l^2}{2} \tag{9.14}$$

Note that $(\Delta\psi)_{cs}$ in this case is a negative value.

Equation (9.9) or (9.10) may also be used to calculate the curvature due to shrinkage at a fully cracked section (compare Equations (3.16) and (7.27)). Here cracking is assumed to have occurred due to loads applied prior to shrinkage. For the fully cracked section, concrete in the tension zone is ignored and the geometric properties of the cracked section assumed unchanged by the effect of shrinkage.

For a cracked simple beam of length l, the deflection due to uniform shrinkage may be determined by interpolation between the limiting values:

$$(\Delta D_1)_{cs} = -\varepsilon_{cs}\kappa_{cs1}\frac{l^2}{8d} \tag{9.15}$$

$$(\Delta D_2)_{cs} = -\varepsilon_{cs}\kappa_{cs2}\frac{l^2}{8d} \tag{9.16}$$

$$\kappa_{cs} = -\frac{A_c d}{\bar{I}} y_c \tag{9.17}$$

The subscripts 1 and 2 are employed with κ_{cs} to refer to uncracked and fully cracked states.

Equations (9.15–16) are derived by combining Equations (9.10) and (9.11) and (9.17) by comparing Equation (9.10) with Equations (3.16) and (7.27); equations for a cantilever can be derived in a similar way. The curvature coefficients κ_{cs1} and κ_{cs2} are to be calculated for the 'determinant' section which is at mid-span for a simple beam and at the fixed end for a cantilever (see Section 9.3).

In statically indeterminate structures, hyperstatic forces develop which tend to reduce the deflection due to shrinkage. Consider as an example the interior span of a continuous beam of equal spans (Fig. 9.3(b)). Assume that the span shown is sufficiently far from the end spans such that the rotations at A and B are zero. Use the force method (see Section 4.2) to calculate the statically indeterminate connecting moments. This gives for a beam of constant cross-section: $M = -\bar{E}_c\bar{I}(\Delta\psi)_{cs}$; where $(\Delta\psi)_{cs}$ represents the curvature if the beam were simply supported. The curvature due to the connecting moments is of constant value equal to $-(\Delta\psi)_{cs}$. Thus, the statically indeterminate beam has no curvature and no deflection due to shrinkage and the concrete stress is uniform tensile of magnitude:

$$(\Delta\sigma)_{cs} = -\varepsilon_{cs}\bar{E}_c\left(1 - \frac{A_c}{A}\right) \tag{9.18}$$

Note that the stress in this case depends only on the sum of the reinforcement areas $(A_s + A'_s)$ not on their locations in the cross-section. For a rectangular section with 1 per cent reinforcement, $\varepsilon_{cs} = -300 \times 10^{-6}$ and $\chi\varphi = 2$, $(\Delta\sigma)_{cs} = 0.45\,\text{MPa}$ (0.065 ksi) (Fig. 9.3(b)).

The statically indeterminate reactions and bending moments caused by uniform shrinkage in continuous beams of constant cross-section having two to five equal spans are given in Fig. 10.7. This figure, intended for the effect of temperature, is also usable for the effect of shrinkage, the only difference is that the multiplier $(\Delta\psi)$ used for the values of the figure represents the change in curvature due to uniform shrinkage of a simple beam (Equation (9.9) or (9.10)). The deflection is largest for the end span and its value at the middle of the span may be expressed as follows:

> Deflection at the centre of a continuous span
> = reduction coefficient × deflection of a simple beam.   (9.19)

The reduction coefficient for an end span is respectively 0.25, 0.40, 0.36 and 0.37 when the number of spans is 2, 3, 4 and 5. The values of the reduction

coefficient given here apply only when the cross-section and the reinforcement are constant within the span; other values for the coefficient are
suggested later in this subsection for the more common case when A_s and
A'_s vary within the span. When A_s and A'_s are constant, the tensile stress at
bottom fibre in a section at the middle of an end span may be approximated
by the average of the values calculated by Equations (9.13) and (9.18):

$$(\Delta\sigma)_{\text{cs bot}} = \bar{E}_c \left[\frac{(\Delta\psi)_{\text{cs}}}{2} y_b - \varepsilon_{\text{cs}} \left(1 - \frac{A_c}{A} \right) \right] \qquad (9.20)$$

Note that $(\Delta\psi)_{\text{cs}}$ is the value of curvature which would occur in a simply
supported beam (Equation (9.9) or (9.10)).

The curvature $(\Delta\psi)_{\text{cs}}$ due to shrinkage depends mainly on $(A_s - A'_s)$. In
actual continuous beams, the bottom reinforcement is larger than the top
reinforcement at mid-span, but the reverse is true at the supports. The curvature $(\Delta\psi)_{\text{cs}}$ of any span, when released as a simple beam (Fig. 9.3(b)), will be
positive at mid-span and negative at the supports. This has the effect of
reducing the absolute value of the statically indeterminate connecting
moment, $|M|$. It can be shown that in the interior span of a continuous beam
of rectangular cross-section (Fig. 9.3(b)), the statically indeterminate con-
necting moments $M = 0$, when the absolute value $|A_s - A'_s|$ is constant, with
the heavier steel at the bottom for only the middle half of the span and at the
top for the remainder of the span. It can also be shown that the deflection in
this case is half the value for a simple beam (Equation (9.11)). For a more
general case, accurate calculation of the value of the connecting moment and
the deflection due to shrinkage must account for the values of A_s and A'_s at
various sections of the span.

As approximation, the change of location of the heavier reinforcement
between top and bottom in a common case may be accounted for by the use
of Equation (9.19) with the reduction coefficient 0.5 for an interior span and
0.7 for an end span. This coefficient is to be multiplied by the deflection of a
simple beam of a constant cross-section, based on the reinforcement at mid-
span. The tensile stress at bottom fibre at the same section may be approximated by Equation (9.13); this implies ignoring the effect of the statically
indeterminate connecting moment.

9.3.3   Total deflection

The deflections due to applied load, including the effects of creep and shrinkage for states 1 and 2 are (by superposition):

$$D_1 = D_1(t_0) + (\Delta D_1)_\varphi + (\Delta D_1)_{\text{cs}} \qquad (9.21)$$

$$D_2 = D_2(t_0) + (\Delta D_2)_\varphi + (\Delta D_2)_{\text{cs}} \qquad (9.22)$$

9.4   Interpolation procedure: the 'bilinear method'

We consider in this section the maximum deflection of a member in flexure, without axial force. The case of a member subjected to axial force combined with bending will be discussed in Section 9.7.

The probable maximum deflection for the member considered in the preceding section (Fig. 9.2) is determined by interpolation between the lower and upper bounds D_1 and D_2. Thus, the deflection due to load, including creep and shrinkage is

$$D = (1 - \zeta)D_1 + \zeta D_2 \tag{9.23}$$

where ζ is the interpolation coefficient, for which the following empirical equation is suggested:

$$\zeta = 1 - \beta_1 \beta_2 \frac{M_r}{M} \tag{9.24}$$

where M is the bending moment at the determinant section due to loading; M_r is the value of bending moment which produces in state 1 a tensile stress f_{ct} at extreme fibre; f_{ct} is the modulus of rupture (tensile strength of concrete in flexure). M_r is given by Equation (8.46), which is repeated here:

$$M_r = f_{ct} W_1 \tag{9.25}$$

where W_1 is the section modulus of the transformed uncracked section at time t_0. As an approximation W_g, the section modulus of the gross concrete area may be employed in Equation (9.25) in lieu of W_1.

The coefficient $\beta_1 = 1.0$ or 0.5 for high-bond reinforcements or plain bars, respectively; β_2 represents the influence of the duration of application and repetition of loading. $\beta_2 = 1$ at first loading and 0.5 for loads applied in a sustained manner or for a large number of load cycles.

Equation (8.45) for the interpolation coefficient used for curvature differs from the interpolation coefficient for deflection (Equation (9.24)) only in the term (M_r/M) which is raised here to the power 1 instead of 2. The two equations are merely empirical and the difference between the two is only justified by a better correlation with test results or with more accurate computation methods.

With the assumptions involved in the calculation of the deflection in states 1 and 2 due to applied load, the two values D_1 and D_2 vary linearly with the applied load or with the value of the free shrinkage ε_{cs}. Thus, Equation (9.23) interpolates between two straight lines. For this reason, the procedure is referred to as the 'bilinear method'[1].

9.5 Effective moment of inertia

An analogous approach for estimation of the instantaneous deflection due to load on a cracked reinforced concrete member is based on calculation of an *effective moment of inertia* I_e, to be assumed constant over the full length of the member. Several empirical expressions have been suggested. The best known is by Branson:[2]

$$I_e = \left(\frac{M_r}{M}\right)^m I_g + \left[1 - \left(\frac{M_r}{M}\right)^m\right] I_2 \qquad \text{with } M \geqslant M_r \qquad (9.26)$$

where

I_e = an effective moment of inertia
I_g = moment of inertia of gross concrete area about its centroidal axis, neglecting reinforcement
I_2 = moment of inertia of transformed fully cracked section (state 2) about its centroidal axis
M = maximum moment in the member at the stage for which the deflection is computed
M_r = The moment which produces cracking.

The power $m = 3$. Branson uses the same equation with $m = 4$ when I_e is intended for calculation of curvature in an individual section.

Example 9.1 Use of curvature coefficients: member in flexure

Figure 9.4 shows a reinforced concrete simple beam of a rectangular cross-section. A uniform load $q = 17\,\text{kN/m}$ is applied at time t_0. Calculate the deflection at time t at mid-span, including effects of creep and shrinkage. The ratio ρ and ρ' for the bottom and top reinforcements are:

$$\rho = \frac{A_s}{bd} = 0.6 \text{ per cent} \qquad \rho' = \frac{A'_s}{bd} = 0.15 \text{ per cent.}$$

Other data are: $E_s = 200\,\text{GPa}$ ($29 \times 10^3\,\text{ksi}$); $E_c(t_0) = 30.0\,\text{GPa}$ ($4350\,\text{ksi}$); $\varphi(t, t_0) = 2.5$; $\chi(t, t_0) = 0.8$; $\varepsilon_{cs}(t, t_0) = -250 \times 10^{-6}$; $f_{ct} = 2.5\,\text{MPa}$ ($0.36\,\text{ksi}$).

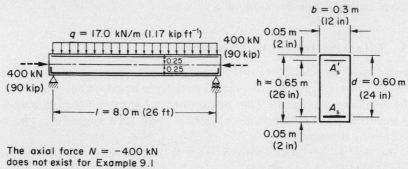

The axial force $N = -400$ kN
does not exist for Example 9.1

Figure 9.4 Beam of Examples 9.1, 9.2 and 9.3.

$$I_g = \frac{bh^3}{12} = \frac{0.3 \times (0.65)^3}{12} = 6.866 \times 10^{-3} \text{ m}^4.$$

Basic deflection

$$D_c = \frac{5}{384} \frac{ql^4}{E_c(t_0)I_g} = \frac{5}{384} \frac{17 \times 10^3 \times 8^4}{30 \times 10^9 \times 6.866 \times 10^{-3}} = 4.40 \text{ mm}.$$

The following curvature coefficients can be read from the graphs in Figs F.1, F.2, F.4, F.7, F.9 and F.10 (or by Equations (F.1–3)):

$$\kappa_{s1} = 0.92 \quad \kappa_{\varphi1} = 0.79 \quad \kappa_{cs1} = 0.27$$

$$\kappa_{s2} = 3.80 \quad \kappa_{\varphi2} = 0.14 \quad \kappa_{cs2} = 0.97$$

Instantaneous deflections in states 1 and 2 (Equations (9.5) and (9.6))

$$D_1(t_0) = 0.92 \times 4.40 = 4.05 \text{ mm}$$

$$D_2(t_0) = 3.80 \times 4.40 = 16.72 \text{ mm}.$$

Changes in deflections in the two states due to creep (Equations (9.7) and (9.8)) are

$$(\Delta D_1)_\varphi = 4.05 \times 2.5 \times 0.79 = 8.00 \text{ mm}$$

$$(\Delta D_2)_\varphi = 16.72 \times 2.5 \times 0.14 = 5.85 \text{ mm}.$$

Changes in deflections due to shrinkage (Equations (9.15) and (9.16)) are

$$(\Delta D_1)_{cs} = 250 \times 10^{-6} \times 0.27 \frac{8^2}{8 \times 0.6} = 0.90\,\text{mm}$$

$$(\Delta D_2)_{cs} = 250 \times 10^{-6} \times 0.97 \frac{8^2}{8 \times 0.6} = 3.23\,\text{mm}.$$

Lower and upper bounds on deflection at time t (Equations (9.21) and (9.22)) are

$$D_1 = 4.05 + 8.00 + 0.90 = 12.95\,\text{mm}$$

$$D_2 = 16.72 + 5.85 + 3.23 = 25.8\,\text{mm}.$$

Value of bending moment which produces cracking in state 1 at mid-span (Equation (9.25)) is:

$$M_r \simeq \frac{bh^2}{6} f_{ct} = \frac{0.3 \times (0.65)^2}{6} \, 2.5 \times 10^6 = 52.8\,\text{kN-m}.$$

(The reinforcement could be included in calculations of section modulus, but this is ignored here.) Actual bending moment at mid-span is

$$M = 17 \times \frac{8^2}{8} = 136.0\,\text{kN-m}.$$

Interpolation coefficient, using $\beta_1 = 1$ assuming high-bond reinforcement and $\beta_2 = 0.5$ for sustained loading (Equation (9.24)) is

$$\zeta = 1 - 1 \times 0.5 \frac{52.8}{136.0} = 0.81.$$

Probable deflection at time t (Equation 9.23) is

$$(1 - 0.81)12.95 + 0.81 \times 25.8 = 23.4\,\text{mm}\ (0.920\,\text{in}).$$

The deflection for the same beam is calculated by a more accurate procedure involving numerical integration in Example 8.4. The answers are almost identical.

9.6   Simplified procedure for calculation of curvature at a section subjected to *M* and *N*

Favre *et al.*[3] suggest the following approximation for the mean curvature at a cracked section subjected to a moment and a normal force.

Consider a reinforced concrete section subjected at time t_0 to a moment M and a normal force N located at the centroid of the gross concrete section (Fig. 9.5). The force N in this figure is assumed to be compressive, but the discussion applies also when the normal force is tensile. The graph represents the variation of instantaneous curvature, excluding creep, when N is kept constant and M increased gradually from zero. The straight line AB represents the curvature $\psi_1(t_0)$ in state 1. In the case when the section has heavier reinforcement at the bottom than at the top, the centroid of the transformed uncracked section at time t_0 is slightly lower than the centroid

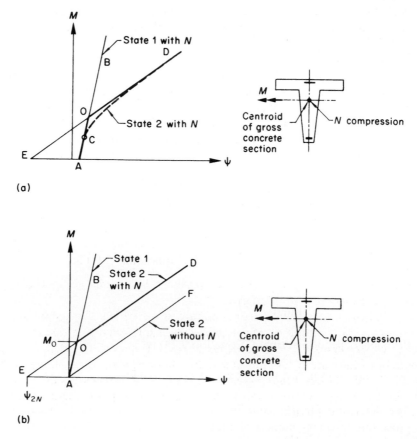

(a)

(b)

Figure 9.5 Moment versus curvature in presence of constant normal force: (a) actual graph; (b) idealized graph.

of the gross concrete section; thus, the compressive force N produces a positive moment about an axis through the centroid of the transformed section. This is why the line AB is slightly shifted from the origin in Fig. 9.5(a).

When M is zero, the neutral axis is outside the section, indicating that all the stress are of one sign (compression in the case considered here). When M reaches a certain level, tensile stress is produced at the bottom fibre; this is represented by point C in Fig. 9.5(a). If M is further increased and concrete in tension ignored, the curvature will follow the broken curve shown. The non-linear behaviour is caused by change in position of the neutral axis, altering the size of the compression zone. Thus, the geometrical properties of the cracked cross-section vary as M changes. However, as M increases, the broken curve in Fig. 9.5(a) gradually approaches the straight line OD, parallel to the line AF of Fig. 9.5(b) which represents the curvature when the cross-section properties of the cracked cross-section are those of a section in state 2 subjected to a bending moment, without a normal force. As an approximation, we accept the two straight lines AO and OD for the curvature in state 2 in lieu of the broken curve.

When a section is subjected to M, without N, the neutral axis in state 2 coincides with the centroid of the transformed fully cracked section (its position can be determined by Equation (7.16)). We further assume that the part of concrete considered effective does not change with time. With these assumptions, the curvature coefficients κ_s, κ_φ and κ_{cs} can be employed to find the instantaneous and time-dependent curvatures in states 1 and 2, accounting for the effects of creep and shrinkage as discussed in Sections 9.3.1 to 9.3.3.

The moment–curvature relation in Fig. 9.5(a) is further simplified in Fig. 9.5(b) by ignoring the small curvature when $M = 0$; thus, the line AB is moved parallel to itself bringing A to the origin. The line AF in this figure represents the curvature in state 2 when the section is subjected to M, without N. Thus, the presence of N has resulted simply in translation of AF, without a change in the slope to OD.

The mean curvature can now be obtained by empirical interpolation between the two straight lines OB and OD.

Fig. 9.5(b) is an idealized representation of M versus the instantaneous curvature $\psi(t_0)$. A graph of M versus the change in curvature due to creep, $(\Delta\psi)_\varphi$, or M versus the instantaneous plus creep curvature $[\psi(t_0) + (\Delta\psi)_\varphi]$ would be of the same form as in Fig. 9.5(a), differing only in the slopes of the lines ED and AB. Let us now consider that Fig. 9.5(b) represents the instantaneous plus creep curvature and write expressions for parameters related to the geometry of the figure:

$$(\text{slope})_{AB} = \frac{E_c(t_0)I_g}{\kappa_{s1}(1 + \phi\kappa_{\varphi1})} \tag{9.27}$$

$$(\text{slope})_{OD} = \frac{E_c(t_0)I_g}{\kappa_{s2}(1 + \phi\kappa_{\varphi 2})} \tag{9.28}$$

The value M_O at the intersection of AB and CD (Fig. 9.5(b)) does not vary with time (in the usual range of variation of $\chi\varphi$). Favre *et al.* employ the following expression for the value of M_O:

$$M_O \simeq -N\lvert y_{12}\rvert \frac{1}{1 - (\kappa_{s1}/\kappa_{s2})} \tag{9.29}$$

where $\lvert y_{12}\rvert$, the absolute value, is the distance between the centroid of the transformed section at time t_0 in state 1 and the centroid of the transformed section at the same time in state 2 subjected to M, without axial force.

The length EA in Fig. 9.5(b) represents a hypothetical curvature, ψ_{2N} is the value of curvature due to the normal force N on a cracked section (state 2), with $M = 0$. From geometry

$$\psi_{2N} = -M_O\left(\frac{1}{(\text{slope})_{OD}} - \frac{1}{(\text{slope})_{AB}}\right) \tag{9.30}$$

Equation (9.29) is applicable when N is tension or compression. According to the sign convention followed throughout this book, N is positive when tensile.

9.7   Deflections by the bilinear method: members subjected to M and N

This section is concerned with the maximum deflection of a reinforced concrete member subjected to a moment M, which may vary over the length of the member, combined with a constant axial force.

Fig. 9.6 represents a simple beam subjected to a normal force N at the centroid of the gross concrete section combined with a uniform load q. A compressive normal force is indicated in the figure, but the discussion applies also when N is tensile. The idealized M-ψ relationship in Fig. 9.5(b) will be used to extend the use of the bilinear method for calculation of the probable maximum deflection in the member shown in Fig. 9.6.

The graphs in Fig. 9.6 represent the variation of M at the determinant section (caused by the variation of q), with the corresponding instantaneous deflection at mid-span. Line AB represents the deflection D_1 in state 1. Line AF represents deflection in state 2 in the absence of the normal force. Line ED represents the deflection D_2 in state 2 due to M and N; the length EA represents the deflection due to a (negative) bending moment equal to $N\lvert y_{12}\rvert$; where y_{12} is the upward shift of the centroid of the transformed section as state 1 is changed to state 2.

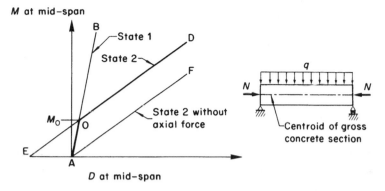

Figure 9.6 Maximum deflection versus bending moment at the determinant section in a simple beam.

If the deflection due to creep is included, the M–D diagram in Fig. 9.6 will not differ in form, but lines AB and ED will have smaller slopes. Inclusion of deflection due to shrinkage will cause the two lines to translate (to the right) without change in slope.

In the bilinear method suggested by Favre *et al.*,[4] two deflection values are calculated:

$D_1 =$ maximum deflection assuming that the member has a constant uncracked cross-section (state 1)

$D_2^* =$ maximum deflection assuming that the member is subjected to bending with no axial force and has a constant fully cracked cross-section (state 2)

The probable maximum deflection is determined by interpolation between these two values, using the equation:

$$D = (1 - \zeta)D_1 + \zeta D_2^* \qquad (9.31)$$

where ζ is the interpolation coefficient given empirically by one of the following four equations.

When $(\beta_1\beta_2 M_r) \geqslant M_O$ (Fig. 9.7(b)),

$$\zeta = \begin{cases} 1 - \beta_1\beta_2 \dfrac{M_r}{M} & \qquad (9.32) \\ 0 \qquad \text{for } M < M_r & \qquad (9.33) \end{cases}$$

When $(\beta_1\beta_2 M_r) < M_O$

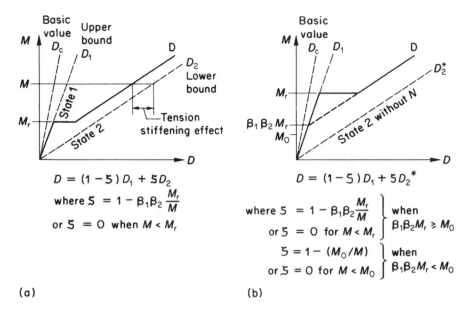

Figure 9.7 Summary of the bilinear method for prediction of maximum deflection of reinforced concrete members: (a) member subjected to bending moment without axial force (see Fig. 9.2); (b) member subjected to bending moment combined with axial force (see Fig. 9.6).

$$\zeta = \begin{cases} 1 - \dfrac{M_O}{M} & \text{(9.34)} \\ 0 \quad \text{for } M < M_O & \text{(9.35)} \end{cases}$$

M_o is given by Equation (9.29); M_r is the value of the bending moment which produces cracking in the presence of the axial force. M_r is given by Equation (8.35) which is repeated here:

$$M_r = \left(f_{ct} - \frac{N}{A_1} \right) W_1 \qquad (9.36)$$

where f_{ct} is the strength of concrete in tension; A_1 and W_1 are respectively the area and section modulus of the transformed uncracked section at time t_0. As an approximation, the area A_g and section modulus W_g of the gross concrete section may be used instead of A_1 and W_1.

The coefficient $\beta_1 = 1$ for high-bond reinforcement and 0.5 for plain bars; $\beta_2 = 1$ for first loading and 0.5 for loads applied in a sustained manner or in a large number of load cycles.

Comparing the equations of this section with Section 9.4, it can be seen that they differ only in the equations for M_r (Equations (9.25) and (9.36)) and for the interpolation coefficient in the case when $(\beta_1\beta_2M_r) < M_O$ (Equations (9.34) and (9.35)).

Figure 9.7 gives a concise presentation of the bilinear method for prediction of probable maximum deflections in reinforced concrete members subjected to a bending moment or a bending moment combined with an axial force. It should be noted that the chosen interpolation equations result in a probable deflection D which varies linearly with the bending moment M at the determinant section. The horizontal distance between the parallel lines D and D_2 in Fig. 9.7(a) represents the stiffening effect of concrete in tension. However, the distance between the lines D and D_2^* represents the tension stiffening combined with the effect of an additional bending moment resulting from the shift of centroid of the transformed section when cracking occurs.

Example 9.2 Use of curvature coefficients: member subjected to
M and N

Consider the same beam of Example 9.1 (Fig. 9.4) subjected at time t_0 to a uniform downward load $q = 17\,\text{kN/m}$ (1.17 kip/ft), combined with an axial compressive force $N = -400\,\text{kN}$ (89.9 kip) at mid-height of the section. It is required to find the maximum deflection at time $t > t_0$ including the effect of creep, but without shrinkage, using the bilinear method. Other data are the same as in Example 9.1.

The calculations are identical to the case of simple bending, without axial force (Example 9.1), except for the cracking moment M_r and the interpolation coefficient ζ. We give here some values calculated in Example 9.1:

Basic deflection = 4.40 mm
Curvature coefficients:

$$\kappa_{s1} = 0.92 \quad \kappa_{\varphi1} = 0.79 \quad \kappa_{s2} = 3.8 \quad \kappa_{\varphi2} = 0.14$$

Instantaneous deflections in states 1 and 2:

$$D_1(t_0) = 4.05\,\text{mm} \quad D_2(t_0) = 16.72\,\text{mm}$$

Changes in deflections in the two states due to creep:

$$(\Delta D_1)_\varphi = 8.00\,\text{mm} \quad (\Delta D_2)_\varphi = 5.85\,\text{mm}$$

Lower bound on deflection:

$$D_1 = 4.05 + 8.00 = 12.05 \, \text{mm}$$

Upper bound on deflection, assuming no axial force

$$D_2{}^* = 16.72 + 5.85 = 22.57 \, \text{mm}$$

Bending moment at the determinant section (mid-span)

$$M = 136.0 \, \text{kN-m}$$

Cracking bending moment (Equation (9.36))

$$M_r = \left(2.5 \times 10^6 - \frac{-400 \times 10^3}{0.3 \times 0.65}\right) \frac{0.3(0.65)^2}{6} = 96.1 \, \text{kN-m}$$

The centroid of the transformed uncracked section at time t_0 is at 331 mm below top edge. In the cracked stage, when the section is subjected to bending without axial force, the depth of the compression zone, $c = 145$ mm (Equation (7.16)). This is also equal to the distance between the top edge and the centroid of the transformed cracked section. Thus, the shift of the centroid as the section changes from state 1 to 2 is:

$$|y_{12}| = 331 - 145 = 186 \, \text{mm}.$$

The value M_O by Equation (9.29) is

$$M_O = - (-400 \times 10^3)0.186 \frac{1}{1 - (0.92/3.8)} = 98.1 \, \text{kN-m}.$$

β_1 and $\beta_2 = 1.0$ and 0.5, the same as in Example 9.1:

$$\beta_1 \beta_2 M_r = 0.5 \times 96.1 = 48.1 \, \text{kN-m} < M_O$$

Interpolation coefficient (Equation (9.34)) is

$$\zeta = 1 - \frac{98.1}{136.0} = 0.28.$$

The probable deflection (Equation (9.31)) is

$$D = (1 - 0.28)12.05 + 0.28 \times 22.57 = 15.0\,\text{mm} \quad (0.59\,\text{in}).$$

If the deflection due to shrinkage is excluded in Example 9.1, the probable deflection will be 20.6 mm (0.810 in). Thus, the compressive force N reduces the deflection in this example by 27 per cent.

9.8 Estimation of probable deflection: method of 'global coefficients'

In the majority of cases in practical design, particularly in preliminary studies, the engineer is only interested in an estimate of the probable deflection. To this effect, Favre *et al.*[5] have prepared graphs based on the bilinear method, permitting a simple and rapid estimation (within ±30 per cent) of long-term deflections due to sustained loads and shrinkage.

9.8.1 Instantaneous plus creep deflection

Equations (9.5–8), (9.21–23) can be combined in one equation for the deflection due to a sustained load including the effect of creep (but not shrinkage):

$$D = D_c[(1 - \zeta)\kappa_{s1}(1 + \varphi\kappa_{\varphi1}) + \zeta\kappa_{s2}(1 + \varphi\kappa_{\varphi2})] \tag{9.37}$$

where ζ is the interpolation coefficient (Equation (9.24)).

Based on parametric study, this equation may be replaced by the following approximation:

$$D \simeq D_c\left[\kappa_t\left(\frac{h}{d}\right)^3 (1 - 20\,\rho')\right] \tag{9.38}$$

This equation was derived for rectangular sections; h is total height; d is the distance between tension reinforcement and extreme compressive fibre; $\rho' = A'_s/bd$; b is the breadth of section and A'_s is the cross-section area of compression reinforcement.

κ_t is a global correction coefficient which depends on the level of loading expressed by the ratio (M_r/M) at the determinant section, creep coefficient φ and the product $\alpha\rho$, with $\alpha = E_s/E_c(t_0)$ and $\rho = A_s/bd$; A_s is the cross-section area of tension reinforcement.

The graphs in Fig. 9.8 give the global correction coefficient κ_t. These were prepared by calculating a value κ_t such that the terms between the square brackets in Equations (9.37) and (9.38) are equal. The following parameters are assumed constants; $d/h = 0.9$; $d'/h = 0.1$; $\alpha = E_s/E_c(t_0) = 7$; $\chi = 0.8$; $\beta_1 = 1$ and

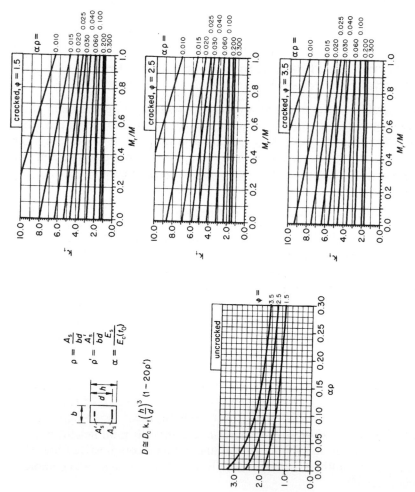

Figure 9.8 Global coefficient κ_t for calculation of instantaneous plus creep deflections of uncracked or cracked members by Equation (9.38).

$\beta_2 = 0.5$ (assuming use of high-bond reinforcement and sustained load). The compression steel reduces the long-term deflection by approximately 5 to 10 per cent. In preparation of the graphs of Fig. 9.8, $\rho' = A'_s/bd$ is considered zero, but the term $(1 - 20\rho')$ approximately accounts for the effect of the compression reinforcement.

Equation (9.38) is applicable for cracked or uncracked members. When the bending moment at the determinant section does not exceed cracking ($M \leqslant M_r$), $\zeta = 0$ and the corresponding graph in Fig. 9.8 may be employed to determine κ_t. Comparison of the values of κ_t for uncracked and cracked members shows that when M is close to M_r it is important to determine whether cracking occurred or not, because the value κ_t and hence the deflection can increase by a factor of 1 to 3 once cracking occurs.

The approximate Equation (9.38) may be employed for members having cross-sections other than rectangular, but with less accuracy. For this purpose, when calculating ρ and ρ' the section is transformed into a rectangle of the same height and with a width calculated such that the moment of inertia of the gross area is the same. Calculation of M_r should be based on section modulus of the actual section.

The tensile reinforcement has a great influence on deflection in the cracked state ($M \geqslant M_r$); on the other hand, its influence is small in the uncracked state. The amount of the tensile reinforcement is accounted for in κ_t and its position is included in Equation (9.38) by the ratio h/d.

The value M_r of the cracking moment at the determinant section and consequently the tensile strength of concrete f_{ct} (see Equations (9.25) and (9.36)), play an important role, particularly when the bending moment in the vicinity of the determinant section is close to M_r, because the deflection may then vary greatly. On the other hand, the influence of f_{ct} diminishes in the cracked stage.

The method of global coefficients was designed for members subjected to flexure, without axial force. If bending is combined with axial compression, produced for example by prestressing, the method may be used but again with less accuracy. The effect of the axial force will be limited to increasing the value M_r (Equation (9.36)).

9.8.2   Shrinkage deflection

Equations (9.15, 16), (9.21–23) may be combined in one equation for the deflection at mid-span of a cracked reinforced concrete simple beam due to shrinkage:

$$(\Delta D)_{cs} = -\varepsilon_{cs} \frac{l^2}{8d} [(1 - \zeta)\,\kappa_{cs1} + \zeta\kappa_{cs2}] \tag{9.39}$$

where ε_{cs} is the value of free shrinkage of concrete (generally a negative quantity), ε_{cs} is assumed uniform. κ_{cs1} and κ_{cs2} are coefficients for the calculation of curvature at the determinant section assumed uncracked and fully cracked, respectively (Equation (9.10)). Values of κ_{cs1} and κ_{cs2} may be determined by Equation (9.17) or the graphs of Figs F.9 and F.10. ζ is an interpolation coefficient given by Equation (9.24) which is repeated below:

$$\zeta = 1 - \beta_1\beta_2 \frac{M_r}{M} \tag{9.40}$$

M_r is the value of the bending moment which produces cracking (Equation (9.25)); M is the bending moment at the determinant section (at mid-span). M is assumed to have been applied before the occurrence of shrinkage.

The term inside the square brackets in Equation (9.39) may be combined in one global coefficient for shrinkage deflection:

$$\kappa_{tcs} = (1 - \zeta)\, \kappa_{cs1} + \zeta\kappa_{cs2} \tag{9.41}$$

The deflection due to shrinkage in a simple beam is

$$(\Delta D)_{cs} = -\varepsilon_{cs} \frac{l^2}{8d} \kappa_{tcs} \tag{9.42}$$

Shrinkage deflection in continuous beams can be predicted by multiplication of the simple-beam deflection calculated by Equation (9.42) by a reduction factor (see Section 9.3.2, near its end).

In a similar way, an equation may be derived for the shrinkage deflection at the free end of a cantilever

$$(\Delta D)_{cs} = -\varepsilon_{cs} \frac{l^2}{2d} \kappa_{tcs} \tag{9.42a}$$

The determinant section in this case is at the fixed end, where the bending moment produces cracking at the top; thus when the graphs in Figs F.9 and F.10 are used the pairs (A_s with d) and (A'_s with d') must refer to the top and bottom reinforcements, respectively. Equations (9.42) and (9.42a) are applicable to uncracked and cracked members.

In the common case when $\beta_1\beta_2 = 0.5$, the interpolation coefficient ζ for a cracked member is a value between 0.5 and 1.0. The graphs in Fig. 9.9 give the values of the global coefficient for shrinkage deflection κ_{tcs} calculated for rectangular sections with the assumptions: $\zeta = 0.5, 0.75$ and 1.0; $d/h = 0.9$ and $d'/h = 0.1$ and $\chi\varphi = 2.0$. The graphs may be used to calculate approximate values of κ_{tcs} for sections other than rectangles or when the values d/h, d'/h or $\chi\varphi$ are different.

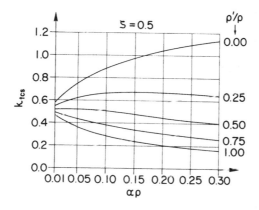

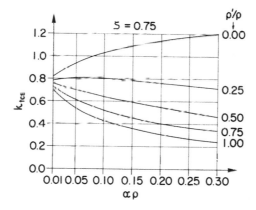

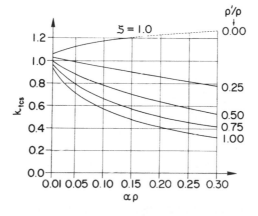

Figure 9.9 Global coefficient κ_{tcs} for calculation of shrinkage deflection of cracked members by Equation (9.42) or (9.42a).

When the member is uncracked, $\zeta = 0$ and $\kappa_{tcs} = \kappa_{cs1}$ (Equation (9.17) or Fig. F.9).

Example 9.3 Non-prestressed beam: use of global coefficients

Estimate the deflection at mid-span for the beam of Example 9.1 (Fig. 9.4) by the method of global coefficients.

The following values calculated in Example 9.1 are required here:

Basic deflection, $D_c = 4.40\,mm$

$$M_r = 52.8\,kN\text{-}m \qquad M = 136.0\,kN\text{-}m \qquad \zeta = 0.81$$

$$a\rho = \frac{200}{30} \times \frac{0.6}{100} = 0.04$$

$$\frac{M_r}{M} = \frac{52.8}{136.0} = 0.39$$

Entering the last two values in the graph for $\varphi = 2.5$ in Fig. 9.8 gives $\kappa_t = 3.8$. The probable instantaneous plus creep deflection (Equation (9.38)) is

$$4.4 \times 3.8 \left(\frac{0.65}{0.6}\right)^3 \left(1 - 20 \times \frac{0.15}{100}\right) = 20.6\,mm.$$

Entering the graph of Fig. 9.9 with $\zeta = 0.81$; $a\rho = 0.04$ and $\rho'/\rho = 0.25$ gives $\kappa_{tcs} = 0.85$. Thus, the deflection due to shrinkage (Equation (9.42)) is

$$(\Delta D)_{cs} = -(-250 \times 10^{-6}) \frac{8^2(0.85)}{8(0.6)} = 2.8\,mm.$$

Estimated value of deflection including effects of creep and shrinkage is

$$D = 20.6 + 2.8 = 23.4\,mm\ (0.94\,in).$$

Example 9.4 Prestressed beam: use of global coefficients

Estimate the deflection at mid-span of the prestressed beam in Fig. 9.10 due to the effects of a sustained load $q = 20\,kN/m$ (1.4 kip/ft) combined

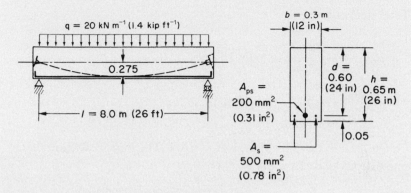

Figure 9.10 Prestressed beam of Example 9.4.

with prestressing. Assume that the effective prestress, after loss, balances 40 per cent of the dead load. Use the method of global coefficients. The beam has a rectangular cross-section as shown; the area of non-prestressed steel at the bottom is $500 \, \text{mm}^2$ ($0.78 \, \text{in}^2$) and at the top $200 \, \text{mm}^2$ ($0.31 \, \text{in}^2$); the area of prestressed steel is $200 \, \text{mm}^2$ ($0.31 \, \text{in}^2$). Other data are: $E_s = 200 \, \text{GPa}$ ($29\,000 \, \text{ksi}$); $E_c(t_0) = 30 \, \text{GPa}$ ($4350 \, \text{ksi}$); $f_{ct} = 2.5 \, \text{MPa}$ ($0.36 \, \text{ksi}$); $\varphi = 2.5$. The prestress duct is grouted after tensioning.

Prestress force necessary to balance 40 per cent of q (at time t after loss) is

$$P(t) = \frac{0.4 \times 20 \times 10^3 \times 8^2}{8 \times 0.275} = 232.7 \, \text{kN}.$$

Bending moment at mid-span due to dead load and effective prestress force is

$$M = 20 \times \frac{60}{100} \times \frac{8^2}{8} = 96.0 \, \text{kN-m}.$$

Value of bending moment producing cracking (Equation (9.36)) is

$$M_r = \left(2.5 \times 10^6 - \frac{-232.7 \times 10^3}{0.3 \times 0.65}\right) \frac{0.3(0.65)^2}{6} = 78.0 \, \text{kN-m}$$

$$\frac{M_r}{M} = \frac{78}{96} = 0.81.$$

Total steel ratio,

$$\rho = \frac{(500 + 200)10^{-6}}{0.3 \times 0.60} = 3.9 \times 10^{-3}$$

$$a = \frac{200}{30} = 6.67 \qquad a\rho = 0.026$$

Graph for $\varphi = 2.5$ in Fig. 9.8 gives $\kappa_t = 4.4$. The basic deflection due to unbalanced load (Equation (C.8)) is

$$\frac{96.0 \times 10^3}{30 \times 10^9 (0.3 \times 0.65^3/12)\, 9.6}\, \frac{8^2}{} = 3.11\,\text{mm}$$

$$\rho' = \frac{200 \times 10^{-6}}{0.3 \times 0.60} = 0.0011.$$

The probable deflection at mid-span (Equation (9.38)) is

$$D \simeq 3.11 \times 4.4 \left(\frac{0.65}{0.60}\right)^3 (1 - 20 \times 0.0011) = 17.1\,\text{mm} \quad (0.671\,\text{in}).$$

Included in the data given in this example is the value of the effective prestress after loss due to creep, shrinkage and relaxation. However, in practice, the initial prestress is known and the effective prestress must be calculated by an estimation of the amount of loss. In this example, the prestress balances only 40 per cent of the dead load, but when the upward load produced by prestressing is of almost the same magnitude as the downward gravity load, the long-term deflection is mainly due to prestress loss. Hence, in such a case the estimate of deflection is largely affected by accuracy in the calculation of prestress loss.

9.9   Deflection of two-way slab systems

This section is concerned with prediction of the maximum deflection in reinforced concrete floor systems taking into account the effects of creep, shrinkage and cracking. The method presented is applicable to slab systems with or without beams between supports. The supports are either columns or walls arranged in a rectangular pattern.

Calculation of the bending moments in two-way slab systems is extensively covered in codes and books on structural design.[6] Tables and other design aids[7] are available for this purpose. In this section, we assume that the bending moment values at the supports and at mid-span are available – in the two directions – at the centre lines of columns and at the centre lines of panels. Also, we assume that the reinforcement has been chosen and it is required to determine the long-term deflection at the centre of the panels.

9.9.1   Geometric relation

The deflection at the centre of a straight member relative to its ends can be calculated from the curvatures at three sections using the equation:

$$\delta = \frac{l^2}{96} (\psi_1 + 10\psi_2 + \psi_3) \tag{9.43}$$

where

> δ = deflection at centre, measured perpendicular to the member from the straight line joining the two ends (see Equation (C.8) and Fig. C.2)
> l = length of member
> ψ_1, ψ_2 and ψ_3 = curvatures at the left end, the centre and the right end of the member.

Equation (9.43) is based on the assumption that the variation of curvature follows a second degree parabola defined by the three ψ-values employed. This geometric relation, which can be proved by double integration, is valid for simply supported and for continuous members. It is of course applicable to a strip of a slab.

In most practical applications, the main loading is the member self-weight producing curvature which varies as second-degree parabola when the flexural rigidity is constant. When cracking occurs, the flexural rigidity is no longer constant and the ψ-values will be changed. Use of Equation (9.43) for calculation of deflection of a member of variable cross-section or for a cracked member results in tolerable error, acceptable in practice, as long as the three ψ-values employed are determined with appropriate account of the flexural rigidity at the respective sections.

Figure 9.11(a) is the top view of a two-way slab with rectangular panels. The deflected shape of a typical panel is shown in Fig. 9.11(b). The deflection D at the centre of the panel can be expressed as the sum of deflections of a strip joining two columns and a strip running along a centre line of the panel. One of the two following equations may be used (see Fig. 9.11(c)):

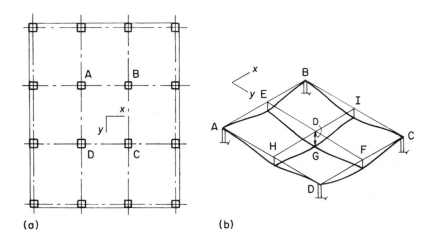

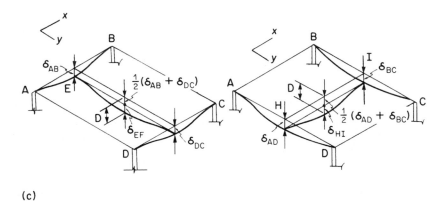

Figure 9.11 Deflection at the centre of a panel of a two-way slab system: (a) top view of system; (b) deflection of a typical panel; (c) definition of symbols employed in Equations (9.44) and (9.45).

$$D = \delta_{EF} + \frac{1}{2}(\delta_{AB} + \delta_{DC}) \qquad\qquad (9.44)$$

or

$$D = \delta_{HI} + \frac{1}{2}(\delta_{AD} + \delta_{BC}) \qquad\qquad (9.45)$$

where D is the deflection at the centre of the panel; δ represents the deflection at the centre of a column or a middle strip with respect to its ends.

The values of δ required in any of the equations may be calculated by Equation (9.43).

Application of Equation (9.44) or (9.45) should theoretically give the same answer for the deflection at the centre; this is in fact a check on compatibility. However, practical application of the two equations results in different answers and it is here suggested that the deflection be considered equal to the average of the two answers. The two answers may differ for the following reasons: (a) the true curvature variation is not parabolic; (b) the curvature values cannot be accurately determined. The curvature is usually calculated using bending moment values based on elastic analysis which does not account for cracking or may account for cracking in an empirical way.

9.9.2 Curvature-bending moment relations

In the elastic state, without consideration of the effects of the presence of the reinforcement, creep, shrinkage or cracking, the curvatures in x and y directions at any point of a slab can be calculated from the bending moments due to the applied load as follows:

$$M_x = E_c I_{g \text{ slab}} (\psi_x + v\,\psi_y); \qquad M_y = E_c I_{g \text{ slab}} (v\psi_x + \psi_y) \qquad (9.46)$$

$$\psi_x = \frac{12}{E_c h^3}(M_x - v\,M_y); \qquad \psi_y = \frac{12}{E_c h^3}(-v\,M_x + M_y) \qquad (9.47)$$

where E_c is the modulus of elasticity of concrete and v its Poisson's ratio (normally close to 0.2); M_x and M_y are bending moment values in a strip of unit width running in the x and y directions; $I_{g \text{ slab}}$ is an effective moment of inertia of the gross concrete area of the strip

$$I_{g \text{ slab}} = \frac{h^3}{12(1 - v^2)} \qquad (9.48)$$

where h is slab thickness.

When the floor system has beams, the curvature of a beam is

$$\psi_{\text{beam}} = \frac{M}{E_c I_{g \text{ beam}}} \qquad (9.49)$$

where $I_{g \text{ beam}}$ is the moment of inertia of the gross concrete cross-section. If the beam is monolithic with the slab, the beam cross-section includes a portion of the slab on each side of the beam of width equal to the projection of the beam above or below the slab. (This width may also be determined by other empirical rules.)

The equations presented in this subsection give curvature values to be substituted in Equation (9.43) to determine δ-values for column and middle strips followed by Equation (9.44) or (9.45) to obtain the deflection at the centre of two-way slab panels. This gives a basic deflection value D_{basic} which does not account for the reinforcement, creep or cracking. The true deflection can be much higher than the value D_{basic} (five to eight times), as will be discussed in the following subsection.

Table 9.2[8] may be employed to find the basic deflection value D_{basic} at the centre of an interior panel and at the centre of column strips for two-way slab systems with or without beams. It should be noted that the deflection values given in this table are based on elastic analysis of an interior panel. Exterior panels usually have larger deflection. Other limitations of Table 9.2 are mentioned with the table.

9.9.3   Effects of cracking and creep

In this subsection an approximate procedure is presented to account for the effect of the reinforcement ratio, cracking and creep on the deflection at the centre of panels of two-way slab systems. The effect of shrinkage will be discussed in the following subsection.

The deflection at the centre of a panel can be considered as the sum of deflection values, δ for column and middle strips (see Equation (9.44) or (9.45) and Fig. 9.11(c)). The symbol δ represents the deflection at the middle of a strip relative to its ends. First, basic δ-values are determined using Equation (9.43) and curvature values based on gross concrete sections, without consideration of cracking or creep. The basic δ-values may also be extracted from Table 9.2 or from alternative sources.

The deflection δ of a strip is largely influenced by the curvature at its mid-span. Thus, the reinforcement ratio, ρ at mid-span section of the strip is used to determine coefficients κ_{s1} and $\kappa_{\varphi1}$ and also κ_{s2} and $\kappa_{\varphi2}$ when cracking occurs. These coefficients are employed as multipliers to the basic δ-values to approximately account for the effects of creep and cracking in the same way as discussed for beams in Section 9.4.

The suggested procedure is more clearly explained in steps given below. The steps are to be followed after the bending moments and the reinforcements in the middle and column strips have been determined.

(1) Calculate curvatures at ends and at mid-span for a column strip running in the x or y direction and for a middle strip running in the perpendicular direction. In this step use the moment values corresponding to the service load for which the deflection is required. Ignore the reinforcement, creep, shrinkage and cracking (Equations (9.46–49)).

(2) Use the curvatures to determine basic deflection δ_{basic} of the two strips relative to their ends (by Equation (9.43)).

Alternatively, steps 1 and 2 may be replaced by a design aid which gives δ-values such as Table 9.2.

Table 9.2 Basic deflections for interior panels of two-way slab systems

Deflection at A, B or C = (coefficient × 10^{-3}) $\dfrac{ql^4}{(EI)_{slab}}$

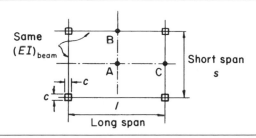

	$(EI)_{beam}$	$c/l = 0.0$			$c/l = 0.1$			$c/l = 0.2$		
s/l	$s(EI)_{slab}$	A	B	C	A	B	C	A	B	C
1.0	0.0	5.81	4.35		4.41	3.04		2.89	1.73	
	0.2	4.38	2.99	same	3.40	2.07	same	2.40	1.22	same
	0.5	3.31	1.98	as	2.71	1.41	as	2.05	0.85	as
	1.0	2.60	1.30	B	2.22	0.92	B	1.79	0.56	B
	2.0	2.06	0.77		1.84	0.54		1.58	0.33	
	4.0	1.74	0.43		1.59	0.30		1.44	0.18	
0.8	0.0	4.20	3.78	2.30	3.01	2.62	1.31	1.89	1.55	0.57
	0.2	3.16	2.71	1.49	2.37	1.92	0.88	1.59	1.16	0.40
	0.5	1.46	1.95	0.99	1.91	1.38	0.59	1.36	0.85	0.28
	1.0	1.91	1.34	0.63	1.54	0.95	0.38	1.17	0.58	0.18
	2.0	1.47	0.83	0.36	1.24	0.58	0.22	1.00	0.36	0.11
	4.0	1.16	0.48	0.19	1.03	0.33	0.12	0.89	0.20	0.06
0.6	0.0	3.27	3.21	0.99	2.34	2.28	0.40	1.43	1.37	0.08
	0.2	2.56	2.46	0.63	1.89	1.78	0.27	1.19	1.08	0.06
	0.5	2.01	1.87	0.40	1.50	1.34	0.17	0.98	0.98	0.04
	1.0	1.53	1.35	0.25	1.16	0.96	0.11	0.79	0.59	0.03
	2.0	1.10	0.87	0.13	0.85	0.61	0.06	0.61	0.37	0.02
	4.0	0.77	0.51	0.07	0.63	0.35	0.03	0.48	0.22	0.01
0.4	0.0	2.84	2.84	0.31	2.04	2.04	0.04			
	0.2	2.31	2.30	0.20	1.66	1.65	0.03			
	0.5	1.83	1.81	0.12	1.31	1.28	0.02			
	1.0	1.37	1.34	0.07	0.98	0.94	0.01			
	2.0	0.93	0.88	0.04	0.66	0.61	0.01			
	4.0	0.59	0.53	0.02	0.42	0.36	0.00			

$(EI)_{slab} = E_c \dfrac{h^3}{12(1 - v^2)}$

q = load intensity; h = slab thickness; v = Poisson's ratio.
Basic deflection values do not account for the effects of the reinforcement, creep or cracking.

(3) For the cross-section at the middle of the two strips, determine the curvature coefficients κ_{s1} and $\kappa_{\varphi 1}$ for a non-cracked section, using graphs of Figs. F.1, F.3–5 or Equations (F.1) and (F.2). For each of the two strips, calculate the instantaneous plus creep value of δ in the uncracked state 1:

$$\delta_1 = \delta_{\text{basic}} \kappa_{s1}(1 + \varphi \kappa_{\varphi 1}) \tag{9.50}$$

Calculate the value M_r of the bending moment which produces cracking. If the bending moment at the centre of a strip is less than or equal to M_r, no cracking occurs and $\delta = \delta_1$; where δ is the deflection at the middle of the strip relative to its ends, a value to be used in step 5.

(4) When cracking occurs at the mid-span section of any of the two strips, determine the curvature coefficients κ_{s2} and $\kappa_{\varphi 2}$ for a fully cracked section, using graphs of Figs. F.2, F.6, F.7 and F.8 or Equations (F.1) and (F.2) and calculate the instantaneous plus creep deflection for a fully cracked strip:

$$\delta_2 = \delta_{\text{basic}} \kappa_{s2}(1 + \varphi \kappa_{\varphi 2}) \tag{9.51}$$

Calculate the interpolation coefficient using Equation (9.24) and determine the δ-value including effects of creep and cracking.

$$\delta = (1 - \zeta)\delta_1 + \zeta \delta_2 \tag{9.52}$$

(5) Add the δ-values of a column and a middle strip according to Equation (9.44) or (9.45) to obtain the deflection at the centre of the panel. For a more reliable answer, two possible patterns of strips may be used and the probable deflection considered equal to the average of the answers from the two patterns.

When the column strips running in one direction have different δ-values, an average value is to be used in Equations (9.44) and (9.45), as shown in Fig. 9.11(c).

Example 9.5 Interior panel

Figure 9.12 is a top view of an interior square panel of a two-way slab supported directly on columns. It is required to calculate the long-term deflection due to a uniform load $8.40 \, \text{kN/m}^2$ ($175 \, \text{lb/ft}^2$), which represents the dead load plus a part of the live load. The bending moments[9] due to this load are indicated in Fig. 9.12(b) for a section at mid-span of a column and a middle strip. The reinforcement cross-section areas[10] at these two locations are given in Fig. 9.12(a). Other data are: slab thickness, $h = 0.20 \, \text{m}$ ($8 \, \text{in}$); average distance from top of slab to centroid of

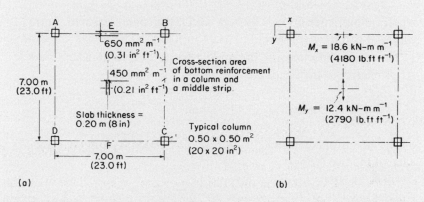

Figure 9.12 Calculation for deflection at the centre of an interior panel of a two-way slab accounting for creep and cracking (Example 9.5): (a) top view of an interior panel; (b) bending moments at mid-span of column and middle strips due to service load 8.4 kN/m² (175 lb/ft²).

bottom reinforcements in x and y directions, $d = 0.16\,\text{m}$ (6.3 in). Modulus of elasticity of concrete at time t_0 when the load is applied; $E_c(t_0) = 25\,\text{GPa}$ (3600 ksi); creep coefficient $\varphi = 2.5$; aging coefficient $\chi = 0.8$; tensile strength of concrete in flexure (modulus of rupture), $f_{ct} = 2.0\,\text{MPa}$ (290 psi), modulus of elasticity of the reinforcement = 200 GPa (29000 ksi).

Effective moment of inertia of the gross concrete section of a strip of unit width (Equation (9.48)) is

$$I_{g\,\text{slab}} = \frac{(0.2)^3}{12(1 - 0.2^2)} = 694 \times 10^{-6}\,\text{m}^4/\text{m}.$$

Poisson's ratio is assumed equal to 0.2.

In this example, the basic deflection can be calculated using coefficients from Table 9.2 which gives:

deflection at centre of panel

$$0.00482\frac{ql^4}{E_c I_{g\,\text{slab}}} = 5.60\,\text{mm}\,(0.221\,\text{in})$$

Deflection at mid-span of column strip

$$0.00342\frac{ql^4}{E_c I_{g\,\text{slab}}} = 3.97\,\text{mm}\,(0.157\,\text{in})$$

where q (load intensity) = 8.4 kN/m^2 and l, the span measured centre to centre of columns, = 7.00 m.

The basic deflections of the middle section of column and middle strips relative to their ends are:

$$\delta_{ABbasic} = 3.97\,\text{mm} \quad \delta_{EFbasic} = 1.63\,\text{mm}.$$

The effects of creep and cracking are calculated separately below for each of the two strips.

Column strip
The following parameters are determined for a section of unit width at the middle of the strip: $b = 1.00\,\text{m}$; $d = 0.16\,\text{m}$; $\rho = A_s/bd = 4.06 \times 10^{-3}$; $\chi\varphi = 2.0$; $\alpha = 8.0$. Curvature coefficients for the section in uncracked state 1 (from graphs of Figs F.1 and F.5 or Equations (F.1) and (F.2)):

$$\kappa_{s1} = 0.98 \quad \kappa_{\varphi1} = 0.93.$$

In the fully cracked state 2, the depth of compression zone is 0.036 m and the curvature coefficients are (Figs F.2 and F.6 or Equations (F.1) and (F.2)):

$$\kappa_{s2} = 7.0 \quad \kappa_{\varphi2} = 0.14.$$

Lower and upper bounds of deflection of the strip, corresponding to states 1 and 2 (Equations (9.50) and (9.51)) are

$$\delta_1 = 3.97(0.98)[1 + 2.5(0.93)] = 12.94\,\text{mm}$$

$$\delta_2 = 3.97(7.0)[1 + 2.5(0.14)] = 37.52\,\text{mm}.$$

Value of the bending moment which produces cracking (Equation (9.25)) is

$$M_r = f_{ct}W_1 = 2.0 \times 10^6 \left(\frac{1.0 \times 0.2^2}{6}\right) = 13.3\,\text{kN-m}$$

(the reinforcement is ignored in calculation of W_1). The interpolation coefficient (Equation (9.24)) is

$$1 - \beta_1 \beta_2 \frac{M_r}{M} = 1 - 1.0(0.5)\,\frac{13.3}{18.6} = 0.64.$$

The deflection at mid-span relative to the ends of the column strip including effects of cracking and creep (Equation (9.52)) is

$$\delta_{AB} = (1 - 0.64)12.94 + 0.64(37.52) = 28.67\,\text{mm}.$$

Middle strip
The value of the bending moment at mid-span does not exceed M_r; thus no cracking occurs. The cross-section has the following parameters:

$$b = 1.00\,\text{m} \quad d = 0.16\,\text{m} \quad p = A_s/bd = 2.81 \times 10^{-3} \quad \chi\varphi = 2.0.$$

Curvature coefficients for the section in the uncracked state 1:

$$\kappa_{s1} = 0.98 \quad \kappa_{\varphi 1} = 0.95$$

The deflection at mid-span relative to the ends of a middle strip including effect of creep (Equation (9.50)) is

$$\delta_{EF} = 1.63(0.98)[1 + 2.5(0.95)] = 5.38\,\text{mm}.$$

Deflection at centre of panel including effects of creep and cracking (Equation (9.44)) is

$$D = 28.67 + 5.38 = 34.05\,\text{mm}\ (1.341\,\text{in}).$$

Example 9.6 Edge panel

Figure 9.13(a) is a typical bay of a two-way slab system of equal spans 7.00 m in the x and y directions. The slab is provided by edge beams running in the y direction. It is required to find the deflection at the centre of an edge panel ABCD due to load 8.40 kN-m^2 (175 lb/ft^2). The corresponding bending moments[11] in column and in middle strips and in the edge beam are indicated in Fig. 9.13(b) and the reinforcements at mid-span sections in Fig. 9.13(c). Other data are the same as in Example 9.5.

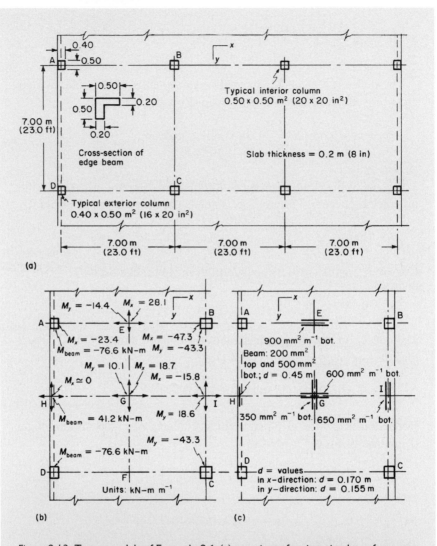

(a)

(b)

(c)

Figure 9.13 Two-way slab of Example 9.6: (a) top view of an interior bay of system; (b) bending moment in slab and in edge beam of panel ABCD; 1 kN-m/m = 225 lb-ft/ft; (c) reinforcement cross-section areas at mid-span of column and middle strips, 1 mm²/m = 0.472 × 10⁻³ in²/ft.

The basic deflection cannot be calculated by the use of a design aid such as Table 9.2 because it does not apply. Thus, we employ the equations of subsections 9.9.1 and 9.9.2.

The effective moment of inertia of the gross concrete section of a strip of unit width (Equation (9.48)) is 694×10^{-6} m⁴/m.

Consider a column strip along AB; the curvature in the x direction at the two ends is:

$$\psi_1 = -\frac{23.4 \times 10^3}{(25 \times 10^9)(694 \times 10^{-6})} = -1349 \times 10^{-6}\,\text{m}^{-1}$$

$$\psi_3 = -\frac{47.3 \times 10^3}{(25 \times 10^9)(694 \times 10^{-6})} = -2726 \times 10^{-6}\,\text{m}^{-1}.$$

Curvature at middle of the strip (Equation 9.46) is

$$\psi_2 = \frac{12[28.1 - 0.2\,(\ 14.4)]\,10^3}{(25 \times 10^9)\,(0.2)^3} = 1859 \times 10^{-6}\,\text{m}^{-1}.$$

Basic deflection at mid-span of the strip relative to its ends (Equation (9.43)) is

$$\delta_{\text{ABbasic}} = \frac{(6.55)^2}{96}(-1349 + 10 \times 1859 - 2726)10^{-6} = 6.49\,\text{mm}.$$

For a middle strip along EF, the basic curvatures at the ends and at the middle are (Equation (9.46)):

$$\psi_1 = \psi_3 = \frac{12[-14.4 - 0.2(28.1)]10^3}{(25 \times 10^9)\,(0.2)^3} = -1201 \times 10^{-6}\,\text{m}^{-1}$$

$$\psi_2 = \frac{12[10.1 - 0.2(18.7)]\,10^3}{(25 \times 10^9)\,(0.2)^3} = 382 \times 10^{-6}\,\text{m}^{-1}.$$

Basic deflection at mid-span relative to ends (Equation (9.43)) is

$$\delta_{\text{EFbasic}} = \frac{(7.00)^2}{96}(-1201 + 10 \times 382 - 1201)\,10^{-6} = 0.72\,\text{mm}.$$

Basic deflection at centre of panel (point G, Equation (9.44)) is

$$D_{\text{basic}} = 6.49 + 0.72 = 7.21\,\text{mm}\ (0.284\,\text{in}).$$

A second pattern of strips may be used to calculate D_{basic} as the average δ-value for column strips BC and AD plus δ-value for the

Table 9.3 Deflection at centre of panel calculated from two strip patterns: Example 9.6

Strip	Basic deflection, δ_{basic} (mm)	Bending moment at mid-span, M (kN-m)	Cracking moment, M_r (kN-m)	Geometrical properties of section at mid-span			Curvature coefficients and deflection in uncracked state			Curvature coefficients and deflection in fully-cracked state			Inter-polation coefficient ζ	Deflection of strip δ (mm)
				A_s (mm²)	d (m)	ρ (10^{-3})	κ_{s1}	$\kappa_{\varphi1}$	δ_1 (mm)	κ_{s2}	$\kappa_{\varphi2}$	δ_2 (mm)		
AB	6.49	28.1	13.3	900	0.170	5.29	0.96	0.88	19.91	4.68	0.16	42.46	0.76	37.05
EF	0.72	10.1	13.3	350	0.155	2.26	0.99	0.97	2.44	–	–	–	–	2.44

First estimate of deflection at centre of panel: $D = \delta_{AB} + \delta_{EF} = 37.05 + 2.44 = 39.49\ \text{mm}$

Strip	Basic deflection, δ_{basic} (mm)	Bending moment at mid-span, M (kN-m)	Cracking moment, M_r (kN-m)	A_s (mm²)	d (m)	ρ (10^{-3})	κ_{s1}	$\kappa_{\varphi1}$	δ_1 (mm)	κ_{s2}	$\kappa_{\varphi2}$	δ_2 (mm)	ζ	δ (mm)
BC	3.55	18.6	13.3	650 200³	0.155 0.04	4.19 0.89	0.98	0.93	11.57	7.48	0.14	35.85	0.64	27.11
AD²	1.46	41.2 kN-m	20.44 kN-m	top 500 bottom 600	0.45	2.22	0.92	0.79	4.00	8.21	0.09	14.68	0.75	12.03
HI	4.51	18.7	13.3	600	0.170	3.53	0.97	0.91	14.33	6.55	0.13	39.14	0.64	30.21

Second estimate of deflection at centre of panel: $D = \frac{1}{2}(\delta_{BC} + \delta_{AD}) + \delta_{HI} = \frac{1}{2}(27.11 + 12.03) + 30.21 = 49.78$

Deflection at centre of panel including effects of creep and cracking $= \dfrac{39.4 + 49.78}{2} = 44.6\ \text{mm}\ (1.76\ \text{in})$

---

[1] When $M \leq M_r$ cracking does not occur; $\delta = \delta_1$ and the columns for κ_{s2}, $\kappa_{\varphi2}$, δ_2 and ζ are left blank.

[2] The edge beam is treated as a T-section with flange width = 0.50 m.

[3] This line gives A'_s, d' and ζ' for the top reinforcement of strip AD.

middle strip along HI. This gives $D_{basic} = 7.02$ mm (0.276 in) (see Table 9.3).

The effects of the presence of the reinforcement and cracking on the initial deflection and the effect of creep on the long-term deflection are calculated using the κ-coefficients as in Example 9.5. A summary of the computations is given in Table 9.3.

9.9.4   Deflection of two-way slabs due to uniform shrinkage

The reinforcement in a slab restrains shrinkage resulting in curvature and stresses which tend to increase the deflection. The deflection due to shrinkage in a two-way slab is of course dependent upon the amount of reinforcement in two directions and on the extent of cracking in the two directions. As an approximation, it is here suggested to calculate the deflection at the centre of a slab panel as the sum of shrinkage deflections at mid-spans of column and middle strips, treated as beams using the equations of Section 9.3.2 (see Fig. 9.3).

As for beams, the deflection is affected by shrinkage in two ways. The first is a direct effect; shrinkage produces curvature which increases deflection. The second effect is indirect; shrinkage produces tensile stresses at bottom fibre at mid-span of the strips and hence enhances cracking. This may be approximately accounted for by an appropriate reduction of the tensile strength of concrete in flexure (the modulus of rupture) f_{ct} when calculating the value M_r of the bending moment which produces cracking (Equation (9.25)), as will be demonstrated in the following example.

Example 9.7 Edge panel

Calculate the deflection in the two-way slab panel of Example 9.5 due to shrinkage, $\varepsilon_{cs} = -300 \times 10^{-6}$.

Column strip AB (Fig. 9.12)
For a section at mid-span, $A_s = 900$ mm^2/m; $d = 0.17$ m; $h = 0.20$ m; $d/h = 0.85$; $a = E_s/E_c = 8$; $a\rho = 0.0424$; $\chi\varphi = 2$. Equation (F.3) or Figs. F.9 and F.10 give for the uncracked and the fully cracked states: $\kappa_{cs1} = 0.30$ and $\kappa_{cs2} = 1.12$.

Deflection due to shrinkage in non-cracked and fully cracked strips when simply supported (Equations (9.15) and (9.16)) is

$$(\Delta D_1)_{cs} = 300 \times 10^{-6}(0.30)\frac{(6.55)^2}{8(0.17)} = 2.88 \text{ mm}$$

$$(\Delta D_2)_{cs} = 300 \times 10^{-6}(1.12)\frac{(6.55)^2}{8(0.17)} = 10.61\,\text{mm}.$$

Interpolation between these two values using $\zeta = 0.76$ (see Table 9.3), gives for the simply supported strip

$$(\Delta D)_{cs} = (1 - 0.76)2.88 + 0.76(10.61) = 8.75\,\text{mm}.$$

The slab is continuous over three equal spans in the direction of strip AB; multiply the simple-beam deflection by 0.7 according to Equation (9.19) to obtain the deflection at mid-span of strips AB or DC relative to their ends

$$\delta_{AB} = \delta_{DC} = 0.7(8.75) = 6.13\,\text{mm}.$$

Middle strip EF (Fig. 9.12)
For a section at mid-span, $A_s = 350\,\text{mm}^2/\text{m}$; $d = 0.155\,\text{m}$; $h = 0.20\,\text{m}$; $d/h = 0.78$; $a = E_s/E_c = 8$; $a\rho = 0.0181$; $\chi\varphi = 2$. Equation (F.3) or Fig. F.9 gives for the uncracked state: $\kappa_{cs1} = 0.10$.

Deflection due to shrinkage in the uncracked state for a simply supported strip is

$$(\Delta D_1)_{cs} = 300 \times 10^{-6}(0.10)\frac{(7.00)^2}{8(0.155)} = 1.19\,\text{mm}.$$

The bending moment due to applied load is not sufficient to produce cracking at mid-span; the fully cracked state need not be considered. The deflection of the strip if it were simply supported is

$$(\Delta D)_{cs} = 1.19\,\text{mm}.$$

For an interior span of a continuous strip, multiply the simple-beam deflection by 0.5 according to Equation (9.19); thus

$$\delta_{EF} = 0.5(1.19) = 0.60\,\text{mm}.$$

The deflection at the centre of the slab panel due to shrinkage (Equation (9.44) with $\delta_{DC} = \delta_{AB}$) is

$$(\Delta D)_{cs} = 6.13 + 0.60 = 6.73\,\text{mm}.$$

Consider an alternative strip pattern: deflection at centre of panel

$= \dfrac{1}{2}(\delta_{AD} + \delta_{BC}) + \delta_{HI}$. Similar calculations as above give $\delta_{AD} = 1.40 \, \text{mm}; \delta_{BC}$

$= 3.95 \, \text{mm}; \delta_{HI} = 5.92 \, \text{mm}$. The deflection at the centre of the panel due to shrinkage is $8.60 \, \text{mm}$.

The average of the two values obtained by the strip patterns considered is the probable deflection at the centre of the panel due to shrinkage and is equal to $7.7 \, \text{mm}$ ($0.30 \, \text{in}$). Addition of this value to the deflection value $44.6 \, \text{mm}$ ($1.76 \, \text{in}$) calculated in Table 9.3 gives the total deflection including the effects of creep, shrinkage and cracking.

For the indirect effect of shrinkage, we determine the tensile stress at bottom fibre at mid-span of strips. In this problem the indirect effect of shrinkage is small and will be calculated for strip AB only. At mid-span, we have: $\bar{E}_c = E_c/(1 + \chi\varphi) = 8.33 \, \text{GPa}; \bar{a} = E_s/\bar{E}_c = 24; A_c = 0.1991 \, \text{m}^2; \bar{A} = 0.2207 \, \text{m}^2$; distance between centroid of $\bar{A}$ (the age-adjusted transformed section) and the bottom fibre, $y_b = 0.093 \, \text{m}$.

Curvature due to shrinkage if the strip were simply supported (Equation (9.10) with $\kappa_{cs1} = 0.30$) is

$$(\Delta\psi)_{cs} = 300 \times 10^{-6} \, \frac{0.30}{0.17} = 536 \times 10^{-6} \, \text{m}^{-1}.$$

Tensile stress at bottom fibre caused by shrinkage (Equation (9.13)) is

$$\Delta\sigma_{bot} = 8.33 \times 10^{9} \left[(536 \times 10^{-6})0.093 - (-300 \times 10^{-6}) \left(1 - \frac{0.1991}{0.2207} \right) \right]$$

$$= 0.660 \, \text{MPa}.$$

The value of tensile strength of concrete in flexure, $f_{ct} = 2.0 \, \text{MPa}$ may now be reduced to $1.340 \, \text{MPa}$. Cracking occurs at a reduced bending moment (Equation (9.25)), $M_r = 8.93 \, \text{kN-m/m}$. The corresponding interpolation coefficient (Equation (9.24)) $\zeta = 0.84$, which is larger than the value $\zeta = 0.76$ calculated in Table 9.3. This means that the indirect effect of shrinkage is to bring the deflection closer to the fully cracked state and gives $\delta_{AB} = 38.85 \, \text{mm}$, instead of $37.05 \, \text{mm}$ calculated in Table 9.3.

From the above, it can be seen that the indirect effect of shrinkage is more conveniently accounted for by estimating a reduced value of f_{ct} and then using it in the calculations for Table 9.3.

9.10   General

The simplified procedures of deflection calculation presented in this chapter are justified by extensive studies[12] comparing the results with more accurate methods using a wide range of the parameters involved.

Codes of various countries specify limits to the maximum deflection, which are not discussed here; but only a brief discussion is made below of problems which may result from excessive deflection.

Visibly large deflections are a cause of anxiety for owners and occupants of structures. However, the human eye is not, generally speaking, very sensitive to deflections and relatively large values can be tolerated. An exception is when the eye can be situated at the same level as the bottom of the member. If appearance is the only concern, one should avoid deflections greater than the span/250.

Excessive deflections can produce cracking in partitions or cause damage to other non-structural elements, e.g. glass panels. A limit on acceptable deflections in such cases is often suggested to be the smaller of: span/500 or 10 mm. However, unacceptable damages have been reported with deflections as small as span/1000.

The age of concrete when various dead loads are applied is generally not known at the time of design. In prediction of deflection, all dead loads may be assumed to be introduced simultaneously at a chosen average age.

Notes

1 See reference mentioned in Note 1, page 302.
2 Branson, D.E. (1977), *Deformation of Concrete Structures*, McGraw-Hill, New York.
3 See reference mentioned in Note 1, page 302.
4 See reference mentioned in Note 1, page 302.
5 See reference mentioned in Note 1, page 302.
6 See, for example, the following references:
   ACI 318–01, Building Code Requirements for Structural Concrete and Commentary, American Concrete Institute, Farmington Hills, Michigan 48333–9094.
   Park, R. and Gamble, W.L. (1980), *Reinforced Concrete Slabs*, Wiley, New York.
7 See, for example, the following references:
   Timoshenko, S. and Woinowsky-Krieger, S. (1959), *Theory of Plates and Shells*, McGraw-Hill, New York.
   Szilard, R. (1974), *Theory of Analysis of Plates, Classical and Numerical Methods*, Prentice Hall, Englewood Cliffs, New Jersey.
8 Table 9.2 is extracted from: Vanderbilt, M.D., Sozen, M.A. and Siess, C.P. (1965), Deflections of multiple-panel reinforced concrete floor slabs. *J. Struct. Div., Am. Soc. Civil Engrs*, **91**, No. ST4, August, 77–101.
9 The bending moment values are determined by the 'Direct Design Method' of the first two references mentioned in Note 6, above.
10 The reinforcement cross-section area approximately corresponds to an ultimate strength design with ultimate load 14.8 kN/m^2 and yield strength of reinforcement = 400 MPa (58 ksi).
11 See Notes 9 and 10, above.
12 See reference mentioned in Note 1, above.

Chapter 10

Effects of temperature

Post-tensioned precast segmental bridge erected by means of a launching truss, Kishwaukee River, Illinois. (Courtesy Prestressed Concrete Institute, Chicago.)

Precast concrete liquid storage tank, Gold Beach, Oregon. (Courtesy Prestressed
Concrete Institute, Chicago.)

10.1   Introduction

It is well known that changes in temperature can produce stresses in concrete
structures of the same order of magnitude as the dead or live loads. However,
the stresses due to temperature are produced only when the thermal expansion or contraction is restrained. High tensile stresses due to temperature
often result in cracking of concrete; once this occurs, the restraint to thermal
expansion or contraction of concrete is gradually removed and its stresses
reduced.

Most design codes require that temperature effects be considered, although
in many cases very little guidance is given on how this can be done. Thermal
stresses can be substantially reduced and the risk of damage caused by temperature eliminated by provision of expansion joints and sufficient well-
distributed reinforcements. For this reason and because of the complexity of
the problem, many structures are designed by following empirical rules for
details (e.g. Equation (E.1)), with virtually no calculation of the effects of
temperature. However, for important structures exposed to large temperature
variations, e.g. structures with members of relatively large depth exposed to
the weather, it is appropriate to have assessment of the magnitude of temperature variations and the corresponding stresses. This chapter attempts to

solve some of the problems involved. Particular attention is given to bridge superstructures.

Bridges are usually provided with expansion joints which allow the longitudinal movement due to temperature expansion in the direction of the bridge axis. Even with these joints, important stresses can develop, particularly when the structure is statically indeterminate. The stresses in the longitudinal direction in a bridge cross-section will be analysed here, treating the structure as a beam.

The first part of this chapter is concerned with temperature distribution in bridge cross-sections; other sections focus on analysis of the corresponding stresses. Effects of creep and cracking on the response of concrete structures to temperature variations will be briefly discussed.

10.2   Sources of heat in concrete structures

The chemical reaction of hydration of cement generates heat over the curing period.[1] A significant rise of temperature may occur in thick members when the dissipation of heat by conduction and convection from the surfaces is at a smaller rate than the liberated heat of hydration. Because the conductivity of concrete is relatively low, steep temperature gradients can occur between the interior of a large concrete mass and the surfaces so that the resulting stresses may produce cracking. A temperature rise of 30 to 50 °C (54 to 90 °F) can be expected in members thicker than 0.5 m (1.6) ft.[2]

The stresses due to heat of hydration occur at an early age and are thus considerably relieved by creep. Prediction of temperature distribution and the corresponding stresses and deformations due to heat of hydration, creep and shrinkage is a complex problem which has been treated only in simplified cases.[3]

Exposed concrete structures, e.g. bridges, continuously lose and gain heat from solar radiation, convection and re-radiation to or from the surrounding air. Analysis of heat flow in a body is generally a three-dimensional problem. For a concrete slab or wall or for a bridge cross-section, it is sufficient to treat it as a one- or two-dimensional problem. The major part of this chapter is concerned with temperature distribution and the corresponding stresses in bridge cross-sections.[4] The temperature at any instant is assumed constant over the bridge length, but variable over the cross-section.

The temperature distribution over a bridge cross-section varies with time and depends upon several variables:

1   geometry of the cross-section;
2   thermal *conductivity, specific heat* and *density* of the material;
3   nature and colour of the exposed surfaces, expressed in terms of solar radiation *absorptivity, emissivity* and *convection* coefficients;

4   orientation of the bridge axis, latitude and altitude of the location;
5   time of the day and the season;
6   diurnal variations of ambient air temperature and wind speed;
7   degree of cloudiness and *turbidity* of the atmosphere.

In daytime, especially in summer, heat gain is greater than heat loss, resulting in a rise of temperature. The reverse occurs in winter nights and the temperature of the structure drops. Figure 10.1 is a schematic representation of heat flow for a bridge deck during daytime in summer. Incident solar radiation is partly absorbed and the rest is reflected. The absorbed energy heats the surface and produces a temperature gradient through the deck. The amount of absorbed radiation depends upon the nature and colour of the surface: the absorptivity is higher in a dark rough surface compared to a smooth surface of light colour. Some of the absorbed heat of radiation is lost to the air by convection and re-radiation from the surface. The amount of convection depends upon wind velocity and the temperatures of the air and the surface.

10.3   Shape of temperature distribution in bridge cross-sections

Bridges are generally provided with bearings which allow free longitudinal translation of the superstructure. A change in temperature, which varies linearly over the cross-section of a simply supported bridge, produces no stresses. When the temperature variation is non-linear, the same bridge will be subjected to stresses, because any fibre, being attached to other fibres, cannot

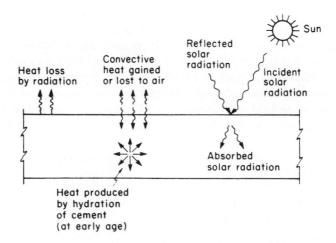

Figure 10.1 Heat transfer processes for a bridge deck in daytime in summer.

exhibit free temperature expansion. Thermal stresses in the cross-section of a statically determinate structure will be referred to as the *self-equilibrating stresses*.

Figure 10.2 shows the strain and stress distribution and the deflection of a simple beam subjected to a rise of temperature which varies linearly or non-linearly over the depth of the section. Two lines are shown for the strain distribution in the case of non-linear temperature variations. The broken line represents the hypothetical strain which would occur if each fibre were free to expand. But, because plane cross-sections tend to remain plane, the actual strain distribution is linear as shown. The difference between the ordinates of the broken line and of the straight line represents expansion or contraction which is restrained by the *self-equilibrating stresses*. Calculation of the actual strain and the self-equilibrating stress in a statically determinate structure was discussed in Section 2.4 and Example 2.1.

In a continuous bridge, a temperature rise varying linearly or non-linearly over the cross-section produces statically indeterminate reactions and internal forces. The stresses due to these forces are referred to as *continuity stresses*.

A change in temperature which is uniform over a bridge cross-section will result in a longitudinal free translation at the bearings without change in stresses or in transverse deflections. Thus, for the purpose of calculation of stresses or deflections, the temperature variation over the cross-section may be measured from an arbitrary datum. Fig. 10.3 represents the distribution

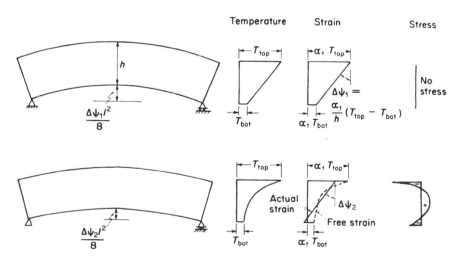

Figure 10.2 Deflection, strain and stress distribution in a simple beam due to a rise of temperature which varies linearly or non-linearly over the depth.

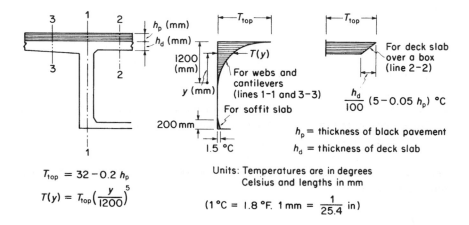

$$T_{top} = 32 - 0.2\,h_p$$

$$T(y) = T_{top}\left(\frac{y}{1200}\right)^5$$

Units: Temperatures are in degrees
Celsius and lengths in mm

$$(1\,°C = 1.8\,°F. \quad 1\,mm = \frac{1}{25.4}\,in)$$

Figure 10.3 Distribution of a rise of temperature suggested by Priestley for design of bridges or T or box sections.

over the cross-section of a bridge of a temperature rise which may be considered for design of box- and T-girders. The distribution is a combination of straight lines and a fifth-degree parabola and is based on finite difference analyses by Priestley.[5] The temperature ordinates shown in Fig. 10.3 are measured from a datum representing the temperature over the zone in the vicinity of mid-height of the web.

The temperature distribution in Fig. 10.3 represents the conditions in the early afternoon of a hot summer day. A temperature distribution of the same form, but reversed in sign (with smaller ordinates), is often suggested for design to consider the conditions in winter during the night or early in the morning.

10.4   Heat transfer equation

With the assumption that the temperature distribution over a bridge cross-section is constant over the bridge length, no heat flow occurs in the longitudinal direction and the following well-known equation applies for the heat flow in the plane of the cross-section:

$$k\frac{\partial^2 T}{\partial x^2} + k\frac{\partial^2 T}{\partial y^2} + Q = pc\frac{\partial T}{\partial t} \qquad (10.1)$$

where

T = the temperature at any point, (x, y) at any instant, t.

k = *thermal conductivity* which is the rate of heat flow by conduction per unit area per unit temperature gradient. The units of k are W/(m °C) (or Btu/(h ft °F)).

Q = amount of heat generated within the body (e.g. by hydration of cement) per unit time per unit volume, W/m³ (or Btu/(h ft³)).

ρ = density, kg/m³ (or lb/ft³).

c = *specific heat capacity*; that is, the quantity of heat required to increase the temperature of the unit mass of the material by one degree, J/(kg °C) (or Btu/(lb °F)).

Heat flow at any instant, at any point on the cross-section boundaries, follows the equation:

$$k\frac{\partial T}{\partial x}n_x + k\frac{\partial T}{\partial y}n_y + q = 0 \tag{10.2}$$

where n_x and n_y are direction cosines of an outwards vector normal to the boundary; q is the amount of heat transfer per unit time per unit area of the boundary in units of W/m² (or Btu/h ft²).

The value of q, which varies with time and with position of the point on the boundary, is the sum of three components:

$$q = q_s - q_c - q_r \tag{10.3}$$

where

q_s = the solar radiation; that is, the heat gain due to sun rays
q_c = the convection due to temperature difference between surface and air
q_r = the re-radiation from the surface to the surrounding air.

The solar radiation can be expressed:

$$q_s = a_a I_s \tag{10.4}$$

where a_a is a dimensionless solar radiation absorptivity coefficient less than 1.0; I_s is the total heat from sun rays reaching the surface per unit area per unit time.

The solar energy incident upon a surface normal to rays of the sun at a point on the outer edge of the earth's atmosphere is almost constant and equal to 1350 W/m² (428 Btu/(h ft²)). However, seasonal variation of the distance between the sun and the earth produces variation in radiation in the range of ± 3 per cent.

Only a portion of this solar radiation reaches the earth's surface, because of the atmosphere which acts like a filter. The amount of radiation which reaches the surface of the earth depends upon the length of the path of the sun's rays through the atmosphere, hence on the latitude and altitude. It also depends upon the air pollution. The angle of incidence of the sun's rays on the surface also affects the amount of solar radiation.[6]

The maximum solar radiation in summer on a horizontal surface in Europe and North America (around latitude 50) is in the order of $800–900 \, W/m^2$ ($250–300 \, Btu/(h \, ft^2)$).

The amount of heat transfer by convection is given by Newton's law of cooling:

$$q_c = h_c(T - T_a) \tag{10.5}$$

where T and T_a are temperatures of the surface and of the surrounding air, respectively; h_c is the convection heat transfer coefficient ($W/(m^2 \, °C)$ or $Btu/(h \, ft^2 \, °F)$). The value h_c depends mainly upon wind speed and to a small degree on the orientation and configuration of the surface and type of material.[7]

The amount of re-radiation from the surface to the air is given by the Stefan-Boltzmann law which may be written in the form:

$$q_r = h_r(T - T_a) \tag{10.6}$$

where h_r is the radiation heat transfer coefficient given by

$$h_r = C_s a_e [(T + T^*)^2 + (T_a + T^*)^2](T + T_a + 2T^*) \tag{10.7}$$

where C_s = Stefan-Boltzmann constant = 5.67×10^{-8} ($W/(m^2 K^4)$) or ($0.171 \times 10^{-8} \, Btu/(h \, ft^2 \, °R^4)$); T^* = constant = 273 used to convert temperature from degrees Celsius (°C) to degrees Kelvin (K) (or = 460 to convert degrees Fahrenheit (°F) to degrees Rankin (°R); a_e is a dimensionless coefficient of *emissivity* of the surface, and takes a value between 0 and 1. The latter value is for an ideal radiator, the *black body*.

Equation (10.7) indicates that h_r can be calculated only when the temperature T of the surface is known. However, because h_r is only slightly affected by T, in a time-incremental solution of Equations (10.1) and (10.2), an approximate value of h_r can be employed, based on earlier values of T.

The convection and re-radiation coefficients h_c and h_r may be combined in one overall heat transfer coefficient for the surface

$$h = h_c + h_r \tag{10.8}$$

and the heat flow by convection and re-radiation, q_{cr}, can be expressed by one equation combining Equations (10.5) and (10.6):

$$q_{cr} = h(T - T_a) \tag{10.9}$$

where

$$q_{cr} = q_c + q_r \tag{10.10}$$

For analysis of temperature distribution over the thickness of a slab or a wall, it is sufficient to employ a simplified one-dimensional form of Equations (10.1) and (10.2), by dropping out the term involving x (or y).

Numerical solution of the differential Equation (10.1), subject to the boundary condition expressed by Equation (10.2), gives the temperature distributions at various time intervals. Finite difference or finite elements[8] methods may be employed.

10.5   Material properties

From the preceding sections, it is seen that a number of values related to thermal properties of the material are involved in heat transfer analyses. For concrete, the material properties vary over wide ranges, depending mainly on composition and moisture content.

Table 10.1 gives several material properties which may be employed for analysis of temperature distribution and the corresponding stresses in bridge cross-sections.

The following values may be employed for the convection heat transfer coefficient h_c (W/(m^2 °C))(or Btu/(h ft^2 °F)), based on a wind speed of 1 m/s (3 ft/s) for all surfaces of a box-section bridge, except for the inner surfaces of the box, where the wind speed is considered zero.

	W/(m^2 °C)	Btu/(h ft^2 °F)
Top surface of concrete deck	8.5	1.5
Asphalt cover	8.8	1.6
Bottom surface of a cantilever	6.0	1.1
Inner surfaces of box	3.5	0.6
Outside box surface	7.5	1.3

10.6   Stresses in the transverse direction in a bridge cross-section

In Section 10.3 we discussed analysis of self-equilibrating and continuity thermal stresses in the direction of the axis of a bridge. Equally important stresses occur in the transverse direction in a closed box cross-section.

Table 10.1 Material properties

	Concrete		Steel		Asphalt	
Thermal conductivity, k W/(m °C) or [Btu/(h ft °F)]	1.5–2.5	(0.87–1.5)	45	(26)	1.0	(0.60)
Specific heat, c J/(kg °C) or [Btu/(lb °F)]	840–1200	(0.20–0.29)	460	(0.11)	920	(0.22)
Density, ρ kg/m³ or (lb/ft³)	2400	(150)	7800	(490)	2100	(130)
Solar radiation absorptivity coefficient, α_a (dimensionless)	0.65–0.80		0.7 (rusted)		0.9	
Radiation emissivity coefficient, α_c (dimensionless)	0.9		0.8		0.9	
Coefficient of thermal expansion, α_t per °C (or per °F)	8.0×10^{-6}	(4.4×10^{-6})	12×10^{-6}	(6.7×10^{-6})	–	

Figure 10.4(a) represents the cross-section of a box-girder bridge subjected to a rise of temperature which is assumed to be constant over the length of the bridge but varies arbitrarily over the cross-section. Stresses in the transverse direction may be calculated by considering a closed-plane frame made up of a strip between two cross-sections of the box a unit distance apart (Fig. 10.4(b)). The method of analysis discussed below is applicable to any plane frame of a variable cross-section subjected to a temperature rise which varies non-linearly over any individual cross-section (Fig. 10.4(b)) and varies from section to section. The material is assumed to be homogeneous and elastic.

If temperature expansion is artificially restrained, the restraining stress at any fibre will be:

$$\sigma_{\text{restraint}} = -\frac{E\alpha_t}{(1 - v^2)} T(y) \tag{10.11}$$

where $T(y)$ is the temperature rise at the fibre considered; α_t is the coefficient of thermal expansion; E is the modulus of elasticity and v is Poisson's ratio. The term $(1 - v^2)$ is included in this equation because the expansion of the strip (of unit width) considered here is restrained by the presence of adjacent identical strips. When considering an isolated plane frame or when expansion in the direction of the bridge axis is free to occur, which is the common case, the term $(1 - v^2)$ should be dropped.

At any section, 1 – 1, the restraining stresses have the following resultants:

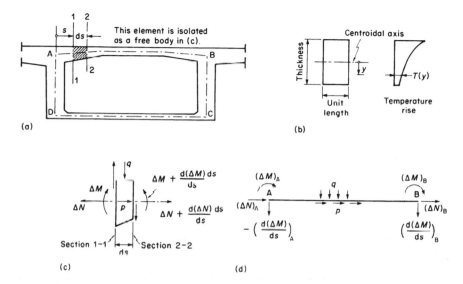

Figure 10.4 Forces necessary for the artificial restraint of the transverse expansion due to temperature in a box-girder bridge: (a) cross-section of a bridge treated as a plane frame ABCD; (b) section 1–1; (c) free-body diagram; (d) a set of self-equilibrating restraining forces for a typical member of a frame.

$$\Delta N = \int_{thickness} \sigma_{restraint}dy \tag{10.12}$$

$$\Delta M = \int_{thickness} y\sigma_{restraint}dy \tag{10.13}$$

ΔN is a normal force at the centroid of the section. Both ΔN and ΔM may vary with s as a result of the variation of temperature or the thickness; where s is a distance measured on the frame from an arbitrary origin as shown in Fig. 10.4. The element ds in Fig. 10.4(a) is isolated as a free body in Fig. 10.4(c). Considering equilibrium, we can see that tangential and transverse forces of magnitudes per unit length equal to p and q must exist; where

$$p = -\frac{d(\Delta N)}{ds} \tag{10.14}$$

$$q = -\frac{d^2(\Delta M)}{ds^2} \tag{10.15}$$

In other words, the loads p and q must be applied in order to restrain artificially the thermal expansion. The set of restraining forces for a typical member represents a system in equilibrium (Fig. 10.4(d)).

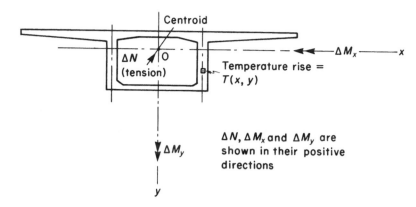

Figure 10.5 Cross-section of a statically determinate bridge; symbols and sign conventions used in Equations (10.17–25).

The artificial restraint must now be eliminated by the application – to all members – of forces equal and opposite to those shown in Fig. 10.4(d). The internal forces and stresses due to this loading on the continuous frame are to be analysed by a conventional method (e.g. the displacement method; see Section 5.2). Let the stress at any fibre obtained from such an analysis be $\Delta\sigma$. The actual stress due to temperature is given by superposition:

$$\sigma = \sigma_{\text{restraint}} + \Delta\sigma \qquad (10.16)$$

10.7   Self-equilibrating stresses

Analysis of the stresses in a direction parallel to the axis of a bridge due to temperature is discussed here and in the following section. As mentioned in Section 10.3, longitudinal stress (referred to as self-equilibrating stresses) occurs in a statically determinate bridge only when the temperature distribution is non-linear. Equation (2.30) can be used to calculate the self-equilibrating stress in a bridge cross-section when the temperature is assumed to vary in the vertical (y) direction. The equations given below are usable when the temperature T varies in both the horizontal and vertical directions.

Consider a statically determinate bridge having the cross-section shown in Fig. 10.5. In general, the temperature distribution varies in both the x and y directions. Consider the stress, strain and curvature caused by a temperature rise with a given temperature distribution $T(x, y)$. If the temperature expansion is artificially restrained, a normal stress will be produced and its magnitude at any fibre will be:

$$\sigma_{\text{restraint}} = - Ea_t T(x, y) \qquad (10.17)$$

and the stress resultants on any section are:

$$\Delta N = \int \int \sigma_{\text{restraint}} \, \mathrm{d}x \, \mathrm{d}y \tag{10.18}$$

$$\Delta M_x = \int \int \sigma_{\text{restraint}} \, y \mathrm{d}x \, \mathrm{d}y \tag{10.19}$$

$$\Delta M_y = \int \int \sigma_{\text{restraint}} \, x \mathrm{d}x \, \mathrm{d}y \tag{10.20}$$

where ΔN is a normal force at the centroid.

To remove the artificial restraint, we apply on the cross-section the forces $-\Delta N$, $-\Delta M_x$ and $-\Delta M_y$, producing at any point the stress

$$\Delta \sigma = -\left(\frac{\Delta N}{A} + \frac{\Delta M_x}{I_x} y + \frac{\Delta M_y}{I_y} x \right) \tag{10.21}$$

where A is the area of the cross-section; I_x and I_y are moments of inertia about centroidal axes x and y.

The self-equilibrating stresses are given by superposition:

$$\sigma = \sigma_{\text{restraint}} + \Delta \sigma \tag{10.22}$$

The normal strain at the centroid O and the curvatures in the yz and xz planes respectively are:

$$\Delta \varepsilon_O = -\frac{\Delta N}{EA} \tag{10.23}$$

$$\Delta \psi_x = -\frac{\Delta M_x}{EI_x} \tag{10.24}$$

$$\Delta \psi_y = -\frac{\Delta M_y}{EI_y} \tag{10.25}$$

Substitution of Equations (10.17–20) in the last three equations would show that the values $\Delta \varepsilon_O$, $\Delta \psi_x$ and $\Delta \psi_y$ are independent of the value of E.

10.8 Continuity stresses

Equations (10.23–25) give the axial strain and the curvatures at any cross-section of a statically determinate beam. These can be used to calculate the displacements at member ends. If these displacements are not free to occur, as for example in a continuous structure, statically indeterminate forces develop, producing continuity stresses which must be added to the self-equilibrating stresses to produce the total stresses at any section. Analysis of the statically indeterminate forces can be performed by the general force or displacement methods (see Sections 4.2 and 5.2).

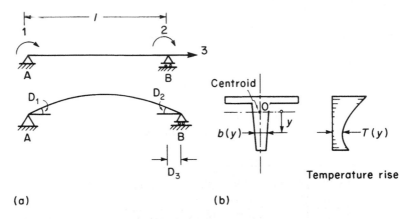

Figure 10.6 Displacements at the ends of a simple beam due to temperature:
(a) coordinate system; (b) temperature distribution.

Consider the simple beam shown in Fig. 10.6(a), which has a constant cross-section subjected to a rise of temperature varying over the depth as shown in Fig. 10.6(b). The displacements at the coordinates shown at the member ends are given by:

$$D_1 = -D_2 = \frac{\Delta\psi l}{2} \tag{10.26}$$

$$D_3 = \Delta\varepsilon_0 l \tag{10.27}$$

where $\Delta\varepsilon_0$ and $\Delta\psi$ are respectively the normal strain at the centroid and the curvature at any section caused by the temperature change and are given by substitution of Equations (10.17–20) into (10.23) and (10.24) (dropping the integral with respect to x):

$$\Delta\varepsilon_0 = \frac{a_t}{A} \int T b \, dy \tag{10.28}$$

$$\Delta\psi = \frac{a_t}{I} \int T b \, y dy \tag{10.29}$$

where $b = b(y)$ is the breadth of the cross-section at any fibre; A and I are the cross-section area and moment of inertia about the centroidal axis.

If the same beam AB is made continuous with an identical span BC as shown for the first beam in Fig. 10.7, the rotation at B cannot occur and a statically indeterminate connecting moment must be produced at B. The

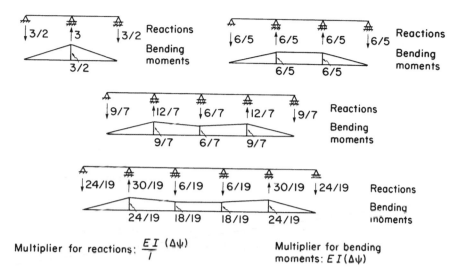

Figure 10.7 Statically indeterminate forces due to temperature rise in continuous beams of equal spans. $\Delta\psi$ = curvature in a statically determinate beam $= (\alpha_t/h) \int T\, by\, dy$ (see Fig. 10.6(b)).

value of the connecting moment may be calculated by the force method (see Example 10.1).

Figure 10.7 gives the statically indeterminate bending moment diagrams and the reactions in continuous beams of constant cross-section, having two to five equal spans, subjected to a rise of temperature which varies over the depth of the section in arbitrary shape (Fig. 10.6(b)). The statically indeterminate values are expressed in terms of the quantity $\Delta\psi$, the curvature at any section when the static indeterminacy is released. The numerical value of $\Delta\psi$ is obtained by evaluation of the integral in Equation (10.29) (see Example 10.1).

Example 10.1 Continuous bridge girder

Find the stress distribution at support B of the continuous bridge shown in Figs 10.8(a) and (b) due to a rise of temperature whose distribution varies over the depth of the cross-section as suggested by Priestley (Fig. 10.3). Consider $E = 30.0\,\text{GPa}$ (4350 ksi) and $\alpha_t = 1.0 \times 10^{-5}$ per °C (5.6×10^{-6} per °F). Ignore rigidity of the pavement.

In accordance with the rules in Fig. 10.3, the temperature rise to be considered in the analysis varies over the top 1.2 m (4 ft) as a fifth-degree parabola (Fig. 10.8(b)).

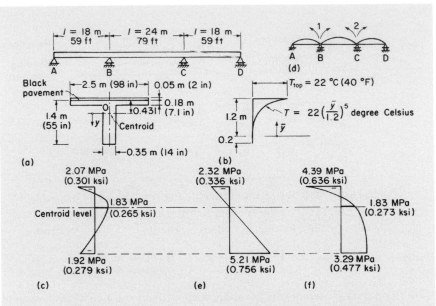

Figure 10.8 Analysis of stress distribution due to temperature in a bridge girder
(Example 10.1): (a) elevation and cross-section; (b) rise of temperature;
(c) self-equilibrating stresses in any section; (d) released structure; (e)
continuity stresses at B; (f) total temperature stresses at B.

The cross-section area A is $0.877\,\mathrm{m}^2$; the centroid O is at $0.969\,\mathrm{m}$ above the soffit; the moment of inertia I about a horizontal axis through the centroid is $0.1615\,\mathrm{m}^4$.

Hypothetical strain that would occur at any fibre if it were free to expand (Equation (2.21)) is:

$$\varepsilon_f = 1.0 \times 10^{-5}\left[22\left(\frac{\bar{y}}{1.2}\right)^5\right]$$

where $\bar{y}$ is a distance in metres measured upwards from a point $0.2\,\mathrm{m}$ above the soffit (see Fig. 10.8(b)).

The stress necessary to prevent this expansion (Equation (2.22)) is

$$\sigma_{\text{restraint}} = -(30 \times 10^9)\left[10^{-5} \times 22\left(\frac{\bar{y}}{1.2}\right)^5\right]$$

$$= -(2.652 \times 10^6)(\bar{y})^5 \ \text{N-m}^2.$$

The resultants of this stress are (Equations (2.23) and (2.24)):

$$\Delta N = -(2.652 \times 10^6) \left(0.35 \int_0^{1.02} \bar{y}^5 \, d\bar{y} + 2.5 \int_{1.02}^{1.20} \bar{y}^5 \, d\bar{y} \right)$$

$$= -2.229 \times 10^6 \text{ N.}$$

$$\Delta M = -(2.652 \times 10^6) \left(0.35 \int_0^{1.02} (0.769 - \bar{y}) \bar{y}^5 \, d\bar{y} \right.$$

$$\left. + 2.5 \int_{1.02}^{1.2} (0.769 - \bar{y}) \bar{y}^5 d\bar{y} \right) = 0.7438 \times 10^6 \text{ N-m.}$$

Release the artificial restraint by application of $(-\Delta N)$ and $(-\Delta M)$ on the cross-section; the resulting axial strain and curvature are (Equation (2.29)):

$$\begin{Bmatrix} \Delta\varepsilon_o \\ \Delta\psi \end{Bmatrix} = \frac{1}{30 \times 10^9} \times \begin{Bmatrix} \dfrac{2.229 \times 10^6}{0.877} \\ -\dfrac{0.7438 \times 10^6}{0.1615} \end{Bmatrix} = \begin{Bmatrix} 84.73 \times 10^{-6} \\ -153.5 \times 10^{-6} \text{ m}^{-1} \end{Bmatrix}$$

The stress in a statically determinate beam (the self-equilibrating stresses, Equation (2.30)) is:

$$\sigma_{\text{self-equilibrating}} = [2.542 - 4.606y - 2.652(0.769 - y)^5] \text{ MPa}$$

$$\text{for } y = -0.431 \text{ to } 0.769 \text{ m}$$

or

$$\sigma_{\text{self-equilibrating}} = (2.542 - 4.606y) \text{ MPa}$$

$$\text{for } y = 0.769 \text{ to } 0.969 \text{ m}$$

where y is the distance in metres measured downwards from the centroid O. The distribution of the self-equilibrating stress is shown in Fig. 10.8(c).

We use the force method for the analysis of the statically indeterminate forces. The structure is released by the introduction of hinges at B and C as shown in Fig. 10.8(d). The displacements of the released structure at the two coordinates indicated are (Equation (10.26)):

$$D_1 = D_2 = -153.5 \times 10^{-6} \left(\frac{18}{2} + \frac{24}{2} \right) = -3224 \times 10^{-6}.$$

The displacements at the two coordinates due to $F_1 = 1$ at coordinate 1 (the flexibility coefficients):[9]

$$f_{11} = \left(\frac{l}{3EI} \right)_{AB} + \left(\frac{l}{3EI} \right)_{BC}$$

$$= \frac{1}{3 \times 30 \times 10^9 \times 0.1615} (18 + 24) = 2.890 \times 10^{-9}$$

$$f_{21} = \left(\frac{l}{6EI} \right)_{BC} = 0.826 \times 10^{-9}.$$

Because of symmetry, one compatibility equation only is necessary to solve for the two redundant forces ($F_1 = F_2$):

$$(f_{11} + f_{12})F_1 = -D_1$$

$$F_1 = -\frac{D_1}{f_{11} + f_{12}} = 867.5 \times 10^3 \text{ N-m.}$$

Thus, the statically indeterminate bending moment at B or C is 867.5 kN-m. The stress distribution due to this bending moment (the continuity stress) is shown in Fig. 10.8(e). The total stress distribution at B due to temperature is the sum of the values in Fig. 10.8(c) and (e); the result is shown in Fig. 10.8(f).

In a prestressed concrete bridge, the prestress force near the top fibre over an interior support is often relatively high, resulting in small or no compressive stress at the bottom fibre in service conditions. The thermal tensile stress in this area (see Fig. 10.8(e)) may cause vertical cracks near the soffit in the vicinity of the support. The consequence of this cracking may be alleviated by provision of reinforcement and reduction of the prestress force (partial prestressing).

10.9   Typical temperature distributions in bridge sections

Concrete bridges of the same depth but with different cross-section shapes have almost the same temperature distribution. However, the temperature distribution and the resulting stresses vary considerably with the cross-section depth. With greater depth, higher temperature stresses occur.

Figure 10.10[10] shows the temperature distributions and the corresponding self-equilibrating stresses in three bridges of the same depth but with the cross-section configurations shown in Fig. 10.9. Fig. 10.11 shows the distributions of temperature and self-equilibrating stresses in five cross-sections varying in depth from 0.25 to 2.25 m (10–89 in).

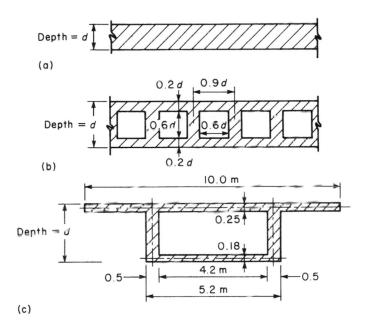

Figure 10.9 Bridge cross-sections analysed by Elbadry and Ghali to study effects of section shape and depth on temperature distribution: (a) solid slab; (b) cellular slab; (c) box girder.

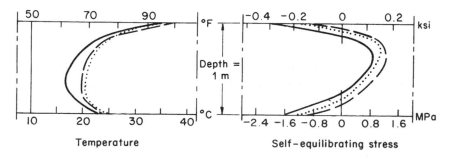

Figure 10.10 Temperature and self-equilibrating stresses in three bridge cross-sections with the same depth (Fig. 10.9) (summer conditions in Calgary, Canada): full curve, solid slab; dotted curve, cellular slab; broken curve, box girder.

Composite cross-sections may exhibit a high temperature difference between concrete and steel (45 °C (81 °F)) when the vertical sides of the webs are exposed to the sun.

10.10   Effect of creep on thermal response

In Sections 10.7 and 10.8 we considered the stresses produced by a temperature rise which varies in an arbitrary manner over the cross-section of a concrete beam. Now we shall consider that the rise of temperature develops gradually with time during a period t_0 to t; where t_0 and t are the ages of concrete at the start and at the end, respectively. Assuming that the temperature expansion is artificially prevented, the normal stress which will be produced at any fibre

$$\sigma_{\text{restraint}} = -\bar{E}aT(x, y) \tag{10.30}$$

where $\bar{E} = \bar{E}_c(t, t_0)$ is the age-adjusted elasticity modulus of concrete as defined by Equation (1.31), which is repeated here for the sake of convenience:

$$\bar{E}_c(t, t_0) = \frac{E_c(t_0)}{1 + \chi\varphi(t, t_0)} \tag{10.31}$$

where

$E_c(t_0)$ = modulus of elasticity of concrete at age t_0
$\varphi(t, t_0)$ = the ratio of creep during the period $(t - t_0)$ to the instantaneous strain due to a stress introduced at age t_0
$\chi = \chi(t, t_0)$ = the aging coefficient.

For values of χ and φ, see Appendix A. The aging coefficient $\chi = 1.0$ when the stress is introduced in its entire value at time t_0 and maintained constant to time t. The value of χ is less than 1.0, when the stress is introduced gradually (see graphs in Figs A.6 to A.45).

The equations derived in Sections 10.7 and 10.8 for the self-equilibrating and the continuity stresses are applicable in the case considered here, with E replaced by $\bar{E}$. The change in temperature due to weather conditions occurs over a period of time (several hours or days), during which some creep occurs. Thus, it may be more appropriate to employ $\bar{E}$, rather than the instantaneous elasticity modulus of concrete. This will generally result in a smaller absolute value of the calculated stresses due to temperature.

Heat of hydration of cement causes a rise of temperature which may develop gradually to a peak over a period of time, for example one week; the

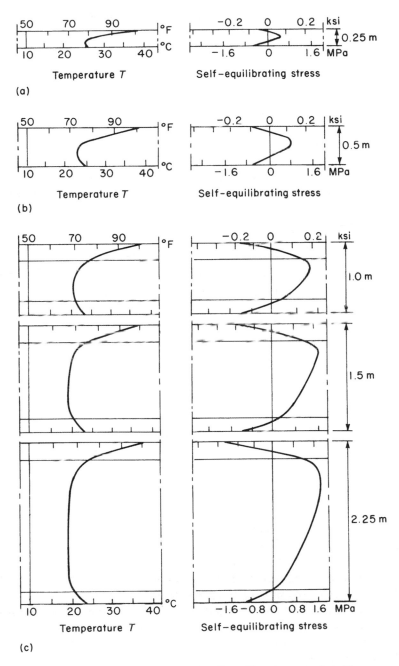

Figure 10.11 Distribution of temperature and self-equilibrating stress in bridge cross-sections of different depths and of shapes shown in Fig. 10.9 (summer conditions in Calgary, Canada): (a) solid slab; (b) cellular slab; (c) box girder.

temperature rise subsequently vanishes slowly over a much longer period. The stresses due to this temperature change may be analysed in steps by dividing the time into intervals and considering that increments of temperature or stresses occur suddenly at the middle of the intervals. For each interval, an appropriate creep coefficient and modulus of elasticity is employed (see Section 5.8). Considering creep in this fashion will result in substantially different stresses from a calculation in which creep and change in modulus of elasticity are ignored.

In fact, considering these time-dependent effects may indicate that the stresses developed at peak temperature reverse signs after a long time when the heat of hydration is completely lost.[11] This can be seen in Example 10.2 which treats the problem using a step-by-step numerical analysis.

A general procedure for a step-by-step procedure of stress analysis of concrete structures is discussed in Section 5.8. Consider here the application of the method for analysis of the self-equilibrating stresses in a cross-section of a concrete member due to a rise of temperature which varies with time. Divide the time, during which the temperature change occurs, into a number of intervals. The symbols $t_{i-\frac{1}{2}}$, t_i and $t_{i+\frac{1}{2}}$ represent the age of concrete at the beginning, middle and end of the ith interval. At the end of any interval i, the strain due to free temperature expansion is the summation:

$$a_t \sum_{j=1}^{i} (\Delta T)_j \qquad\qquad (10.32)$$

This strain is prevented artificially by the introduction of stress $(\Delta\sigma_{\text{restraint}})_j$ at the middle of the intervals. The combined strain caused by temperature and these stress increments is zero. For the end of the ith interval, we can write

$$a_t \sum_{j=1}^{i} (\Delta T)_j + \sum_{j=1}^{i} \frac{(\Delta\sigma_{\text{restraint}})_j}{E_c(t_j)} [1 + \varphi(t_{i+\frac{1}{2}}, t_j)] = 0 \qquad\qquad (10.33)$$

where $E_c(t_j)$ is the modulus of elasticity of concrete at the middle of the jth interval; $\varphi(t_{i+\frac{1}{2}}, t_j)$ is the ratio of creep occurring between the middle of the jth interval and the end of the ith interval to the instantaneous strain when a stress is introduced at t_j. The summation in the second term of the equation represents the instantaneous strain plus creep caused by the stress increments during the intervals 1, 2, . . . , i.

In a step-by-step analysis, when Equation (9.33) is applied at any interval i, the stress increments are known for the earlier intervals. Thus, the equation can be solved for the stress increment in the ith interval, giving:

$$(\Delta\sigma_{\text{restraint}})_i = -\frac{E_c(t_i)}{1 + \varphi(t_{i+\frac{1}{2}},\, t_i)}\left(a_t \sum_{j=1}^{i}(\Delta T)_j \right.$$

$$\left. + \sum_{j=1}^{i-1}\frac{(\Delta\sigma_{\text{restraint}})_j}{E_c(t_j)}[1 + \varphi\,(t_{i+\frac{1}{2}},\, t_j)]\right) \qquad (10.34)$$

The resultant of the stress increment $(\Delta\sigma_{\text{restraint}})_i$ for the ith interval is to be integrated over the area of the cross-section to determine the corresponding stress resultants $(\Delta N)_i$, $(\Delta M_x)_i$ and $(\Delta M_y)_i$. Equal and opposite forces are applied on the cross-section to remove the artificial restraint; the corresponding stress, strain or curvature are derived using Equations (10.21–25), employing a modulus of elasticity $E = E_c(t_i)$. The analysis in this way gives the changes in the self-equilibrating stresses in the individual increments and these may be summed up to find the stress at any time.

Example 10.2 Wall: stress developed by heat of hydration

Figure 10.12(a) represents a typical distribution of the temperature rise developed by the heat of hydration in a concrete wall of thickness h. Let T_c be the difference of temperature rise between the middle surface of the wall and its faces; assume the distribution to be parabolic at all times. The value of the temperature rise T_c is assumed to be zero at the age of casting, reaches a peak value $T_{c\ \max}$ at the age of 6 days and drops to $0.06\ T_{c\ \max}$ at the age of 50 days. It is required to determine the self-equilibrating stress distributions at ages 6 and 50 days. Assume that temperature expansion of the wall is free to occur.

Use three time intervals for which the interval limits are: 0, 2, 6 and 50 days, and assume the temperature increments at the wall centre line in the three intervals $T_{c\ \max}\{0.53, 0.47, -0.94\}$. Assume the moduli of elasticity of concrete at the middle of the three intervals $E_c(28)\ \{0.44, 0.73, 1.00\}$; where $E_c(28)$ is the modulus of elasticity at age 28 days (see Equation (A.37)).

The following creep coefficients are required in the step-by-step analysis:

$\varphi(2, 1) = 0.56$ $\varphi(6, 1) = 0.60$ $\varphi(50, 1) = 0.82$

$\varphi(6, 4) = 0.55$ $\varphi(50, 4) = 0.76$ $\varphi(50, 28) = 0.48$.

Successive application of Equation (10.34), with $i = 1$, 2 and 3 gives

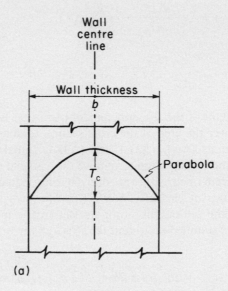

Wall
centre
line

Wall thickness
b

T_c

Parabola

(a)

Multiplier:

$\alpha_t\, T_{c\,max}\, E_c(28)$

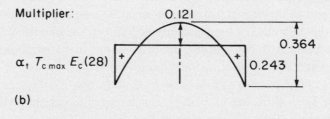

0.121

0.364

0.243

(b)

Multiplier:

$\alpha_t\, T_{c\,max}\, E_c(28)$

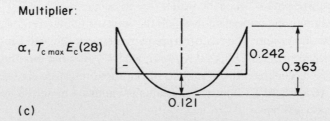

0.242

0.363

0.121

(c)

Figure 10.12 Self-equilibrating stresses caused by heat of hydration of cement in a thick concrete wall (Example 10.2): (a) assumed distribution of temperature at any time; (b) self-equilibrating stress at age 6 days (time at which $T_{c\,max}$ occurs); (c) self-equilibrating stress at age 50 days (time when the temperature rise is almost lost).

the following values of the stress increments at wall centre line if the temperature expansion is artificially prevented:

$$(\Delta\sigma_{\text{restraint}})_1 = a_t T_{c\,\text{max}} E_c(28) \left(-\frac{0.44}{1+0.56}\right)(0.53) = -0.150[a_t T_{c\,\text{max}} E_c(28)]$$

$$(\Delta\sigma_{\text{restraint}})_2 = a_t T_{c\,\text{max}} E_c(28) \left(-\frac{0.73}{1+0.55}\right)\left((0.53+0.47)\right.$$

$$\left. -\frac{0.150}{0.44}(1+0.60)\right)$$

$$= -0.214[a_t T_{c\,\text{max}} E_c(28)]$$

$$(\Delta\sigma_{\text{restraint}})_3 = a_t T_{c\,\text{max}} E_c(28) \left(-\frac{1.00}{1+0.48}\right)$$

$$\times \left((0.53+0.47-0.94) - \frac{0.150}{0.44}(1+0.82) - \frac{0.214}{0.73}\right.$$

$$\left. (1+0.76)\right)$$

$$= 0.727\,[a_t T_{c\,\text{max}} E_c(28)].$$

Summation of the increments gives the following values of stresses at ages 6 and 50 days:

$$\Delta\sigma_{\text{restraint}}(6) = -0.364[a_t T_{c\,\text{max}} E_c(28)]$$

$$\Delta\sigma_{\text{restraint}}(50) = 0.363[a_t T_{c\,\text{max}} E_c(28)].$$

The increments of self-equilibrating stress may be calculated separately for each interval and then the increments added to give the stress distributions shown in Figs 10.12(b) and (c) at ages 6 and 50 days, respectively. The same results will be reached if the change in temperature is considered instantaneous and the modulus of elasticity $E = E_c(28)$ and the temperature distributions parabolic with values at the wall centre line of $0.364\,T_{c\,\text{max}}$ and $-0.363\,T_{c\,\text{max}}$ for the stresses at 6 and 50 days, respectively.

For any symmetrical temperature distribution T, as considered in this example, the self-equilibrating stress for an elastic material may be calculated by the equation:

$$\sigma = -Ea_t T + (Ea_t)\,T_{\text{average}} \qquad (10.35)$$

where T_{average} is the average temperature.

Figures 10.12(b) and (c) indicate that the stresses at the outer fibres of the wall are tensile at age 6 days but they become compressive at age 50 days. To have an idea about the magnitude of the self-equilibrating stress at the surfaces, assume the following values: $T_{c\,max} = 30\,°C\,(54\,°F)$; wall thickness $b = 1.0\,m$ (3.3 ft); ($E_c(28) = 30.0\,GPa$ (4350 ksi); $a_t = 1 \times 10^{-5}$ per °C (0.6 × 10^{-5} per °F). This gives the following stresses at the outer surface: 2.19 MPa (0.317 ksi) at age 6 days and −2.18 MPa (−3.17 ksi) at age 50 days. Note that the dimension b does not directly affect the stress, but of course it affects the value $T_{c\,max}$ and the creep coefficients.

The stress reversal at the older age may be explained as follows. A stress introduced at an early age causes a relatively large strain because of a smaller modulus of elasticity and larger creep. A rise of temperature at an early age can be restrained by a stress smaller in absolute value than the corresponding stress for a drop of temperature of the same magnitude but occurring at an older age. Thus, the self-equilibrating stress developed while the temperature is rising is more than offset by the self-equilibrating stress produced by the subsequent cooling.

In this example, we assumed that the wall is free to expand and we calculated the self-equilibrating stresses. Much larger stress would occur if the wall edges were not free; this can be seen by comparing the magnitude of $\sigma_{restrained}$ with the self-equilibrating stress in Fig. 10.12(c).

It can be seen from this example that for the analysis of the stresses due to heat of hydration, it is necessary to know the temperature distribution and its history, as well as the mechanical properties of concrete at various ages starting from the time of hardening. The information required for the analysis is not usually easy to obtain.

It is well known that creep of concrete is of a larger magnitude when the temperature is higher. This is of importance when concrete is subjected to elevated temperature, e.g. in power plants. The step-by-step analysis discussed above and in Section 5.8 may be applied, using values of creep coefficients that are functions of the temperature.

10.11   Effect of cracking on thermal response

In general, the absolute values of stresses caused by temperature in a cracked reinforced concrete cross-section are smaller than in an uncracked section.

Calculation of stresses caused by temperature on cracked structures is complex. Simplifying assumptions are necessary to make the calculations reasonably simple. In the following, we shall consider that cracking is produced by loads other than temperature and assume that the depth of the compression zone remains unchanged by the effect of temperature. With these assumptions, the analysis of the self-equilibrating stresses in a statically determinate structure or the continuity stresses in an indeterminate structure may be performed in the same way as discussed in Sections 10.7 and 10.8. But the actual cross-section of the members must be replaced by a transformed section composed of the area of concrete plus a times the area of steel; where $a = E_s/E_c$ is the ratio of the modulus of elasticity of steel to that of concrete.

For qualitative assessment of the effect of cracking, we consider the cross-section in Fig. 10.13(a) of a statically determinate structure and calculate the self-equilibrating stresses and the strain due to a temperature rise which

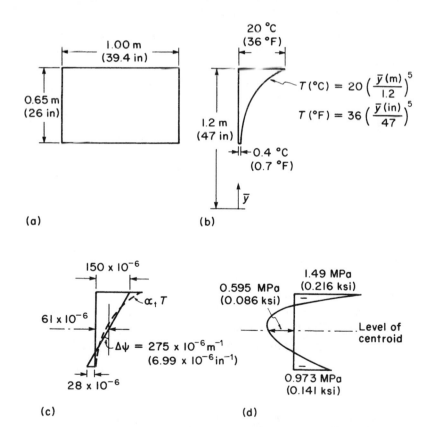

Figure 10.13 Distributions of strain and self-equilibrating stresses due to temperature in a homogeneous elastic cross-section: (a) concrete cross-section; (b) distribution of temperature rise; (c) strain; (d) stress.

varies over the depth as shown in Fig. 10.13(b). Figures 10.13(c) and (d) show the distributions of strain and stress if the section is considered of homogeneous elastic material. The values shown may be checked by the method presented in Section 10.7 with the following data:

$$E_c = 30.0\,\text{GPa}\ (4350\,\text{ksi});\ a_t = 1 \times 10^{-5}\ \text{per}\ ^\circ\text{C}\ (0.6 \times 10^{-5}\ \text{per}\ ^\circ\text{F}).$$

The same cross-section is considered cracked at the bottom or at the top and provided with 1 per cent reinforcement ($6000\,\text{mm}^2$ ($9.30\,\text{in}^2$)) at the cracked face (Figs 10.14(a) and (d)). Concrete is ignored over a cracked zone of depth $0.467\,\text{m}$ ($18.4\,\text{in}$). The distributions of strain and the self-equilibrating stresses due to the temperature rise in Fig. 10.13(b) are shown in Figs 10.14(b) and (c) when the cracking is at the bottom and in Figs 10.14(e) and (f) when the cracking is at the top. The values shown in these figures are obtained by application of Equations (2.21), (2.22), (2.29) and (2.30) and employing the following properties of the transformed section: $a = E_s/E_c = 6.67$; area, $A = 0.223\,\text{m}^2$ ($346\,\text{in}^2$); moment of inertia about centroidal axis, $I = 9.00 \times 10^{-3}\,\text{m}^4$ ($21\,600\,\text{in}^4$).

Comparison of the stress values in Figs 10.13(a), 10.14(c) and 10.14(f) indicates that the self-equilibrating stresses caused by temperature are generally smaller in the cracked section. However, the corresponding strain values and particularly the curvatures, $\Delta\psi$ are not much different (Figs 10.13(c), 10.14(b) and 10.14(e)). It follows that the strains and hence the displacements

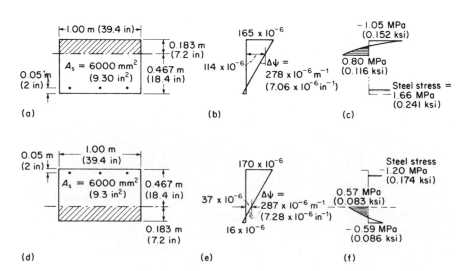

Figure 10.14 Distributions of strain and self-equilibrating stresses in a cracked reinforced concrete section due to a temperature rise shown in Fig. 10.13(b): (a) cracking at bottom face; (b) strain; (c) stress; (d) cracking at top face; (e) strain; (f) stress.

(e.g. elongations or rotations at member ends) due to temperature in a statically determinate cracked structure may be approximated by considering the cross-sections to be homogeneous, elastic and with no cracks.

It also follows that calculation of the statically indeterminate forces produced by temperature, using the force method (see Section 4.2), can be simplified by calculation of the displacements $\{D\}$ of a statically determinate uncracked structure and considering cracking only when calculating the flexibility matrix $[f]$. If, for example, the continuous beam in Fig. 10.8(a) (Example 10.1) was cracked, the flexibility coefficients would be (see Equation (3.32))

$$ f_{11} = \int \frac{M_{u1}^2 \, \mathrm{d}l}{E_c I} \quad f_{12} = \int \frac{M_{u1} M_{u2} \mathrm{d}l}{E_c I} \quad f_{22} = \int \frac{M_{u2}^2 \, \mathrm{d}l}{E_c I} $$

where M_{u1} and M_{u2} are bending moments due to unit couples applied at coordinates 1 and 2 on a released structure (Fig. 10.8(d)). I is the moment of inertia of a transformed cracked section (or uncracked where no cracking occurs). The displacements $\{D\}$ calculated in Example 10.1 may be employed without change in the case of a cracked continuous beam. The result of such calculation will generally give smaller statically indeterminate forces $\{F\}$ when cracking is considered.

It should be mentioned that when the concrete in the cracked zone is completely ignored, as suggested above, the flexibility coefficients f_{ij} are generally overestimated; hence, the calculated statically indeterminate forces will be somewhat lower than the true values.

The above discussion indicates that the stresses or the internal forces produced by temperature depend upon the extent of cracking caused by other loading. Thus, at service conditions when no or little cracking occurs, the stresses or internal forces induced by a temperature increment are large compared to the effect of the same increment when introduced at a higher level close to the ultimate strength of the structure. This may be seen by considering the identical moment-curvature diagrams shown in Fig. 10.15, which are typical for a reinforced concrete cross-section. An increment $(\Delta \psi)$ in the curvature due to temperature is introduced in Fig. 10.15(a) in the service condition near the linear part of the graph. The same increment, introduced near the ultimate strength of the section (Fig. 10.15(b)), produces a smaller increment in moment compared to the increment in moment in Fig. 10.15(a). Thus, the effect of temperature is of less significance at the ultimate load than at service conditions. The effects of temperature must be considered in design for service conditions and sufficient reinforcement provided to ensure that the cracks are closely distributed and the crack width within acceptable limits, as opposed to wide cracks far apart.

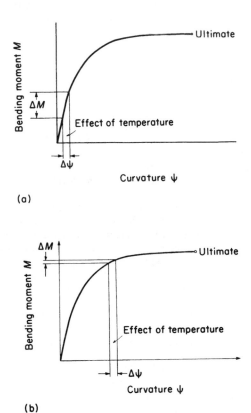

Figure 10.15 Comparison of bending-moment increments corresponding to an increment of curvature introduced at service conditions (a) or near ultimate limit state (b).

10.12 General

Design of concrete structures for the effects of temperature is complex. The temperature distribution is not easy to predict and is variable with time. The combination of the effect of temperature and that of other loadings is not clearly specified in design codes. The stresses developed due to temperature are affected by the age of concrete and by creep. A further complication results from cracking which limits the validity of superposition. This chapter by no means provides a complete solution to all these problems.

The stresses due to temperature are generally smaller in a cracked structure compared to a structure without cracks. This favours the use of limited pre-stressing and provision of non-prestressed steel to furnish the necessary strength with allowance for cracking, as opposed to a design in which the total strength is provided for by prestressed reinforcement without cracking.

It is believed that with the first design, the structure is less vulnerable to damages at peak temperatures; when sufficient and well-distributed non-prestressed reinforcement is provided, the cracks will be of small width.

Notes

1 For information on the amount and rate of heat generated and additional references, see Neville, A.M., (1997), *Properties of Concrete*, 4th ed., Wiley, New York.
2 See Leonhardt, F. (1982), *Prevention of Damages in Bridges: Proceedings of the Ninth International Congress of the FIP*, Stockholm, Commission Reports, Vol. 1, June.
3 See Thurston, S.J., Priestley, N. and Cooke, N. (1980), Thermal analysis of thick concrete sections. *American Concrete Institute Journal*, No. 77–38, Sept.–Oct., 347–57.
4 For more details and references, see Elbadry, M.M. and Ghali, A. (1983), Nonlinear temperature distribution and its effects on bridges. *International Association of Bridge and Structural Engineering Proceedings*, pp. 66/83, Periodica 3/1983.
5 Priestley, M.J.N. (1976), Design of thermal gradients for concrete bridges. *New Zealand Engineering*, **31**, No. 9, September, 213–19.
6 Extensive discussion and equations for evaluation of the solar radiation can be found in Duffie, J.A. and Beckmann, W.A. (1974), *Solar Energy Thermal Processes*, Wiley, New York.
7 See Kreith, F. (1983), *Principles of Heat Transfer*, 3rd edn, Intext Educational Publishers, New York.
8 See Elbadry, M.M. and Ghali, A. (1983), Temperature variations in concrete bridges. *Proc. Am. Soc. Civil Engrs. J. Structural Div.*, **109**, No. 10, 2355–74.
   Also see Elbadry, M.M. and Ghali, A. (1982), User manual and computer program FETAB: finite element thermal analysis of bridges. *Research Report No. CE82–10*, Department of Civil Engineering, The University of Calgary, Canada, October.
9 See Appendix B of the reference mentioned in Note 3, page 99.
10 Figures 10.9–10.11 are taken (with permission) from the last reference mentioned in Note 4, above. The values shown in the last two figures are based on heat transfer analyses with climatic conditions representing a summer day in Calgary, Canada (air temperature extremes in 24 hours: 10 and 30 °C (50 and 86 °F)). The temperature and stress graphs represent the variations when maximum absolute values occur (in the afternoon).
11 See Zienkiewicz, O.C. (1961), Analysis of visco-elastic behaviour of concrete structures with particular reference to thermal stresses. *J. Am. Concrete Inst., Proc.*, **58**, No. 4, October, 383–94.

Chapter 11

Control of cracking

Viaduc de Sylans-France. (Courtesy Bouygues Contractor, France).

11.1   Introduction

Cracking occurs in a concrete member when the stress in concrete reaches the tensile strength, f_{ct}. The value of f_{ct} depends upon several parameters. Choice of the appropriate value of f_{ct} is the first difficulty in the analysis of crack prediction, to be discussed in this chapter.

When a statically determinate member is subjected to external applied load of sufficient magnitude to produce cracking, the member stiffness drops and an increase in displacement occurs. As long as the external applied load is sustained, there will be no change in the internal forces. On the other hand, when cracking of a statically indeterminate structure is due to temperature variation, volumetric change or settlement of supports, a reduction of stiffness occurs and the magnitude of the internal forces drops from the values existing before cracking. Cracking in the first and second cases will be referred to as force-induced and displacement-induced cracking and analysis of the two types of cracking will be discussed.

Provision of bonded reinforcement of sufficient magnitude and appropriate detailing can effectively limit the mean crack width to any specified value. The amount of reinforcement required for crack control is discussed below. The equations presented include two parameters which have to be predicted by empirical expressions. The two parameters are the mean crack spacing s_{rm} and the coefficient ζ used to account for the additional stiffness which concrete in tension provides to the state of full cracking where concrete in tension is ignored (see Equation 8.48). The empirical expressions used here for ζ and s_{rm} are adopted from codes (see Equation 8.45 and Appendix E); the accuracy of the predicted crack width depends upon these empirical expressions.

Concrete of very high strength reaching 80–100 MPa (12 000–15 000 psi) is increasingly used in practice. The increase in f_{ct} can prevent cracking. But if cracking takes place the crack width will generally be smaller because of the improved bond between the concrete and the reinforcing bars. Cracking of high-strength concrete members will be discussed.

11.2 Variation of tensile strength of concrete

The value of the normal force, N_r and/or the bending moment M_r at which cracking of a section occurs, is directly proportional to the tensile strength of concrete, f_{ct}. It is important to use an appropriate value of f_{ct} to predict whether or not cracking will occur and to account for the effect of cracking in the calculations of the probable deflections. The minimum reinforcement required for control of cracking also depends upon the value f_{ct}, which will be further discussed in Section 11.5.

The values of f_{ct} determined in tests for a given concrete composition can differ from the average value f_{ctm} by plus or minus 30 per cent. The value of f_{ct} in a member varies from section to section; as a result, cracks do not all form at the same load level. Furthermore, in a structure, the value of f_{ct} is generally smaller than the value measured by the testing of cylinders made out of the same concrete; the difference between the two values is larger in members of larger size. This can be attributed to microcracking and to surface shrinkage cracking resulting from the rapid loss of moisture in freshly placed concrete.

Appendix A includes equations for f_{ct} according to codes and technical committee reports.

11.3   Force-induced and displacement-induced cracking

Figures 11.1(a) and (b) show the variation of displacement D with the axial force N in two experiments in which a reinforced concrete member is subjected to an axial force or imposed end displacement. The two graphs are identical in the range $0 \leqslant N \leqslant N_r$; where N_r is the force producing the first crack. After cracking, the behaviour depends upon the way the experiment is conducted. Figure 11.1(a) represents the case when the force N is controlled during the experiment; specified increments of N are applied and the corresponding D is measured. In Figure 11.1(b), the displacement D is controlled by imposing specified increments and measuring the corresponding value of N.

In the force-controlled test, the occurrence of a crack is accompanied by a sudden increase in D, without change in N. In the displacement-controlled test, formation of a crack is accompanied by a sudden drop in the value of N.

The cracks in both tests (Figs 11.1(a) and 11.1(b)) correspond to the same values of N: N_{r1}, N_{r2} . . ., N_{rn}. The first crack occurs at the weakest section, when the stress in concrete reaches its tensile strength f_{ct1} (the corresponding strain $\varepsilon_c \simeq 0.001$). The second crack occurs at the second weakest section, when the stress reaches a value f_{ct2}, slightly greater than f_{ct1}. The distance between the two cracks cannot be smaller than the crack spacing s_r. At a crack, the stress in the concrete is zero; the distance s_r is necessary for transmission, by bond, of the force from the reinforcement to the concrete until the stress σ_c again reaches the tensile strength. This is discussed in Appendix E, which mentions the parameters that affect the crack spacing and gives empirical equations for estimation of the mean crack spacing s_{rm}. The maximum number of cracks that can occur, n, is equal to the integer part of the quotient (l/s_{rm}); a subsequent increase of N or D causes a widening of the existing cracks.

The formation of each crack is accompanied by a reduction of the member stiffness (increase in flexibility); this is demonstrated in Fig. 11.1(b) by a reduction in the slope of the N–D diagram.

The force-controlled test represents the effects of external applied forces on a statically determinate structure; the cracking in this case is referred to as force-induced cracking. The behaviour of a statically indeterminate structure subjected to external applied forces is more complex, because of the changes in the statically indeterminate internal forces due to cracking. The behaviour in the displacement-controlled test can occur in statically indeterminate structures due to the effects of temperature variation, shrinkage of concrete or settlement of supports.

With force-induced cracking, the stabilized cracking is reached when $N > N_{rn}$. Because N_{rn} is not substantially larger than the force N_{r1}, we can expect stabilized cracking in most cases when the cracking force is exceeded.

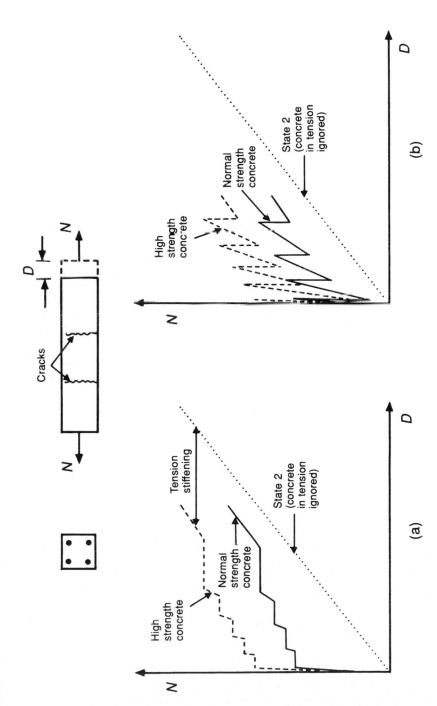

Figure 11.1 A reinforced concrete member subjected to: (a) axial force N; (b) imposed end displacement D.

On the other hand, stabilized cracking rarely occurs with displacement-induced cracking. If the first crack is formed at a displacement ΔD_1, subsequent cracks require increments $\Delta D_2, \ldots, \Delta D_n$; the magnitude of each increment is greater than the preceding one. This is further explained in the following example of a beam subjected to a bending moment produced by a temperature gradient.

11.3.1   Example of a member subjected to bending[1]

A member representing an interior span of a continuous beam of infinite number of equal spans is shown in Fig. 11.2(a). The member has a rectangular cross-section and is subjected to a rise of temperature which varies linearly over the depth, with a difference of ΔT degrees between the top and bottom fibres. It is required to study the variation of the bending moment M

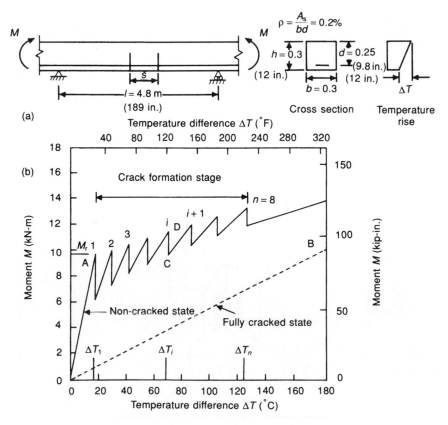

Figure 11.2 Development of cracks and statically indeterminate moment due to temperature gradient: (a) an interior span of a continuous beam; (b) variation of M with ΔT (Elbadry (1988), see Note 1, page 406).

with ΔT as ΔT increases from zero to a value ΔT_n causing stabilized crack formation. Assume the average crack spacing $s_{rm} = 0.6\,\text{m}$ (24 in). Consider the tensile stress of concrete at which the successive cracks form to vary as:[2]

$$f_{cti} = f_{ct1}[1 + 350a_t(\Delta T_i - \Delta T_1)] \qquad (11.1)$$

where f_{cti} is the tensile strength of concrete at the ith crack; ΔT_i is the value of ΔT at which the ith crack is formed; a_t is the coefficient of thermal expansion.

The cross-section geometrical data are given in Fig. 11.2(a). Other data are: $f_{ct1} = 2.1\,\text{MPa}$ (0.30 ksi); $a_t = 10 \times 10^{-6}$ per °C (5.6×10^{-6} per °F); moduli of elasticity of concrete and steel are $E_c = 25\,\text{GPa}$ (3600 ksi) and $E_s = 200\,\text{GPa}$ (29 000 ksi).

Assume that yielding of the reinforcement does not occur when the last crack occurs, the nth crack, with $n = 8$.

The value of M at which the ith crack occurs is:

$$M_{ri} = f_{cti}W_1 \qquad (11.2)$$

where W_1 is the section modulus in the non-cracked state 1.

For simplicity we assume that the beam does not lose symmetry about the centre of the span as a result of crack formation. This assumption makes the structure statically indeterminate to the first degree, with the indeterminate force being a bending moment, M whose magnitude is constant over the length of the span. Before cracking, M is given by (line OA in Fig. 11.2(b)):

$$M = a_t E_c I_1 \frac{\Delta T}{h} \quad (0 \le \Delta T \le \Delta T_1) \qquad (11.3)$$

where I_1 is the second moment of area about the centroidal axis of the transformed non-cracked section (state 1).

Setting $M = M_{r1}$ and $\Delta T = \Delta T_1$ in Equation (11.3) and solving, gives the value of ΔT at which the first crack occurs:

$$\Delta T_1 = \frac{M_{r1}h}{a_t E_c I_1} \qquad (11.4)$$

After occurrence of the ith crack and before formation of the next crack the span l can be considered to be composed of a non-cracked part of length $(l - i\,s_{rm})$ with flexural rigidity $E_c I_1$ and a cracked part of length $i\,s_{rm}$ with a mean flexural rigidity $E_c I_{rm}$ given by the equation:

$$\frac{1}{E_c I_{rm}} = (1 - \zeta)\frac{1}{E_c I_1} + \zeta\left(\frac{1}{E_c I_2}\right) \qquad (11.5)$$

where ζ is the coefficient of interpolation between the curvatures in states 1 and 2 (Equation (8.45)); I_2 is the second moment of area about the centroidal axis of the transformed fully cracked section (state 2). Equation (11.5) is derived from Equation (8.40) by substitution for ψ_m, ψ_1 and ψ_2 by $1/(E_c I_{rm})$, $1/(E_c I_1)$ and $1/(E_c I_2)$ respectively; each of these quantities represents curvature due to a unit moment.

Solution of Equation (11.5) gives:

$$I_{rm} = \frac{I_1 I_2}{\zeta I_1 + (1 - \zeta) I_2} \tag{11.6}$$

If we assume that high-bond deformed bars are used, the interpolation coefficient, just after cracking, is $\zeta = 0.5$ (by Equation (8.41), substituting $M = M_r$ and assuming $\beta_1 = 1$, $\beta_2 = 0.5$).

After formation of the ith crack and before occurrence of the next crack, the value of M is represented by (line CD in Fig. 11.2(b)):

$$M = \frac{a_t \Delta T}{h} E_c I_1 \left[1 + \frac{i \, s_{rm}}{l} \left(\frac{I_1}{I_{rm}} - 1 \right) \right]^{-1} \text{ (with } \Delta T_i \leqslant \Delta T \leqslant \Delta T_{i+1}) \tag{11.7}$$

This equation can be derived by the force method (Section 4.2).

Equation (11.7) can be solved for the value M_{ri} at which the ith crack is formed (by substituting ΔT_i for ΔT and $i - 1$ for i):

$$M_{ri} = \frac{a_t \Delta T_i}{h} E_c I_1 \left[1 + \frac{(i - 1) s_{rm}}{l} \left(\frac{I_1}{I_{rm}} - 1 \right) \right]^{-1} \tag{11.8}$$

The values of M_{ri} calculated by Equations (11.2) and (11.8) depend upon ΔT_i, whose value can be determined by elimination of M_{ri} and solving the two equations.

The mean crack width may be calculated by Equation (8.48):

$$w_m = s_{rm} \zeta \varepsilon_{s2} \tag{11.9}$$

where s_{rm} is the mean crack spacing (see Appendix E); ε_{s2}, the steel strain in a fully cracked section (state 2), may be calculated by:

$$\varepsilon_{s2} = \frac{M}{y_{CT} A_s E_s} \tag{11.10}$$

where y_{CT} is the distance between the tension steel and the resultant of compression on the section; A_s and E_s are the cross-section area and modulus of elasticity of the tension steel.

The value of M varies in the stage of crack formation as shown in Fig. 11.2(b). Just before formation of the second crack $M = M_{r2} = 10.1$ kN-m; substituting this value in Equations (11.9) and (11.10) gives ($y_{CT} \approx 0.9d = 0.225$ m):

$$\varepsilon_{s2} = 1496 \times 10^{-6} \qquad w_m = 0.45\,\text{mm} \ (0.018\,\text{in}).$$

The value of w_m can be reduced to any specified limit by increasing the steel area, A_s. On the other hand, the mean crack width can become much larger if A_s is reduced below a minimum at which $\sigma_{s2} = E_s\varepsilon_{s2} = f_y$; where f_y is the yield strength of the steel. The minimum value of the steel ratio required to avoid this situation is discussed in Section 11.6.

This procedure can be employed to determine ΔT_i for $i = 2, 3, \ldots, n$. Substitution of the value of ΔT_i in Equation (11.8) gives the larger of two M-ordinates corresponding to ΔT_i, required to construct the graph in Fig. 11.2(b); Equation (11.7) gives the lesser ordinate. The values of ΔT_1 and the two M-ordinates corresponding to the first crack can be determined by Equations (11.2), (11.4) and (11.7).

The results of the above analysis are plotted in Fig. 11.2(b); for comparison, the dashed line OB is included to represent the case when concrete in tension is ignored. The values of ΔT_i and the corresponding ordinates are listed below:

Crack number	1	2	3	4	5	6	7	8
M_{ri} (larger ordinate kN-m)	9.7	10.1	10.5	11.0	11.6	12.2	12.9	13.6
M (lesser ordinate kN-m)	5.8	7.2	8.2	9.0	9.8	10.6	11.4	12.2

11.3.2   Example of a member subjected to axial force (worked out in British units)

It is required to study the variation of N versus $\varepsilon(= D/l)$ for a member of length l subjected to an imposed end displacement D (Fig. 11.3) in the range $0 \le \varepsilon \le \varepsilon_s$; where $\varepsilon_s = D_s/l$ with D_s being the displacement at which stabilized cracking occurs. Assume that yielding of the reinforcement does not occur in this range. Consider average crack spacing $s_{rm} = 12$ in (300 mm); the value of the tensile strength of concrete at which successive cracks form is:[3]

$$f_{cti} = f_{ct1}[1 + 350(\varepsilon_i - \varepsilon_1)] \qquad (11.11)$$

where f_{cti} is the tensile strength of concrete at the location of the ith crack; $\varepsilon_i = D_i/l$ with D_i being the imposed displacement at which the ith crack is formed.

The cross-section geometrical data are given in Fig. 11.3. Other data are: $f_{ct1} = 0.35$ ksi (2.4 MPa); $E_c = 4150$ ksi (28.6 GPa); $E_s = 29000$ ksi (200 GPa).

The equations derived in Section 11.3.1 apply for a member subjected to an

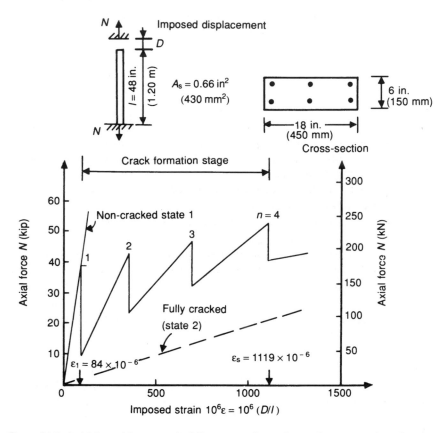

Figure 11.3 Axial force N versus ε (= D/l) in a member subjected to imposed axial end displacement.

imposed axial end displacement, by changing some of the parameters, as given below:

$$N_{ri} = f_{cti}A_1 \tag{11.12}$$

$$N = E_cA_1\varepsilon \ (0 \leqslant \varepsilon \leqslant \varepsilon_1) \tag{11.13}$$

$$\varepsilon_1 = \frac{N_{r1}}{E_cA_1} \tag{11.14}$$

$$\frac{1}{E_cA_{rm}} = (1-\zeta)\frac{1}{E_cA_1} + \zeta\left(\frac{1}{E_cA_2}\right) \tag{11.15}$$

$$A_{rm} = \frac{A_1A_2}{\zeta A_1 + (1-\zeta)A_2} \tag{11.16}$$

$$N = \varepsilon E_c A_1 \left[1 + \frac{i \, s_{\mathrm{rm}}}{l} \left(\frac{A_1}{A_{\mathrm{rm}}} - 1 \right) \right]^{-1} \quad (\text{with } \varepsilon_i \leq \varepsilon \leq \varepsilon_{i+1}) \tag{11.17}$$

$$N_{ri} = \varepsilon_i E_c A_1 \left[1 + \frac{(i-1) s_{\mathrm{rm}}}{l} \left[\frac{A_1}{A_{\mathrm{rm}}} - 1 \right] \right]^{-1} \tag{11.18}$$

where

A_1 and A_2 = areas of transformed sections in non-cracked and in fully cracked states; $A_1 = A_c(1 + a\rho)$; $A_2 = A_c a\rho$; $a = E_s/E_c$; A_c = area of concrete; $\rho = A_s/A_c$.

$\quad A_{\mathrm{rm}}$ = mean transformed cross-section area.

$\quad N$ – axial normal force.

$\quad N_{ri}$ = value of N just before formation of the ith crack.

$\quad s_{\mathrm{rm}}$ = mean crack spacing.

$\quad\quad \varepsilon = D/l$; where D is imposed displacement.

$\quad\quad \varepsilon_i = D_i/l$; where D_i is the imposed displacement at which the ith crack is formed.

The transformed section area $A_1 = 112.6 \, \mathrm{in}^2$; $A_2 = 4.61 \, \mathrm{in}^2$. Using $\zeta = 0.5$, Equation (11.16) gives $A_{\mathrm{rm}} = 8.86 \, \mathrm{in}^2$.

The number of cracks at crack stabilization, $n = l/s_{\mathrm{rm}} = 4$ cracks.

Equations (11.12) and (11.14) give: $N_{r1} = 39.4 \, \mathrm{kip}$; $\varepsilon_1 = 84 \times 10^{-6}$. Substituting the value of ε_1 in Equation (11.17) gives $N = 10.0 \, \mathrm{kip}$; this is the lower ordinate plotted for $\varepsilon = \varepsilon_1$.

Setting $i = 2$, 3 and 4 and solving Equations (11.11), (11.12) and (11.18) for ε_i and substitution of this value in Equations (11.17) and (11.18) give all the values required for plotting the graph in Fig. 11.3. The following is a list of the values of ε_i and the corresponding ordinates for $i = 1, 2, \ldots, 4$:

Crack number i	1	2	3	4
ε_i	84×10^{-6}	362×10^{-6}	700×10^{-6}	1119×10^{-6}
N_{ri}(kip)	39.4	43.0	47.6	53.4
N (lesser ordinate, kip)	10.0	24.6	33.4	41.1

Discussion of results

If the same example is analysed with a reduced value of the steel area, A_s, the vertical drops in the N-value at each crack formation will be larger and the degradation of the slope of the N–ε graph will be faster with the successive crack formations. Furthermore, the value given for s_{rm} should be increased because of the reduction in A_s (see Appendix E). As a result, the number of cracks will be smaller and the cracks will be wider.

When the steel ratio $\rho = A_s/A_c$ is reduced below a limiting value $\rho_{\mathrm{min, \, y}}$ the

N–ε graph (Fig. 11.3) will exhibit a large drop in the value of N at the formation of the first crack (at $\varepsilon = \varepsilon_1$ and $N = N_{r1}$). Subsequent increase in N will occur, at a relatively low rate, to reach a limiting value equal to $N_y = A_s f_y < N_{r1}$; where f_y is the yield strength. At this point, the mean crack width $w_m = s_{rm} \zeta f_y / E_s$. Any further increase of the imposed displacement, D will be accompanied by an increase of the same magnitude in the crack width, while the value of N remains constant equal to N_y and no further cracks develop. Thus, when the steel ratio $\rho \leq \rho_{min,\,y}$ a single, usually excessively wide, crack occurs. Equations will be derived in Section 11.5 for $\rho_{min,\,y}$ in reinforced concrete sections with or without prestressing.

Experimental verification
Jaccoud[4] conducted experiments on reinforced concrete prisms subjected to an imposed axial end displacement (Fig. 11.4). The geometrical and material data are given in the figure; all parameters have values approximately equal to the values employed in the above example, with the exception of A_s which is reduced, but is still sufficient to avoid yielding. Fig. 11.4 compares the graphs of σ_c ($= N/A_1$) versus ε obtained by experiment and by analysis setting $f_{cti} =$ constant (as observed for this specimen) and using Equations (11.12) to (11.18).

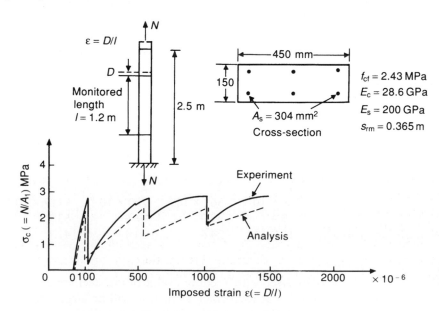

Figure 11.4 Comparison of analysis with experimental results of Jaccoud (see Note 2, page 406). A prism subjected to imposed end displacement.

11.4   Advantage of partial prestressing

Cracking results in considerable reduction of the statically indeterminate forces caused by an imposed displacement or restraint of volumetric changes due to temperature or shrinkage. Without cracking, the graph in Fig. 11.2(b) would be the straight line OA extended upwards. This would be the case in prestressed structures in which cracking is not allowed. However, a design which does not allow tensile stresses requires a high prestressing level. In addition to being costly, the higher compression due to prestressing increases the losses due to creep and can produce excessive deformations. Thus, it is beneficial to use partial prestressing, allowing cracking to occur, while controlling the width of cracks by provision of adequate non-prestressed steel. The amount of steel required for crack control is discussed in the following sections.

11.5   Minimum reinforcement to avoid yielding of steel

If the reinforcement in a cross-section of a member is below a minimum ratio, $\rho_{min, y}$, yielding of the reinforcement occurs at the formation of the first crack; such a crack will be excessively wide and formation of several cracks with limited width does not take place. This is true when cracking is induced by applied forces or imposed displacements. The minimum reinforcement cross-section area $A_{s\,min,\,y}$ and the corresponding steel ratio $\rho_{min,\,y}$ to ensure that wide isolated cracks do not occur due to yielding are determined below.

Consider a section subjected to axial tension, N. The value of N just sufficient to produce cracking is:

$$N_r = f_{ct} A_1 \tag{11.19}$$

where f_{ct} is the tensile strength of concrete and $A_1 = A_c(1 + \alpha\rho)$ is the area of the transformed non-cracked section (in state 1); $\alpha = E_s/E_c$, where E_c is the modulus of elasticity of concrete at the time considered and E_s the modulus of elasticity of the reinforcement; $\rho = A_s/A_c$.

Immediately after cracking, the force N_r is resisted entirely by the reinforcement; thus $\sigma_s = N_r/A_s$. Setting in this equation $\sigma_s = f_y$ and $A_s = A_{s\,min,\,y}$ gives the minimum steel ratio to ensure non-yielding at cracking:

$$\rho_{min,\,y} = \frac{f_{ct}}{f_y}\left[\frac{1}{1 - \alpha(f_{ct}/f_y)}\right] \tag{11.20}$$

where $\rho_{min,\,y} = A_{s\,min,\,y}/A_c$.

The term inside the square brackets in Equation (11.20) is approximately equal to unity; thus the equation is frequently written in the simpler form: $\rho_{min, y} = f_{ct}/f_y$.

Derivation of Equation (11.20) implies that the normal force is equal to N_r just before and just after formation of the first crack. However, in the case of displacement-induced cracking (defined in Section 11.3), a sudden drop of the value of the normal force takes place once the crack is formed. Subsequent increase in the imposed displacement will increase the normal force to a value $N \leq N_r$.

When the section is prestressed, the first crack occurs when $N_2 = N_r$; where $N_2 = N - N_1$ with N_1 being the decompression force (Equation 7.40). Equation (11.20) can be used substituting for f_y the yield stress of the non-prestressed steel; the resulting value $\rho_{min, y}$ will be equal to $(A_{ps} + A_{s\ min,\ y})/A_c$, where A_{ps} is the area of the prestressed steel and $A_{s\ min,\ y}$ is the minimum area of the non-prestressed steel required to ensure no yielding. Use of Equation (11.20) in this way implies the assumption that a is the same for all reinforcements.

The case when cracking is produced by a normal force N applied at a reference point O and a bending moment M about an axis through O is considered below. Assuming that the pair N and M are just sufficient to produce cracking, we can write:

$$f_{ct} = \frac{N}{A_1} + \frac{M}{W_1} \tag{11.21}$$

where A_1 and W_1 are respectively the transformed non-cracked cross-section area and section modulus (in state 1). Equation (11.21) applies only when the reference point O is at the centroid of the transformed non-cracked section; if this is not the case, the statical equivalents of the normal force and the moment must be determined to be used in the equation. The stress in steel at the crack can be calculated and equated to f_y to give $\rho_{min, y}$:

$$\rho_{min, y} = \frac{(Ne_s/y_{CT}) + N}{f_y bd} \tag{11.22}$$

where $\rho_{min, y} = (A_{s\ min,\ y}/bd)$; y_{CT} is an absolute value equal to the distance between resultant tension and resultant compression when the concrete in tension is ignored (Fig. 11.5); b and d are defined in Fig. 11.5; e_s is the eccentricity of the resultant of M and N measured downward from the centroid of the tension steel; $e_s = (M/N) - y_s$, with y_s being the y coordinate of the centroid of the tension steel.

Calculation of y_{CT} will involve determination of the depth c of the compression zone by solving Equation (7.10) or (7.9).

When the cross-section is prestressed, Equations (11.21) and (11.22) can be applied by substituting N_2 and M_2 for N and M; where $N_2 = N - N_1$ and $M_2 = M - M_1$, with N_1 and M_1 being the decompression forces (Equations (7.40)

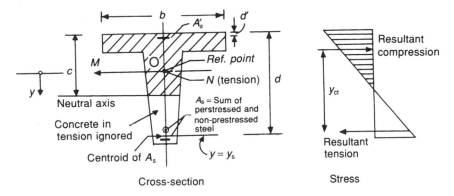

Figure 11.5 Stress distribution in a cracked prestressed section. Positive sign convention for N, M and y.

and (7.41)). Again the value of f_y should be equal to the yield strength of the non-prestressed steel and the resulting value $\rho_{\min, y} = (A_{ps} + A_{s\ \min,\ y})/(bd)$, with $A_{s\ \min,\ y}$ being the minimum non-prestressed steel area.

In the case of a reinforced rectangular section without prestressing, subjected to bending moment without axial force, Equations (11.21) and (11.22) give:

$$\rho_{\min,\ y} = \frac{A_{s\ \min,\ y}}{bd} \approx 0.24 \frac{f_{ct}}{f_y} \tag{11.23}$$

which is derived by assuming $d \approx 0.9h$ and $y_{CT} = 0.9d$.

It is to be noted that in Equations (11.21) and (11.22), the values of N and M are just sufficient to produce the first crack. The two equations apply to reinforced concrete sections with or without prestressing, subjected to any combination of N and M, satisfying Equation (11.21). Thus, for the case of axial tension, $M = 0$ and $e_s = 0$; Equations (11.21) and (11.22) give the same results as Equations (11.19) and (11.20).

11.6 Early thermal cracking

In many cases cracking of concrete structures occurs at an early age due to heat of hydration of cement. When the heat of hydration is generated at a higher rate than heat dissipation, rise of temperature and expansion of the concrete occurs. The expansion is followed by contraction as the concrete cools down to the ambient temperature. These volumetric changes are in most cases partially restrained and stresses result; the magnitude of stress may be assumed proportionate to the modulus of elasticity of concrete, E_c.

The value of E_c when the contraction occurs is large compared with the corresponding value at earlier stages of hydration, when the expansion occurs. As a result, the tensile stress during contraction exceeds the compressive stress which has occurred during expansion. The difference is frequently sufficient to exceed the tensile strength, f_{ct} of the young concrete.

In a member without reinforcement, one wide crack is induced. Provision of reinforcement controls the crack so that the member remains serviceable. The amount of reinforcement required for this purpose may be calculated by the equations in Section 11.5, substituting an appropriate (relatively low) value[5] of f_{ct}, representing the tensile strength of concrete at an early age (see Appendix A).

11.7   Amount of reinforcement to limit crack width

The width of force-induced or displacement-induced cracks can be limited to any specified value by provision of sufficient area of bonded reinforcement in the tension zone. The same objective can be achieved by limiting the steel strain at a cracked section. The following equations can be used for this purpose for a reinforced concrete section with or without prestressing. It is assumed that the section is subjected to normal force N at a reference point O and a bending moment about a horizontal axis through the reference point (Fig. 11.5).

$$w_m = s_{rm} \zeta \varepsilon_{s2} \tag{11.24}$$

$$e_s = \frac{M}{N} - y_s \quad \text{(with } N \neq 0\text{)} \tag{11.25}$$

$$\varepsilon_{s2} = \frac{(Ne_s/y_{CT}) + N}{E_s A_s} \tag{11.26}$$

$$\sigma_{s2} = E_s \varepsilon_{s2} \tag{11.27}$$

$$A_s = \frac{(Ne_s/y_{CT}) + N}{\sigma_{s2}} \tag{11.28}$$

where

A_s = sum of the cross-section areas of prestressed and non-prestressed steel resisting the tension.

E_s = modulus of elasticity of steel, assumed the same for the prestressed and the non-prestressed steel.

e_s = eccentricity of the resultant of N and M, measured downward from the centroid of A_s. Equation (11.26) implies the assumption that the resultant tension is at the centroid of A_s; ε_{s2} for any reinforcement layer can be more accurately calculated by the procedure given in Chapter 8.

N and M = normal force at reference point O and bending moment about a horizontal axis through O; the sign conventions for N and M are defined in Fig. 11.5. It is assumed that the stress in concrete $\sigma_c = 0$ prior to the introduction of N and M. If this is not the case, substitute N and M in Equations (11.25), (11.26) and (11.28) by $N_2 = N - N_1$ and $M_2 = M - M_1$; where N_1 and M_1 are the 'decompression forces' (Equations (7.40) and (7.41)).

s_{1m} = crack spacing (see Appendix E).

w_m = mean crack width.

y_{CT} = absolute value equal to the distance between resultant tension and compression forces.

ε_{s2} = strain increment in steel (prestressed or non-prestressed) due to N and M, with concrete in tension ignored.

σ_{s2} = stress increment in steel (prestressed or non-prestressed) due to N and M, with concrete in tension ignored.

ζ = coefficient of interpolation between state 1 where cracking is disregarded and state 2 where concrete in tension is ignored.

The product (Ne_s) in Equation (11.26) is to be replaced by the value of M when $N = 0$. Calculation of y_{CT} is commonly preceded by determination of the depth c of the compression zone (Equation (7.10) or (7.9)).

11.7.1   Fatigue of steel

Fatigue can occur in the non-prestressed steel or in the prestressed steel when cyclic change in steel stress, σ_{s2}, is relatively large. Equations (11.26) and (11.27) can be used to calculate the magnitude of σ_{s2} when the values of N and M producing the cyclic change in stress are known. For given values of N and M, the change in steel stress, σ_{s2}, can be critical for fatigue when the total steel ratio, prestressed and non-prestressed, is small. This can be the case in partially prestressed structures, where a smaller area of steel is used compared to the case of no prestressing.

The allowable limits of σ_{s2} to avoid fatigue failure are given in various codes. Approximate values are 125 MPa (18 ksi) for non-prestressed deformed bars and 10–12 per cent of the ultimate strength for prestressed steel.

11.7.2   Graph for the change in steel stress in a rectangular cracked section

Equations (11.26) and (11.27) are used to derive the graph in Fig. 11.6 for rectangular sections subjected to a normal force N at O, at mid-height of the section and a bending moment M about a horizontal axis through O. To use

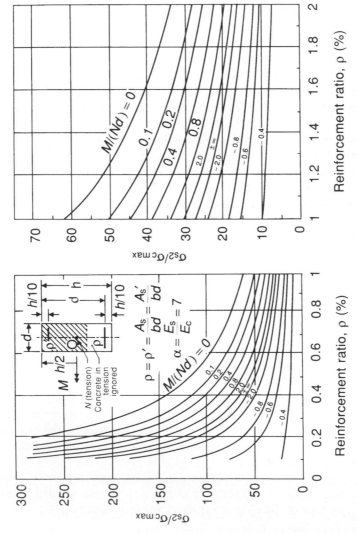

Figure 11.6 Change in steel stress, σ_{s2} in a rectangular cracked section due to a normal force N and a bending moment M. The value of σ_{s2} is equal to the value read from the graph multiplied by $\sigma_{c\,max}$; where $\sigma_{c\,max}$ is the extreme fibre stress ignoring cracking. Positive sign convention for N and M are indicated.

the graph, enter the dimensionless parameters $\rho(= A_s/bd)$ and (M/Nd) and read the ordinate value, representing the ratio $\sigma_{s2}/\sigma_{c\ max}$; where σ_{s2} is stress increment in steel (prestressed and non-prestressed) due to N and M with concrete in tension ignored; $\sigma_{c\ max}$ is the extreme fibre stress due to N and M with cracking ignored. The value of $\sigma_{c\ max}$ may be calculated by Equations (2.17) and (2.20), using properties of the transformed non-cracked section.

When the graph is used for a prestressed section, A_s represents the sum of the cross-section area of the prestressed and the non-prestressed steel; this implies an approximation by the assumption that the resultant tension is at the centroid of A_s. In all cases, it should be noted that N and M represent the values of the normal force and bending moment after deduction of the decompression forces (see definition of N and M given below Equation (11.28)).

Assumed parameters, which have small influence, used in preparing the graph are: $a = E_s/E_c = 7.0$; $A'_s = A_s$; $d = 0.9h$; $d' = 0.1h$ (Fig. 11.6).

For given values of N and M, the graph in Fig. 11.6 may be used to calculate the steel ratio $\rho(= A_s/(bd))$ required to limit the mean crack width, w_m, to a specified value. For this purpose, determine the extreme fibre stress $\sigma_{c\ max}$ in the non-cracked section and calculate ζ by Equation (8.49). Using an assumed crack spacing s_{rm} (Appendix E), determine ε_{s2} and σ_{s2} by Equations (11.26) and (11.27). Enter the graphs with the values of $(\sigma_{s2}/\sigma_{c\ max})$ and $[M/(Nd)]$ and read the value of ρ.

Example 11.1

What is the change in steel stress σ_{s2} and the mean crack width due to a bending moment $= 40\,\text{kN-m}$ (350 kip-in) and an axial force $= 0$, applied at time t on the section shown in Fig. 11.7? The free shrinkage occurring prior to age t is $\varepsilon_{cs}(t, t_0) = -400 \times 10^{-6}$; where t_0 is the age of concrete when curing is stopped. Use the following data: $E_s = 200\,\text{GPa}$ (29 000 ksi); $a = E_s/E_c(t) = 7$; creep coefficient $\varphi = 2.5$; aging coefficient $\chi = 0.8$; $E_c(t_0) = E_c(t)$; mean crack spacing $s_{rm} = 400\,\text{mm}$ (16 in); interpolation coefficient $\zeta = 0.8$.

Determine the bottom steel area required to limit the mean crack width to 0.2 mm.

The age-adjusted modulus of elasticity of concrete, Equation (1.31), $\bar{E}_c(t, t_0) = 9.52\,\text{GPa}$. Properties of the age-adjusted transformed section are: $\bar{A} = 0.3720\,\text{m}^2$; $\bar{B} = 0$; $\bar{I} = 3.287 \times 10^{-3}\,\text{m}^4$. The stress in concrete at time t due to shrinkage is constant over the section and its value is $\sigma_c(t) = 0.774\,\text{MPa}$ (Equations (3.15) and (3.19)).

Properties of the transformed non-cracked section at time t are: $A =$

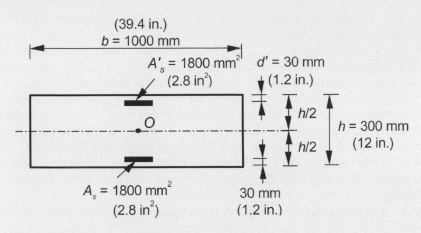

Figure 11.7 Cross-section subjected to shrinkage and a bending moment, analysed to determine crack width (Example 11.1).

$0.3216 \, \text{m}^2$; $B = 0$; $I = 2.561 \times 10^{-3} \, \text{m}^4$. The decompression forces are $M_1 = 0$ and by Equation (7.43):

$$N_1 = -0.3216(0.774 \times 10^6) = -249 \, \text{kN}.$$

Forces producing cracking, after deducting the decompression forces are:

$$N = 0 - (-249) = 249 \, \text{kN} \quad M = 40 \, \text{kN-m}.$$

Stress at the extreme fibre, ignoring cracking is:

$$\sigma_{c \, max} = \frac{249}{0.3216} + \frac{40(0.15)}{2.561 \times 10^{-3}} = 3.12 \, \text{MPa}.$$

The steel ratio $\rho = 1800/(1000 \times 270) = 0.67$ per cent; $M/(Nd) = 40/(249 \times 0.27) = 0.60$. Entering the graph in Fig. 11.6 with the values of ρ and $M/(Nd)$ gives: $(\sigma_{s2}/\sigma_{c \, max}) = 51$; thus,

$$\sigma_{s2} = 51(3.12) = 160 \, \text{MPa} \, (23.2 \, \text{ksi}).$$

The mean crack width (Equations (11.24) and (11.27)):

$$w_m = 400(0.80) \frac{160}{200 \times 10^3} = 0.26 \, \text{mm} \, (0.010 \, \text{in}).$$

To limit the mean crack width to 0.2 mm, σ_{s2} is to be reduced to $\sigma_{s2} =$ 125 MPa (18.1 ksi). Enter the graph in Fig. 11.6 with ordinate = 125/ 3.12 = 40 and $M/(Nd) = 0.6$ to read $\rho \simeq 1.04$ per cent. The increase in steel ratio will change $\sigma_c(t)$ and $M/(Nd)$ values; hence iteration is required. Repetition of the analysis will give a more accurate value of $\rho = 1.0$ per cent, or $A_s = 2700 \, \text{mm}^2$ (4.19 in^2).

11.8 Considerations in crack control

This section discusses the motives and the most important measures for control of cracking.

There are three motives for crack control: durability by reducing risk of corrosion of reinforcement, aesthetic appearance and functional requirements such as gas or liquid tightness or hygiene (cracks can be the focus of development of pathogenic microbes). The three motives are discussed separately below.

Corrosion of reinforcement
There is no general agreement on the influence of width of cracks on corrosion of reinforcing steel. Some research[6] indicates that intensity of observed corrosion is not dependent upon width of cracks, w, as long as $w_{maximum}$ is limited to 0.3–0.5 mm (0.01–0.2 in); the lower limit is for cracks running parallel to reinforcing bars, producing a higher risk of corrosion. Only the length of time (a few years) before initiation of corrosion is influenced by crack width; but this period is relatively short and has little influence on the longevity of structures.

On the other hand, there is agreement that thickness and porosity of concrete covering the reinforcement are important parameters influencing corrosion. Improving the quality of concrete (mainly by limiting the water–cement ratio) and at the same time controlling the crack width are, at present, considered important to control cracking. Thus, it is prudent to specify the crack width dependent upon the aggressiveness of the environment.[7] Also, stricter requirements should be applied to prestressed structures, because the prestressed steel is more susceptible to corrosion than ordinary reinforcing bars [$w_{maximum} \leqslant 0.2 \, \text{mm}$ (0.008 in) at the exposed concrete surface and $w = 0$ at the level of the prestressed steel].

Aesthetic appearance
Cracks of width smaller than 0.3 mm (0.01 in) generally are not of much concern to the public. However, owners and users of structures are in general sensitive to the aesthetic damage of appearance when wide cracks develop. Obviously, the crack width of tolerable appearance is subjective and dependent on many factors, such as the distance between the crack and the observer, the lighting and the roughness of the surface.

Gas or liquid tightness
The need for tightness depends upon the nature of gas or liquid to be contained in the structure. It is theoretically possible to specify and expect a structure with no cracks. It is more realistic, however, to specify a limit for crack width. Research and experience have shown that water-retaining structures can be water tight with crack width, $w = 0.1$ to 0.2 mm (0.004–0.009 in). Such a crack, even when it traverses the full thickness of a wall, may allow moisture to penetrate after first occurrence of the crack; but 'healing' and stopping of leakage occurs within a few days.

Measures for control of cracking
Controlled cracks of width, $w = 0.10$ to 0.30 mm (0.004 to 0.012 in) generally do not undermine the use, the durability or the appearance of concrete structures. On the other hand, uncontrolled wide cracks ($w > 0.5$ mm) must be avoided. This objective may be achieved by:

1. reducing the risk of cracking by measures such as use of appropriate mix, curing, casting sequence and construction joints of concrete, provision of temporary or permanent expansion joints, prestressing, etc.
2. provision of minimum bonded reinforcement in all parts of reinforced concrete or prestressed structures when cracking is probable during construction or during use of the structure. Thus designers should consider the combinations of direct and indirect loads (settlement of supports, temperature and volumetric changes, etc.) which can produce tensile stress close to or exceeding the tensile strength of concrete.
3. limiting the steel stress, calculated with concrete in tension ignored. This design check is commonly done considering only the quasi-permanent loads and allowing wider partly reversible cracks to occur due to additional transient loads.
4. provision of prestressing, even at a low level, can be effective in reducing crack width. This is particularly the case when cracking is caused by flexure. When cracking is caused by a normal force, provision of prestressing is effective only when the element considered is free to shorten.

11.9 Cracking of high-strength concrete

The effects of use of high-strength concrete (HSC) on cracking of reinforced concrete structures with or without prestressing are briefly discussed here. For this purpose, consider a reinforced concrete member subjected to an axial force, N, as shown in Fig. 11.1(a). If concrete of higher strength is used for the member, cracks will occur at higher N values, because of the increase of the tensile strength (see the dashed lines in Fig. 11.1(a)). Also, because of the improved bond, slip between the concrete and the reinforcing bars occurs at higher bond stress; this decreases the crack spacing (see Appendix E) and also increases the effect of the tension stiffening, thus reducing the width of cracks. Therefore, use of HSC may prevent the cracking, or when N is greater than the cracking value the crack width will be smaller compared with an identical member with the same reinforcement but lower concrete strength.

On the other hand, when N is caused by an imposed displacement (Fig. 11.1(b)), the first crack will be formed at a higher N value and higher steel stress will occur at the crack. This means that a larger steel ratio is necessary to avoid yielding of steel at cracking (see Equation (11.20)). In spite of the higher steel stress at the crack, the crack width will increase only slightly, because of the increase of tension stiffening effect due to improved bond.

The effects of use of HSC on cracking of members subjected to bending is not different from what is discussed above. This is evident in Fig. 11.8 which summarizes the results of long-term tests[8] on simply supported slabs

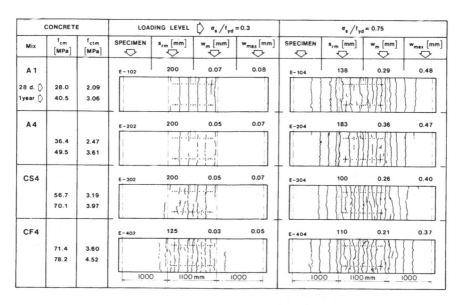

Figure 11.8 Mean crack width w_m and mean crack spacing s_{rm} observed in tests on reinforced concrete slabs of varying concrete strength.

subjected to two symmetrically located equal forces. The results are given for two load levels represented by the ratio σ_{s2}/f_y; where σ_{s2} is steel stress at a cracked section in the central part of the span (zone of constant bending) and f_y is the yield strength of the steel (460 MPa).

The empirical equations given in some codes to predict crack spacing or to account for tension stiffening do not accurately represent structures made of HSC. This status will no doubt change because concrete strength higher than 50 MPa (7000 psi) and reaching up to 80 or 100 MPa (12 000 or 15 000 psi) is increasingly used in modern structures.

11.10   Examples worked out in British units

Example 11.2 Prestressed section: crack width calculation

Figure 11.9 shows the stress distribution in a prestressed concrete section, at time t, after occurrence of creep, shrinkage and relaxation (the same cross-section was analysed in Example 2.2, Fig. 2.6). Find the crack width after application of live-load bending $M = 7000$ kip-in. (790 kN-m) about an axis through reference point O. Use the following data: $\sigma_{cO}(t) = -0.360$ ksi (-2.48 MPa); $\gamma(t) = -6.38 \times 10^{-3}$ ksi/in (-1.73 MPa/m); $E_s = 29\,000$ ksi (200 GPa), for all reinforcements, $\alpha(t) = E_s/E_c(t) = 7$; mean crack spacing $s_{rm} = 16$ in (400 mm); interpolation coefficient, $\zeta = 0.9$.

Properties of the transformed non-cracked section at time t:

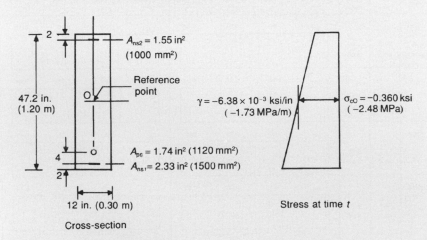

Figure 11.9 Prestressed cross-section analysed to determine crack width after application of live-load bending moment (Example 11.2).

$A = 598.3 \, \text{in}^2$ $B = 270.4 \, \text{in}^3$ $I = 118 \, 500 \, \text{in}^4$.

The decompression forces are (Equations (7.40) and (7.41)):

$N_1 = 213.2 \, \text{kip}$ $M_1 = 835.4 \, \text{kip-in.}$

Forces producing cracking, after deducting the decompression forces;

$N_2 = -213.2 \, \text{kip}$; $M_2 = 7000 - 835.4 = 6164.6 \, \text{kip-in.}$

Stress at the extreme fibre, ignoring cracking (Equations (2.19) and (2.17)), $\sigma_{c \, max} = 0.868 \, \text{ksi}$.

The tension is resisted by the prestressed and the bottom non-prestressed steel; the total steel area resisting tension = 1.74 + 2.33 = 4.07 in^2 and its centroid is at depth $d = 43.6 \, \text{in}$. The steel ratio $\rho = 4.07/(12 \times 43.6) = 0.78$ per cent. The value $M/(Nd) = 6164.6/(-213.2 \times 43.6) = -0.663$. Entering the graph in Fig. 11.6 with the values of ρ and $M/(Nd)$ gives: $(\sigma_{s2}/\sigma_{c \, max}) = 21.3$. Thus,

$\sigma_{s2} = 21.3(0.868) = 18.5 \, \text{ksi}$.

The mean crack width (Equations (11.24) and (11.27)) is:

$$w_m = 16(0.9) \frac{18.5}{29 \, 000} = 0.0092 \, \text{in} \, (0.23 \, \text{mm}).$$

A more accurate analysis, using the equation of Chapter 8 gives: $c = 20.3 \, \text{in}$; σ_{s2} in the bottom non-prestressed steel = 17.8 ksi; mean crack width = 0.0088 in (0.22 mm).

Example 11.3 Overhanging slab: reinforcement to control thermal cracking

Figure 11.10 represents top view and section in a reinforced concrete slab extending as a cantilever from the floor of a building. Transverse cracks can occur in the cantilever due to temperature difference between the outside air and the interior heated building. It is required to calculate the concrete stress which would occur, ignoring cracking, due

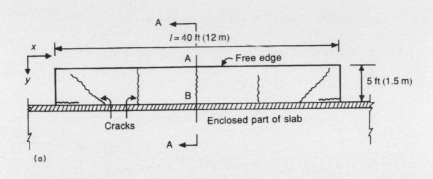

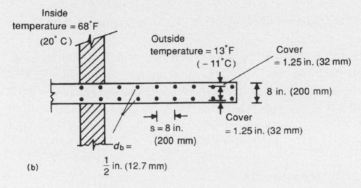

(b)

$\dfrac{1}{2}$ in. (12.7 mm)

Figure 11.10 Minimum reinforcement requirement to control cracking due to temperature in an overhanging reinforced concrete slab in a building (Example 11.3): (a) top view; (b) section A–A.

to a temperature drop of 55 °F (31 °C) of the cantilever below the temperature of the interior part of the floor. Determine the reinforcement required to avoid yielding of the steel. Consider the coefficient of thermal expansion, $a_t = 4.4 \times 10^{-6}$ per degree Fahrenheit (8.0×10^{-6} per degree Celsius); $E_c = 2900 \, \text{ksi} \, (20 \, \text{GPa})$; $f_y = 60 \, \text{ksi} \, (410 \, \text{MPa})$; $E_s = 29 \, 000 \, \text{ksi} \, (200 \, \text{GPa})$.

Strain if thermal contraction were free to occur:

$$\varepsilon_{\text{free}} = 4.4 \times 10^{-6} \, (-55) = -242 \times 10^{-6}$$

Assuming the strain is completely prevented, the stress in concrete will be:

$$\sigma_c = -E_c \varepsilon_{\text{free}} = -2900(-242 \times 10^{-6}) = 702 \, \text{psi} \, (4.84 \, \text{MPa}).$$

Cracking will no doubt occur before this hypothetical high stress value is reached. The minimum reinforcement ratio necessary to avoid yielding (Eq. (11.20)) is:

$$\rho_{min, y} = \frac{360}{60 \times 10^3}\left[\frac{1}{1 - \frac{29000}{2900}\left(\frac{360}{60 \times 10^3}\right)}\right] = 0.0063.$$

This steel ratio is approximately provided by using $\frac{1}{2}$ in bars (A_s/bar = 0.20 in^2 (130 mm^2)) with spacing s = 8 in at top and bottom. The corresponding steel ratio is:

$$\rho = \frac{A_s/\text{bar}}{st_s/2} = \frac{0.20}{(8 \times 8)/2} = 0.0063$$

where t_s = slab thickness.

The ACI 318–89[9] assumes an effective tension area to be used in crack analysis (Fig. E. 3(b)). Accordingly, the effective tension area per bar is 24 in^2; if this value is used to replace the quantity ($st_s/2$), the above equation gives: A_s/bar = 0.0063(24) = 0.15 in^2 (98 mm^2).

The MC-90 and the EC-91 assume an effective tension area defined in Fig. E.2(c); accordingly the effective tension area will be 30 in^2 and the required cross-section area per bar will be 0.0063 (30) = 0.19 in^2 (120 mm^2).

Commentary

In the above example the stress σ_x ignoring cracking is based on the assumption that the volumetric change in the x direction is fully restrained. This simple analysis is sufficient to give an approximate value of the stress σ_x and to conclude that f_{ct} will be exceeded; thus cracking will occur. Analysis based on elastic theory shows that the stress σ_x at section AB varies between the approximate value calculated above (assuming complete restraint) and a smaller value at point A, the tip of the cantilever. When l/b = 2, 4 or 8 the value of stress at A is respectively equal to 9, 55 or 96 per cent of the stress calculated assuming complete restraint. In this example l/b = 8; thus the stress σ_x is approximately constant over the section A–B.

Typical crack pattern is shown in Fig. 11.10(a); the width and spacings of cracks depend upon the reinforcement provided mainly in the x direction. The type of stress distribution and crack pattern described above occurs in practice in walls where the volumetric change of the wall due to temperature or shrinkage is restrained by the wall footings. Another example where similar crack patterns can occur is in bridge superstructures,[10] where parts of the cross-section are cast separately, using longitudinal casting joints (for example parapets or overhanging parts of the deck cast separately to the main deck).

11.11 General

This chapter discusses the parameters that influence crack width and gives equations to determine the steel area of bonded reinforcement required to limit the width of cracks to a specified value. The analysis is approximate because it includes empirical parameters: s_{rm} and ζ to predict crack spacing and to account for tension stiffening. The empirical coefficients for s_{rm} and ζ are to be determined from codes (some code expression are given in Appendix E and Equation (8.45).

Notes

1 This example is based on: Elbadry, M.M. (1988), *Serviceability of Reinforced Concrete Structures*, Ph.D. thesis, Department of Civil Engineering, University of Calgary, Calgary, Canada, 294 pp.
2 Equation (11.1) is derived from experiments reported in: Jaccoud, J.P. (1987), *Armature minimale pour le contrôle de la fissuration des structures en béton*, Ph.D. thesis, Département de Génie Civil, École Polytechnique Fédérale de Lausanne, Lausanne, Switzerland, 195 pp.
3 Equation (11.11) is derived from experiments reported in Jaccoud (1987); see reference mentioned in Note 2, above.
4 See reference mentioned in Note 2, above.
5 A suggested equation can also be found in: Department of Transport Highways and Traffic (1987), Department Advice Note BA24/87, *Early Thermal Cracking of Concrete*, 16 pp., DOE/DTp Publication Sales Unit, Bldg. 1, Victoria Rd., 5, Ruislip, Middlesex HA4 0NZ, UK.
6 Schiessl, P. (1985), *Mindestbewehrung zur Vermeidung klaffender Risse*, Institut fur Betonstahl und Stahlbetonbau, Munich, Bericht 284.
7 Beeby, A.W. (1983), 'Cracking, cover and corrosion of reinforcement', *Concrete International*, **28**, (2), Feb., pp. 35–40.
8 See Jaccoud, J.-P, Charif, H. and Farra, B. (1993), *Cracking Behaviour of HSC Structures and Practical Considerations for Design*, Publication No. 139, IBAP, Swiss Federal Institute of Technology, Lausanne, Switzerland.
9 ACI 318–89 Building Code Requirements for Reinforced Concrete Institute, Farmington Hills, Michigan 48333–9094. The clause referred to here has been dropped out in subsequent issues of the Code.
10 See reference mentioned in Note 5, above.

Chapter 12

Design for serviceability of prestressed concrete

Viaduct at Gruyère, Switzerland

12.1 Introduction

Prestressed concrete structures commonly contain non-prestressed reinforcement to control cracks that develop before introduction of the prestressing or when the structure is in service. The appropriate design of the non-prestressed reinforcement and the prestressing forces can control deflections and limit the opening of cracks. This chapter[1] discusses the choice of the level of prestressing forces and the amount of non-prestressed reinforcement to achieve these objectives.

12.2   Permanent state

Durability of concrete structures is closely linked to their serviceability in the permanent state. This is defined here as the state of the structures subjected to sustained loads such as the prestressing, the self-weight and the superimposed dead loads and the quasi-permanent live loads. A prestressed structure may be designed such that cracks occur only under the effect of exceptional live load combined with temperature variation. Such cracks open and close in each cycle that the loads are applied. However, cyclic loading of cracked structures produces residual opening of cracks and residual deflections which are discussed in this chapter. Some bridges exhibit increasing deflections after several decades in service. This is attributed partly to the irreversible curvature which adds to the deflections due to the effects of creep, shrinkage and relaxation.

The approach adopted in this chapter to achieve satisfactory serviceability is to limit the tensile stress in the permanent state to a specified value. This can be achieved by designing the prestressing such that its effect combined with the sustained loads produce stress in concrete not exceeding the specified value. In the permanent state the structure has no or only limited cracks; thus any elastic analysis ignoring cracking is considered adequate to calculate the stress in concrete due to the sustained loads and the prestressing combined. In addition to the prestressing, the structure should have non-prestressed reinforcement, the design of which is discussed below.

12.3   Balanced deflection factor

The balanced deflection factor β_D is defined as the ratio between the elastic deflections at mid-span due to prestressing and due to sustained quasi-permanent loads:

$$\beta_D = -\frac{D(P_m)}{D(q)} \tag{12.1}$$

where $D(P_m)$ is the deflection at mid-span due to prestressing. In the calculation of this deflection the prestressing force in a tendon is taken equal to the mean of the initial prestressing force excluding friction loss and the force remaining after losses due to creep and shrinkage of concrete and relaxation of prestressed steel. $D(q)$ is the elastic (immediate) deflection at mid-span due to permanent and quasi-permanent load.

A parabolic tendon having constant prestressing force exerts on a prismatic concrete member a uniform upward load. If, in addition, the permanent load is uniform downward, the balanced deflection factor, β_D is the same as the well-known balancing load factor, which is equal to minus the ratio of the intensities of the upward and the downward loads. However, use of the balanced deflection factor is preferred here because it applies with any tendon profile and with members having variable depth.

The significance of the balanced deflection factor is explained by Fig. 12.1, which depicts the strain distribution in a section of members having $\beta_D = 0$ and 1. In the former the strain at the centroid, $\varepsilon_O = 0$ and the curvature, $\psi \neq 0$; in the latter $\varepsilon_O \neq 0$ and $\psi = 0$. We recognize that with $\beta_D = 0$, the member is non-prestressed and with $\beta_D = 1$, the prestressing is just sufficient to eliminate the deflection.

In determining β_D by Equation (12.1), the deflection is calculated using the cross-sectional area properties of gross concrete sections and an estimated reduction of the prestressing forces to account for the time-dependent losses due to creep, shrinkage and relaxation. Because the analysis is concerned with the behaviour of the structure during its service life, it is suggested that the prestressing force used in calculating β_D be the average of the values before and after the time-dependent losses.

12.4   Design of prestressing level

In the design of a prestressed structure the level of prestressing, expressed by the balanced deflection factor β_D, can be a means of controlling cracks in service condition. For this purpose the structure can be designed such that the stress at a specified fibre due to prestressing combined with sustained quasi-permanent loads, σ_{perm} satisfy the condition:

$$\sigma_{\text{perm}} \leqslant \sigma_{\text{allowable}} \tag{12.2}$$

where $\sigma_{\text{allowable}}$ is an allowable stress value depending upon the width of cracks that can be tolerated and the amount of non-prestressed reinforcement that is

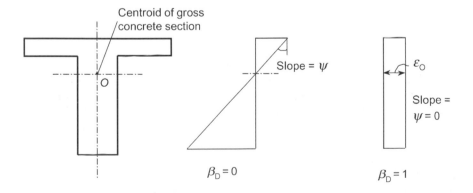

Figure 12.1 Strain distribution in a cross-section of members having the balanced deflection factors $\beta_D = 0$ and 1.

provided. For example, at the extreme top fibre of a bridge deck, exceptional live load combined with temperature variation can produce transient stress $\Delta\sigma$ and cause cracking when:

$$\Delta\sigma + \sigma_{\text{perm}} \geqslant f_{\text{ct}} \tag{12.3}$$

where f_{ct} is the tensile strength of concrete. Thus, when $f_{\text{ct}} = 3\text{MPa}$ and σ_{perm} is equal to an allowable value, say $\sigma_{\text{allowable}} = -2\text{MPa}$, cracking occurs when $\Delta\sigma \geqslant 5\text{MPa}$. The cracks partially close when the live and the thermal loads are removed. With the repetition of these loads, the cracks will have some residual opening and the structure will exhibit residual deformations (see Sections 12.7 and 12.11).

Equation (12.8) is derived below to give for a continuous or a simple beam subjected to uniform permanent downward load q/unit length, a design value of the balanced deflection factor β_D. This value represents the prestress force P_m that satisfies Equation (12.2) at any fibre of a specified section. The bridge cross-sections shown in Fig. 12.2 will be used as examples for the application of Equation (12.8) and to study the sensitivity of the design to the choice of β_D.

Figure 12.3 shows a typical parabolic tendon profile for a simple or a continuous span. The tendon exerts on the concrete a uniform load whose intensity is:

$$q_{\text{prestress}} = -8\ P_m f_0 / l^2 \tag{12.4}$$

where P_m is the absolute value of the mean prestress force, an average of the values before and after the loss due to creep, shrinkage and relaxation; l is the span; f_0 is the distance between the chord joining the tendon ends over the supports and the tendon profile at mid-span; this distance is measured in the direction of the normal to the centroidal axis of the beam. The value of P_m is assumed to be constant within each span; at a simply supported end the tendon has zero eccentricity; the value ($q_{\text{prestress}}/q$) is assumed to be the same in all spans; thus, when the spans are unequal, f_0 is assumed to vary such that $q_{\text{prestress}}$ is the same for all spans (Equation (12.4)). The intensity q of the permanent or quasi-permanent load is assumed to be constant and equal in all spans. With these assumptions, the balanced deflection factor, β_D is the same as the balanced load factor; thus,

$$\beta_D = -\frac{q_{\text{prestress}}}{q} \tag{12.5}$$

$$M_{\text{prestress}} = -\beta_D M_q \tag{12.6}$$

where $M_{\text{prestress}}$ and M_q are bending moments at any section due to the

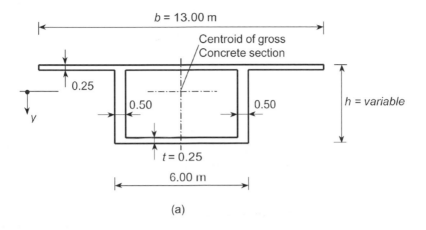

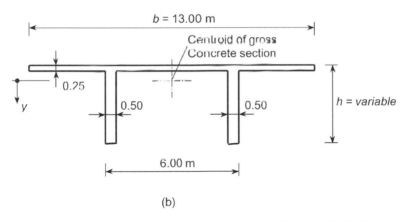

Figure 12.2 Bridge cross-sections considered in Examples 12.1 to 12.3: (a) closed section; (b) open section.

prestressing force P_m and due to the sustained load q, respectively. The permanent stress at any fibre due to P_m and q combined is:

$$\sigma_{\text{perm}} = M_q \frac{y}{I}(1 - \beta_{\text{D}}) - \frac{P_m}{A} \qquad (12.7)$$

where A and I are the area and the second moment of area about centroidal axis of the gross concrete section; y is the coordinate of the fibre considered, measured downward from the centroidal axis (Fig. 12.2). Substituting Equation (12.5) to (12.7) in Equation (12.2) and solving for the balanced deflection factor gives:

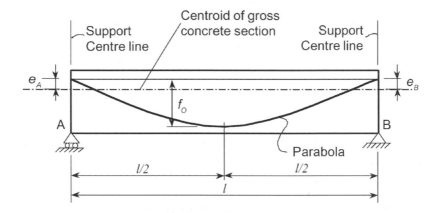

Figure 12.3 Typical cable profile in a span of a simple or a continuous beam. Assumption used in derivation of Equation 12.8. When end A or B is simply supported, the tendon eccentricity e_A or e_B is zero.

$$\beta_D \geq \frac{1 - \sigma_{\text{allowable}} / \sigma_q}{1 + I/(8\alpha_q A f_0 y)} \tag{12.8}$$

where σ_q is a hypothetical value of the stress that would occur at the fibre considered if q were applied without prestressing and the section is homogeneous noncracked:

$$\sigma_q = M_q \frac{y}{I} \tag{12.9}$$

α_q is a dimensionless coefficient defined as:

$$\alpha_q = M_q / (ql^2) \tag{12.10}$$

The mean value of the prestressing force required is (by Equations (12.4) and (12.5)):

$$P_m = \beta_D \frac{ql^2}{8f_0} \tag{12.11}$$

For a simple span, Equation (12.8) is to be applied at the section at mid-span to give β_D and the result substituted in Equation (12.11) to give P_m. In a continuous beam the two equations should be applied for critical sections over the interior supports and at (or close to) mid-spans. The largest P_m thus obtained should be adopted in design.

For a chosen value of β_D, the permanent stress is (Equations (12.7) and (12.11)):

$$\sigma_{perm} = \sigma_q[1 - \beta_D\,(1 + I/\,(8a_qAf_0y)] \tag{12.12}$$

12.5 Examples of design of prestress level in bridges

Figures 12.2(a) and 12.2(b) represent cross-sections of bridges that will be used in design examples and in parametric studies. The thickness h will be varied as well as the span to thickness ratio l/h.

Example 12.1 Bridges continuous over three spans

Consider a bridge deck having a constant cross-section shown in Fig. 12.2(a), continuous over three spans $0.7l$, l and $0.7l$, with $l = 60$ m and $h = 3$ m. The cross-sectional area properties are: $A = 7.25\,\text{m}^2$; $I = 9.51\,\text{m}^4$; the y coordinate of the top-fibre is $y = -1.168$ m. In addition to its self-weight ($24\,\text{kN-m}^3$), the deck carries a sustained dead load of $32.5\,\text{kN-m}$. Thus, the total permanent load is: $q = 32.5 + 24\,(7.25) = 206.5\,\text{kN-m}$. Assume a parabolic tendon profile (Fig. 12.3) with $f_0 = h - 0.1$ m. Determine the balanced deflection factor β_D and the required mean prestressing force, P_m such that the permanent stress at top fibre over the two interior supports equal the allowable stress, $\sigma_{allowable} = -2\,\text{MPa}$.

For a continuous beam of the specified spans, the bending moment over the two interior supports is (by elastic analysis):

$$M_q = -0.0763\ ql^2; \text{ thus } a_q = -0.0763$$

The value a_q can be more accurately calculated by considering the fact that over a short distance over the supports, the actual profile of the tendon should be convex to avoid sudden change in direction.

The hypothetical stress value at top fibre if q were applied without prestressing (Equation (12.9)):

$$\sigma_q = -0.0763\,(206.5 \times 10^3)\,(60)^2\,\frac{(-1.168)}{9.51} = 6.97\,\text{MPa}$$

The balanced deflection factor is (Equation (12.8)):

$$\beta_D \geq \frac{1-(-2.0)/6.97}{1+9.51/[8(-0.0763)(7.25)(2.9)(-1.168)]} = 0.79$$

The absolute value of mean prestressing force should exceed (Equation (12.11)):

$$P_m = 0.79 \frac{(206.5 \times 10^3)(60)^2}{8(2.9)} = 25\,200\,\text{kN}.$$

Table 12.1 gives the variation of β_D for the same structure when the allowable stress, $\sigma_{allowable}$ is varied between +2.4 and −4.4 MPa. The same table indicates the sensitivity of the permanent stress, σ_{perm} to the variation of the prestressing force. The table shows the typical result that the absolute value of the sustained compressive stress drops rapidly with the decrease of β_D. In this example a change of β_D from 1.0 to 0.6 varies σ_{perm} from −4.4 to 0.1 MPa. We recall that $\beta_D = 1.0$ corresponds to zero curvature (Fig. 12.1), or zero deflection. With $\beta_D = 0.6$ the curvature, the deflection and the cracking can be critical, particularly when the transient stress due to live load and temperature, $\Delta\sigma$ is high. Table 12.1 also demonstrates the typical result that the reserve compressive stress σ_{perm} can be substantially eroded when the actual time-dependent prestress loss is greater than estimated (e.g. when $P_{m\,actual} = 0.9\,P_m$). This can cause σ_{perm} to become small compressive or even tensile, causing the residual cracking and the residual deflection to be critical, after cyclic application of exceptional live load and temperature variation.

The values on the last line of Table 12.1 are calculated for the open cross-section shown in Fig. 12.2(b), with $h = 3$ m and area properties: $A = 6.00\,\text{m}^2$; $I = 5.10\,\text{m}^4$; $y_t = -0.813$ m; $q = 176.5$ kN-m; other data are the

Table 12.1 Variation of the permanent stress, σ_{perm} with the balanced deflection factor β_D. The table also gives variation of the required β_D for a given allowable stress, $\sigma_{allowable}$. Three-span bridge of Example 12.1, Fig. 12.2(a) and (b) with $h = 3$ m

Balanced deflection factor, β_D	0.4	0.5	0.6	0.7	0.8	0.9	1.0
Top fibre stress σ_{perm} or $\sigma_{allowable}$ (MPa); closed box section, Fig. 12.2(a)	2.41	1.27	0.13	−1.00	−2.14	−3.28	−4.42
Top fibre stress σ_{perm} or $\sigma_{allowable}$ (MPa); open section, Fig. 12.2(b)	2.81	1.58	0.35	−0.88	−2.11	−3.33	−4.56

same as above. The same remarks made above about the sensitivity of σ_{perm} to the choice of β_D apply to the open cross-section.

For comparison, we give below the results when the above calculations are repeated for the open cross-section in Fig. 12.2(b), with $q = 176.5$ kN-m and $\sigma_{allowable} = -2$ MPa (unchanged):

$$\sigma_q = 7.73 \text{ MPa}$$

$$\beta_D \geqslant 0.79$$

$$P_m = 21\ 600 \text{ kN}$$

Example 12.2 Simply supported bridges

Determine the balanced deflection factor β_D and the required mean force, P_m using the same data as in Example 12.1 but for a simply supported span $l = 60$ m and; $h = 3$ m; $\sigma_{allowable} = -2$ MPa at the bottom extreme fibre of mid-span section.

(a) Closed cross-section (Fig. 12.2a): $q = 206.5$ kN-m; $y = 1.832$ m; $A = 7.25$ m²; $I = 9.51$ m⁴.

$$M_q = 0.125\ ql^2; \ a_q = 0.125$$

The hypothetical stress value at bottom fibre ($y = 1.832$ m) if q were applied without prestress:

$$\sigma_q = 1.25(206.5 \times 10^3)(60)^2 \frac{1.832}{9.51} = 17.90 \text{ MPa}$$

The balanced deflection factor is (Equation (12.8)):

$$\beta_D \geqslant \frac{1 - (-2.0)/17.90}{1 + 9.51/[8(0.125)(7.25)(2.9)(1.832)]} = 0.89$$

$$P_m = 0.89 \frac{(206.5 \times 10^3)(60)^2}{8(2.9)} = 28\ 500 \text{ kN}$$

(b) Open cross-section (Fig. 12.2(b)): $q = 176.5$ kN-m; $y = 2.187$ m;

$A = 6.00 \, \text{m}^2$; $I = 5.10 \, \text{m}^4$. Repetition of the above calculation using this data gives:

$\sigma_q = 34.05 \, \text{MPa}$

$\beta_D = 0.85$

$P_m = 23\,300 \, \text{kN}$

Example 12.3: Effects of variation of span to thickness ratio on β_D

For the closed bridge cross-section shown in Fig. 12.2(a), determine the value of β_D required for an allowable stress, above the interior supports, $\sigma_{\text{allowable}} = -2 \, \text{MPa}$. Assume that the bridge deck is continuous over three spans of lengths: $0.7l$, l and $0.7l$. Consider $l = 30$, 60 and 90 m and $l/h = 20$, 25 and 30. In all cases use $f_0 = h - 0.1 \, \text{m}$ and q = self-weight plus $32.5 \, \text{kN/m}$; specific weight of concrete $24 \, \text{kN-m}^3$.

Calculations similar to Example 12.1 give the results in Table 12.2 which indicate that β_D varies between 0.94 and 0.69; the lower value is approached with the increase in l or in $\frac{1}{h}$ The values of the mean prestress force P_m for each case are also given in the same table.

Table 12.2 Variation of the required balanced deflection factor β_D with the span l and span to thickness ratio l/h. Bridge deck with spans $0.7l$, l and $0.7l$; closed cross-section (Fig. 12.2a); sustained load q = self-weight plus $32.5 \, \text{kN-m}$; allowable permanent stress at top fibre above interior supports, $\sigma_{\text{allowable}} = -2 \, \text{MPa}$

l/h	$l = 30$		$l = 60 \, \text{m}$		$l = 90 \, \text{m}$	
	β_D	P_m (kN)	β_D	P_m (kN)	β_D	P_m (kN)
20	0.94	12 800	0.79	25 400	0.75	41 600
25	0.87	14 500	0.73	27 500	0.71	45 500
30	0.83	16 400	0.72	31 100	0.69	49 700

12.6   Transient stresses

The transient stress in concrete, $\Delta\sigma$ caused by variable actions such as live load and temperature variation when added to the sustained stress σ_{perm} can produce irreversible opening of cracks and irreversible deformations. The values of $\Delta\sigma$ and σ_{perm} will be used below (Section 12.9) to give the non-prestressed reinforcement ratio required for control of residual crack opening. Again, homogenous uncracked sections are assumed in calculating $\Delta\sigma$ and σ_{perm}.

The magnitude of the transient stress $\Delta\sigma$ is different from structure to structure, depending upon the type of live load and the climate. As example, consider a bridge continuous over three spans: $0.7l$, l and $0.7l$ and having a box or open cross-section shown in Fig. 12.2(a) or (b). A temperature rise varying over the depth as shown in Fig. 12.4, produces at the section at the middle of the interior span the stress distributions[2] shown in Fig. 12.5. Here the span is varied between $l = 30$ and 90 m, maintaining the span to thickness ratio $l/h = 20$. It can be seen that in this case, the maximum tensile stress due to temperature rise is of the order 2 MPa (0.3 ksi) and occurs near the bottom fibre. The temperature distribution in Fig. 12.4 may be representative of the condition in the afternoon of a summer day in moderate climate. A distribution of the same shape, but with half the temperature values and the sign reversed (representing drop of temperature) may appear in the night or in the early morning in winter. This drop in temperature produces at the section over the interior supports the stress distributions shown in Fig. 12.5, with the stress values multiplied by –0.5. Here the maximum tensile stress is again close to 2 MPa (0.3 ksi), occurring at the top fibre. It is to be noted that because of the roller supports, the constant part of the temperature rise shown in Fig. 12.4 produces no stress. In calculation of the stresses presented in Fig. 12.5, the coefficient of thermal expansion is taken equal to 10×10^{-6} per degree Celsuis (5.6×10^{-6} per degree Fahrenheit) and the modulus of elasticity is considered equal to 30 GPa (4400 ksi). The stresses presented are the sum of self-equilibrating stresses and stresses due to statically indeterminate moment. (See Example 10.1.)

In the same bridges the maximum stress $\Delta\sigma_{traffic}$ due to exceptionally heavy load[3] (convoy weighing 4000 kN (900 kip)) is between 1.8 and 4.7 MPa (0.26

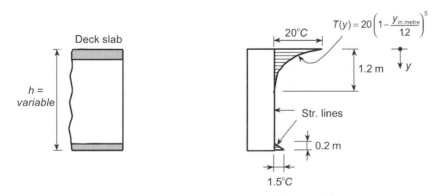

Figure 12.4 Distribution of temperature rise over the height of a bridge cross-section used in the analyses whose results are shown in Fig. 12.5.

and 0.68 ksi), in the bottom fibre at the section in the middle of the interior span. The lower stress value is for interior span $l = 30$ m and closed cross-section (Fig. 12.2(a)); the upper value is for $l = 90$ m and open cross-section (Fig. 12.2(b)). At the section over the interior support the maximum value is $\Delta\sigma_{traffic} = 1.5$ to 2 MPa (0.2 to 0.3 ksi), with the lower value being for the closed section.

The thermal stresses given above are calculated for bridges continuous over supports that allow free axial elongation. Thus, the distribution of temperature rise is divided into a constant part and a variable part (hatched) in Fig. 12.4; only the latter part produces thermal stresses. Cracking is ignored in the calculation of the thermal stresses presented in Fig. 12.5. The statically indeterminate bending moments due to temperature variation are proportionate to the flexural rigidity EI. In a cracked zone, the flexural rigidity is smaller than in the uncracked zone; if this variation in flexural rigidity is taken into account (as discussed in Chapter 13), the calculated statically indeterminate bending moments can be much smaller than what is obtained with linear analysis. This fact makes it advantageous to allow cracking, by not choosing a too high value of the balanced deflection factor β_D, particularly in climates with temperature extremes. Details of the linear analysis that give the thermal stress distributions in Fig. 12.5 are given in Section 5.9 in Ghali & Neville.[4]

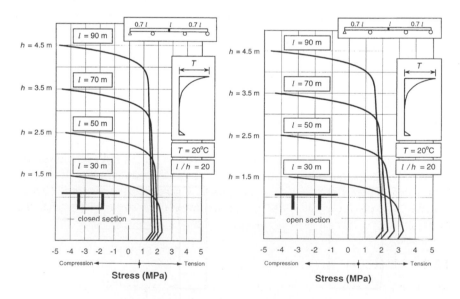

Figure 12.5 Distribution of stress due to temperature rise (Fig. 12.4) in a bridge continuous over three spans. Cross-section at the middle of the central span. Constant cross-section with height $h = l/20$, $l =$ variable. The cross-section is either closed or open (Fig. 12.2). 1 MPa = 0.145 ksi; 1 m = 3.28 ft; 1 °C = 1.8 °F.

12.7   Residual opening of cracks

Consider a prestressed concrete member containing non-prestressed reinforcement. While the sustained stress, σ_{perm} at extreme fibre of a cross-section is compressive, assume that the member is subjected to cyclic loads producing cyclic stress change at the same fibre. Assume that the maximum value, $\Delta\sigma$ represents in practice the transient effect of live load and/or temperature variation. It is assumed here that the value $\Delta\sigma$ is calculated for a noncracked section. Experiments and observations have shown that in each load cycle the crack opens and closes and the width varies between w_{max} and w_{res}; where the former is the maximum width and the latter is a *residual crack width*, caused by a permanent damage of the bond between the concrete and the reinforcement.

Experiments[5] on a prestressed member subjected to axial tension produced by a displacement controlled actuator show that the residual crack width, w_{res} is highly dependent upon the sustained compressive stress, σ_{perm}. Figure 12.6 shows the results of a series of tests in which σ_{perm} is varied between 0 and -3.5 MPa. It can be seen that as the absolute value of σ_{perm} is varied from zero (reinforced non-prestressed member) to 3.5 MPa, w_{res} varies from 0.12 mm to almost zero. The ordinates plotted in the graph are crack widths measured after 9000 cycles; the widths measured after one cycle are not substantially different. The experiments repeated with different non-prestressed steel ratio ρ_{ns} $(= A_{\text{ns}}/\text{gross concrete cross-sectional area})$ show that w_{res} is slightly influenced by ρ_{ns}. This is contrary to what is generally observed for reinforced non-prestressed elements.

The specified concrete strength and the diameter of the non-prestressed bars (for a given ρ_{ns} value) have an important effect on the maximum width of crack at the peak of the transient stress; however, the same two parameters have negligible influence on the residual crack width w_{res}.

The residual crack width may be considered an important parameter in design, because it is closely related to the durability of the structure. Characteristic limit values, $w_{k\ \text{res}} = 0.2$ to 0.05 mm are recommended or required by some codes in the design of prestressed concrete structure; the lower value is for the case when water-tightness is required, as for example the deck slab of a bridge. The characteristic value $w_{k\ \text{res}}$ is assumed to be equal to 1.5 times the mean value obtained in experiments or by analysis.

12.8   Water-tightness

Avoiding or limiting leakage is one of the reasons of controlling cracking in structural members that are permanently or occasionally in contact with water. Examples of such structures are water tanks, tunnels, parking floors and bridge decks. The repeated flow of contaminated water, often with de-icing salts has negative impact on the structures' durability.

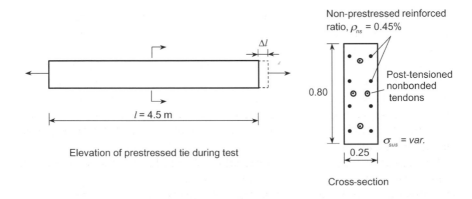

Elevation of prestressed tie during test

Cross-section

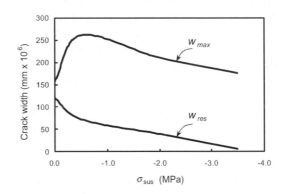

Figure 12.6 Variation of residual crack with, w_{res} with the sustained prestress, σ_{sus}. Cyclic imposed elongation $\Delta l / l$ varying between 200×10^{-6} and 400×10^{-6}. Specified concrete strength $f'_c \approx 30\,\text{MPa}$ (4400 psi). Reference is mentioned in Note 1 of this chapter.

For a specified liquid, the rate of flow through a crack in a floor or a wall (mass per unit time per unit length) is proportional to the gradient of liquid pressure and inversely proportional to the thickness of the structural element, but more importantly to the cube of the crack width.[6] The crack width is thus the governing factor for liquid-tightness.

When water is permanently in contact with a cracked concrete surface, it will cause swelling that tends to close cracks having widths less than 0.2 mm. However, this self-sealing process is hampered when the crack is repeatedly widened and the increase in width exceeds 0.1 mm.

12.9   Control of residual crack opening

The experimental work on prestressed concrete members subjected to axial tension may be considered representative of the upper or the lower slabs of a prestressed box cross-section subjected to bending moment that produce cracking. The results of the experiments have been used to verify an analytical model that calculates the width of residual cracks, w_{res}. Involved in the analysis are equations for the bond stress and the tensile stress in the vicinity of a crack under the effect of cyclic loading. The analytical model is then used in parametric studies and in the development of the design charts[7] presented in Figs. 12.7 (a) to (e).

For a specified characteristic residual crack width, w_{res}, the solid lines in the charts give the non-prestressed steel ratio ρ_{ns} as functions of the permanent stress, σ_{perm} and the maximum transient stress, $\Delta\sigma$. The dashed straight lines in the graphs give a non-prestressed steel ratio that must be exceeded to avoid yielding of this reinforcement when the maximum transient loading is applied. The specified yield stress of the non-prestressed steel is assumed equal to 400, 500 or 800 MPa. The residual crack width, w_{res} is slightly influenced by the specified concrete strength, or by the diameter of the non-prestressed steel bar; thus, these two parameters are absent from the design charts.

The charts give the recommended value for ρ_{ns} from the input parameters σ_{perm}, $\Delta\sigma$, $w_{k\ res}$ and $\sigma_{s\ max}(\equiv f_y)$. The symbols are redefined below and some remarks are given on the calculation of the parameters involved:

f_y = specified yield stress of non-prestressed reinforcement.

$w_{k\ res}$ = specified characteristic residual crack width, a value equal to 1.5 times the mean residual crack width.

$\Delta\sigma$ = maximum change in tensile stress in concrete due to transient actions (e.g. live load and temperature variation). In calculation of $\Delta\sigma$, cracking is ignored.

$\rho_{ns} = A_{ns}/A_{cef}$; where A_{ns} is cross-sectional area of non-prestressed reinforcement within an effective concrete area A_{cef}. The charts are derived for sections subjected to axial tension. Their use is recommended without adjustment for a slab, whose thickness is less than 0.3 m, in the tension face of a box section in flexure; in this case A_{cef} is equal to the cross-sectional area of the slab; it is also empirically recommended that σ_{perm} and $\Delta\sigma$ be calculated at a distance 0.1 m away from the extreme fibre. Use of the same charts is extended to other cases by empirically taking A_{cef} as specified in Fig. E.2 (see Appendix E).

σ_{perm} = permanent stress at a fibre due to dead loads and other quasi-permanent loads combined with prestress. The value of σ_{perm} is calculated for a gross concrete section (uncracked) at a fibre that will subsequently be in tension when the transient load is applied. The fibre considered may be empirically taken at 0.1 m away from the extreme fibre of a slab of a box section; alternatively at the centroid of A_{cef}. The prestressing forces to be used in calculating σ_{perm} is an average of the values of the prestressing forces before and after the time-dependent losses.

12.10   Recommended[8] longitudinal non-prestressed steel in closed-box bridge sections

Figure 12.8 represents top view of the top and bottom slabs, respectively of a bridge having the closed cross-section in Fig. 12.2(a). The two slabs are divided into zones labelled 1, 2, 3 and 4. Use of the charts in Figs. 12.7(a) to (e) is recommended to give ρ_{ns}; but this ratio should be greater or equal to 0.4 per cent. Near the tip of the cantilevers, over a width equal to one third of the width of the overhang, b_c and a length equal the full length of the deck, ρ_{ns} is empirically recommended to be not less than 0.6 per cent. This is because the slab edges are vulnerable to cracking caused by shrinkage of the tip, which is often cast subsequent to the casting of the deck. The edges are also subjected to stresses due to temperature gradient in the horizontal direction, often not considered in calculation of the transient stress, $\Delta\sigma$. In the longitudinal direction for a length equal $l/5$ on either side of an interior support (zones 1 and 4), ρ_{ns} is to be controlled by σ_{perm} and $\Delta\sigma$ at the section over the support; for the remainder of the length of the deck (zones 2 and 3), ρ_{ns} is to be controlled by σ_{perm} and $\Delta\sigma$ at the section at mid-span.

Commonly the thickness of the top and the bottom slabs are variable in the transverse direction (as opposed to the simplified bridge cross-sections in Fig. 12.2). It is suggested that the cross-sectional area of the non-prestressed reinforcement per unit width of the slab be equal to the value of ρ_{ns} as recommended above for each zone multiplied by the minimum slab thickness in the zone; in this way the diameter of the bars and their spacing can be constant in each zone.

12.11   Residual curvature

The graph in Fig. 8.5 represents the relationship between the bending moment, M and the curvature, ψ for a reinforced non-prestressed section subjected to no normal force. The same figure is shown in Fig. 12.9(a) for comparison with the M-ψ graph when the section is subjected to a normal force $N < 0$, caused by prestressing. Here, the coefficient $\beta = \beta_1\beta_2 = 0.5$, representing the case of repetitive loading and the use of high bond bars. The graph AFG in Fig. 12.9(b) represents M versus ψ_1, or M versus mean curvature ψ_m after cracking, for a section subjected to a constant normal

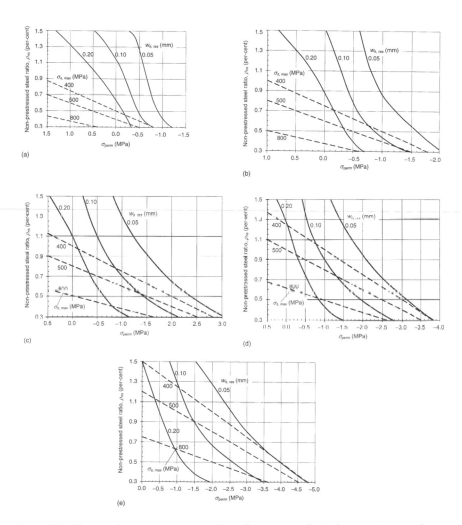

Figure 12.7 Choice of the non-prestressed steel ratio, ρ_{ns} to limit characteristic crack width $w_{k\ res}$ for variable maximum transient stress $\Delta\sigma$. The charts can also give the mean residual crack width (= $w_{k\ res}/1.5$) for given values of ρ_{ns} and $\Delta\sigma$; (a) to (e) $\Delta\sigma$ = 2, 3, 4, 5 and 6 MPa, respectively. 1 MPa = 0.145 ksi.

force $N < 0$ (compressive) and variable M. The graph AED in the same figure represents the case when $N = 0$ (same as AED in Fig. 12.9(a)). It can be seen that the effect of prestressing is to move ED upward to FG; thus the value $M_r \sqrt{0.5}$ the upper limit of the noncracked state is increased; also, after cracking, ψ_m is reduced. We recall that M_r is the value of the moment just sufficient to cause a virgin section to crack; M_r can be determined from the equation:

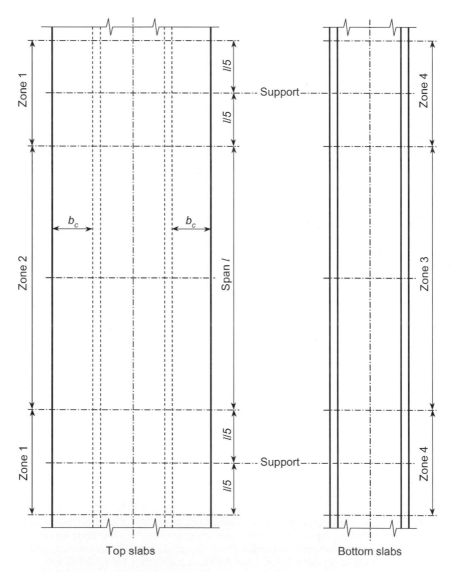

Figure 12.8 Division of top and bottom of slabs of a closed-box section into zones in which the recommended non-prestressed steel ratio, ρ_{ns} is constant.

$$f_{ct} = \frac{N}{A_1} + \frac{M}{W_1} \qquad (12.13)$$

where f_{ct} is the tensile strength of concrete, A_1 and W_1 are the area and the section modulus of the transformed uncracked section, respectively.

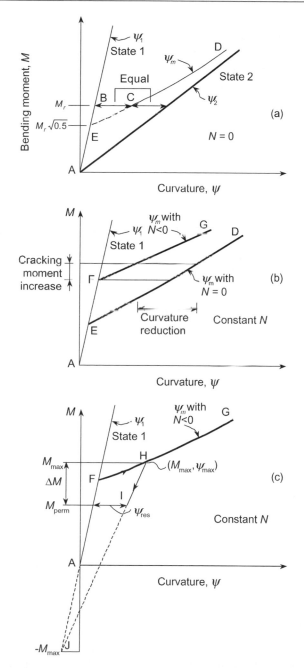

Figure 12.9 Moment curvature relationship: (a) $N = 0$; (b) comparison of the mean curvatures, ψ_m with $N = 0$ and with $N < 0$ (compressive); (c) residual curvature in a prestressed section subjected to transient moment increment ΔM producing cracking.

Figure 12.9(c) represents a prestressed section for which the moment M_{perm} due to prestressing and sustained load is below M_r. A transient moment increment ΔM brings the moment level to M_{max} and the curvature to ψ_{max}. Removal of ΔM does not fully recover the curvature and the section is left with a residual curvature ψ_{res} as the moment level returns to M_{perm}. The descending branch HI of the M-ψ curve is defined by the straight line HIJ, where J is a point whose ordinate is $-M_{max}$ and situated on the straight line FAJ. This graphical construction that gives ψ_{res} is based on extensive experiments.[9] The residual curvature can be calculated by the equation:

$$\psi_{res} = \frac{M_{max} + M_{perm}}{2M_{max}}\left(\psi_{m,\,M_{max}} - \frac{M_{max}}{EI_1}\right) \tag{12.14}$$

where $\psi_{m,\,M_{max}}$ is the mean curvature corresponding to M_{max}; the value of the mean curvature is given by Equation (8.21).

The residual curvature can cause an increase in deflection, which is additional to the deformation caused by the time-dependent effects of creep, shrinkage and relaxation. Studies of the bridge cross-sections shown in Fig. 12.2 and discussed in Section 12.5 show that the residual curvature increases as the balanced deflection factor β_D is decreased. Significant residual curvature and deflection, that should be avoided, occurs when $\beta_D \leq 0.6$ and $l \geq$ 60 m. For structures expected to have long service life, such as bridges, it is recommended to adopt β_D between 0.8 and 1.0. It is to be noted that for the same value of β_D, the permanent stress, σ_{perm} is a compressive stress of smaller absolute value for smaller spans. This can be seen by setting a value of β_D in Equation (12.8) and solving for the allowable stress. Thus, cracking occurs at a smaller moment increment, ΔM when the span is smaller. For this reason, β_D should be closer to 1.0 for smaller spans.

12.12   General

This chapter is concerned with prestressed structures in which cracking occurs under transient live load and/or temperature variation. These repetitious actions produce residual opening of cracks and residual curvature associated with residual deflections that are additional to the time-dependent effects of creep, shrinkage and concrete. Structures should be designed such that in the permanent state the opening of cracks and the deflections are not excessive.

A balanced deflection factor, β_D is defined as equal to the absolute value of the permanent deflection due to prestressing divided by the deflection due to permanent load. Choice of the parameters β_D and the corresponding amount of prestressing are important factors in limiting the residual opening of cracks and the residual deflections.

The residual width of cracks also depends upon the amount of the non-prestressed steel. The charts presented can be used to give the non-prestressed steel ratios that limit the residual crack widths to specified values. Choice of low value for β_D requires higher non-prestressed steel ratio.

The value of β_D may be selected between 0.7 and 1.0, depending upon the type of the structure. The selected value of β_D can be 1.0 or even more in structures exposed to large variable loads, such as railway bridges. For bridges built by incremental launching, β_D between 0.5 and 0.6 is possible, because of the favourable effect of high axial prestressing force needed for launching. (Such an axial force does not produce deflection, thus is not accounted for by the parameter β_D.) In prestressed concrete floor slabs, β_D can be as low as 0.5.

Notes

1 For further reference on bridge design, relating to concepts in this chapter, see Laurencet, P., Rotilio, J.-D., Jaccoud, J.-P., and Favre, R., *Influence des actions variables sur l'état permanent des ponts en béton précontraint*, Swiss Federal Institute of Technology, IBAP, Lausanne, Switzerland, May 1999, 168 pp.
2 Slightly different stress values are presented in a figure by Rotilio, J.-D., *Contributions des actions variables aux déformations à long terme des ponts en béton*. Doctorate thesis No.1870, Swiss Federal Institute of Technology, IBAP, Lausanne, Switzerland, 1997.
3 See the reference in Note 2, above.
4 See the reference in Note 3, page 99.
5 See Note 1, above.
6 Mivelaz, P., Jaccoud, J.-P. and Favre, R., 'Experimental Study of Air and Water Flow through Cracked Concrete Tension Members', *4th International Symposium on Utilization of High-Strength/High-Performance Concrete, Paris*, 1996, Publication IBAP No. 173, Swiss Federal Institute of Technology, Lausanne, Switzerland, January 1996.
7 The charts are taken from the reference mentioned in Note 1, above.
8 The recommendations are taken from the reference in Note 1, above.
9 Details of the experiments and their results are given in the reference in Note 1, above.

Chapter 13

Non-linear analysis of plane frames

Arch bridge in Switzerland

13.1 Introduction

In statically indeterminate structures, the reduction in member stiffness due to cracking is accompanied by changes in the reactions and internal forces. The changes can result in an increase in deflection. An example of this situation is the so-called 'redistribution' of moments associated with cracking in continuous beams and slabs. The relatively large negative moments over interior supports produce cracking accompanied by a drop in the absolute value of the negative moment and an increase in the positive moment and in the deflection. Cracking also causes considerable reduction in the stresses and internal forces induced by temperature variations, support movements or any other type of imposed deformations (see Chapter 11).

The present chapter is concerned with the analysis of reinforced concrete plane frames, with or without prestressing, accounting for the effects of cracking. The general displacement method of analysis (Section 5.2) is used for this purpose. The structures are here considered in service condition, in which the stress–strain relation for concrete can be assumed linear. Cracking, however, causes non-linearity; therefore, the analysis requires iterative computations using a computer. Analysis beyond service condition up to failure requires consideration of the non-linear stress–strain relationship for the concrete and the reinforcement; this is beyond the scope of this book. Shear deformations are ignored in the analysis presented below.

When the structure is prestressed, it is assumed that the stresses in the structure are known after accounting for the effects of creep and shrinkage of concrete and relaxation of the prestressing reinforcement. Furthermore, it is assumed that no cracking has occurred at this stage and the analysis is required to find the effects of additional loading (e.g., live loads or temperature variations) that may produce cracking.

13.2   Reference axis

In the analysis which will follow, we will use A_1, B_1 and I_1, which are the properties of transformed uncracked sections composed of the area of concrete plus the area of the reinforcement, A_{ns} or A_{ps}, multiplied by α_{ns} or α_{ps}; where $\alpha_{ns} = (E_{ns}/E_c)$ or $\alpha_{ps} = (E_{ps}/E_c)$, with E_{ns} and E_{ps} being the moduli of elasticity of non-prestressed and prestressed reinforcement, respectively. We will also use A_2, B_2 and I_2, the properties of transformed cracked section calculated in the same manner, but the concrete in tension is ignored. The symbols A, B and I represent, respectively, the cross-sectional area, its first and second moment about an axis through a reference point. Obviously, the uncracked and the cracked transformed sections do not have the same centroid. Thus, it is more suitable for frame analysis to place the nodes on a fixed non-centroidal axis of members. The axis is through reference point O, which may be chosen at any fibre, for example, at the top fibre. However, choice of the reference axes at, or close to, the centroid of the gross concrete section may be advantageous in some cases; thus, for practice, it is recommended to select the reference axis at – or close to – this centroid in all cases.

13.3   Idealization of plane frames

Figure 13.1(a) shows a typical reinforced concrete plane frame with or without prestressing. The frame is idealized as an assemblage of straight non-prismatic beam elements connected at the joints (nodes); see Fig. 13.1(b). Each node has three displacement components: two translations and a rotation, in direction of arbitrarily chosen global axes, x, y and z.

For each member, a local system of axes x^*, y^* and z^* is defined in Fig.

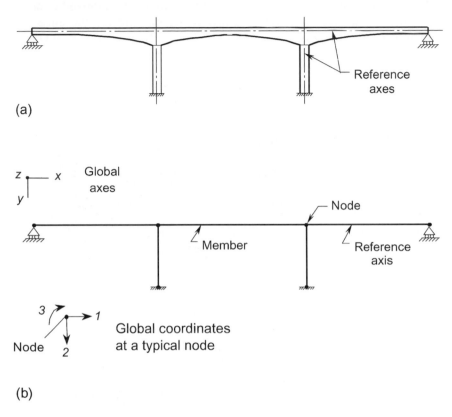

(a)

(b)

Figure 13.1 Idealization of a plane frame: (a) a typical plane frame; (b) idealized structure.

13.2(a). The axis x^* coincides with the reference axis of the member and is directed from end node O_1 to end node O_2. The y^*- and z^*-axes are mutually perpendicular to the x^*-axis, with y^* lying in the plane of the frame. The purpose of the analysis is to determine the changes in nodal displacements in global directions and, for each member, the internal forces at its ends along the six local coordinates 1^* to 6^*. Define a number of cross-sections, arbitrarily spaced; the first section is at node O_1 and the last section is at node O_2 (Fig. 13.3(a)). For each cross-section, the geometry (including the areas A_{ns} and/or A_{ps} of the reinforcement layers and their depths) and the parameters $\{\sigma_O, \gamma\}_{in}$, defining the distribution of the initial stress (Fig. 13.3(b)) are part of the given data. External applied loads on an individual member are given at the same sections in directions of the local axes. Also, external forces may be given as nodal forces in the directions of the global axes. Figure 13.3(b) shows the positive sign convention for σ_O and γ and the normal force N and the bending moment M.

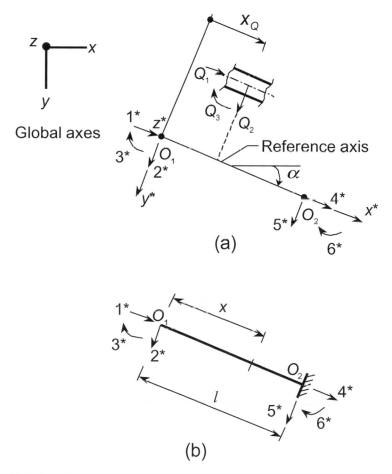

Figure 13.2 Coordinate system for plane frame analysis: (a) local coordinates representing displacements $\{D^*\}$ or forces $\{F^*\}$ at ends of a typical member; (b) system of forces in equilibrium caused by a displacement introduced at coordinate $1^*, 2^*$ or 3^*, while end O_2 is totally fixed.

13.4   Tangent stiffness matrix of a member

The tangent stiffness matrix of a typical member (Fig. 13.2(a)) relates the forces and the displacements in local coordinates:

$$[S^*] \{D^*\} = \{F^*\} \qquad (13.1)$$

where $\{F^*\}$ and $\{D^*\}$ represent *small* increments of nodal forces and nodal

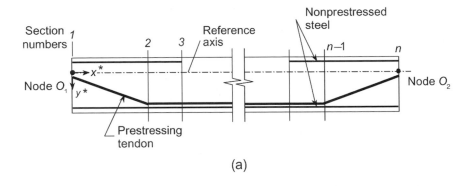

(a)

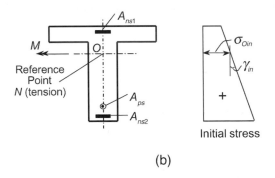

(b)

Figure 13.3 Input data defining the geometry and initial stresses: (a) a number of sections defined on a typical member; (b) cross-section geometry and initial stress distribution at a typical section; also, positive sign convention for σ_O, γ, N and M.

displacements. A typical element S_{ij}^* of the tangent stiffness matrix is equal to the force increment at coordinate i due to a small unit displacement at coordinate j. The cross-sectional area properties A, B and I are assumed variable; where A, B and I are the area, its first moment and its second moment about an axis through the reference point O (Fig. 13.3(a)). The variation of the cross-sectional area properties can be caused by variation of geometry or by cracking.

The tangent stiffness matrix $[S^*]$ can be partitioned into 3×3 submatrices:

$$[S^*] = \begin{bmatrix} [S_{11}^*] & [S_{12}^*] \\ [S_{21}^*] & [S_{22}^*] \end{bmatrix} \tag{13.2}$$

The submatrices in the first row contain forces at coordinates 1*, 2* and 3* at node O_1. The equilibrants of these forces at coordinates 4*, 5* and 6*, at node O_2, form the elements of the submatrices in the second row. Thus, the three elements of any column of $[S_{11}^*]$ and the three elements of the same column of $[S_{21}^*]$ represent a system of forces in equilibrium (Fig. 13.2(b)). This equilibrium relationship may be expressed as:

$$[S_{21}^*] = [R]\,[S_{11}^*] \tag{13.3}$$

where

$$[R] - \begin{bmatrix} -1 & 0 & 0 \\ 0 & -1 & 0 \\ 0 & l & -1 \end{bmatrix} \tag{13.4}$$

with l being the member's length.

Because of this equilibrium relationship and symmetry of $[S^*]$, Equation (13.2) may be rewritten as:

$$[S^*] = \begin{bmatrix} [S_{11}^*] & [S_{11}^*]\,[R]^T \\ [R][S_{11}^*] & [R]\,[S_{11}^*]\,[R]^T \end{bmatrix} \tag{13.5}$$

The matrix $[S_{11}^*]$ can be determined by:

$$[S_{11}^*] = [f^*]^{-1} \tag{13.6}$$

where $[f^*]$ is the flexibility matrix of the member when it is treated as a cantilever fixed at node O_2 (Fig. 13.2(b)). Any element f_{ij}^* of the flexibility matrix is equal to the change in displacement at coordinate i due to a small unit increment of force at coordinate j. Using virtual work (unit load theory), a typical element of $[f^*]$ is given by:

$$f_{ij}^* = \int_0^l N_{ui}\varepsilon_{Ouj}\,dx + \int_0^l M_{ui}\psi_{uj}\,dx \tag{13.7}$$

where N_{ui} and M_{ui} are the normal force and bending moment at any section at a distance x from end O_1, due to a virtual unit force at coordinate i, with $i = 1$, 2 or 3; ε_{Ouj} and ψ_{uj} are the changes in the strain, in the same section, at reference point O and in the curvature produced by a small unit force applied at coordinate j, with $j = 1$, 2 or 3. For $F_1^* = 1$, $N_{u1} = -1$ and $M_{u1} = 0$; for $F_2^* = 1$, $N_{u2} = 0$ and $M_{u2} = -x$; and for $F_3^* = 1$, $N_{u3} = 0$ while $M_{u3} = 1$. Substitution in Equation (13.7) gives:

$$f^*_{1j} = -\int_0^l \varepsilon_{Ouj} dx; \quad f^*_{2j} = -\int_0^l \psi_{uj} x \, dx; \quad f^*_{3j} = \int_0^l \psi_{uj} dx \text{ with } j = 1, 2, 3 \quad (13.8)$$

The integrals in this equation are evaluated numerically (Section 13.8) using values of ε_O and ψ, determined by Equation (2.19) at a number of sections for which the geometry and cross-sectional area of reinforcement are known. For cracked sections, ε_O and ψ represent mean values determined by Equations (8.43) and (8.44). This requires that the depth c of the compression zone and the interpolation coefficient ζ (defined in Section 8.3) be known. The two parameters depend upon the stresses existing before introducing the increments in the forces at the ends. Thus, the tangent stiffness depends upon the stress level and the state of cracking of the member.

In order to generate the tangent stiffness matrix for the structure, the tangent stiffness matrices, $[S^*]$ of individual members must be transformed from the local coordinate systems to the global system:

$$[S_{\text{member}}] = [T]^T [S^*] [T] \quad (13.9)$$

where $[S_{\text{member}}]$ is the member stiffness matrix in global coordinates; $[T]$ is a transformation matrix given by:

$$[T] = \begin{bmatrix} [t] & [0] \\ [0] & [t] \end{bmatrix} \quad ; \quad [t] = \begin{bmatrix} c & s & 0 \\ -s & c & 0 \\ 0 & 0 & 1 \end{bmatrix} \quad (13.10)$$

where $c = \cos a$ and $s = \sin a$, with a being the angle between the global x-direction and the local x^*-axis (Fig. 13.2(a)). The matrix $[T]$ can be used for transformation of member end forces and displacements from local to global or vice versa:

$$\{D^*\} = [T]\{D\} \quad ; \quad \{F\}_{\text{global}} = [T]^T \{F^*\} \quad (13.11)$$

13.5   Examples of stiffness matrices

Example 13.1 Stiffness matrix of an uncracked prismatic cantilever

Derive the stiffness matrix with respect to non-centroidal coordinates, shown in Fig. 13.4(a), for an uncracked cantilever having a constant cross-section with properties, A, B and I. What are the displacements at the three coordinates due to a downward force P applied at the free end?

Assume that the member has a rectangular cross-section of width b and height h.

The normal strain $\varepsilon_{\text{O}uj}$ and curvature ψ_{uj} at any section are obtained by the application of a unit force at each of the three coordinates at the end O_1 and the use of Equation (2.19):

For $F_1^* = 1$, $N_{u1} = -1$ and $M_{u1} = 0$,

$$\varepsilon_{\text{O}u1} = -\frac{I}{E(AI - B^2)} \quad \text{and} \quad \psi_{u1} = \frac{B}{E(AI - B^2)} \tag{13.12}$$

For $F_2^* = 1$, $N_{u2} = 0$ and $M_{u2} = -x$,

$$\varepsilon_{\text{O}u2} = \frac{Bx}{E(AI - B^2)} \quad \text{and} \quad \psi_{u2} = -\frac{Ax}{E(AI - B^2)} \tag{13.13}$$

For $F_3^* = 1$, $N_{u3} = 0$ and $M_{u3} = 1$,

$$\varepsilon_{\text{O}u3} = -\frac{B}{E(AI - B^2)} \quad \text{and} \quad \psi_{u3} = \frac{A}{E(AI - B^2)} \tag{13.14}$$

Substitution of the above expressions for $\varepsilon_{\text{O}u}$ and ψ_u in Equation (13.8) gives:

$$[f^*] = \frac{l}{E(AI - B^2)} \begin{bmatrix} I & -\dfrac{Bl}{2} & B \\[2ex] -\dfrac{Bl}{2} & \dfrac{Al^2}{3} & -\dfrac{Al}{2} \\[2ex] B & -\dfrac{Al}{2} & A \end{bmatrix} \tag{13.15}$$

Application of Equations (13.6), (13.4) and (13.5) gives the stiffness matrix corresponding to the six local coordinates in Fig. 13.2(b):

$$[S^*] = E \begin{bmatrix} \dfrac{A}{1} & & & & \text{symmetrical} & \\ 0 & \dfrac{12(AI - B^2)}{Al^3} & & & & \\ -\dfrac{B}{1} & \dfrac{6(AI - B^2)}{Al^2} & \dfrac{4AI - 3B^2}{Al} & & & \\ \hline -\dfrac{A}{1} & 0 & \dfrac{B}{1} & \dfrac{A}{1} & & \\ 0 & -\dfrac{12(AI - B^2)}{Al^3} & -\dfrac{6(AI - B^2)}{Al^2} & 0 & \dfrac{12(AI - B^2)}{Al^3} & \\ \dfrac{B}{1} & \dfrac{6(AI - B^2)}{Al^2} & \dfrac{2AI - 3B^2}{Al} & -\dfrac{B}{1} & -\dfrac{6(AI - B^2)}{Al^2} & \dfrac{4AI - 3B^2}{Al} \end{bmatrix} \quad (13.16)$$

When O is chosen at the centroid of the cross-section, B will be equal to zero and the matrix $[S^*]$ in Equation (13.16) will reduce to the conventional form of the stiffness matrix for a plane frame member.[1] The 3 × 3 submatrix at the top left-hand corner of this matrix is the stiffness matrix of the cantilever in Fig. 13.4(a).

For the cantilever with a rectangular section, the area properties, with the reference point O at the top fibre, are: $A = b\,h$; $B = \dfrac{b\,h^2}{2}$; $I = \dfrac{b\,h^3}{3}$. Substitution in Equation (13.16) gives:

$$[S^*]_{\text{cantilever}} = \frac{Ebh}{l} \begin{bmatrix} 1 & 0 & -\dfrac{h}{2} \\ 0 & \dfrac{h}{l^2} & \dfrac{h^2}{2l} \\ -\dfrac{h}{2} & \dfrac{h^2}{2l} & \dfrac{7h^2}{12} \end{bmatrix} \quad (13.17)$$

The displacements at the free end due to the applied force P are:

$$\{D^*\} = [S^*]^{-1} \begin{Bmatrix} 0 \\ P \\ 0 \end{Bmatrix} = \frac{l}{Ebh^3} \begin{bmatrix} 4h^2 & -3hl & 6h \\ -3hl & 4l^2 & -6l \\ 6h & -6l & 12 \end{bmatrix} \begin{Bmatrix} 0 \\ P \\ 0 \end{Bmatrix} = \frac{Pl^2}{Ebh^3} \begin{Bmatrix} -3h \\ 4l \\ -6 \end{Bmatrix} \quad (13.18)$$

If the procedure followed in this example is redone with the reference axis chosen through the cross-sectional centroid, D_1^* would be zero, with the other two displacements unchanged. The top fibre at the tip of the cantilever would move horizontally outwards at a distance equal to $h/2$ multiplied by D_3^*, which is the same answer as obtained above.

Example 13.2 Tangent stiffness matrix of a cracked cantilever

Find the tangent stiffness matrix corresponding to the coordinates in Fig. 13.4(a) for the cantilever of Example 13.1, assuming that it has a constant concrete cross-section, reinforced with non-prestressed steel of areas $A_{ns1} = A_{ns2} = 0.01\ bh$ (Fig. 13.4(b)). Also find the three displacements at the three coordinates due to a downward force P applied at the free end. Assume that initially the cantilever has been cracked due to a negative bending moment having the same value at all sections, such that c and ζ are constant. Given data: The elasticity moduli of steel and concrete are $E_s = 200\ \text{GPa}$ and $E_c = 25\ \text{GPa}$; $c = 0.275\ h$ and $\zeta = 0.8$; the given c value is determined by Equation (7.16); the compression zone is at the bottom of the section; the transformed cross-sectional area properties are: $A_1 = 0.6840\ h^2$; $B_1 = 0.3420\ h^3$; $I_1 = 0.2344\ h^4$; $A_2 = 0.2549\ h^2$; $B_2 = 0.1849\ h^3$; $I_2 = 0.1582\ h^4$; where the subscripts 1 and 2 refer to the uncracked and the fully cracked states, respectively. Give the answer in terms of P and E_c. Assume that P is small such that it causes negligible change in the value of ζ.

At any section of the cantilever, the strain parameters ε_0 and ψ are calculated assuming that the sections are uncracked (using A_1, B_1, and I_1), and again assuming that all sections are fully cracked (using A_2, B_2, and I_2). Then the mean strain parameters are determined using $\zeta = 0.8$

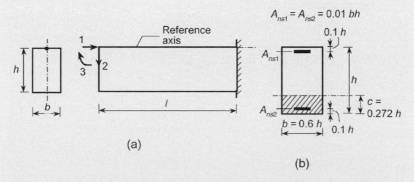

(a)

(b)

Figure 13.4 The cantilever considered in Examples 13.1 and 13.2: (a) coordinate system; (b) cracked reinforced section considered in Example 13.2.

Table 13.1 Strain parameters at any section of the cantilever of Example 13.2 due to unit force $F_1{}^*, F_2{}^*$ or $F_3{}^*$ applied at the free end

Force applied	Strain parameters	Uncracked	Fully cracked	Mean	Multiplier
$F_1^* = 1$	ε_{Ou1}	−5405	−25780	−21700	$10^{-3}(h^2 E_c)^{-1}$
	ψ_{u1}	7886	30130	25680	$10^{-3}(h^3 E_c)^{-1}$
$F_2^* = 1$	ε_{Ou2}	7886x	30130x	25680x	$10^{-3}(h^3 E_c)^{-1}$
	ψ_{u2}	−15770x	−41530x	−36380x	$10^{-3}(h^4 E_c)^{-1}$
$F_3^* = 1$	ε_{Ou3}	−7886	−30130	−25680	$10^{-3}(h^3 E_c)^{-1}$
	ψ_{u3}	15770	41530	36380	$10^{-3}(h^4 E_c)^{-1}$

in Equations (8.43) and (8.44). The results of these calculations are presented in Table 13.1. We give below, as example, the calculations for $F_1^* = 1$.

For, $F_1^* = 1$, $N_{u1} = -1$ and $M_{u1} = 0$ at any section. Apply Equation (13.12) for uncracked section:

$$\varepsilon_O = -\frac{0.2344}{E_c h^2 [0.6840(0.2344) - (0.3420)^2]} = -5405 \times 10^{-3} (E_c h^2)^{-1}$$

$$\psi = \frac{0.3420}{E_c h^3 [0.6840(0.2344) - (0.3420)^2]} = 7886 \times 10^{-3}(E_c h^3)^{-1}$$

Apply the same equation for a fully cracked section:

$$\varepsilon_O = -\frac{0.1582}{E_c h^2 [0.2549(0.1582) - (0.1849)^2]} = -25780 \times 10^{-3}(E_c h^2)^{-1}$$

$$\psi = \frac{0.1849}{E_c h^3 [0.2549(0.1582) - (0.1849)^2]} = 30130 \times 10^{-3}(E_c h^3)^{-1}$$

Mean parameters (with $\zeta = 0.8$):

$$\left\{\begin{matrix} \varepsilon_O \\ \psi \end{matrix}\right\}_{mean} = (1 - \zeta)\left\{\begin{matrix} \varepsilon_O \\ \psi \end{matrix}\right\}_{uncracked} + \zeta\left\{\begin{matrix} \varepsilon_O \\ \psi \end{matrix}\right\}_{fully-cracked} = \left\{\begin{matrix} -21700\, h \\ 25680 \end{matrix}\right\}(E_c h^3)^{-1}$$

The flexibility coefficients are determined by Equation (13.8), with the integrals evaluated explicitly giving:

$$[f^*] = \frac{10^{-3}l}{E_c h^2} \begin{bmatrix} 21700 & -\dfrac{25680}{h}\left(\dfrac{l}{2}\right) & \dfrac{25680}{h} \\[2ex] -\dfrac{25680}{h}\left(\dfrac{l}{2}\right) & \dfrac{36380}{h^2}\left(\dfrac{l^2}{3}\right) & -\dfrac{36380}{h^2}\left(\dfrac{l}{2}\right) \\[2ex] \dfrac{25680}{h} & -\dfrac{36380}{h^2}\left(\dfrac{l}{2}\right) & \dfrac{36380}{h^2} \end{bmatrix}$$

The stiffness matrix is:

$$[S^*] = [f^*]^{-1} = \frac{E_c h^2}{l} \begin{bmatrix} 0.2799 & 0 & -0.1976h \\[2ex] 0 & 0.3295\,\dfrac{h^2}{l^2} & 0.1647\,\dfrac{h^2}{l} \\[2ex] -0.1976h & 0.1647\,\dfrac{h^2}{l} & 0.2493\,h^2 \end{bmatrix}$$

The displacements at the free end are:

$$\{D^*\} = [S^*]^{-1} \left\{ \begin{matrix} 0 \\ P \\ 0 \end{matrix} \right\} = \frac{Pl^2}{E_c h^4} \left\{ \begin{matrix} -12.84h \\ 12.13l \\ -18.19 \end{matrix} \right\}$$

13.6   Fixed-end forces

When the external forces are applied at intermediate sections away from the nodes (Fig. 13.2(a)), the analysis by the displacement method involves calculation of the actions $\{A_r\}$. These are the values of the end forces due to the applied loads with the member ends totally fixed. The vector $\{A_r\}$ may be partitioned into two 3×1 vectors:

$$\{A_r\} = \left\{ \begin{matrix} \{A_r\}_{O1} \\ \{A_r\}_{O2} \end{matrix} \right\} \tag{13.19}$$

Consider the case when a nonprismatic member is subjected to the three force components $\{Q\}$ shown in Fig. 13.2(a) at one section located at a distance x_Q from the member end O_1. The values of $\{A_r\}_{O1}$ can be determined by the force method (Section 4.2) using, as the released structure, the cantilever in Fig. 13.2(b). This gives:

$${A_r}_{O1} = -[f^*]^{-1} {D_s^*} \tag{13.20}$$

where $[f^*]$ is the flexibility matrix derived by Equation (13.8); ${D_s^*}$ is vector of the displacements of the released structure; these are given by virtual work:

$$D_{1s}^* = -\int_0^l \varepsilon_O\, dx; \qquad D_{2s}^* = -\int_0^l \psi\, x\, dx; \qquad D_{3s}^* = \int_0^l \psi\, dx \tag{13.21}$$

where ε_O and ψ are the strain at the reference point O and the curvature produced in the released structure at any section at a distance x from O_1. Here again, Equation (2.19) is to be used to determine ε_O and ψ at different sections and the integrals are evaluated numerically (Section 13.8).

The forces at end O_2 can be determined by equilibrium:

$${A_r}_{O2} = [R] {A_r}_{O1} - \begin{Bmatrix} Q_1 \\ Q_2 \\ Q_3 - Q_2(l - x_Q) \end{Bmatrix} \tag{13.22}$$

where $[R]$ is the matrix defined in Equation (13.4).

Distributed loads in the transverse or the tangential direction of a member can be replaced by statical equivalent concentrated forces in the same direction. The following equation can be used for this purpose:

$$\begin{Bmatrix} Q_A \\ Q_B \end{Bmatrix} = \frac{s}{6} \begin{bmatrix} 2 & 1 \\ 1 & 2 \end{bmatrix} \begin{Bmatrix} q_A \\ q_B \end{Bmatrix} \tag{13.23}$$

where q_A and q_B are intensities of the distributed applied load at two sections A and B spaced at a distance s; Q_A and Q_B are statical equivalent forces, assuming that q varies linearly over the spacing s. Replacement of the distributed load by concentrated forces (rather than concentrated forces combined with moments) in this way involves an error which is considered here negligible when s is small compared to the member length (say $s \leqslant l/8$). The fixed-end forces due to the external forces applied at more than one section can be obtained by summation, using Equations (13.19) to (13.22) for each section where forces are applied.

13.7   Fixed-end forces due to temperature

Consider a nonprismatic uncracked member subjected to temperature rise, which varies over the length of the member and also over the height of its cross-section (Fig. 13.5(c)). Such temperature distribution can occur in practice on a summer day in bridges of variable cross-section. We will consider

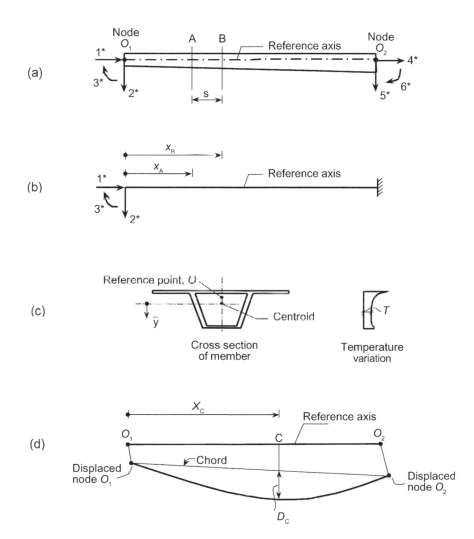

Figure 13.5 Analysis of fixed-end forces and numerical integration: (a) typical nonprismatic member; (b) released structure; (c) temperature rise that varies nonlinearly over the height of a typical cross-section; (d) deflected shape.

here the fixed-end forces $\{A_r\}$ at the six coordinates shown in Fig. 13.2(a) caused by the temperature rise.

Apply the force method (Section 4.2), using as a released structure a cantilever fixed at the right-hand end O_2 (Fig. 13.5(b)). Equations (13.19) to (13.21), and Equation (13.22) (with Q_1 to Q_3 set equal to zero), can be used to give the required fixed-end forces; the values of ε_O and ψ at any section, to be substituted in Equation (13.21), can be determined by:

$$\varepsilon_O = \varepsilon_{\text{centroid}} + \bar{y}_O \psi \qquad\qquad (13.24)$$

$$\varepsilon_{\text{centroid}} = \frac{a_t}{A_1} \int T \, dA_1 \qquad ; \qquad \psi = \frac{a_t}{I_{1\,\text{centroidal}}} \int T \bar{y} \, dA_1 \qquad (13.25)$$

$$\bar{y}_O = -\frac{B_1}{A_1} \qquad ; \qquad I_{1\,\text{centroidal}} = I_1 - \frac{B_1^2}{A_1} \qquad (13.26)$$

where a_t is the thermal expansion coefficient, assumed the same for the concrete and the reinforcement; T is the temperature rise at any fibre (Fig. 13.5(c)); $\bar{y}$ is the coordinate of any fibre measured downward from the centroid of the transformed section; $\bar{y}_O$ is the $\bar{y}$-coordinate of the reference point O; A_1, B_1 and I_1 are the area of the transformed uncracked section and its first and second moments about an axis through the reference point O; $I_{1\,centroidal}$ is the second moment of area of the uncracked transformed section about an axis through its centroid.

A temperature rise that varies over the depth as a straight line, defined by $T_O = \varepsilon_O/a_t$ and $dT/dy = \psi/a_t$, produces the same fixed-end forces as the nonlinear distribution having the same values of ε_O and ψ, determined by Equations (13.24) and (13.25). However, only when the variation of temperature is nonlinear, self-equilibrating stresses given by Equation (2.30) occur. The nonlinear analysis presented in this chapter does not explicitly account for the self-equilibrating stresses. Presence of these stresses can cause cracking at a section to occur at lower load level. As approximation, the value of f_{ct} may be reduced by an estimated value of the tensile stress at the extreme fibre estimated by Equation (2.30).

13.8   Numerical integration

Consider a typical member (Fig. 13.5(a)) and define a number of its cross sections (say ≥ 9). In evaluating the integrals involving ε_O and ψ at the sections it will be considered sufficiently accurate to assume that the parameters vary linearly over a typical spacing AB. Thus, the well-known trapezoidal rule can be employed to determine the areas below the broken lines representing the variation of ε_O and ψ over the member length; this gives the value of the first and the last integrals in each of Equations (13.8) and (13.21). The remaining integral in each of the two equations can be evaluated over the spacing s of two consecutive sections by (Fig. 13.5(b)):

$$\int_s \psi x \, dx = \frac{s}{6} (2\psi_A x_A + \psi_A x_B + \psi_B x_A + 2\psi_B x_B) \qquad (13.27)$$

where the subscripts A and B refer to values of ψ and x at the ends of the spacing, s. This equation can be employed for each spacing; the sum of the results gives the value of the integral over the member length.

When the curvatures ψ have been determined at a number of specified sections on a member, Equations (13.28) and (13.29) can be used to determine the deflection D_C; where D_C is the transverse distance between the chord and the deflected member at any section C (Fig. 13.5(c)). The chord is the straight line joining the nodes O_1 and O_2 in their displaced position. Again, the contribution of the curvature over a typical spacing, s can be calculated separately and the contributions of all spacings can be summed up to give D_C.

Contribution of spacing AB to D_C (Fig. 13.5(c)):

$$\frac{l-x_C}{l}\int_s \psi\, x\, dx = \frac{l-x_C}{l}\frac{s}{6}(2\psi_A\, x_A + \psi_A\, x_B + \psi_B\, x_A + 2\psi_B\, x_B)$$

$$\text{when } x_B \leqslant x_C \quad (13.28)$$

$$\frac{x_C}{l}\int_s \psi(l-x)dx = \frac{x_C}{l}\left[\frac{l\, s(\psi_A+\psi_B)}{2} - \frac{s}{6}(2\psi_A\, x_A + \psi_A\, x_B\right.$$

$$\left. + \psi_B\, x_A + 2\psi_B\, x_B)\right] \quad \text{when } x_A \geqslant x_C \quad (13.29)$$

where x_C is the distance between node O_1 and the point considered.

13.9   Iterative analysis

The analysis described below applies the displacement method in iterative cycles. Each cycle starts with known values of the parameters σ_O, γ, c and ζ at each section of individual members; where σ_O is the stress at reference point O; γ is the slope of the stress diagram (Fig. 13.3(b)); c here means depth of the part of the section in which the concrete is not ignored; thus, for a cracked section, c is the depth of the compression zone, but for an uncracked section, $c = h$, with h being the full height of the section; ζ is the interpolation coefficient. In each iteration, these values are updated. For the analysis of a non-prestressed reinforced concrete frame, the initial stresses are assumed null and the sections are assumed uncracked; thus at all sections, $\sigma_O = 0$; $\gamma = 0$; $c = h$; $\zeta = 0$. For a prestressed frame, the initial stresses are defined by the given parameters σ_{Oin} and γ_{in} and again the sections are assumed uncracked; thus $c = h$ and $\zeta = 0$. The cycles of analysis are repeated until the residual vector $\{F\}_{residual}$ becomes approximately equal to $\{0\}$; generation of the vector $\{F\}_{residual}$ is explained below. The analysis cycle is completed in three steps:

Step 1   Determine by conventional linear analysis the nodal displacements and the member end forces. This involves: generation of stiffness matrices, $[S^*]$ of individual members, transformation of $[S^*]$ from local member directions to global directions, assemblage of the transformed matrices ($[T]^T [S^*] [T]$) to obtain the stiffness matrix, $[S]$, of the structure, adjustment of the stiffness matrix according to the support conditions of the structure, and solution of the equilibrium equations:

$$[S] \{D\} = -\{F\} \tag{13.30}$$

where $\{F\}$ is a vector of forces that can restrain the displacements at the nodes; $\{F\}$ is generated by summing up the forces applied at the nodes with reversed sign to the transformed fixed-end forces ($[T]^T\{A_r\}$) for each member; where $[T]$ is a transformation matrix defined by Equation (13.10). Solution of Equation (13.30) gives the nodal displacements, $\{D\}$ in the global directions. These are used to find the member end forces for individual members:

$$\{A\} = \{A_r\} + [S^*] \{D^*\} \tag{13.31}$$

$\{D^*\}_{6\times1}$ is a vector generated by transformation of three displacement components at each of the two nodes at the member ends (Equations (13.11)).

In the first cycle, $\{A_r\}$ is determined by the force method, ignoring cracking (see Sections 13.6 and 13.7). In other cycles, $\{A_r\}$ is given by Equation (13.34). In each cycle, the member's tangent stiffness matrix $[S^*]$ is calculated considering the updated c and ζ values for the sections (values existing at the end of the preceding cycle).

Step 2   Update the nodal displacements $\{D\}$ and the member end forces $\{A\}$, by adding the values determined in step 1 of this cycle to the values existing at end of the preceding cycle. For each member, the forces at the ends and the forces at intermediate sections represent a system in equilibrium. Determine the values of normal force and the bending moment in all sections. Use these to update c, ζ, σ_O and γ and calculate the strain parameters ε_O and ψ. Apply Equation (13.21) to determine $\{D^*_s\}$, a vector of the three displacements at node O_1 relative to node O_2. Calculate for each member the error in nodal displacements of node O_1 relative to node O_2 by:

$$\{D^*\}_{error} = [H] \{D^*\} - \{D^*_s\} \tag{13.32}$$

where

$$[H] = \begin{bmatrix} 1 & 0 & 0 & -1 & 0 & 0 \\ 0 & 1 & 0 & 0 & -1 & l \\ 0 & 0 & 1 & 0 & 0 & -1 \end{bmatrix} \tag{13.33}$$

$\{D^*\}$ are the displacements, in local directions, at the two ends of the member. The elements of $\{D^*\}$ are obtained by transformation, using Equation (13.11), of the displacements of the two nodes associated with the member; these are part of the updated global displacements $\{D\}$.

Step 3   Calculate the residual member end forces by the equation:

$$
\{A\}_{\text{residual}} = \begin{Bmatrix} [S^*_{11}] \{D^*\}_{\text{error}} \\ [R] [S^*_{11}] \{D^*\}_{\text{error}} \end{Bmatrix} \tag{13.34}
$$

where $[R]$ is the matrix defined by Equation (13.4); $[S^*_{11}]$ is a 3×3 matrix, the top left-hand part of the partitioned matrix in Equation (13.2). Generation of $[S^*_{11}]$ is to be based on the updated c and ζ determined in step 2. Transform $\{A\}_{\text{residual}}$ to global directions (Equation (13.11)) and sum up for all members to generate a vector of residual nodal forces, $\{F\}_{\text{residual}}$. If $\{F\}_{\text{residual}}$ is smaller than a specified tolerance (Section 13.10) terminate the analysis; otherwise, set $\{F\} = \{F\}_{\text{residual}}$ and go to step 1 to start a new cycle.

13.10   Convergence criteria

The iteration cycles discussed above may be terminated when:

$$
\left(\{F\}^T \{F\} \right)^{1/2}_{\text{residual}} \leq a_{\text{tolerance}} \left(\{F\}^T \{F\} \right)^{1/2}_{\text{cycle } 1} \tag{13.35}
$$

where $a_{tolerance}$ is a small specified value, say between 0.01 and 0.001. This criterion ensures that the residual forces are small compared to the nodal forces applied on the structure.

When the analysis is for the effect of prescribed displacements, $\{F\}$ can be null, while the nonzero forces are the support reactions $\{R\}$. In this case, the convergence criterion may be:

$$
\left(\{F\}^T \{F\} \right)^{1/2}_{\text{residual}} \leq a_{\text{tolerance}} \left(\{R\}^T \{R\} \right)^{1/2}_{\text{cycle } 1} \tag{13.36}
$$

where $\{R\}_{\text{cycle}}$ is a vector of the reaction components determined in step 1 of cycle 1.

The elements of $\{F\}$ or $\{R\}$ have either the unit of force or force-length. The elements with the latter units may be derived by an arbitrary length before application of Equation (13.35) or (13.36). The arbitrary length may be chosen equal to the larger overall dimension of the frame in the global x and y direction.

The objective of the iteration cycles discussed in Section 13.9 is to bring to a tolerable level $\{D^*\}_{\text{error}}$ (the members' displacements calculated by Equation (13.32)). It is possible that the error in one of the three displacements changes sign in two consecutive iterations, with no convergence. This can occur when at one section of the member a new crack is developed due to a small change in the internal forces, thus causing a sudden change in the mean strain parameters. This iteration problem can be avoided by adopting Equations

(8.43) and (8.44) in the calculation of ε_{Om} and ψ_m for a section when the stress at the extreme fibre exceeds $\sqrt{\beta} f_{ct}$, instead of f_{ct}; where $\beta = \beta_1\beta_2$, defined in Section 8.3. When the section is subjected to M without N, this change will make the moment-curvature graph in Fig. 8.5 follow the curve AED, instead of $ABCD$ (see Section 8.4.1).

13.11   Incremental method

In this section, we discuss the simpler technique, known as the incremental method. The load vector $\{F\}$, generated as described in Section 13.9 is divided into m increments: $\delta_i \{F\}$, with $i = 1$ to m; where δ is a dimensionless increment multiplier. The load increments are applied, one at a time, and an elastic analysis is performed. For each increment, the following equilibrium equation is solved:

$$([S]\{\Delta D\})_i = \delta_i \{F\} \qquad\qquad (13.37)$$

The tangent stiffness matrix $[S]_i$ depends upon the state of cracking (c and ζ) of all sections reached in the preceding increment ($i - 1$). The increments of displacements and the stresses calculated for each load increment are used to update these parameters for each node and for each section. The solution of the problem is achieved at the end of calculations in the last increment, m.

A typical plot of the displacement at any coordinate versus the corresponding nodal force is shown in Fig. 13.6, from which it is seen that the error for any increment exceeds the preceding ones. However, the accuracy can be improved by using smaller increments, thus increasing the computation; this is not a hindrance with most computers and most structures. The advantages of the incremental method are its simplicity and robustness (no convergence criterion has to be satisfied).

Because in the analysis considered in this chapter, the stress–strain relationships of the concrete and the reinforcement are assumed linear, the structure behaves linearly until cracking occurs at one section. Thus, the first load increment multiplier, δ_1 can represent the estimated load level that produces first cracking.

For each load increment, a new analysis is performed; this allows for changes in the support conditions if necessary (e.g. to represent a construction stage). It should be noted that a section cracked in the ith increment remains cracked in a later increment j, even when N combined with M become insufficient to produce cracking. In such a case, a part or all of the tension zone becomes in compression and the crack closes. If in a subsequent increment, N combined with M produce compression at the extreme fibre, the crack closes; further change of N and/or M will cause the crack to reopen when the stress at the same extreme fibre is greater than zero, and not

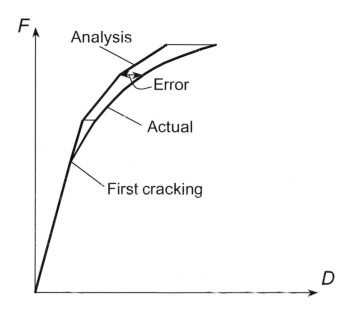

Figure 13.6 Variation of displacement with force at a typical coordinate. Typical result of the incremental method.

necessarily greater than f_{ct}. This should be observed in the storage and in the use of the parameters representing the state of cracking of the sections.

13.12   Examples of statically indeterminate structures

Example 13.3 Demonstration of the iterative analysis

Perform two iteration cycles to determine the bending moment, M_B at the interior support of the reinforced concrete beam shown in Fig. 13.7 due to concentrated load $Q = 60.0\,\text{kN}$ (13.5 kip) at mid-span. Consider $f_{ct} = 2.5\,\text{MPa}$ (0.36 ksi), $E_c = 30.0\,\text{GPa}$ (4350 ksi) and $E_s = 200\,\text{GPa}$ (29000 ksi); $\beta_1 = 1$; $\beta_2 = 0.5$ (definitions of β_1 and β_2 are given in Section 8.3).

The reference axis is chosen at mid-height of cross-section. The structure has only one member, and only three unknown displacements at coordinates 1, 2 and 3 at node A need to be considered (of which $D_2 = 0$). The local coordinates 1*, 2*and 3*of member AB are the same as

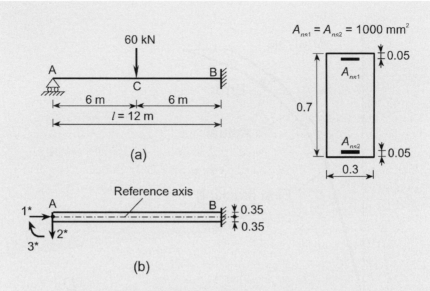

(a)

(b)

Figure 13.7 Reinforced concrete beam analysed in Example 13.3: (a) beam
dimensions; (b) coordinate system for member AC with end C fixed.

the global coordinates. The cross-sectional area properties before cracking are:

$$A_1 = 221.3 \times 10^{-3} \, \text{m}^2 \quad ; \quad B_1 = 0 \quad ; \quad I_1 = 9.595 \times 10^{-3} \, \text{m}^4$$

Eleven sections equally spaced at $0.1l$ are considered in the analysis. The
positive or negative moment that is just sufficient to produce cracking
is:

$$M_r = \pm f_{ct} W_1 = \pm 68.5 \, \text{kN-m}$$

The units used in all calculations given below are Newton and metre.

Iteration cycle 1
In this cycle, the structure is assumed uncracked. A linear analysis gives:

$$M_{A \, \text{cycle} \, 1} = 0 \quad ; \quad M_{B \text{cycle} \, 1} = -135 \times 10^3 \quad ;$$
$$M_{C \, \text{cycle1}} = 112.5 \times 10^3 \quad \text{(a)}$$

The corresponding displacements at the three coordinates at end A
(Fig. 13.7(b)) are:

$$\{D\}_{\text{cycle 1}} = \{D^*\}_{\text{cycle 1}} = 10^{-6} \begin{Bmatrix} 0 \\ 0 \\ 938.0 \end{Bmatrix} \tag{b}$$

Treat AB as a cantilever, fixed at B and subjected to a bending moment diagram composed of two straight lines joining the above three M-values. The magnitudes of these moments indicate that cracking occurs at several sections in the negative and in the positive moment zones. Calculate c_0 and ψ at the eleven sections and apply Equation (13.21) to obtain:

$$\{D_s^*\}_{\text{cycle 1}} = 10^{-6} \begin{Bmatrix} -1119 \\ 6922 \\ 1530 \end{Bmatrix} \tag{c}$$

For the member considered, end B is fixed; thus, the displacements of end A with respect to end B is the same as $\{D^*\}$. From Equation (13.32), determine $\{D^*\}_{\text{error}}$:

$$\{D^*\}_{\text{error cycle 1}} = \{\{D^*\}-\{D_s^*\}\} = 10^{-6} \begin{Bmatrix} 1119 \\ -6922 \\ -592.2 \end{Bmatrix} \tag{d}$$

The flexibility matrix of the cracked cantilever is (Equation (13.8)):

$$\{f^*\}_{\text{cycle 1}} = 10^{-9} \begin{bmatrix} 5.569 & -16.790 & -1.685 \\ -16.790 & 4839 & -569.2 \\ -1.685 & -569.2 & 82.07 \end{bmatrix} \tag{e}$$

Inversion of $[f^*]$ gives the stiffness matrix of the cracked member:

$$[S^*]_{\text{cycle 1}} = [f^*]^{-1}_{\text{cycle 1}} = 10^6 \begin{bmatrix} 216.2 & 6.908 & 52.35 \\ 6.908 & 1.343 & 9.454 \\ 52.35 & 9.454 & 78.83 \end{bmatrix} \tag{f}$$

The residual member end forces (Equation (13.34)):

$$\{A_{O1}\}_{\text{residual cycle 1}} = [S^*]\,\{D^*_{\text{error}}\} = 10^3 \begin{Bmatrix} 163.1 \\ -7.165 \\ -53.57 \end{Bmatrix} \quad ;$$

$$\{A_{O2}\}_{\text{residual cycle 1}} = 10^3 \begin{Bmatrix} -163.1 \\ 7.165 \\ -32.42 \end{Bmatrix} \qquad \text{(g)}$$

Thus, the residual forces are:

$$\{F\}_{\text{residual cycle 1}} = \{A_{O\,1}\}_{\text{residual cycle 1}} = 10^3 \begin{Bmatrix} 163.1 \\ -7.165 \\ -53.57 \end{Bmatrix} \qquad \text{(h)}$$

Iteration cycle 2

The nodal forces to be used in this cycle are $\{F\} = \{F\}_{\text{residual}}$. Because this structure has only one member, $[S]$ to be used in this cycle is the same as $[S^*]$ in the preceding cycle; however, with the support conditions of the actual structure, the stiffness matrix becomes:

$$[S^*]_{\text{cycle 1}} = 10^6 \begin{bmatrix} 216.2 & 6.908 & 52.35 \\ 6.908 & 1.343\times10^6 & 9.454 \\ 52.35 & 9.454 & 78.83 \end{bmatrix} \qquad \text{(i)}$$

This is the same as in Equation (f) above, but with S^*_{22} multiplied by a large number (10^6) to cause the displacement at coordinate 2 to be zero.

The equilibrium equations and their solutions are:

$$[S]\{D\}_{\text{cycle 2}} = -\{F\}_{\text{residual cycle 1}} = -10^3 \begin{Bmatrix} 163.1 \\ -7.165 \\ -53.57 \end{Bmatrix}$$

$$\{D\}_{\text{cycle 2}} = 10^{-6} \begin{Bmatrix} -1095 \\ 0 \\ 1406 \end{Bmatrix} \quad ; \quad \{D\}_{\text{updated cycle 2}} = 10^{-6} \begin{Bmatrix} -1095 \\ 0 \\ 2344 \end{Bmatrix}$$

The member end forces (Equation (13.31)):

$$\{A\}_{\text{cycle 2}} = \{A_r\}_{\text{cycle 2}} + [S^*]_{\text{cycle 1}}\{D\}_{\text{cycle 2}} \quad ;$$
$$\{A_r\}_{\text{cycle 2}} = \{A\}_{\text{residual cycle 1}}$$

$$\{A_{O1}\}_{cycle\ 2} = 10^3 \begin{Bmatrix} 163.1 \\ -7.165 \\ -53.57 \end{Bmatrix} + [S^*] \begin{Bmatrix} -1095 \\ 0 \\ 1406 \end{Bmatrix} 10^{-6} = 10^3 \begin{Bmatrix} 0 \\ -1.430 \\ 0 \end{Bmatrix}$$

$$\{A_{O2}\}_{cycle\ 2} = 10^3 \begin{Bmatrix} 0 \\ 1.430 \\ -17.17 \end{Bmatrix}$$

Update the member end forces by adding the end forces to the end forces in cycle 1. This gives $M_B = (135 + 17.17)\ 10^3 = -117.8\,\text{kN-m}$. The corresponding bending moments at A and C are: $M_A = 0$ and $M_C = 121.1\,\text{kN-m}$.

Again, the bending moment varies linearly between A and C and between C and B. Calculation of ε_0 and ψ at the eleven sections and integration using Equation (13.21) gives:

$$\{D^*_s\}_{cycle\ 2} = 10^{-6} \begin{Bmatrix} -1402 \\ -9574 \\ 3904 \end{Bmatrix}$$

Apply Equation (13.22) and note that $\{D\}$ and $\{D^*\}$ are the same:

$$\{D^*\}_{error\ cycle\ 2} = 10^{-6} \left\{ \begin{Bmatrix} -1095 \\ 0 \\ 2344 \end{Bmatrix} - \begin{Bmatrix} -1402 \\ -9574 \\ 3904 \end{Bmatrix} \right\} = 10^{-6} \begin{Bmatrix} 307.4 \\ 9574 \\ -1559 \end{Bmatrix}$$

The computations are proceeded similar to cycle 1, giving at the end of cycle 2 the residual nodal forces for use in the next cycle:

$$\{F\}_{residual\ cycle\ 2} = 10^3 \begin{Bmatrix} 31.17 \\ 0.340 \\ -11.45 \end{Bmatrix}$$

Performing more iterations will give a more accurate value, $M_B = -126.3\,\text{kN-m}$.

Example 13.4:   Deflection of a non-prestressed concrete slab

Figure 13.8(a) represents a concrete slab continuous over two equal spans, $l = 7\,\text{m}$. The cross-section of a strip of width 1 m is shown in the figure. The bottom reinforcement covers the full length of the spans; while the top reinforcement covers a distance of 2.1 m on both sides of support B. It is required to calculate the immediate deflection (without the effects of creep or shrinkage of concrete or temperature variation) due to uniform load $q = 7.5\,\text{kN/m}$, treating the strip as a continuous beam. Given data: $E_c = 20\,\text{GPa}$ (2900 ksi); $E_s = 200\,\text{GPa}$ (29000 ksi); $f_{ct} = 2.5\,\text{MPa}$ (0.36 ksi); $\beta_1 = 1$ and $\beta_2 = 0.5$.

The moment that is just sufficient to produce cracking at top fibre at B or at bottom fibre near mid-span is:

$$(M_r)_{top \ or \ bot} = f_{ct} W_{top \ or \ bot} = -28.6 \text{ or } 27.6\,\text{kN-m}$$

where W is the section modulus; the subscripts *top* and *bot* refer to the extreme top and bottom fibres, respectively. The first cracking occurs at the top fibre at a load intensity, $q_r = 4.7\,\text{kN/m}$. The corresponding deflection at mid-span, $D_r = 2.1\,\text{mm}$. Ignoring cracking in calculation of the statically indeterminate moment would give $M_B = -0.125\,ql^2 = -45.9\,\text{kN-m}$ and the maximum positive moment in the spans $= 25\,\text{kN-m}$. Because the latter value is lower than M_r, one would conclude, based on this calculation, that cracking occurs only in the negative moment zone. The corresponding deflection at mid-span is equal to 3.2 mm.

An analysis using a computer program that accounts for cracking (with sections placed at one tenth of the span) gives $M_B = -38.8\,\text{kN-m}$ and indicates that the values of M_r are exceeded at B and also in the vicinity of the point of maximum moment in the span. The corresponding deflection $= 6.5\,\text{mm}$, which indicates that ignoring cracking in calculation of M_B leads to underestimation of deflection.

The value of the dimensionless coefficient, $[-M_B/(ql^2)]$ is 0.125 before cracking. When cracking develops in the negative moment zone, and before occurrence of cracking in the positive moment zone, the value of this coefficient drops. For the load intensity considered above, $q = 7.5\,\text{kN/m} = 1.6\,q_r$, the dimensionless coefficient is equal to $[-(-38.8)/(7.5 \times 7^2)] = 0.106$. It is to be noted that the value of this coefficient depends upon the relative flexural rigidity of the negative and positive moment zones. As cracking develops in the positive zone, by the increase of the

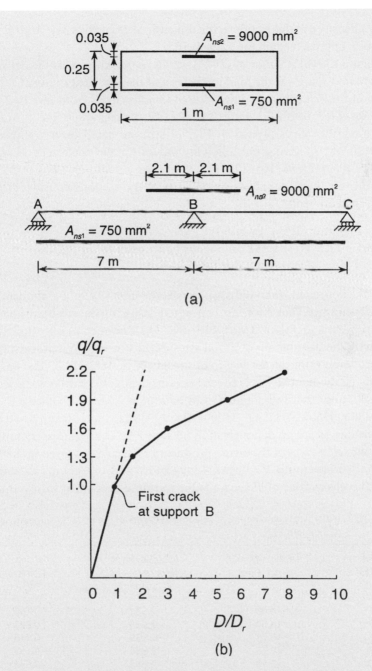

Figure 13.8 Slab continuous over two spans analysed in Example 13.4: (a) slab geometry; (b) variation of the deflection at mid-span with the load intensity.

load intensity q in this example, the dimensionless coefficient $[-M_B/(ql^2)]$ tends to approach again the value 0.125.

Figure 13.8(b) is a plot of (D/D_r) versus (q/q_r); where D is the deflection at mid-span; q is the load intensity and the subscript r refers to the state of the first cracking. The value of (q/q_r) is varied from zero to 2.2, in steps of 0.3; before cracking $(0 \leq (q/q_r) \leq 1.0)$ the graph is linear. After cracking, the values plotted in Fig. 13.8(b) are based on a nonlinear analysis of M_B, giving the following values for the coefficient: $[-M_B/(ql^2)] = 10^{-3} \{125, 98, 105, 122, 118\}$ for $\{q/q_r\} = \{1.0, 1.3, 1.6, 1.9, 2.2\}$, respectively. In the analyses reported here the sections considered in each span are spaced at $l/10$.

Example 13.5 Prestressed continuous beam analysed by the incremental method

Plot the deflection, D at mid-span versus the intensity, q of a uniformly distributed live load covering two equal spans of the post-tensioned beam shown in Figs. 13.9(a) and (b). The parameters $\sigma_{Oin.}$ and $\gamma_{in.}$ defining the distribution of initial stresses, at various sections, existing before application of the live load are given in Table 13.2. The variations of the parameters between sections 1 and 16 and between sections 17 and 22 are parabolic. Given data: $f_{ct} = 2.5$ MPa (0.36 ksi); $E_c = 30.0$ GPa (4350 ksi); $E_{ns} = E_{ps} = 200$ GPa (29000 ksi); $\beta_1 = 1.0$ and $\beta_2 = 0.5$; definitions of β_1 and β_2 are given in Section 8.3. The prestressing force and the self weight of the beam, producing $\sigma_{O in.}$ and $\gamma_{in.}$, given in Table 13.2 are, respectively, $P = 2200$ kN (494.6 kip) and 18.0 kN/m (1.23 kip/ft). The given value of P accounts for the time-dependent losses; thus

Table 13.2 Initial stress parameters at selected sections of the post-tensioned beam of Example 13.5

Section number	Distance from support A (m)	Initial stress at reference point O, σ_{Oin} (MPa)	Initial slope of stress diagram, γ_{in} (MPa/m)
1	0.00	−2.939	0.150
9	10.00	−2.939	0.843
16	18.75	−2.939	0.918
17	18.75	−2.855	−0.152
20	22.50	−2.857	−0.832
22	25.00	−2.856	−1.414

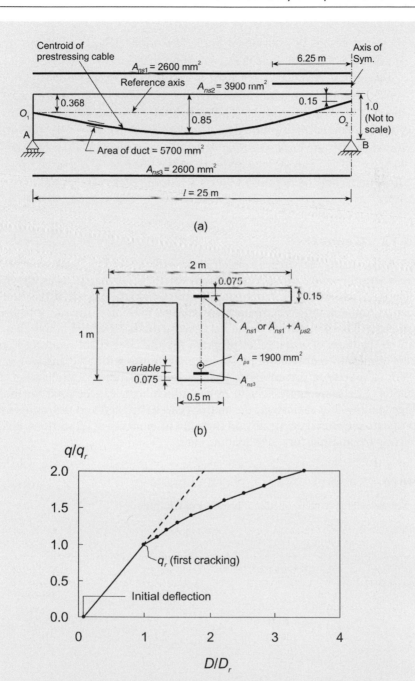

Figure 13.9 Prestressed continuous beam analysed in Example 13.5: (a) beam
dimensions; (b) cross-section; (c) variation of the deflection at mid-span
with the load intensity.

these are not to be considered here. The initial deflection before the application of the live load is $D_{in.} = 1.1$ mm (0.043 in.).

Because of symmetry, one span is considered. Twenty-two equally-spaced sections are used, of which two sections (16 and 17) at zero distance apart are assumed at C. The reference axis is chosen at 0.368 m (14.5 in.) below the top fibre. First cracking occurs at $q_r = 13.4$ kN/m (0.921 kip/ft); the corresponding deflection at mid-span, $D_r = 12.8$ mm (0.5 in.). After cracking, the intensity of the live load is increased up to $2q_r$ in steps of $0.1q_r$. Figure 13.9(c) is a plot of (D/D_r) versus (q/q_r).

13.13   General

At service load, the stress in concrete is sufficiently low, such that linear stress–strain relationship can be used for uncracked concrete; thus, the non-linearity of the analysis is required only after cracking. In spite of this simplification, the amount of computations is large, making it necessary to use a computer. The incremental or the iterative methods presented in this chapter can be the basis of a computer program for the analysis of reinforced concrete frames, with or without prestressing, accounting for cracking.

For simplicity of presentation, a single loading stage is considered in this chapter. The same analysis, with minor adjustment, can be used for multi-stage loading. For each stage, the analysis can be applied and the results used to update the stress, the strain and the state of cracking at all sections, before starting the analysis for a new loading stage.

Note

1 See the reference mentioned in footnote 3 of Chapter 3.

Chapter 14

Serviceability of members reinforced with fibre-reinforced polymers

Fibre-reinforced polymer bars

14.1 Introduction

Corrosion of steel reinforcement increases its volume and causes spalling and deterioration of concrete. To avoid corrosion, stainless steel can be used as reinforcement for concrete. Fibre-reinforced polymers (FRP) products do not corrode; thus they are used in lieu of steel. Several FRP products in the shapes of bars, cables or grids are in use. They have high tensile strength which can exceed that of steel. But many FRP bars have smaller modulus of elasticity compared to that of steel. This makes members reinforced with FRP more vulnerable to excessive deflection and wide cracks when the members are non-prestressed. This chapter[1] discusses design of the amount of FRP reinforcement and choice of span to thickness ratio of concrete members to avoid excessive deflections and wide cracks.

14.2   Properties of FRP reinforcements for concrete

FRP bars are made of continuous fibres bonded by impregnating them with matrices, such as epoxy resins and vinyl ester resin. Three main types of fibres are used to produce FRP bars: carbon, aramid and glass. Carbon fibres are made from petroleum or coal pitch and polyacrilic nitril. The stress–strain relationships for FRP bars in tension is linear up to failure by rupture. Values of the tensile strengths, f_{fu} and the moduli of elasticity, E_f of carbon, aramid and glass FRP bars are compared in Table 14.1 with the nominal yield stress and modulus of elasticity of steel reinforcing bars. The values given in the table for FRP are approximate; the mechanical properties depend upon the constituents as well as the manufacturing process; the manufacturers commonly provide the mechanical properties of their products.

FRP bars have almost no adhesion to concrete. The force in FRP bars embedded in concrete is developed by interlocking with deformations on the surface of the bars (similar to deformations on steel bars). Sand-coated bars, braided bars and strands are FRP products intended to have, when embedded in concrete, resistance to slipping comparable to that of steel reinforcements.

Thermal expansion coefficients of steel and concrete are 12×10^{-6} and 8×10^{-6} per degree Celcius (6.7×10^{-6} and 4.4×10^{-6} per degree Fahrenheit), respectively. The difference between the coefficients of thermal expansion of concrete and FRP products is generally greater than the difference between concrete and steel. Furthermore, FRP bars have substantially larger coefficient of thermal expansion in the radial direction than that of concrete (21×10^{-6} to 23×10^{-6} per degree Celsius for glass and carbon FRP and 60×10^{-6} to 80×10^{-6} per degree Celsius for aramid FRP). In the longitudinal direction, the thermal expansion coefficient is 6×10^{-6} to 8×10^{-6} for glass FRP, while carbon and aramid FRP have zero or small negative thermal expansion coefficient. The incompatibility of thermal expansion coefficients of concrete and FRP products has some adverse effect. The high thermal expansion

Table 14.1 Properties of types of reinforcements for concrete structures

Reinforcement type	Modulus of elasticity		Tensile strength of FRP or yield stress for steel	
	GPa	10^3 ksi	MPa	ksi
Glass FRP	40	6	550	80
Aramid FRP	80	12	1200	170
Carbon FRP	150	22	2000	300
Steel	200	29	400 and 500	60 and 73

coefficient in radial direction of FRP bars can cause hoop tensile stress in the concrete adjacent to the bar, and produce radial hair cracks when the temperature rises. Once cracking occurs, the tensile stresses are relieved; thus, the radial hair cracks do not extend to the surface of concrete member when the cover is not overly small. It should be noted that the thermal expansion coefficients of FRP vary with the method of production, the type of fibre and the resin matrix. The values mentioned above are only approximate. Again, the manufacturer commonly provides information on thermal properties.

The compressive strength of FRP bars is relatively low and their contribution to ultimate strength as compression reinforcement in concrete sections is often not considered. Also, because the modulus of elasticity of FRP is relatively low, particularly in compression, the contribution of FRP bars situated in the compression zone to the flexural rigidity of cracked members is ignored.

Carbon FRP has tensile strength that exceeds the tensile strength of steel used for prestressing. To avoid very wide cracks the high strength of FRP cannot be fully used in non-prestressed concrete members. When used for prestressing, information about relaxation of carbon FRP is required; this data should take into account the temperature and the ratio of the initial tensile stress to the tensile strength. At 20 degree Celsius and initial stress 70 to 80 percent of the tensile strength, the relaxation of FRP in 0.5 million hours (57 yrs) is approximately 15 percent, regardless of the type of fibre. The manufacturers of FRP for prestressing should provide relaxation data.

Certain FRP products are vulnerable to rupture when they are subjected to sustained tensile stress. This phenomenon, referred to as creep rupture, occurs in a shorter time when the ratio of the sustained stress to the tensile strength is larger. To control width of cracks in non-prestressed members, the permissible strain in FRP in service should be relatively low compared to the tensile strength. The permissible strain in service proposed in Section 14.3 for non-prestressed FRP is below the strain that can produce creep rupture. However, when FRP is used for prestressing, the ratio of stress at transfer to the tensile strength should be small compared to the permissible ratio for prestressing steel.

The basic assumptions in analysis of stresses, strains and displacements of steel-reinforced concrete structures in service are also adopted when FRP is employed. Thus, concrete and reinforcement are assumed to have linear stress–strain relationships. Sections that are plane before deformation remain plane after deformation. Concrete in tension in a cracked section is ignored; the tension stiffening effect is accounted for empirically by interpolation between the uncracked state and the state of full cracking. The analysis procedures and equations presented in the remainder of the book for structures reinforced or prestressed with steel can be applied with FRP, using the appropriate characteristic material properties. However, because of some of the differences of properties of FRP and steel, particularly in the moduli of elasticity, the design of sections with FRP may be governed by serviceability requirements (control of deflection and crack width), rather than by ultimate strength.

14.3   Strain in reinforcement and width of cracks

Widths of cracks in flexural members depend upon crack spacing, quality of bond between concrete and reinforcing bars and, above all, upon strain in reinforcement. For steel-reinforced sections codes explicitly or implicitly limit the stain in steel in service to approximately 1200×10^{-6}. The corresponding mean crack width to be expected is 0.4 mm (0.01 or 0.02 in.). Corrosion of steel is one of the reasons to control crack width. Because FRP bars do not corrode, wider cracks are commonly tolerated. In this chapter a permissible strain, $\varepsilon_{fservice}$ in service in FRP bars, is taken equal to 2000×10^{-6}. The anticipated mean crack width where this strain is reached is (2000/1200) times the width when steel is employed; that is 0.7 mm (0.02 or 0.03 in.). The corresponding permissible stress in the FRP in service is $\sigma_{fservice} = 80$, 160 and 300 MPa (12, 24 and 44 ksi) for glass, aramid and carbon FRP, respectively. When these values are treated as the permissible service stresses, the design of the required cross-sectional area, A_f of the FRP will be governed by this serviceability requirement, rather than by ultimate strength. This is so, because the three values of $\sigma_{fservice}$ mentioned above do not exceed 15 percent of the tensile strengths of the three types of FRP (see Table 14.1).

14.4   Design of cross-sectional area of FRP for non-prestressed flexural members

The equations presented below give the cross-sectional area, A_f of FRP required for a non-prestressed section (Fig. 14.1) subjected in service to a moment $M_{service}$ or a moment $M_{service}$ combined with a normal force $N_{service}$, acting at the centroid of the tension reinforcement. The cross-sectional area, A_f is calculated such that the strain in FRP in service be equal to a specified value, $\varepsilon_{fservice}$. It is assumed that concrete dimensions of the section have been selected. Choice of the overall height, h_f of the sections is discussed in Section 14.7.

For a specified value of the permissible strain, $\varepsilon_{fservice}$ in FRP in service, the stress is: $\sigma_{fservice} = E_f \varepsilon_{fservice}$; where E_f is modulus of elasticity of FRP. Ignoring concrete in tension, the cross-sectional area of FRP required, when the section is subjected to a bending moment without normal force, is (Fig. 14.1(a)):

$$A_f = M_{service}/(\sigma_{fservice} y_{CT}) \qquad (14.1)$$

where $M_{service}$ is the moment in service; y_{CT} is the distance between the resultants of tensile and compressive stresses in the section. The depth of

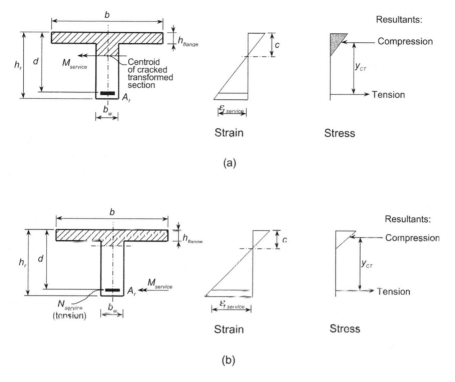

Figure 14.1 Strain and stress distributions in a section with FRP: (a) bending moment $M_{service}$ without normal force; (b) bending moment $M_{service}$ combined with normal force $N_{service}$.

compression zone c can be calculated by Equations (6.12) or (6.16); here the reinforcement in the compression zone is ignored and the section has no prestressed reinforcement. For T- or rectangular sections having one layer of tension reinforcement, and subjected to $M_{service}$ only, without $N_{service}$ (Fig. 14.1(a)), Equation 6.12 reduces to:

$$c = \frac{1}{2}\left(-a_1 + \sqrt{a_1^2 + 4a_2}\right) \qquad \text{(when } N_{service} = 0\text{)} \qquad (14.2)$$

$$a_1 = \frac{2h_{flange}}{b_w}(b - b_w) + \frac{2a\,A_f}{b_w} \qquad \text{(with } b_w = b \text{ when } c \leqslant h_{flange}\text{)} \qquad (14.3)$$

$$a_2 = \frac{h_{flange}^2}{b_w}(b - b_w) + \frac{2a\,A_f d}{b_w} \qquad \text{(with } b_w = b \text{ when } c \leqslant h_{flange}\text{)} \qquad (14.4)$$

where $a = E_f/E_c$, with E_c being the modulus of elasticity of concrete, b, b_w, d

and h_{flange} are defined in Fig. 14.1(a). When $c \le h_{\text{flange}}$ or when the section is rectangular, set $b = b_w$ in Equations (14.3) and (14.4).

The distance y_{CT} between the resultants of tensile and compressive stresses is given by Equation 14.5 or 14.6, which apply when the section is subjected to M_{service} only or when M_{service} is combined with N_{service}:

$$y_{CT} = d - \frac{c}{3} \qquad \text{with } c \le h_f \qquad (14.5)$$

When c is substantially greater than h_{flange}, use Equation 14.6 for y_{CT}:

$$y_{CT} = d - \frac{c}{3}\left[\frac{bc^2 - (b - b_w)(c - h_{\text{flange}})^2\,(c + 2h_{\text{flange}})/c}{bc^2 - (b - b_w)(c - h_{\text{flange}})^2}\right] \text{ (with } c \ge h_{\text{flange}}) \ (14.6)$$

For a given reinforcement ratio, $\rho_{\text{f}}\ (=A_f/bd)$ and specified value for $\varepsilon_{\text{f service}}$, the curvature at a cracked section, due to M_{service} or due to M_{service} combined with N_{service}, can be calculated by (see Equation (11.28)):

$$\psi_{2\text{service}} = \frac{\varepsilon_{\text{f service}}}{d - c} \qquad (14.7)$$

When the section is subjected in service to normal force N_{service} combined with moment M_{service}, the required cross-sectional area of FRP is given by (Fig. 14.1(b)):

$$A_f = \frac{M_{\text{service}}/y_{CT} + N_{\text{service}}}{\sigma_{\text{f service}}} \qquad (14.8)$$

The depth c of the compression zone can be determined by Equation (6.10) or (6.16); then Equations (14.5) to (14.7) can be applied to give y_{CT} and $\psi_{2\text{ service}}$. In Equation (14.8), $M_{\text{service}} = e_f N_{\text{service}}$; where e_f is the eccentricity of the normal force measured downwards from the centroid of A_f.

To calculate A_f by Equation (14.1) or by Equation (14.8), y_{CT} must be known. But y_{CT} depends upon A_f. Thus iteration is required; first a value of y_{CT} is assumed to obtain an approximate value of A_f. In the second iteration (which is often sufficient), the approximate value of A_f is used to calculate y_{CT} and Equation (14.1) or (14.8) is reused.

The above equations are based on the assumption that the amount of tension reinforcement required is governed by the allowable strain in the tension reinforcement, $\varepsilon_{\text{f service}}\ (= \sigma_{\text{f service}}/E_f)$ in service. It will be shown below that deflection of a member is also governed by the value $\varepsilon_{\text{f service}}$ at mid-length section.

14.5   Curvature and deflection of flexural members

In this section we consider the moment–mean curvature relationship for members reinforced with FRP; the mean curvatures calculated at various sections can be used to give deflections in service (e.g. by the equations in Appendix C). For simplicity, the subscript 'service' is omitted in Sections 14.5 to 14.8, which are concerned with deflections or curvatures in service condition. Section 7.4 presents equations that give the mean curvature, ψ_m of a member reinforced with non-prestressed steel bars subjected to bending moment, $M > M_r$, with M_r being the moment that produces first cracking. The equations are repeated here with the symbols adjusted for use when FRP is employed:

$$\psi_m = (1 - \zeta) \psi_1 + \zeta \psi_2 \tag{14.10}$$

$$\psi_1 = \frac{M}{E_c I_1} \quad ; \quad \psi_2 = \frac{M}{E_c I_2} \tag{14.11}$$

$$\zeta = 1 - \beta \left(\frac{M_r}{M}\right)^2 \tag{14.12}$$

or

$$\zeta = 1 - \beta \left(\frac{f_{ct}}{\sigma_{1\,max}}\right)^2 \tag{14.13}$$

where I_1 is the second moment of areas of a transformed area consisting of A_c plus αA_f, with A_c and A_f being the cross-sectional areas of concrete and FRP (ignoring the bars in compression) and $\alpha = E_f/E_c$; where E_f and E_c are the moduli of elasticity of FRP and concrete, respectively; f_{ct} is tensile strength of concrete; $\sigma_{1\,max}$ is stress at extreme fibre in state 1, where cracking is ignored. The coefficient β replaces the product of β_1 and β_2 which account for the bond quality and the effect of repetitious loading; with FRP bars a value of $\beta = 0.5$ is recommended. This recommendation is based on comparisons of published experimental deflections of numerous beams reinforced with different FRP types with the values of deflections calculated from curvatures using the above equations.[2]

Equation (14.10) can be rewritten in the form:

$$\psi_m = \frac{M}{EI_{em}} \tag{14.14}$$

where I_{em} is an effective second moment of area for use in calculation of mean curvature in members subjected to bending moment, without axial force,

$$I_{em} = \frac{I_1 I_2}{I_2 + \left[1 - 0.5\left(\dfrac{M_r}{M} \right)^2 \right](I_1 - I_2)} \tag{14.15}$$

For a cracked member subjected to a normal force N combined with moment M, the mean axial strain ε_{om} and the mean curvature ψ_m can be calculated by Equations (7.36) and (7.40). Validity of these equations, again with $\beta_1 \beta_2$ replaced by $\beta = 0.5$, for prestressed members using carbon FRP has been verified[3] by comparison with published experimental data of several beams.

14.6   Relationship between deflection, mean curvature and strain in reinforcement

Consider a straight non-prestressed concrete member having steel or FRP reinforcement. The deflection, D_{centre} at mid-length section varies almost linearly with the strain ε_s or ε_f in the reinforcement at the same section. The symbol D_{centre} represents the transverse deflection measured from the chord joining the two ends of a continuous or simply-supported member. The linear relationship of D_{centre} versus ε_f is demonstrated below for a simply supported beam reinforced with FRP and will be used in the following section in developing an equation for recommended span to thickness ratio.

Using virtual work, the deflection D_{centre} can be expressed as:

$$D_{centre} = \frac{1}{2} \int_0^{l/2} \psi x \, dx + \frac{1}{2} \int_{l/2}^{l} \psi(l - x)dx \tag{14.16}$$

This is a geometric relationship applicable for any straight member. Here the symbol ψ stands for ψ_m in the cracked part of the length and for ψ_1 in the uncracked part; l is length of member. When ψ is assumed to vary as a second-degree parabola between ψ_{end1}, ψ_{centre} and $\psi_{end\,2}$ (curvature at the two ends and at mid-length), Equation (14.16) gives:

$$D_{centre} = \frac{l^2}{96} (\psi_{end\,1} + 10\psi_{centre} + \psi_{end\,2}) \tag{14.17}$$

Equation (14.16) is employed to calculate D_{centre} for a simple beam reinforced with glass FRP and having the cross-section shown in Fig. 14.2(a). The integrals in the equation are evaluated numerically assuming ψ to vary linearly over short segments of length $0.025\,l$. Figure 14.2(b) is a plot of the deflection, D_{centre} versus the curvature, $\psi_{m\,centre}$ at the centre of a simple beam subjected to uniform load, whose intensity q is varied between zero and $4q_r$; where q_r is the value just sufficient to produce cracking at mid-span section.

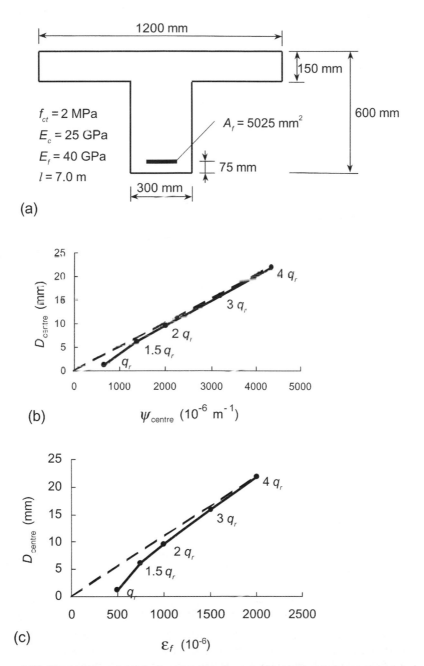

Figure 14.2 Simple beam example showing that D_{centre} is almost proportionate to ψ_m and ε_f at mid-length section: (a) cross-section of beam; (b) deflection versus curvature at mid-length; (c) deflection versus strain in FRP at mid-length section.

The following data are adopted: $f_{ct} = 2\,\mathrm{MPa}$; $E_c = 25\,\mathrm{GPa}$; $E_f = 40\,\mathrm{GPa}$ and $l = 7\,\mathrm{m}$; where f_{ct} and E_c are tensile stress and modulus of elasticity of concrete; E_f is modulus of elasticity of glass FRP. The dashed line in Fig. 14.2(b) is plotted by the equation:

$$D_{centre} = \frac{5}{48}\psi_{m\ centre}\, l^2 \tag{14.18}$$

This is the same as Equation (14.17) applied to a simple beam having zero curvatures at the ends. The difference between the ordinates of the solid line and the dashed line indicates the error in deflection calculation when Equation (14.18) is used; this equation overestimates the deflection because it ignores the fact that the parts of the simple beam adjacent to the ends are uncracked. It can be seen that this simplification overestimates the deflection by a small margin, particularly when the load intensity q is higher, but not close to q_r. This example shows that the deflection at the centre of a simple beam is almost linearly dependent upon the mean curvature at the same section.

When Equation (14.17) is applied to a continuous member, D_{centre} will depend on the curvatures $\psi_{end\ 1}$ and $\psi_{end\ 2}$ at the ends, but will mainly depend upon the curvature ψ_{centre} at mid-length (because this value is multiplied by 10). In Section 14.7, we will adopt the assumption that the deflection at the centre of reinforced concrete cracked members, continuous or simply supported, is proportionate to the mean curvature $\psi_{m\ centre}$ at mid-length.

Figure 14.2(c) shows the variation of D_{centre} with the strain, ε_f in the tension reinforcement at mid-span section of the same simple beam considered above. The value ε_f is calculated, ignoring the concrete in tension, as the load intensity is increased from q_r to $4q_r$. The dashed straight line in Fig. 14.2(c) connects the origin to the ordinate corresponding to the load intensity $4q_r$. From this example it is also concluded that D_{centre} is almost linearly dependent upon ε_f. Again, in Section 14.7 we will adopt the assumption that the deflection at the centre of reinforced concrete cracked members is proportionate to the strain in the tension reinforcement at a cracked section at mid-length. The two assumptions, based on the above study, will be used below to develop an empirical equation for the ratio span to minimum thickness of members reinforced with FRP.

14.7   Ratio of span to minimum thickness

In design of concrete members the overall depths (here referred to as thickness) of members, e.g. thicknesses of slabs, are often selected based on codes or guidelines which give a ratio of span to minimum thickness, (l/h) that normally avoids excessive deflection. The codes and the guidelines are intended for steel-reinforced members; an adjustment to the ratio recommended by the codes and the guidelines will be derived below for use for members with FRP. The subscripts f and s are employed to refer to FRP and steel, respectively.

The deflection, D_{centre} at centre of a member is considered here (based on Section 14.6) to be almost proportional to the strain, ε calculated in the tension reinforcement at a cracked section at mid-length. Thus for a given thickness, when the strain, ε_f in service in FRP is allowed to be greater than the permissible strain, ε_s in service in steel, D_{centre} will be greater when FRP is employed. This can be avoided by adopting a minimum thickness $h_f > h_s$, such that D_{centre} become the same when FRP or steel are used; where h_f and h_s are the minimum thickness of member reinforced with FRP and steel, respectively. Equation 14.25, presented below, gives a recommended value for $(l/h)_f$ in terms of the values specified in codes or guidelines for $(l/h)_s$ and the permissible strain in service ε_f in FRP. When the equation is used in design, the ratio of length to deflection, l/D_{centre} will be the same regardless of the reinforcement material.

Different types of FRP have different elasticity moduli. Thus, identical members with different FRP types have different deflections. However, when the thickness is determined by Equation (14.25) and the amount of reinforcement is designed such that, ε_f at mid-length section in service is equal to a specified value, the deflection will be the same with all FRP types.

14.7.1   Minimum thickness comparison between members reinforced with steel and with FRP

Figures 14.1(a) and (b) show the strain variation over the depth of a cracked reinforced concrete section. With any reinforcement material, the curvature may be expressed by the equation:

$$\psi_2 = \frac{\varepsilon_s \text{ or } \varepsilon_f}{\bar{\beta}h} \tag{14.19}$$

where

$$\bar{\beta} = \frac{d-c}{h} \tag{14.20}$$

where c is depth of compression zone (Equation (14.2)). The mean curvature (Equation (14.10)) may be expressed as:

$$\psi_m = \eta\psi_2 \tag{14.21}$$

where

$$\eta = 1 - 0.5 \left(\frac{f_{ct}}{\sigma_{1\,max}}\right)^2 \left(1 - \frac{I_2}{I_1}\right) \tag{14.22}$$

Use of Equations (14.18), (14.19) and (14.21) gives the deflections at mid-length of a cracked simply-supported concrete member as function of strain, $\varepsilon_{s\,or\,f}$ in the reinforcement at mid-length section:

$$D_{centre} = \frac{5}{48} \frac{l^2}{h} \left(\frac{\eta}{\bar{\beta}}\right)_{s\,or\,f} \varepsilon_{s\,or\,f} \tag{14.23}$$

Application of this equation to a member reinforced with FRP and to a conjugate member reinforced with steel and equating (l/D_{centre}) for the two beams gives:

$$\left(\frac{l}{h}\right)_f = \left(\frac{l}{h}\right)_s \left[\frac{(\eta/\bar{\beta})_s}{(\eta/\bar{\beta})_f}\right] \left(\frac{\varepsilon_s}{\varepsilon_f}\right) \tag{14.24}$$

This equation can be used in design to select thickness, h_f for a member reinforced with FRP guided by $(l/h)_s$ for steel-reinforced members; for this purpose substitute $\varepsilon_s = 1200 \times 10^{-6}$ (or a not substantially different value), representing the commonly allowable value of strain in steel in service. The value of ε_f may be taken equal to 2000×10^{-6}, or a different value depending upon the acceptable crack width (see Section 14.3). The term between square brackets in Equation (14.24) is a dimensionless parameter which can be expressed empirically as a function of $(\varepsilon_s/\varepsilon_f)$ as discussed in Section 14.7.2.

14.7.2   Empirical equation for ratio of length to minimum thickness

The empirical Equation (14.25) given below (as approximation of Equation (14.24)) is based on a parametric study[4] of T and rectangular sections reinforced with FRP varying the thickness h_f, h_{flange}, (b/b_w) and ε_f; where h_f and h_{flange} are the overall height and the flange thickness; b and b_w are width of flange and web and ε_f is the permissible strain in FRP in service. The ratio $(l/h)_f$, length to minimum thickness of a member reinforced with FRP is:

$$\left(\frac{l}{h}\right)_f = \left(\frac{l}{h}\right)_s \left(\frac{\varepsilon_s}{\varepsilon_f}\right)^{a_d} \tag{14.25}$$

where $a_d = 0.5$ for rectangular sections; for T-sections a_d is:

$$a_d = 0.5 + \frac{3b}{100b_w} - \frac{b}{80h_s} \tag{14.26}$$

When $(l/h)_f$ for FRP-reinforced member satisfies Equation (14.25), its ratio l/D_{centre} will be approximately equal to that of a conjugate steel-reinforced member having its length to thickness ratio equal to $(l/h)_s$. The upper limit of the difference between $(l/D_{centre})_f$ and $(l/D_{centre})_s$ will be approximately 11 per cent of the latter ratio. Equation (14.25) is intended to give the minimum thickness for FRP-reinforced concrete members using codes or guidelines that recommend $(l/h)_s$ values for steel-reinforced members. Equation (14.25) can be employed with any type of FRP. The permissible strain in service ε_f used in Equation (14.25) is to be also used in calculating the required cross-sectional area of FRP (Section 14.4).

The design of members reinforced with FRP, avoiding excessive deflection and crack opening can be done by the steps outlined below. Given member length l, cross-sectional dimensions except h_f and internal forces in service $M_{service}$ with or without $N_{service}$:

1   Select the minimum thickness, h_s for a conjugate steel-reinforced member, using code or guideline; the corresponding permissible strain in steel in service, $\varepsilon_s = 1200 \times 10^{-6}$, or a value not substantially different explicitly given or implied by the same code or guideline.
2   Apply Equation (14.25) to obtain minimum thickness, h_f of the member with FRP; the permissible strain in FRP in service may be taken $\varepsilon_f = 2000 \times 10^{-6}$ or different value depending upon the tolerable crack width (Section 14.3).
3   Apply Equation (14.1) or (14.8) to calculate the required cross-sectional area of FRP.

14.8   Design examples for deflection control

In the following examples a simple beam is designed following the steps of the preceding section; then the ratio l/D_{centre} for the beam with FRP is compared with the same ratio for a conjugate steel-reinforced beam.

Example 14.1   A simple beam

Determine the minimum thickness, h_f and the cross-sectional area of glass FRP required for a simple beam of span 8 m to carry a uniform service load $q = 16$ kN-m. Given data: $E_c = 25$ GPa; $E_f = 40$ GPa; assume a T-section with $b = 2$ m; $b_w = 0.4$ m; $h_{flange} = 0.15$ m and $d = 0.85 h_f$; the permissible strain in FRP in service $\varepsilon_f = 2000 \times 10^{-6}$. For the conjugate steel-reinforced beam, take the permissible strain in steel in service, $\varepsilon_s = 1200 \times 10^{-6}$ and $(l/h)_s = 16$.

If the beam is reinforced with steel, the thickness would be:

$$h_s = \frac{8\,m}{16} = 0.5\,\text{m}$$

Application of Equations (14.26) and (14.25) gives

$$a_d = 0.5 + \frac{3(2)}{100(0.4)} - \frac{2}{80(0.5)} = 0.60$$

$$\left(\frac{l}{h}\right)_f = 16 \left(\frac{1200\times10^{-6}}{2000\times10^{-6}}\right)^{0.6} = 11.8$$

Thus, the minimum thickness for beam with FRP is:

$$h_f = \frac{8}{11.8} = 0.68\,\text{m}; \qquad \text{take } h_f = 0.7\text{ m}; \ d = 0.6\,\text{m}$$

The bending moment at mid-span in service,

$$M = (16\,\text{kN/m})\,\frac{(8\,\text{m})^2}{8} = 128.0\,\text{kN.m}$$

The permissible stress in FRP in service is $\sigma_f = E_f \varepsilon_f = 40\,\text{GPa}$ (2000 × 10^{-6}) = 80 MPa. Substitution of this value in Equation (14.1) with an estimated value $y_{CT} = 0.9d = 0.54\,\text{m}$ gives:

$$A_f \approx \frac{128\,\text{kN} - \text{m}}{80\,\text{MPa}\,(0.54)} = 2.96\times10^{-3}\,\text{m}^2 = 2960\,\text{mm}^2$$

Application of Equations (14.2) and (14.5) gives $c = 51$ mm and $y_{CT} = 0.583\,\text{m}$. Substitution of this value in Equation (14.1) gives a more accurate value $A_f = 2740\,\text{mm}^2$.

Example 14.2 Verification of the ratio of span to deflection

Compare l/D_{centre} for the beam designed in Example 14.1 with that of a conjugate steel-reinforced beam carrying the same load intensity and having the same thickness $h_s = h_f = 0.7\,\text{m}$ and $d = 0.6\,\text{m}$, but $(l/h)_s = 16$.

Additional data: $f_{ct} = 2\,MPa$; $E_s = 200\,GPa$; $\varepsilon_s = 1200 \times 10^{-6}$. Other data are the same as in Example 14.1.

(a) Deflection of beam with FRP: The following is calculated at mid-spare section by Equations (14.11), (14.13), (14.10) and (14.18) (with $A_f = 2740\,mm^2$ and $M = 128\,kN\text{-}m$):

$I_1 = 21.9 \times 10^{-3}\,m^4$; $c = 0.049\,m$; $I_2 = 1.41 \times 10^{-3}\,m^4$

$\psi_1 = 234 \times 10^{-6}\,m^{-1}$; $\psi_2 = 3633 \times 10^{-6}\,m^{-1}$

$\sigma_{1\,max} = 2.78\,MPa$ $\zeta = 0.742$

$\psi_m = 2754 \times 10^{-6}\,m^{-1}$

$D_{centre} = 18.4\,mm$

$$(l/D_{centre})_f = \frac{8.0\,m}{0.0184\,m} = 435$$

(b) Beam with steel: With steel, the beam has a longer span $l_s = 16(0.7) = 11.2\,m$. The bending moment at mid-span,

$$M = (16\,kN\text{-}m)\frac{(11.2)^2}{8} = 250.9\,kN\text{-}m$$

$A_s = 1830\,mm^2$; $\varepsilon_s = 1200 \times 10^6$

$I_1 = 23.4 \times 10^{-3}\,m^4$; $c = 0.087\,m$; $I_2 = 4.29 \times 10^{-3}\,m^4$

$\psi_1 = 428 \times 10^{-6}$; $\psi_2 = 2338$

$\sigma_{1\,max} = 5.01\,MPa$; $\zeta = 0.920$

$\psi_m = 2186 \times 10^{-6}\,m^{-1}$

$D_{centre} = 28.6\,mm$

$$\left(l/D_{centre}\right)_s = \frac{11.2\,m}{0.0286\,m} = 392$$

14.9   Deformability of sections in flexure

The discussion in this section is limited to non-prestressed sections. Failure of steel-reinforced sections by flexure is ductile, exhibiting large curvature before the ultimate moment is reached. Unlike steel, FRP continues to exhibit linear stress–strain relationship up to rupture, without yielding or strain hardening. For this reason, FRP-reinforced sections in flexure should have sufficiently large curvature before failure by rupture of the FRP; this requirement is here considered satisfied when the section has a *deformability factor* ≥ 4. The deformability factor is defined as the ratio of the products of moment multiplied by curvature at ultimate and at service. This factor is an approximate indicator of the ratio of strain energy values per unit length of the flexural member at ultimate and at service. Parametric studies[5] show that steel-reinforced sections have deformability factor greater than 4, except when the steel ratio $\rho_s = A_s/(bd)$ is greater than the balanced ratio; in which case the deformability factor is slightly below 4.0. The parametric studies also show that design of cross-sectional areas, A_f in flexural sections with FRP based on a permissible strain in service, $\varepsilon_{f\,service}$ (as discussed in Section 14.4) will normally result in sections having deformability factors greater than 4. Thus, A_f is governed by the serviceability requirement and there is no need to check the deformability, except in the unusual case when the FRP ratio, $\rho_f = A_f/(bd)$ is greater than $0.15 f'_c/\sigma_{f\,service}$; where b is width of section at extreme compressive fibre; f'_c is specified concrete strength; $\sigma_{f\,service}$ $(= E_f \varepsilon_{f\,service})$ is permissible stress in FRP in service. In the parametric studies referred to here, the strain in FRP in service is assumed: $\varepsilon_{f\,service} = 2000 \times 10^{-6}$ and the modulus of elasticity of the FRP, $E_f = 40\,\text{GPa}$ to $150\,\text{GPa}$.

14.10   Prestressing with FRP

In non-prestressed sections, the stress in FRP reinforcement in service is a relatively small fraction of the tensile strength because of control of width of cracks. The high strength of FRP, particularly with carbon fibres, can be more efficiently utilized when the FRP is employed for prestressing. Appropriate permissible stresses at jacking and at transfer should be adopted, accounting for the vulnerability of FRP to creep rupture (e.g. 70 and 60 per cent of the tensile strength, at jacking and at transfers, respectively). Fatigue rupture should also be considered in setting the permissible stresses. The deformability should also be considered, noting that the discussion in the preceding section does not apply.

With FRP types that have low moduli of elasticity compared to steel, the loss of prestress force in the tendons due to creep and shrinkage of concrete is relatively small. An appropriate value of the intrinsic relaxation depending upon the type of the FRP should be used. The procedure in Appendix B for calculating the relaxation reduction coefficient, χ_r can be used with FRP, but not the graph and the empirical equation that give χ_r for prestressing steel.

14.11   General

Properties of FRP for use as reinforcement in concrete vary with the type of fibres, the resin and the manufacturing process. For the design using these materials, their properties should be established with certainty. The difference of modulus of elasticity of FRP from that of steel and its influence on the design for serviceability are presented in this chapter. The basic assumptions and equations adopted in calculation of stresses, strains and displacements of steel-reinforced concrete structures apply when FRP is used.

Notes

1 For further reading on properties of FRP and its design for concrete members see: Japan Society of Civil Engineers (1993), *State-of-the-Art Report on Continuous Fiber Reinforcing Materials*, ed. A. Machida, Concrete Engg. Series 3. See also: ISIS Canada (2001), *Reinforcing New Concrete Structures with Fibre Reinforced Polymers*, Design Manual No. 3, ISIS Canada, Intelligent Sensing for Innovative Structures, A Canadian Network of Centres of Excellence, 227 Engineering Building, University of Manitoba; American Concrete Institute Committee (2001), report ACI 440.1R-01, *Guide for the Design and Construction of Concrete Reinforced with FRP Bars*, 41 pp.

2 Hall, Tara S. (2000), *Deflections of Concrete Members Reinforced with Fibre Reinforced Polymer (FRP) Bars*, M.Sc. Thesis, Department of Civil Engineering, University of Calgary, Calgary, Canada.

3 Ariyawardena, N. (2000), *Prestressed Concrete with Internal or External Tendons: Behaviour and Analysis*, Ph.D. Thesis, Department of Civil Engineering, University of Calgary, Calgary, Canada.

4 Ghali, A., Hall, T. and Bobey, W. (2001), 'Minimum Thickness of Concrete Members Reinforced with Fibre Reinforced Polymer (FRP) Bars', *Canadian J. of Civil Engg.*, **28**, No. 4, pp. 583–592.

5 Newhook, J., Ghali, A. and Tadros, G. (2002), "Concrete Flexural Members Reinforced with FRP: Design for Cracking and Deformability", Canadian J. of Civil Engg., *29*, No. 1.

Appendix A

Time functions for modulus of elasticity, creep, shrinkage and aging coefficient of concrete

The equations and graphs presented below are based on the requirements of the CEB-FIP, *Model Code for Concrete Structures, 1990* (MC-90) and ACI Committee 209, *Prediction of Creep, Shrinkage and Temperature Effects in Concrete Structures, 1992*.[1] It is expected that the requirements of the codes will change in future editions and it is for this reason that this material is presented in an appendix rather than in the body of the text. Thus, all equations and methods of analysis included in the body of the text are independent of the time functions to be used for E_c, φ and ε_{cs}, the modulus of elasticity, creep coefficient and free shrinkage of concrete. Requirements of Eurocode 2–1991[2] (EC2–91) and British Standard BS8110–1997[3] are also discussed.

A.1 CEB-FIP Model Code 1990 (MC-90)

In this code the symbol φ is used differently from the way it is used in this book; for this reason, we adopt the symbol φ_{CEB} for the creep coefficient employed in MC-90. Equation (1.2) expresses the total strain at time t, instantaneous plus creep, due to a constant stress $\sigma_c(t_0)$ introduced at time t_0 as follows:

$$\varepsilon_c(t) = \frac{\sigma_c(t_0)}{E_c(t_0)} [1 + \varphi(t, t_0)] \tag{A.1}$$

where $E_c(t_0)$ is the modulus of elasticity at age t_0; $\varphi(t, t_0)$ is the ratio of creep to the instantaneous strain.

The strain $\varepsilon_c(t)$ is expressed in MC-90 by the equation:

$$\varepsilon_c(t) = \frac{\sigma_c(t_0)}{E_c(t_0)} \left[1 + \frac{E_c(t_0)}{E_c(28)} \varphi_{CEB}(t, t_0) \right] \tag{A.2}$$

where $E_c(28)$ is the modulus of elasticity at age 28 days. Comparison of Equations (A.1) and (A.2) indicates that

$$\varphi = \varphi_{\text{CEB}} \frac{E_c(t_0)}{E_c(28)} \qquad (A.3)$$

Thus, the numerical values of the creep coefficient φ_{CEB} calculated according to MC-90 must be multiplied by the ratio $E_c(t_0)/E_c(28)$ to obtain the value of the creep coefficient for use in the equations of this book. The graphs and equations for the creep coefficient presented in this appendix include this adjustment; thus, they can be used directly in the equations of the book without further adjustment.

A.1.1   Parameters affecting creep

Creep depends upon the age at loading t_0 and the length of the period t_0 to t; where t is the instant at which the value of creep is considered. In the equations which will follow, t_0 and t are in days. Creep also depends upon the relative humidity, RH (per cent) and the notional size h_0 (mm) defined by:

$$h_0 = \frac{2A_c}{u} \qquad (A.4)$$

where A_c and u are the area and perimeter in contact with the atmosphere of the cross-section of the considered member.

The value of creep coefficient is inversely proportional to $\sqrt{f_{\text{cm}}}$; where f_{cm} (MPa) is the mean compressive strength of concrete at age 28 days. The value f_{cm} may be estimated by:

$$f_{\text{cm}} = f_{\text{ck}} + 8\,\text{MPa} \qquad (A.5)$$

f_{ck} (MPa) is characteristic compressive strength of cylinders, 150 mm in diameter and 300 mm in height stored in water at $20 \pm 2\,°C$, and tested at the age of 28 days. The value f_{ck} is the strength below which 5 per cent of all possible strength measurements may be expected to fall.

The graphs presented in this appendix give creep coefficient $\varphi(t, t_0)$ and aging coefficient $\chi(t, t_0)$ for selected combinations of the parameters f_{ck}, RH and h_0. The graphs are based on the code equations given below, which are valid for mean temperature of $20\,°C$, taking into account seasonal variations between $-20\,°C$ and $+40\,°C$.

A.1.2.   Effect of temperature on maturity

When prevailing temperature is higher or lower than 20 degrees Celsius, the effect of temperature on the maturity of concrete may be accounted for by the use of adjusted age t_T in lieu of t_0 or t in the equations or graphs presented below. The adjusted age is given by:

$$t_T = \sum_{i=1}^{n} \left[\Delta t_i \exp \left(13.65 - \frac{4000}{273 + T(\Delta t_i)} \right) \right] \tag{A.6}$$

where t_T is the concrete age adjusted for temperature; Δt_i is the number of days in which a temperature $T(\Delta t_i)$ degree Celsius prevails. For application of (A.6), the age t_0 or t is to be divided into n intervals and a prevailing temperature is to be assumed for each.

A.1.3   Modulus of elasticity

The modulus of elasticity of concrete, $E_c(28)$ (MPa), at age 28 days, for normal-weight concrete can be estimated by:

$$E_c(28) = 21\ 500\ (f_{cm}/f_{cm0})^{\frac{1}{3}} \tag{A.7}$$

where $f_{cmo} = 10\ \text{MPa}$.
When the mean compressive strength f_{cm} MPa is not known, $E_c(28)$ may be estimated from the characteristic compressive strength, f_{ck} (for MPa) at 28 days by the equation:

$$E_c(28) = 21\ 500[(f_{ck} + \Delta f)/f_{cm0}]^{\frac{1}{3}} \tag{A.8}$$

where $\Delta f = 8\ \text{MPa}$.
   Equations (A.7) and (A.8) apply when quartzitic aggregates are used. For other aggregates, multiply $E_c(28)$ by a factor varying between 0.7 and 1.2.
   Equations (A.7) and (A.8) give the tangent modulus of elasticity, which is equal to the slope of the stress–strain diagram at the origin. This modulus is the value to be employed with the creep coefficient given by Equation (A.16) and the graphs presented in this section to calculate the strain at any time (see Equation (A.1)).
   When the modulus of elasticity is for use in an elastic analysis, without considering creep, the value $E_c(28)$ should be reduced by a factor of 0.85 to account for the quasi-instantaneous strain, which occurs shortly (within one day) after loading.

A.1.4   Development of strength and modulus of elasticity with time

The mean concrete strength $f_{cm}(t)$ at age t (days) may be estimated from the strength f_{cm} at 28 days by:

$$f_{cm}(t) = \beta_{cc}(t) f_{cm} \tag{A.9}$$

where

$$\beta_{cc}(t) = \exp[s(1 - \sqrt{28/t})] \qquad (A.10)$$

with s being a coefficient depending on type of cement; s is equal to 0.2, 0.25 and 0.38, respectively, for rapidly hardening high-strength cements, for normal and rapidly hardening cements, and for slowly hardening cements.

The modulus of elasticity of concrete at age t may be estimated by:

$$E_c(t) = \beta_E(t)E_c(28) \qquad (A.11)$$

with

$$\beta_E(t) = \sqrt{\beta_{cc}(t)} \qquad (A.12)$$

A.1.5 Tensile strength

The tensile strength of concrete may be subject to large variation by environmental effects. Upper and lower values of the characteristic axial tensile strength f_{ctk} (MPa) may be estimated by:

$$f_{ctk, \ min} = 0.95(f_{ck}/f_{ck0})^{2/3} \qquad (A.13)$$

$$f_{ctk, \ max} = 1.85(f_{ck}/f_{ck0})^{2/3} \qquad (A.14)$$

where $f_{ck0} = 10\,\text{MPa}$.

Caution should be taken when the tensile strength of concrete is used in analysis of displacements. The value of the tensile strength assumed in such analysis will indicate whether cracking occurred or not. Cracking can substantially increase displacements. Thus, when the displacements are critical, their analysis should be based on the minimum value of tensile strength (Equation (A.13)).

A.1.6 Creep under stress not exceeding 40 per cent of mean compressive strength

We recall the definition of the creep coefficient $\varphi(t, t_0)$ as the ratio of creep at time t to the instantaneous strain due to a constant stress introduced at time t_0. MC-90 gives a coefficient $\varphi_{CEB}(t, t_0)$, which is equal to $\varphi(t, t_0)$ divided by $\beta_E(t_0)$ (see Equations (A.3) and (A.11)), where

$$\beta_E(t_0) = \frac{E_c(t_0)}{E_c(28)} \qquad (A.15)$$

The equation given below for $\varphi(t, t_0)$ is valid for compressive stress not exceeding $0.40 f_{cm}(t_0)$, relative humidity RH = 40 to 100 per cent, mean temperature 5 to 30 degrees Celsius and f_{ck} between 12 and 80 MPa. The same equation applies when the stress is tensile.

The equation given by MC-90 for φ_{CEB} is adjusted below to give the creep coefficient $\varphi(t, t_0)$ as defined above:

$$\varphi(t, t_0) = \varphi_0 \beta_c(t - t_0)\beta_E(t_0) \tag{A.16}$$

where β_c is a coefficient describing development of creep with time after loading; φ_0 is a notional creep coefficient given by:

$$\varphi_0 = \varphi_{RH}\beta(f_{cm})\beta(t_0) \tag{A.17}$$

$$\varphi_{RH} = 1 + \frac{1 - (RH/100)}{0.46(h_0/h_{ref})^{1/3}} \tag{A.18}$$

where $h_{ref} = 100$ mm.

$$\beta(f_{cm}) = \frac{5.3}{\sqrt{f_{cm}/f_{cmo}}} \tag{A.19}$$

where $f_{cm0} = 10$ MPa.

$$\beta(t_0) = \frac{1}{0.1 + t_0^{0.2}} \tag{A.20}$$

The symbol h_0 (mm) is the notional size of member defined by Equation (A.4). Development of creep with time is expressed by:

$$\beta_c(t - t_0) = \left(\frac{t - t_0}{\beta_H + t - t_0}\right)^{0.3} \tag{A.21}$$

β_H (mm) is a function of the notional size h_0 (mm) and the relative humidity, RH (per cent):

$$\beta_H = \frac{150h_0}{h_{ref}}[1 + (0.012\,RH)^{18}] + 250 \leq 1500\,mm \tag{A.22}$$

where $h_{ref} = 100$ mm.

A.1.7   Effect of type of cement on creep

Creep of concrete depends on the degree of hydration needed at the age of loading t_0 and thus on the type of cement. This effect can be accounted for by modifying t_0, using equation:

$$t_0 = t_{0,\,\mathrm{T}} \left(\frac{9}{2 + (t_{0,\,\mathrm{T}})^{1.2}} + 1 \right)^a \geq 0.5 \qquad (A.23)$$

where $t_{0,\,\mathrm{T}}$ (days) is the age of concrete at loading adjusted by Equation (A.6) for substantial deviation of prevailing temperature from 20 degrees Celsius; a is coefficient equal to -1.0, 0, or 1.0, respectively, for slowly hardening cement, for normal or rapidly hardening cement and for rapidly hardening high-strength cements.

A.1.8   Creep under high stress

Creep under compressive stress in the range (0.4 to 0.6) $f_{cm}(t_0)$ can be calculated by Equation (A.16) replacing φ_0 by φ_{0k} given by:

$$\varphi_{0k} = \varphi_0 \exp[1.5(k - 0.4)] \qquad (A.24)$$

where k is the applied stress divided by $f_{cm}(t_0)$.

A.1.9   Shrinkage

Shrinkage starts at time t_s (days) when curing is stopped. On the other hand, concrete immersed in water at time t_s starts to swell. The shrinkage or the swelling at any time t (days) may be estimated by:

$$\varepsilon_{cs}(t, t_s) = \varepsilon_{cs0}\beta_s(t - t_s) \qquad (A.25)$$

where $\beta_s(t - t_s)$ is a function describing the development of shrinkage or swelling with time, given by:

$$\beta_s(t - t_s) = \left(\frac{t - t_s}{350(h_0/h_{ref})^2 + t - t_s} \right)^{0.5} \qquad (A.26)$$

where h_0(mm) is the notional size defined by Equation (A.4) and $h_{ref} = 100\,\mathrm{mm}$.
ε_{cs0} is the notional shrinkage given by:

$$\varepsilon_{cs0} = \varepsilon_s(f_{cm})\beta_{RH} \qquad (A.27)$$

where

$$\varepsilon_s(f_{cm}) = 10^{-6}[160 + 10\beta_{sc}(9 - f_{cm}/f_{cm0})] \qquad (A.28)$$

with β_{sc} equalling 4, 5 or 8, respectively, for slowly hardening cements, for normal or rapidly hardening cements, and for rapidly hardening high-strength cements; $f_{cm0} = 10\,\mathrm{MPa}$.

$$\beta_{RH} = -1.55\left[1 - \left(\frac{RH}{100}\right)^3\right] \text{ for } 40\% \leq RH < 99\% \qquad (A.29)$$

$$\beta_{RH} = +0.25 \text{ for } RH \geq 99\% \text{ (immersed in water).} \qquad (A.30)$$

Positive β_{RH} indicates swelling. RH (per cent) is relative humidity.

A.2   Eurocode 2–1991 (EC2–91)

The values of the parameters discussed in the preceding section for MC-90 will not be much different when estimated in accordance with EC2–91. Some of the differences between the two codes are summarized below.

EC2–91 gives Equation (A.31) for estimation of secant modulus of elasticity (MPa) for normal weight concrete at age t_0 days. The secant modulus is defined as stress divided by strain at a stress level $= \sigma_c(t_0) = 0.4\,f_{ck}(t_0)$ (see Fig. 1.1):

$$E_c(t_0) = 0.95\,\{21\,500[(f_{ck}(t_0) + 8)/10]^{\frac{1}{3}}\} \qquad (A.31)$$

where $f_{ck}(t_0)$ (MPa) is characteristic compressive stress at age t_0.

EC2–91 contains a table of creep coefficients for normal-weight concrete at $t = \infty$ due to compressive stress not exceeding $0.45\,f_{ck}(t_0)$ introduced at age t_0; the value $t_0 = 1, 7, 28, 90$ and 365 days. The EC2–91 values are here adjusted by multiplication by $\beta_E(t_0)$ (given by Equations (A.12) and (A.10) and the resulting coefficients are presented in Table A.1. The values $\varphi(\infty\ t_0)$ given in Table A.1 can be used with the secant modulus of elasticity (Equation (A.31) and Equation (A.1)) to calculate the total strain, instantaneous plus ultimate creep after a very long time.

Table A.2 from EC2–91 gives final shrinkage values of normal-weight concrete $(\varepsilon_{cs}(\infty, t_s))$; where t_s is time when curing is stopped.

The values given in Tables A.1 and A.2 apply for a range of mean temperatures between 10 and 20 degrees Celsius (taking into account seasonal variations between −20 and +40 degrees).

As a complement to Tables A.1 and A.2, EC2–91 gives, for use when more accuracy is required, the same equations as MC-90 for creep coefficient and shrinkage, which are presented above in Section A.1.

Table A.1 Final creep coefficients $\varphi(\infty, t_0)$ of normal-weight concrete subjected to compressive stress not exceeding 0.45 $f_{ck}(t_0)$ (based on EC2–91)

Age at loading t_0 (days)	Notional size h_0 (mm), defined by Equation (A.4)					
	50	150	600	50	150	600
	Dry atmosphere (inside) (RH = 50 per cent)			Humid atmosphere (outside) (RH = 80 per cent)		
1	3.2	2.6	2.1	2.0	1.8	1.5
7	3.4	2.8	2.2	2.2	1.9	1.7
28	3.2	2.5	2.0	1.9	1.7	1.5
90	2.8	2.3	1.7	1.7	1.5	1.3
365	2.2	1.8	1.3	1.3	1.1	1.1

Table A.2 Final shrinkage strain ε_{cs} of normal-weight concrete (based on EC2–91)

Location of member	Relative humidity per cent	Notional size h_0 (mm), defined by Equation (A.4)	
		≤150	≥600
Inside	50	-600×10^{-6}	-500×10^{-6}
Outside	80	-330×10^{-6}	-280×10^{-6}

The quantity inside the curly brackets in Equation (A.31) is the tangent modulus of elasticity according to MC-90 (slope of stress–strain diagram at the origin; see Equation (A.8)). Thus, the EC2–91 allows calculation of the strain $\varepsilon(t)$ by Equation (A.1), using the creep coefficient $\varphi(t, t_0)$ given by Equation (A.16) and the tangent modulus of elasticity. The graphs for creep and aging coefficients presented in Section A.6 are based on Equation (A.16); thus, they are in accordance with EC2–91. At age 28 days, EC2–91 considers that the tangent modulus of elasticity is equal to 1.05 the secant modulus.

A.3   ACI Committee 209

A large number of variables affect the magnitude of creep and shrinkage which is discussed in some detail in the report of the American Concrete Institute Committee 209.[4] The following equations are considered applicable in 'standard conditions'. The term 'standard conditions' is defined in the report by ranges for a number of variables related to the material properties, the climate and the sizes of members. Reference must be made to the mentioned report when the conditions are different from what is specified below.

A.3.1   Creep

The coefficient for creep at time t for age at loading t_0 is given by:

$$\varphi(t, t_0) = \frac{(t - t_0)^{0.6}}{10 + (t - t_0)^{0.6}} \varphi_u \qquad (A.32)$$

where

$$\varphi_u = \varphi(t_\infty, t_0) \qquad (A.33)$$

φ_u is the ultimate creep, after a very long time (10 000 days) for age at loading, t_0. The value φ_u is given by:

$$\varphi_u = 2.35\, \gamma_c \qquad (A.34)$$

where γ_c is a correction factor, the product of several multipliers depending upon ambient relative humidity, average thickness of the member or its volume-to-surface ratio and on the temperature. For relative humidity of 40 per cent, average thickness 6 in (0.15 m) or volume-to-surface ratio of 1.5 in and temperature 70 °F (21 °C), all the multipliers are equal to unity. In this case, γ_c may be calculated as a function of the age at loading t_0:

$$\gamma_c = 1.25\, t_0^{-0.118} \qquad (A.35)$$

or

$$\gamma_c = 1.113\, t_0^{-0.094} \qquad (A.36)$$

Equations (A.35) and (A.36) are respectively applicable for moist-cured concrete and for 1–3 days steam-cured concrete. The two equations give $\gamma_c \simeq 1.0$ when $t_0 = 7$ and 3 days, respectively.

The ratio of the modulus of elasticity at any age t_0 days to the value at age 28 days:

$$\frac{E_c(t_0)}{E_c(28)} = \left(\frac{t_0}{a + \beta t_0}\right)^{\frac{1}{2}} \qquad (A.37)$$

The coefficients a and β are constants depending upon the type of cement and curing used. For cement Type I, $a = 4$ and $\beta = 0.85$.

The value of E_c to be employed with the equations presented in this section may be estimated by the ACI318(1989) Code[5] equation:

$$E_c = w_c^{1.5} 33 \sqrt{f_c'} \qquad (A.38)$$

where E_c (psi) and f'_c (psi) are the modulus of elasticity of concrete and its specified compressive strength; w_c (lb per cu ft) is the unit weight of concrete. For normal-weight concrete, E_c (psi) may be taken as $57\,000\,\sqrt{f'_c}$. Equation (A.38) may be rewritten using SI units:

$$E_c = w_c^{1.5}\,0.043\,\sqrt{f'_c} \tag{A.39}$$

with E_c (MPa), and f'_c (MPa) and w_c (kg/m^3); the corresponding value of E_c (MPa) for normal-weight concrete is $4730\,\sqrt{f'_c}$ (MPa).

Equation (A.38) or (A.39) gives the secant modulus of elasticity, which is the slope of the secant drawn from the origin to a point corresponding to 0.40 f'_c on the instantaneous (1–5 minutes) stress–strain curve.

Use of Equation (A.38) or (A.39) will overestimate E_c when f'_c is higher than 6000 psi (40 MPa), in which case the following equation is suggested[6] for normal-weight concrete:

$$E_c = 40\,000\,\sqrt{f'_c} + 10^6 \text{ psi} \tag{A.40}$$

$$E_c = 3300\,\sqrt{f'_c} + 7000\,\text{MPa} \tag{A.41}$$

From Equations (A.32–34), the ratio of the creep coefficient $\varphi(t_\infty, t_0)$ to $\varphi(t_\infty, 7)$ for moist-cured concrete may be expressed as:

$$\frac{\varphi(t_\infty, t_0)}{\varphi(t_\infty, 7)} = 1.25 t_0^{-0.118} \tag{A.42}$$

Bazant[7] employs Equations (A.31), (A.32), (A.36) and (A.41) to calculate $\varphi(t, t_0)$ and uses a numerical procedure similar to the method in Section 1.10 to calculate the values of the aging coefficient $\chi(t, t_0)$ given in Table A.3.

A.3.2   Shrinkage

For moist-cured concrete, the free shrinkage which occurs between $t_0 = 7$ days and any time t

$$\varepsilon_{cs}(t, t_0) = \frac{t - t_0}{35 + (t - t_0)}\,(\varepsilon_{cs})_u \text{ with } t_0 = 7 \tag{A.43}$$

and for steam-cured concrete, the shrinkage between $t_0 = 1$ to 3 days and any time t

$$\varepsilon_{cs}(t, t_0 = 1 \text{ to } 3) = \frac{t - (1 \text{ to } 3)}{55 + (t - 1 \text{ to } 3)}\,(\varepsilon_{cs})_u \tag{A.44}$$

Table A.3 Aging coefficient $\chi(t, t_0)$ calculated by Bazant

| $(t - t_0)$ | $\varphi(t_\infty, 7)$ | Value of χ | | | | $\dfrac{\varphi(t, t_0)}{\varphi(t_\infty, t_0)}$ |
		$t_0 = 10$	$t_0 = 10^2$	$t_0 = 10^3$	$t_0 = 10^4$	
10 days	0.5	0.525	0.804	0.811	0.809	
	1.5	0.720	0.826	0.825	0.820	0.273
	2.5	0.774	0.842	0.837	0.830	
	3.5	0.806	0.856	0.848	0.839	
10^2	0.5	0.505	0.888	0.916	0.915	
days	1.5	0.739	0.919	0.932	0.928	0.608
	2.5	0.804	0.935	0.943	0.938	
	3.5	0.839	0.946	0.951	0.946	
10^3	0.5	0.511	0.912	0.973	0.981	
days	1.5	0.732	0.943	0.981	0.985	0.857
	2.5	0.795	0.956	0.985	0.988	
	3.5	0.830	0.964	0.987	0.990	
10^4	0.5	0.461	0.887	0.956	0.965	
days	1.5	0.702	0.924	0.966	0.972	0.954
	2.5	0.770	0.940	0.972	0.976	
	3.5	0.808	0.950	0.977	0.980	
$\dfrac{\varphi(t_\infty, t_0)}{\varphi(t_\infty, 7)}$		0.960	0.731	0.558	0.425	
$\dfrac{E_c(t_0)}{E_c(28)}$		0.895	1.060	1.083	1.089	

where $(\varepsilon_{cs})_u$ is the ultimate free shrinkage corresponding to t_∞ (say at 10 000 days). The ultimate free shrinkage is given by:

$$(\varepsilon_{cs})_u = -780 \times 10^6 \gamma_{cs} \tag{A.45}$$

where γ_{cs} is a correction factor, the product of a number of multipliers which depends upon the same factors mentioned above for γ_c. The correction factor $\gamma_{cs} = 1.0$ when the period of initial moist curing is 7 days, the relative humidity of the ambient air is 40 per cent, the average thickness is 6 in (0.15 m) or the volume-to-surface ratio is 1.5 in.

The free shrinkage between any two ages t_0 and t can be calculated as the difference of shrinkage for the periods $(t - 7)$ and $(t_0 - 7)$:

$$\varepsilon_{cs}(t, t_0) = \varepsilon_{cs}(t, 7) - \varepsilon_{cs}(t_0, 7) \tag{A.46}$$

Equation (A.43) is applicable for each of the two terms in Equation (A.46). In a similar way, Equation (A.44) can be employed to calculate $\varepsilon_{cs}(t, t_0)$ for steam-cured concrete.

A.4   British Standard BS 8110[8]

Part 2 of BS 8110 gives equations for modulus of elasticity, creep and shrinkage of concrete that result in level of accuracy greater than that in BS 8110: Part 1. The equations presented below are taken from Part 2.

A.4.1   Modulus of elasticity of concrete

A mean value of the elasticity modulus of normal-weight concrete is given by.

$$E_c(28) = K_0 + 0.2 f_{cu}(28) \tag{A.47}$$

where 28 is the age of concrete in days; E_c is the static modulus of elasticity; f_{cu} is the characteristic cube strength, below which 5 per cent of all possible test results would be expected to fall; $K_0 = 20\,\text{GPa}$, a constant related to the modulus of elasticity of the aggregate. For lightweight aggregate concrete, the value of elasticity modulus given by Equation A.47 should be multiplied by $(w/2400)^2$; where w is the density of concrete in kg/m^3. When E_c is for calculation of deflections that are of great importance, BS 8110 states that tests should be carried out on concrete made with the aggregate to be used in the structure. With unknown aggregate at the design stage, the standard advises to consider a range of $E_c(28)$, based on $K_0 = 14$ to $26\,\text{GPa}$.

Variation of the elasticity modulus with the age of concrete, t is expressed by:

$$E_c(t) = E_c(28) \left(0.4 + \frac{0.6 f_{cu}(t)}{f_{cu}(28)} \right) \text{ with } t \geq 3 \text{ days} \tag{A.48}$$

The value of $f_{cu}(t)$ to be used in this equation is given in Table A.4. For lightweight concrete having density w, multiply the values in Table A.4 by $[w(\text{kg/m}^3)/2400]^2$.

A.4.2   Tensile strength of concrete

The British Standard BS 8110 does not specify tensile strength of concrete. However, non-prestressed sections subjected to bending moment can be considered uncracked when the stress in concrete at the level of the tension reinforcement is less than 1 MPa. When the section is considered cracked, the stress in concrete in tension is assumed to vary linearly over the tension zone;

Table A.4 Variation of cube strength in MPa with age of concrete according to British Standard BS 8110

Grade	Characteristic strength $f_{cu}(28)$	Cube strength at an age of:				
		7 days	2 months	3 months	6 months	1 year
20	20.0	13.5	22.0	23.0	24.0	25.0
25	25.0	16.5	27.5	29.0	30.0	31.0
30	30.0	20.0	33.0	35.0	36.0	37.0
40	40.0	28.0	44.0	45.5	47.5	50.0
50	50.0	36.0	54.0	55.5	57.5	60.0

the value of the tensile stress is zero at the neutral axis and, at the level of the tension reinforcement, the concrete stress is equal to 1 MPa or 0.55 MPa in short term and in long term, respectively (see Fig. E.4, Appendix E).

For prestressed sections, class 2, flexural tensile stress is permitted without visible cracks; the allowable tensile stress is $0.45\sqrt{f_{cu}}$ or $0.36\sqrt{f_{cu}}$ for pretensioned and post-tensioned members, respectively. The allowable tensile stress may be increased by up to 1.7 MPa in certain conditions.

A.4.3 Creep

Final creep is assumed to occur in 30 years. The creep coefficient φ (30 yrs, t_0) is given by the graph in Fig. A.1; where t_0 is age of concrete at loading in days. In this figure the effective thickness is twice the cross-sectional area of a member divided by the exposed perimeter. When concrete is exposed to constant relative humidity, 40, 60 and 80 per cent, the final creep is assumed to develop in the first month, at six months and 30 months, respectively.

A.4.4 Shrinkage

British Standard BS 8110 gives the graph in Fig. A.2 for an estimate of drying shrinkage of plain normal-weight concrete as function of the relative humidity and the effective thickness (defined the same as in the preceding section). The graph is for concrete made without water-reducing admixtures (water content about 190 litre/m³). Shrinkage is considered proportional to water content in the range 150 to 230 litre/m³.

A.5 Computer code for creep and aging coefficients

The computer code in FORTRAN described below employs the step-by-step procedure given in Section 1.10 and Equations (1.23), (1.25), (1.27) and (1.29)

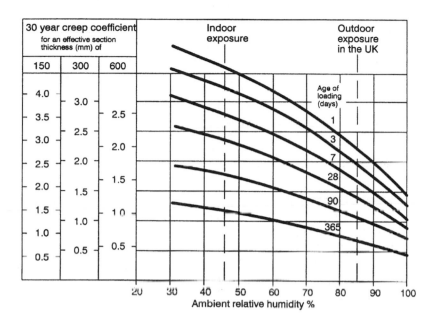

Figure A.1 Creep coefficient φ (30 yrs, t_0); where t_0 is age of loading. Reproduced from BS 8110: Part 2, 1985, with kind permission of BSI.

to calculate the relaxation function $r(t, t_0)$ and the aging coefficient $\chi(t, t_0)$. The values of $E_c(t)$ and $\varphi(t, t_0)$ required in the analysis are based on the equations of MC-90 (see Section A.2).

Figure A.3 is a listing of subroutine named *Chicoef*, for which the input data are f_{ck}, h_0, RH, t and t_0, where f_{ck} (MPa) is characteristic compressive strength (at 28 days); h_0 (mm) is notional size (Equation (A.4)); RH (per cent) is relative humidity; t_0 and t are ages of concrete at the start and at the end of the loading (or relaxation) period. The output gives the relaxation function $r(\tau, t_0)$, varying τ between t_0 and t and the aging coefficient $\chi(t, t_0)$.

The subroutine *Chicoef* employs a subroutine named *Phicoef*, (see Fig. A.4), which calculates $\varphi(t_2, t_1)$ as a function of f_{ck}, h_0, RH, t_1 and t_2; where t_1 is the age of concrete at loading and t_2 is the age at the end of a period in which the load is sustained. The computer programs provided on the Internet for this book include FORTRAN files for the subroutines *Chicoef* and *Phicoef*; thus they can be revised as may become necessary. The present subroutines are employed to produce an executable file, also included on the disc, that can perform the calculations on a micro-computer (see Appendix G).

Figure A.5 is an example plot of the relaxation function $r(t, t_0)$ prepared by the above computer codes with $t_0 = 3$ days and 120 days, $f_{ck} = 30$ MPa, $h_0 = 400$ mm and RH = 50 per cent. The broken line on the same graph represents the variation of E_c with time.

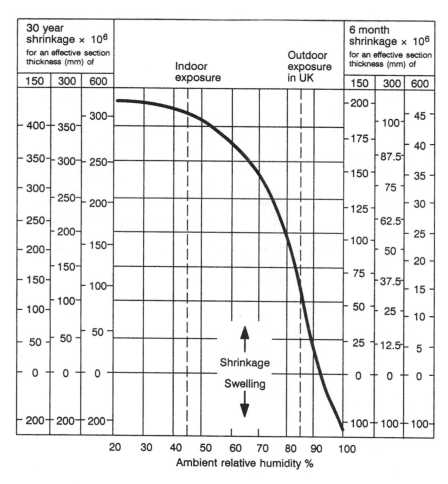

Figure A.2 Free shrinkage of normal-weight concrete. Reproduced from BS 8110: Part 2, 1985, with kind permission of BSI.

A.6   Graphs for creep and aging coefficients

The graphs in Figs A.6 to A.45, based on MC-90, give the values of the creep coefficient $\varphi(t, t_0)$ and the aging coefficients for selected sets of characteristic compressive stress f_{ck}, notional size h_0 (Equation (A.4)) and relative humidity RH. The coefficient $\varphi(t, t_0)$ is the ratio of creep in the period t_0 to t divided by the instantaneous strain $\varepsilon_c(t_0)$. The value of $E_c(t_0)$ to be used in calculating $\varepsilon_c(t_0)$ is the tangent elasticity modulus given by Equations (A.8) and (A.11). As mentioned earlier, the graphs are in accordance with EC2–91.

Table A.4 lists the values of f_{ck} (MPa), h_0 (mm) and RH (per cent) selected for the graphs.

```
       Subroutine Chicoef(fck,ho,RH,to,t,Chi)
       DIMENSION DSIGMA(54), ti(55)
       Ec=2.15e4*((fck+8.)/10.)**0.333333333333
       BetaEto=(   EXP(0.25*(1.-SQRT(28./to) ))    )**.5
       Ecto=Ec*BetaEto
c      Division of the period (t-to) into 53 intrvals selected in accordance
c      with Chiorino,M. A. and Lacidogna,G. , "Design Aids for Creep Analysis
c      of Concrete Structures (CEB Model 1990 for Concrete)", Atti 30,Dep. di
c      Ingeneria Strutturale, Politecnico di Torino, Italy, Oct.,1991.
       ti(1)=to
       ti(2)=to
       ti(3)=ti(1)+0.01
       DO 20 K=4,55
       q=10.**0.125
       ti(K)=(ti(K-1)-ti(1))*q+ti(1)
20     IF(ti(K).GT.t)ti(K)=t
       DSIGMA(1)=Ecto
       DO 30 I=2,54
       tihalf=ti(I+1)
       SUM=0.
       DO 40 J=1,I-1
       tj=(ti(J)+ti(J+1))/2.
       Ectj=Ec*(   EXP(0.25*(1.-SQRT(28./tj)))    )**.5
       CALL Phicoef(fck,ho,RH,tj,tihalf,Phi)
       SUM=SUM+DSIGMA(J)*(1.+Phi)/Ectj
40     CONTINUE
       tii=(ti(I)+ti(I+1))/2.
       Ecti=Ec*(   EXP(0.25*(1.-SQRT(28./tii)) )    )**.5
       CALL Phicoef(fck,ho,RH,tii,tihalf,Phi)
       DSIGMA(I)= (Ecti/(1.+Phi))*(1.-SUM)
30     CONTINUE
```

This figure is continued on next page.

A.7   Approximate equation for aging coefficient

It can be seen from any of the aging coefficient graphs in Figs A.6 to A.45 that for a specified age at start of loading t_0, the value of $\chi(t, t_0)$ is almost constant when $(t - t_0) \geq 1$ year. In this case, the aging coefficient can be approximated by the empirical equation.[9]

$$\chi(t, t_0) \simeq \chi(30 \times 10^3, t_0) \simeq \frac{\sqrt{t_0}}{1 + \sqrt{t_0}} \tag{A.49}$$

The error in this equation is less than ± 10 per cent when $t_0 > 28$ days and can

```
          WRITE(6,300)to,0.0,1.0,Ecto
          DO 50 I=1,54
          tihalf=ti(I+1)
          SUM=0.
          DO 70 J=1,I
70        SUM=SUM+DSIGMA(J)
          rthalf=SUM
c         In exceptional cases, the step-by-step analysis used with MC-90 equations
c         for Phi and Ec,results in a small negative value for r(t,to). Such  value
c         is not accepted and r(t,to) is simply set to equal zero.
          IF(rthalf.LT.0.0)rthalf=0.0
          IF(I.EQ.1)Chi=0.0
          IF(I.EQ.1)GO TO 72
          CALL Phicoef(fck,ho,RH,to,tihalf,Phi)
          Chi=(1./(1.-rthalf/Ecto))-1./Phi
c         The above equation cannot be used in interval 1 (of zero duration). Print
c         a dummy zero value for Chi.
72        r=rthalf/Ecto
          WRITE(6,300)tihalf,Chi,r,rthalf
300       FORMAT(3X,F10.2,3X,F6.3,9X,F6.4,6X,E11.5)
50        CONTINUE
          WRITE(6,400)to,Ecto,  t,to,Chi,    t,to,rthalf
400       FORMAT(/,3X, 'Ec(',F6.0,'days)=',E11.5,' MPa',/,
         +/,3X,'Aging coefficient Chi(',F6.0,',',F6.0,')=',F6.4,/,
         +/,3X,'Relaxation function r(',F6.0,',',F6.0,')=',E11.5,' MPa')
          CALL Phicoef(fck,ho,RH,to,t,Phi)
          WRITE(6,600)t,to,Phi
600       FORMAT(/,3X,'Creep coefficient Phi(',F6.0,',',F6.0,')=',F6.4)
          RETURN
          END
```

Figure A.3 Computer code in FORTRAN for relaxation function $r(t, t_0)$ and aging coefficient $\chi(t, t_0)$.

reach $\pm$ 20 per cent when $t_0 = 3$ days. The equation underestimates the value of χ when creep is high; that is when f_{ck}, RH and h_0 are relatively small. More accuracy can be achieved by replacing the constant 1.0 in the denominator on the right-hand side of Equation (A.49) by a variable[9] $\simeq$ 0.5 to 2.0, which is a function of f_{ck}, RH and h_0.

```
        SUBROUTINE Phicoef(fck,ho,RH,t1,t2,Phi)
c       Calculation of creep coefficient Phi(t2,t1)
        fcm=fck+8.
        Bfcm=5.3/SQRT(fcm/10.)
        PhiRH=1.+(1.-(RH/100.) )/(0.46* ((ho/100.)**0.3333333333333) )
        Betato=1./(0.1+t1**0.2)
        Phio=PhiRH*Bfcm*Betato
        Betah=1.5*(1.+(0.012*RH)**18.)*ho+250.
        IF(Betah.GT.1500.)Betah=1500.

        Betac=( (t2-t1)/(Betah+t2-t1)   )**0.3
        PhiCEB=Phio*Betac
c       Assume normal or rapid-hardening cement; s=0.25
        s=0.25
        Betacc=exp(s* (1.- SQRT(28./t1) ))
        BetaE=SQRT(Betacc)
        Phi=PhiCEB*BetaE

c       write(6,550)t2,t1,Phi
550     FORMAT(3X,'Creep coefficient Phi(',F8.2,',',F8.2,')=',F6.4)
        RETURN
        END
```

Figure A.4 Computer code in FORTRAN for calculation of creep coefficient $\phi(t_2, t_1)$ according to MC-90.

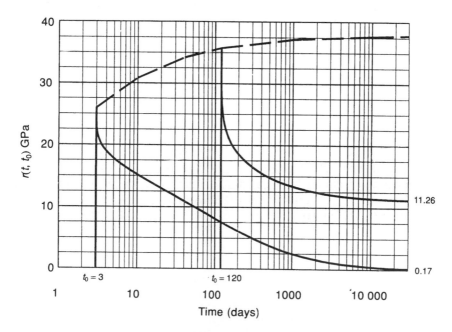

Figure A.5 Example of relaxation function $r(t, t_0)$ and variation of E_c with time: $f_{ck} = 30$ MPa (4500 psi); RH = 50 per cent; $h_0 = 400$ mm (16 in).

Table A.5 List of graphs for $\varphi(t, t_0)$ and $\chi(t, t_0)$ in Figs A.6 to A.45

Characteristic Compressive f_{ck}		Relative humidity RH (per cent)	Notional size h_0 (Eq. A.4)		Figure number
MPa	psi		mm	in	
20	3000	50	100	4	A.6
			200	8	A.7
			400	16	A.8
			1000	40	A.9
		80	100	4	A.10
			200	8	A.11
			400	16	A.12
			1000	40	A.13
30	4500	50	100	4	A.14
			200	8	A.15
			400	16	A.16
			1000	40	A.17
		80	100	4	A.18
			200	8	A.19
			400	16	A.20
			1000	40	A.21
40	6000	50	100	4	A.22
			200	8	A.23
			400	16	A.24
			1000	40	A.25
		80	100	4	A.26
			200	8	A.27
			400	16	A.28
			1000	40	A.29
60	9000	50	100	4	A.30
			200	8	A.31
			400	16	A.32
			1000	40	A.33
		80	100	4	A.34
			200	8	A.35
			400	16	A.36
			1000	40	A.37
80	12000	50	100	4	A.38
			200	8	A.39
			400	16	A.40
			1000	40	A.41
		80	100	4	A.42
			200	8	A.43
			400	16	A.44
			1000	40	A.45

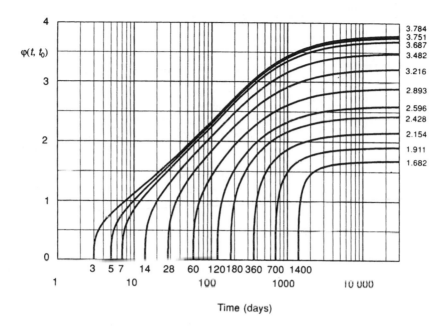

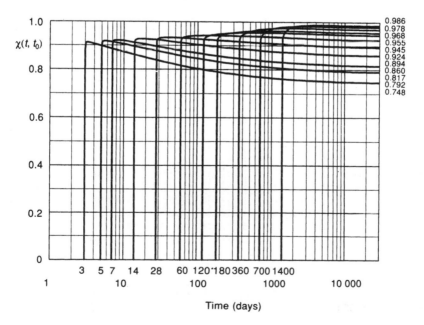

Figure A.6 Creep and aging coefficients of concrete when: $f_{ck} = 20\,\text{MPa}$ (3000 psi); RH = 50 per cent; $h_0 = 100\,\text{mm}$ (4 in).

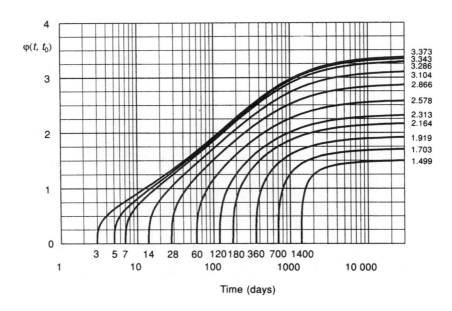

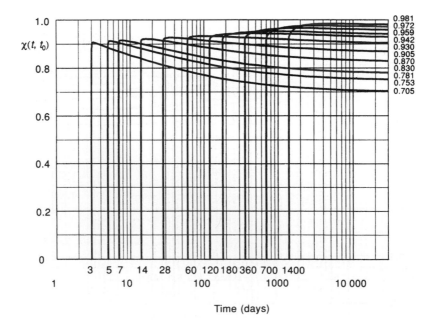

Figure A.7 Creep and aging coefficients of concrete when: $f_{ck} = 20\,\text{MPa}$ (3000 psi); RH = 50 per cent; $h_0 = 200\,\text{mm}$ (8 in).

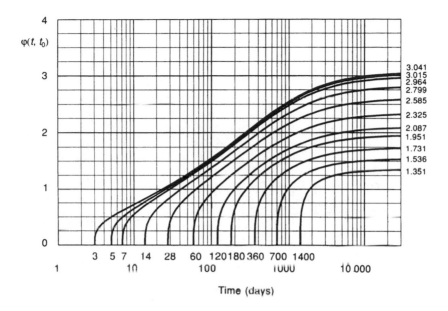

3.041
3.015
2.964
2.799
2.585
2.325
2.087
1.951
1.731
1.536
1.351

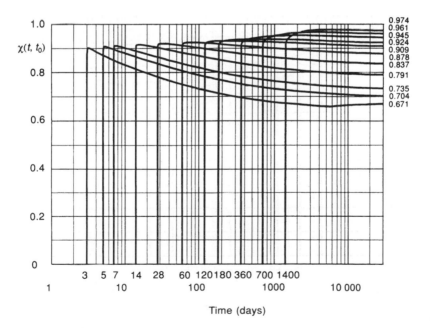

0.974
0.961
0.945
0.924
0.909
0.878
0.837
0.791
0.735
0.704
0.671

Figure A.8 Creep and aging coefficients of concrete when: $f_{ck} = 20\,MPa$ (3000 psi); RH = 50 per cent; $h_0 = 400\,mm$ (16 in).

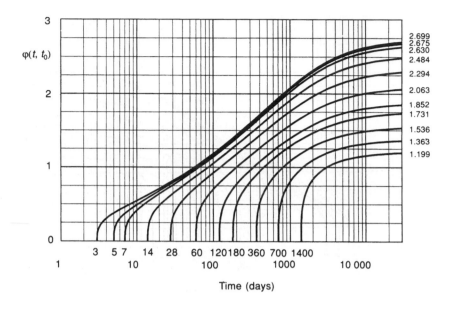

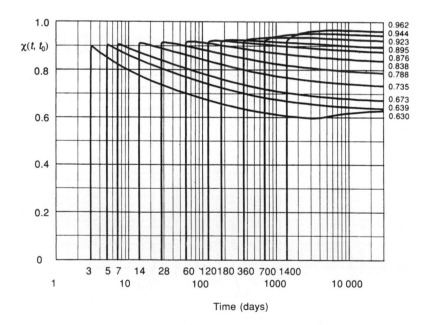

Figure A.9 Creep and aging coefficients of concrete when: f_{ck} = 20 MPa (3000 psi); RH = 50 per cent; h_0 = 1000 mm (40 in).

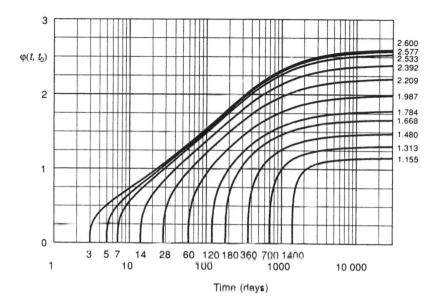

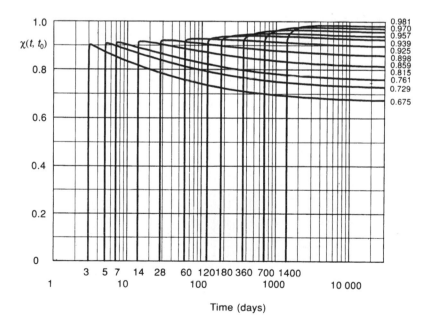

Figure A.10 Creep and aging coefficients of concrete when: $f_{ck} = 20\,\mathrm{MPa}$ (3000 psi); RH = 80 per cent; $h_0 = 100\,\mathrm{mm}$ (4 in).

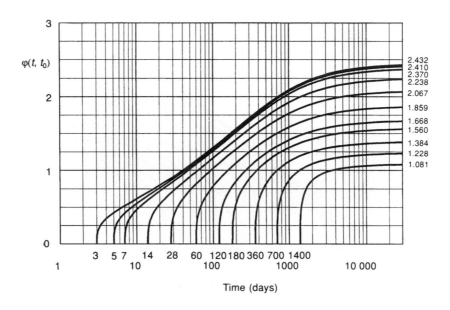

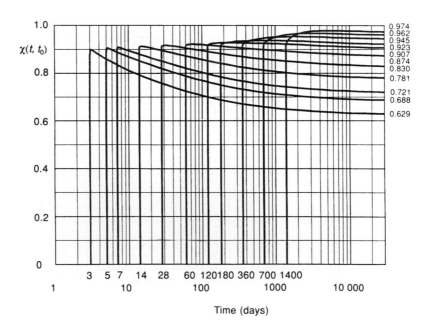

Figure A.11 Creep and aging coefficients of concrete when: $f_{ck} = 20\,\text{MPa}$ (3000 psi); RH = 80 per cent; $h_0 = 200\,\text{mm}$ (8 in).

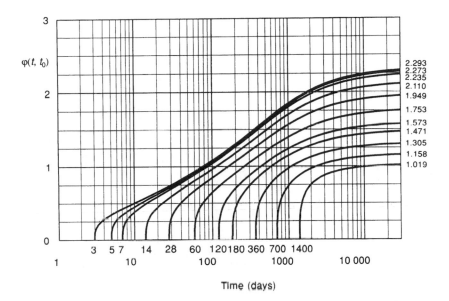

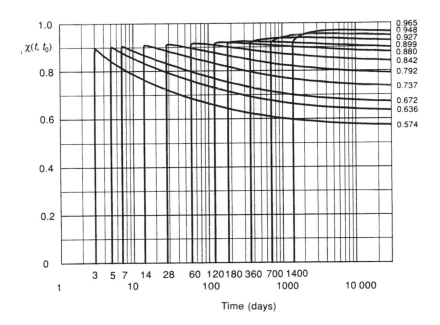

Figure A.12 Creep and aging coefficients of concrete when: $f_{ck} = 20\,\text{MPa}$ (3000 psi); RH = 80 per cent; $h_0 = 400\,\text{mm}$ (16 in).

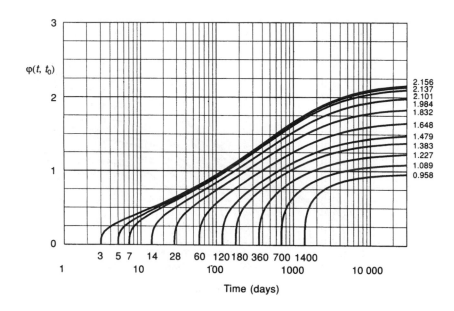

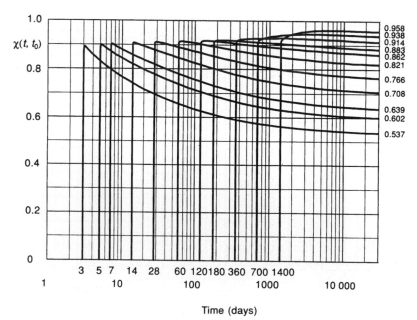

Figure A.13 Creep and aging coefficients of concrete when: $f_{ck} = 20\,MPa$ (3000 psi); RH = 80 per cent; $h_0 = 1000\,mm$ (40 in).

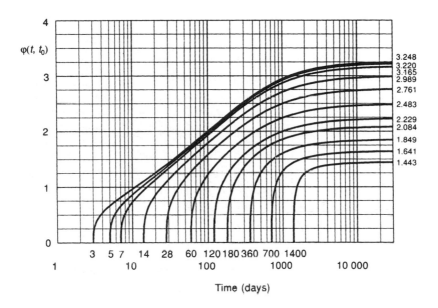

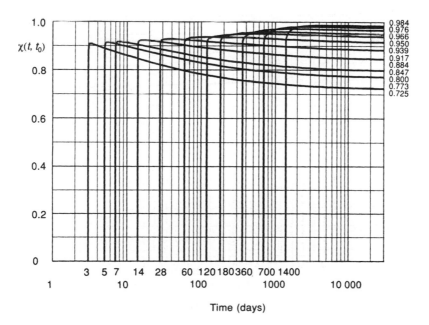

Figure A.14 Creep and aging coefficients of concrete when: $f_{ck} = 30\,\text{MPa}$ (4500 psi); RH = 50 per cent; $h_0 = 100\,\text{mm}$ (4 in).

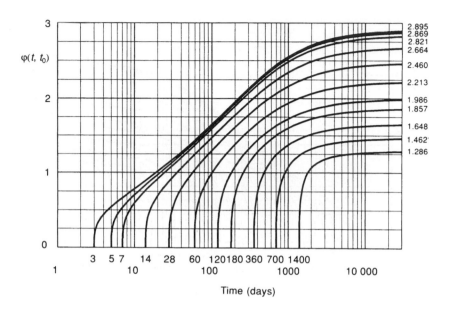

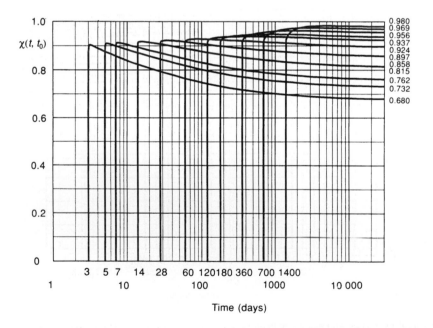

Figure A.15 Creep and aging coefficients of concrete when: $f_{ck} = 30\,\text{MPa}$ (4500 psi); RH = 50 per cent; $h_0 = 200\,\text{mm}$ (8 in).

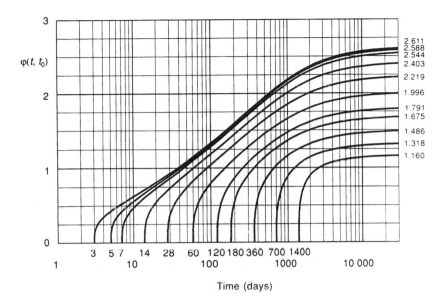

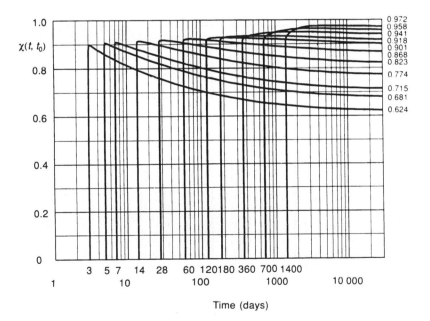

Figure A.16 Creep and aging coefficients of concrete when: $f_{ck} = 30\,\text{MPa}$ (4500 psi); RH = 50 per cent; $h_0 = 400\,\text{mm}$ (16 in).

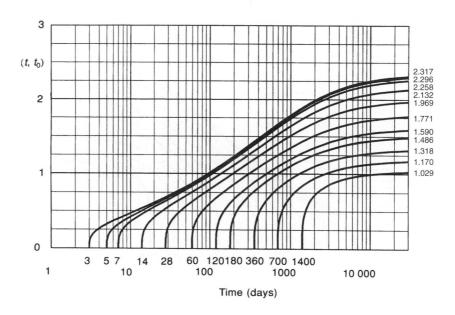

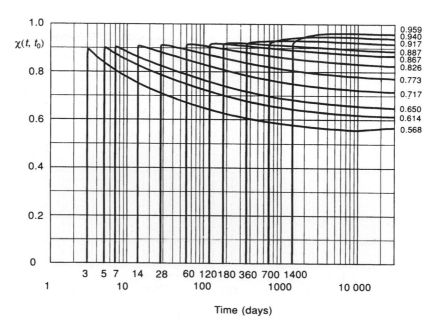

Figure A.17   Creep and aging coefficients of concrete when: $f_{ck} = 30\,\text{MPa}$ (4500 psi); RH = 50 per cent; $h_0 = 1000\,\text{mm}$ (40 in).

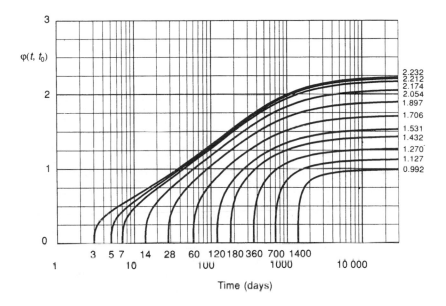

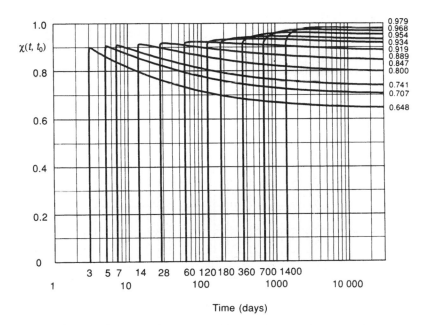

Figure A.18 Creep and aging coefficients of concrete when: $f_{ck} = 30\,\text{MPa}$ (4500 psi); RH = 80 per cent; $h_0 = 100\,\text{mm}$ (4 in).

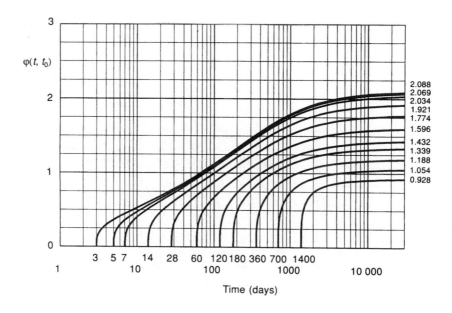

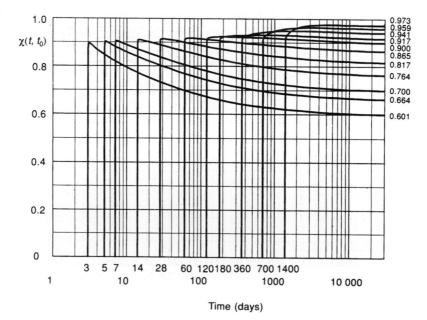

Figure A.19 Creep and aging coefficients of concrete when: $f_{ck} = 30\,\text{MPa}$ (4500 psi); RH = 80 per cent; $h_0 = 200\,\text{mm}$ (8 in).

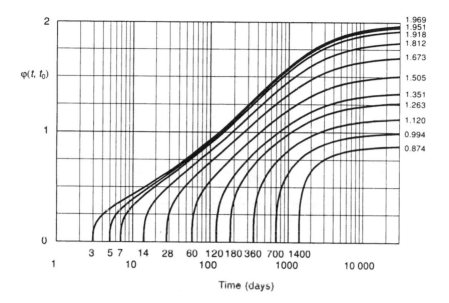

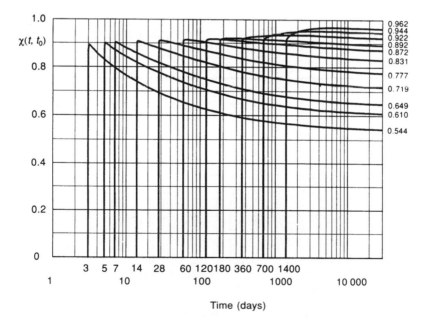

Figure A.20 Creep and aging coefficients of concrete when: $f_{ck} = 30\,\text{MPa}$ (4500 psi); RH = 80 per cent; $h_0 = 400\,\text{mm}$ (16 in).

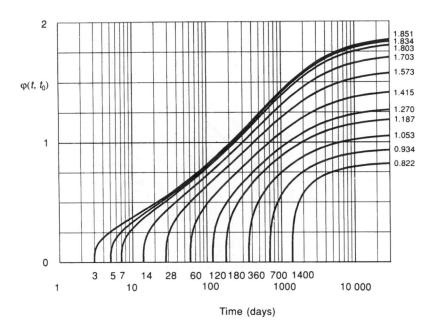

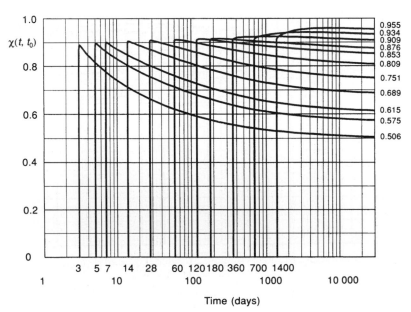

Figure A.21 Creep and aging coefficients of concrete when: $f_{ck} = 30\,\mathrm{MPa}$ (4500 psi); RH = 80 per cent; $h_0 = 1000\,\mathrm{mm}$ (40 in).

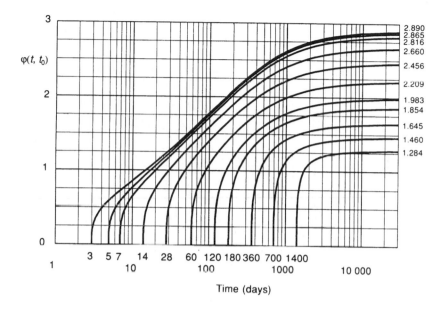

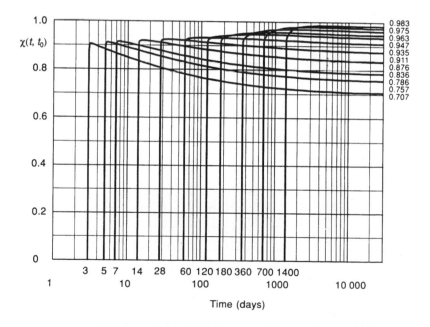

Figure A.22 Creep and aging coefficients of concrete when: $f_{ck} = 40\,MPa$ (6000 psi); RH = 50 per cent; $h_0 = 100\,mm$ (4 in).

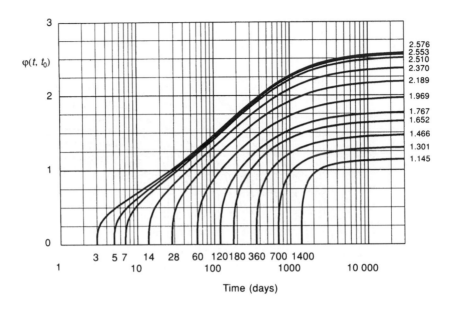

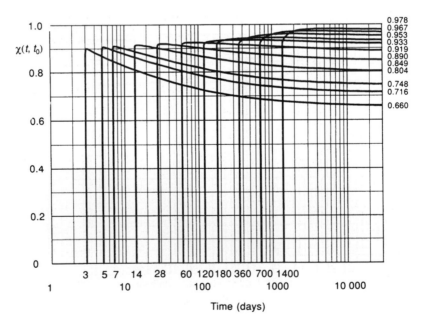

Figure A.23 Creep and aging coefficients of concrete when: $f_{ck} = 40\,MPa$ (6000 psi); RH = 50 per cent; $h_0 = 200\,mm$ (8 in).

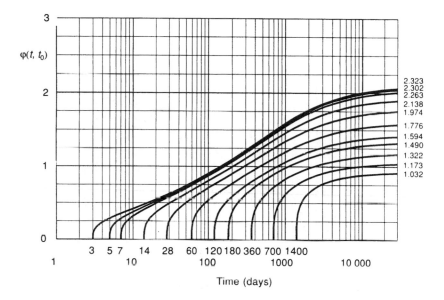

Time (days)

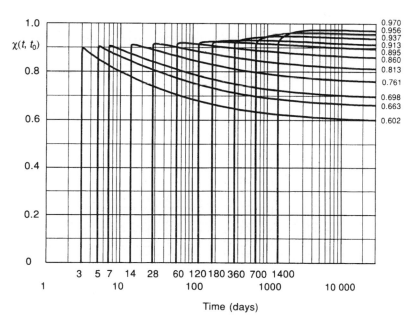

Time (days)

Figure A.24 Creep and aging coefficients of concrete when: $f_{ck} = 40\,MPa$ (6000 psi); RH = 50 per cent; $h_0 = 400\,mm$ (16 in).

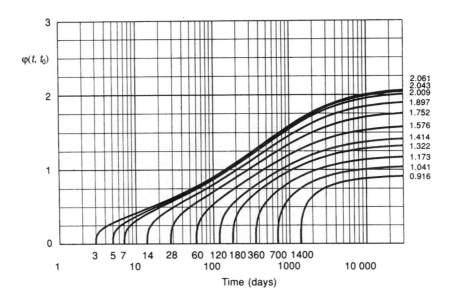

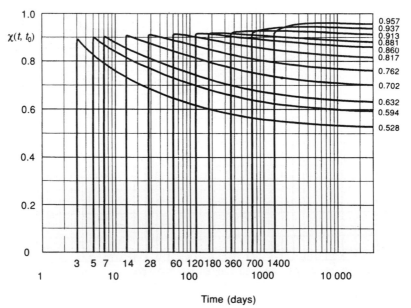

Figure A.25 Creep and aging coefficients of concrete when: $f_{ck} = 40\,\text{MPa}$ (6000 psi); RH = 50 per cent; $h_0 = 1000\,\text{mm}$ (40 in).

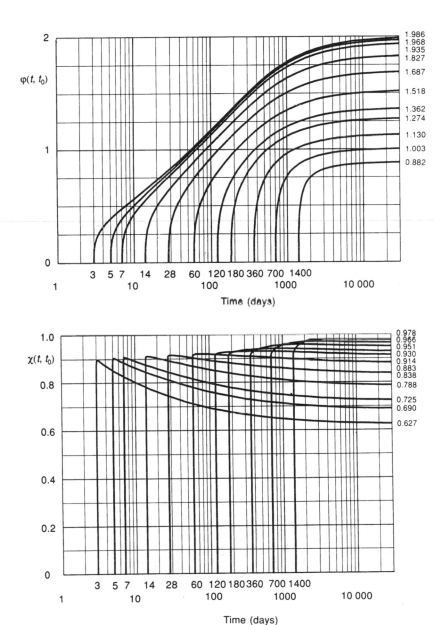

Figure A.26 Creep and aging coefficients of concrete when: $f_{ck} = 40\,\text{MPa}$ (6000 psi); RH = 80 per cent; $h_0 = 100\,\text{mm}$ (4 in).

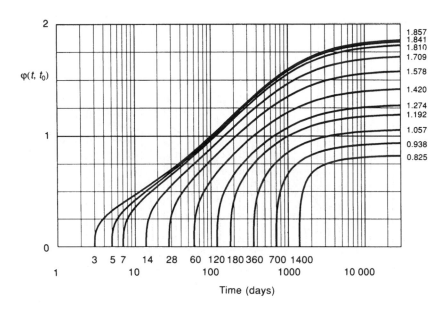

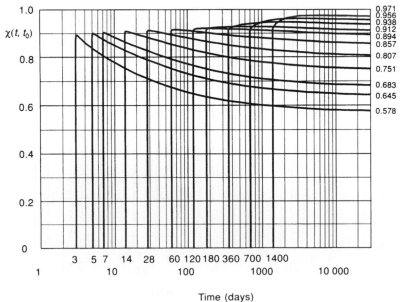

Figure A.27 Creep and aging coefficients of concrete when: $f_{ck} = 40\,\text{MPa}$ (6000 psi); RH = 80 per cent; $h_0 = 200\,\text{mm}$ (8 in).

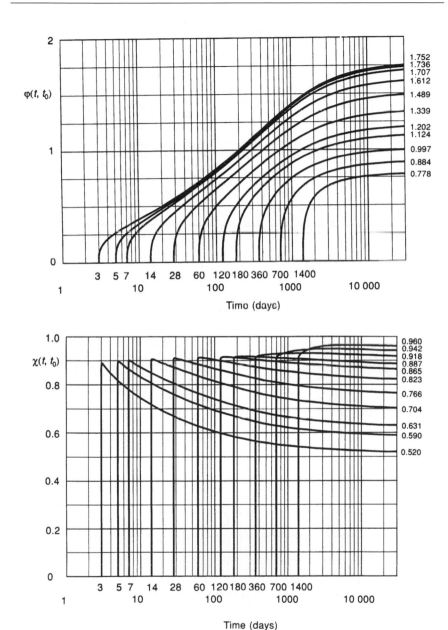

Figure A.28 Creep and aging coefficients of concrete when: $f_{ck} = 40\,\text{MPa}$ (6000 psi); RH = 80 per cent; $h_0 = 400\,\text{mm}$ (16 in).

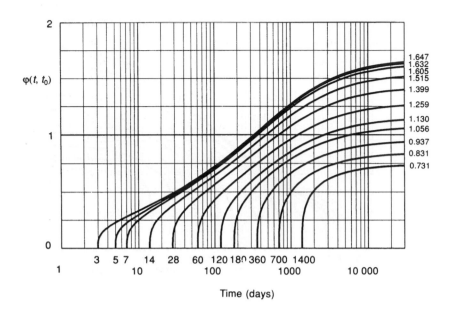

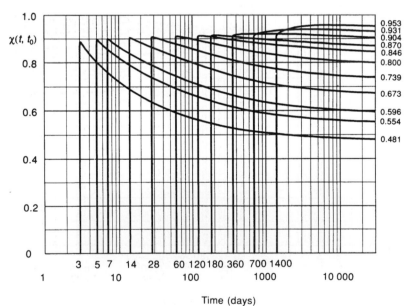

Figure A.29 Creep and aging coefficients of concrete when: $f_{ck} = 40\,\text{MPa}$ (6000 psi); RH = 80 per cent; $h_0 = 1000\,\text{mm}$ (40 in).

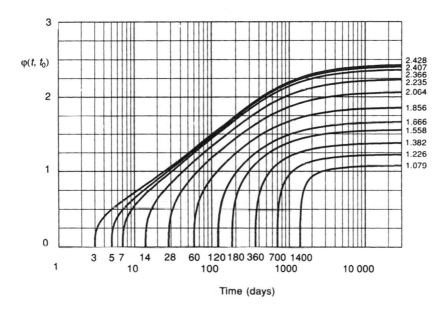

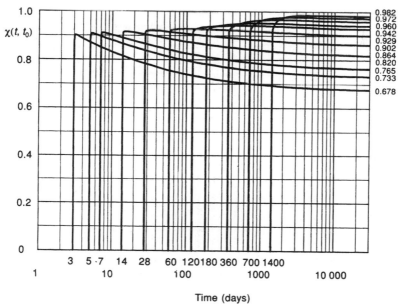

Figure A.30 Creep and aging coefficients of concrete when: $f_{ck} = 60$ MPa (9000 psi); RH = 50 per cent; $h_0 = 100$ mm (4 in).

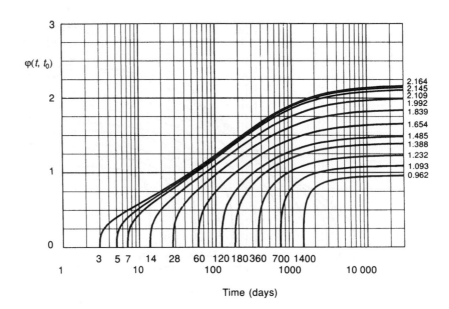

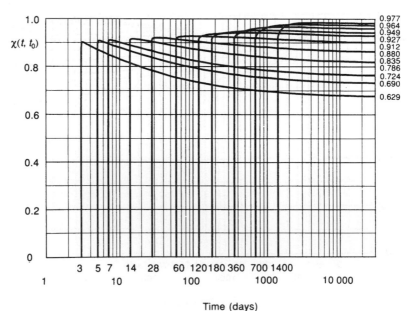

Figure A.31 Creep and aging coefficients of concrete when: $f_{ck} = 60$ MPa (9000 psi); RH = 50 per cent; $h_0 = 200$ mm (8 in).

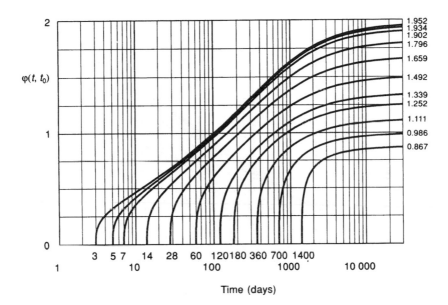

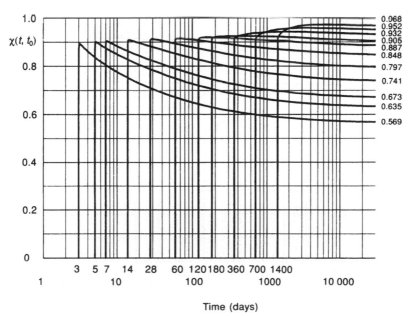

Figure A.32 Creep and aging coefficients of concrete when: $f_{ck} = 60\,\text{MPa}$ (9000 psi); RH = 50 per cent; $h_0 = 400\,\text{mm}$ (16 in).

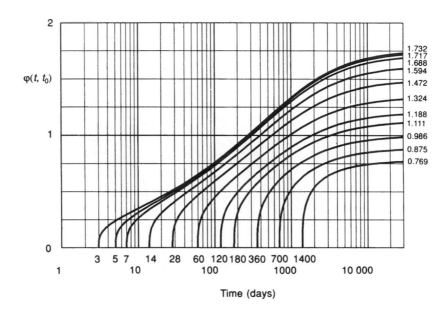

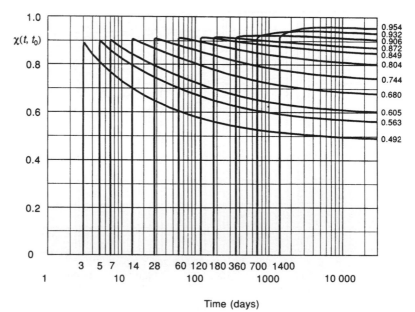

Figure A.33 Creep and aging coefficients of concrete when: f_{ck} = 60 MPa (9000 psi); RH = 50 per cent; h_0 = 1000 mm (40 in).

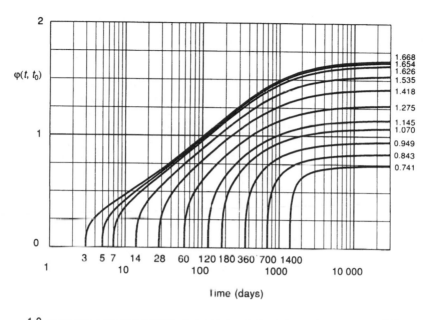

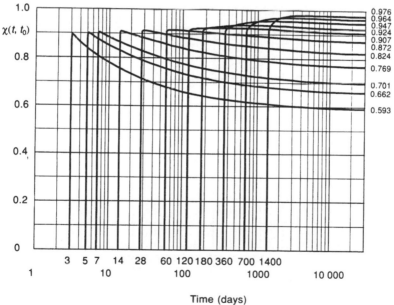

Figure A.34 Creep and aging coefficients of concrete when: f_{ck} = 60 MPa (9000 psi); RH = 80 per cent; h_0 = 100 mm (4 in).

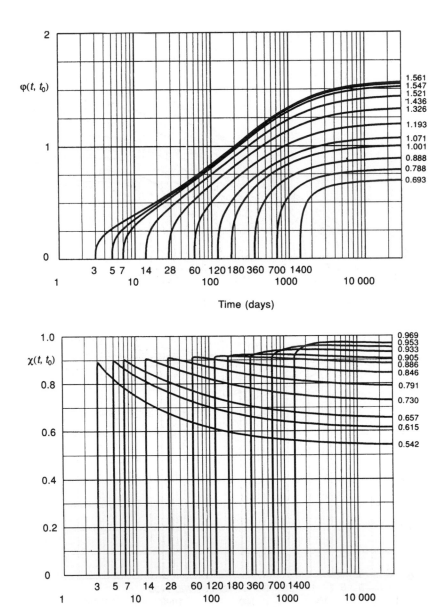

Figure A.35 Creep and aging coefficients of concrete when: $f_{ck} = 60\,\text{MPa}$ (9000 psi); RH = 80 per cent; $h_0 = 200\,\text{mm}$ (8 in).

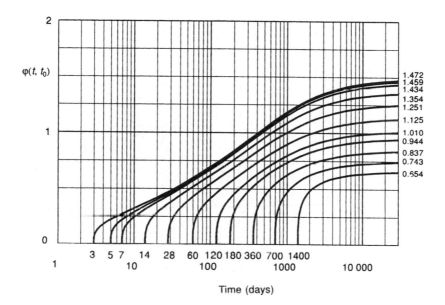

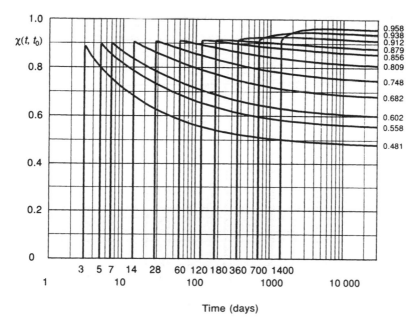

Figure A.36 Creep and aging coefficients of concrete when: $f_{ck} = 60\,\text{MPa}$ (9000 psi); RH = 80 per cent; $h_0 = 400\,\text{mm}$ (16 in).

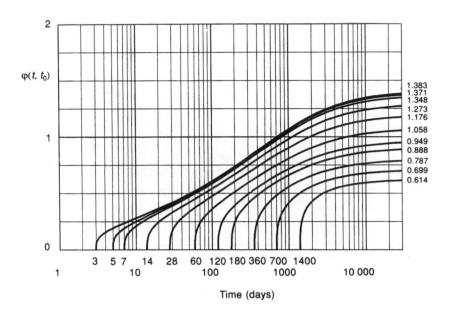

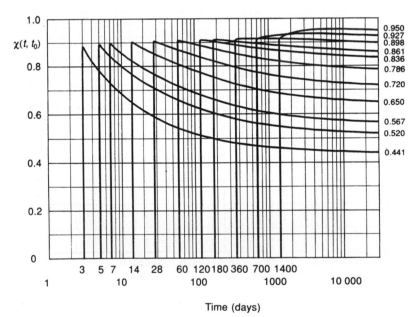

Figure A.37 Creep and aging coefficients of concrete when: $f_{ck} = 60\,\text{MPa}$ (9000 psi); RH $=$ 80 per cent; $h_0 = 1000\,\text{mm}$ (40 in).

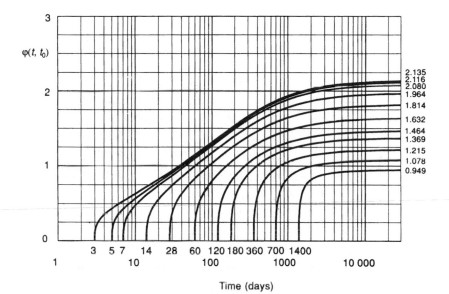

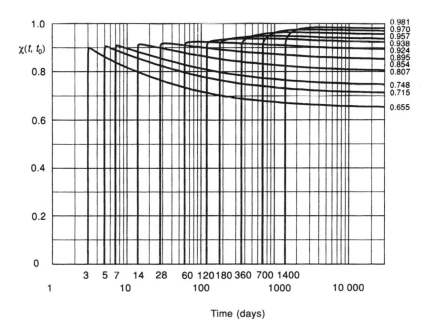

Figure A.38 Creep and aging coefficients of concrete when: $f_{ck} = 80\,$MPa ($12\,000\,$psi); RH = 50 per cent; $h_0 = 100\,$mm ($4\,$in).

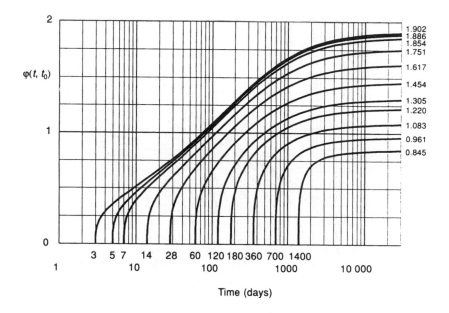

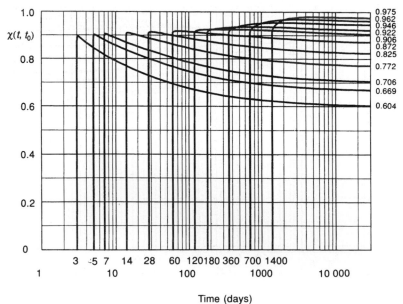

Figure A.39 Creep and aging coefficients of concrete when: $f_{ck} = 80\,\mathrm{MPa}$ (12 000 psi); RH = 50 per cent; $h_0 = 200\,\mathrm{mm}$ (8 in).

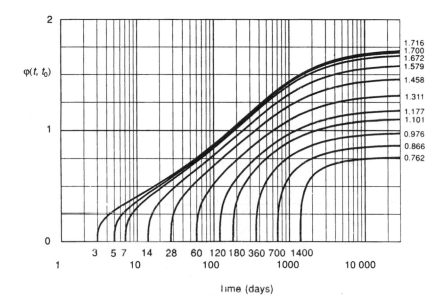

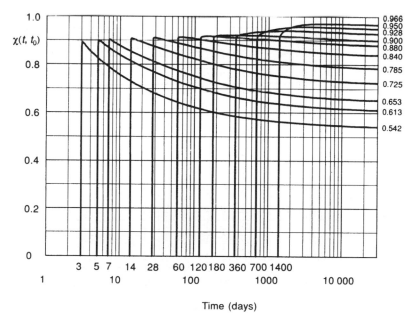

Figure A.40 Creep and aging coefficients of concrete when: $f_{ck} = 80$ MPa (12000 psi); RH = 50 per cent; $h_0 = 400$ mm (16 in).

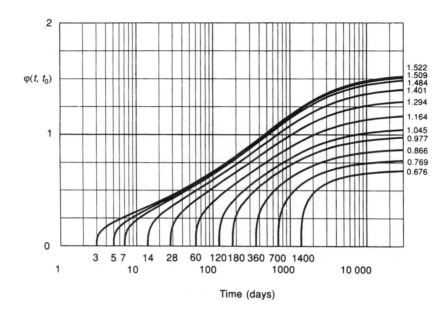

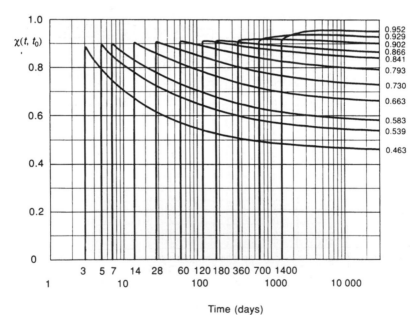

Figure A.41 Creep and aging coefficients of concrete when: $f_{ck} = 80\,\text{MPa}$ (12 000 psi); RH = 50 per cent; $h_0 = 1000\,\text{mm}$ (40 in).

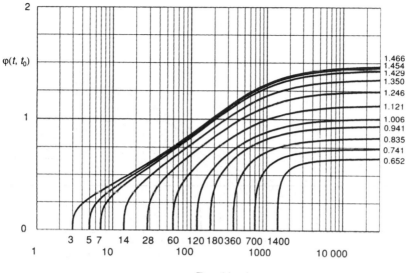

Time (days)

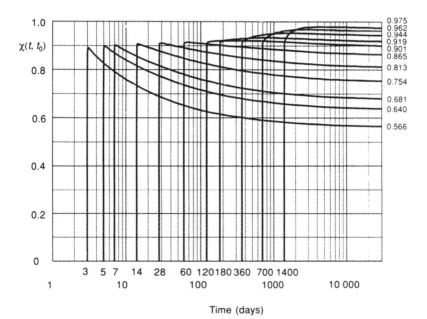

Time (days)

Figure A.42 Creep and aging coefficients of concrete when: $f_{ck} = 80$ MPa (12 000 psi); RH = 80 per cent; $h_0 = 100$ mm (4 in).

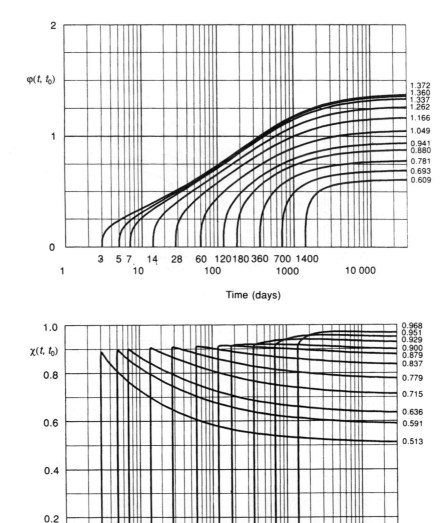

Figure A.43 Creep and aging coefficients of concrete when: $f_{ck} = 80\,\text{MPa}$ (12 000 psi); RH = 80 per cent; $h_0 = 200\,\text{mm}$ (8 in).

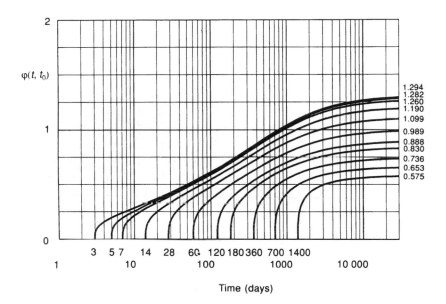

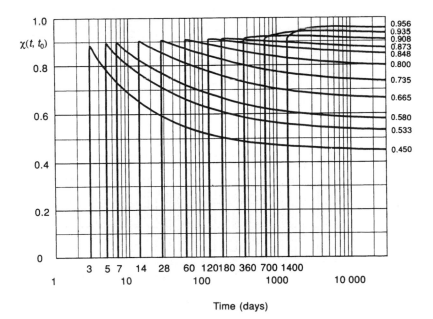

Figure A.44 Creep and aging coefficients of concrete when: $f_{ck} = 80\,\text{MPa}$ (12 000 psi); RH = 80 per cent; $h_0 = 400\,\text{mm}$ (16 in).

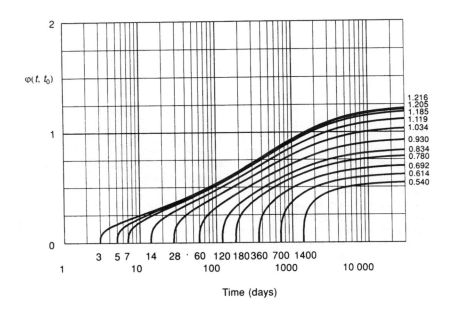

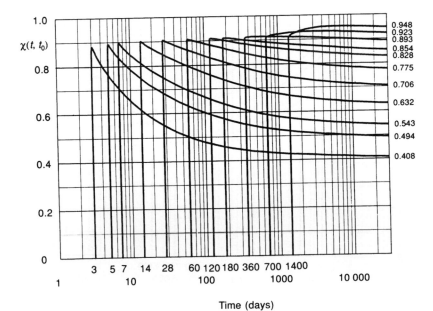

Figure A.45 Creep and aging coefficients of concrete when: $f_{ck} = 80\,\text{MPa}$ (12 000 psi); RH = 80 per cent; $h_0 = 1000\,\text{mm}$ (40 in).

Notes

1 See references mentioned in Note 2, page 19.
2 See reference mentioned in Note 5, page 19.
3 British Standard BS 8110: Part 1: 1997, and Part 2: 1985, *Structural Use of Concrete*, British Standards Institute, 2 Park Street, London W1A 2BS. Part 1 is reproduced by Deco, 15210 Stagg St., Van Nuys, Ca. 91405-1092, USA.
4 See reference in Note 2, page 19.
5 ACI 318(2001), *Building Code Requirements for Reinforcements for Reinforced Concrete*, American Concrete Institute, Detroit, Michigan 48219.
6 Carrasquillo, R.L., Nilson, A.H. and Slate, F.O. (1981), Properties of high-strength concrete subject to short-term loads, *ACI Journal*, **78** (3), 171–8.
7 Bazant, Z.P. (1972), Prediction of concrete creep effects using age-adjusted effective modulus method, *J. Proc. Amer. Concrete Inst.* **69** (4), 212–17.
8 See Note 3, above.
9 Chiorino, M.A. and G. Lacidogna (1991), Approximate values of the aging coefficient for the age-adjusted effective modulus method in linear analysis of creep structures, Report No. 31, Department of Structural Engineering, Politecnico di Torino, Turin, Italy. See also Report No. 35 (1992) by the second author for the more accurate version of Equation (A.49).

Appendix B

Relaxation reduction coefficient χ_r

In a concrete structure, the relaxation of prestressed steel is commonly smaller than the intrinsic relaxation that would occur in a constant-length test with the same initial stress. The coefficient χ_r is a multiplier to be applied to the intrinsic relaxation to obtain a reduced relaxation value to be employed in calculation of the loss of prestress in a prestressed concrete cross-section (see Sections 2.5.2 and 3.2). Values of χ_r are given by the graphs in Fig. 1.4 and Table 1.1 or by Equation (B.11). The equations used for preparation of the graphs and the table are here derived.

Consider a tendon stretched at time t_0 and its length kept constant up to time t. Let $\Delta\sigma_{pr}(t)$ be the intrinsic relaxation during the period $(t - t_0)$; its value depends on the quality of steel and the initial tension σ_{p0} (see Equation (1.5)):

$$\Delta\sigma_{pr}(t) = \begin{cases} -\eta_t\sigma_{p0}(\lambda - 0.4)^2 & \lambda \geq 0.4 \\ 0 & \lambda < 0.4 \end{cases} \tag{B.1}$$

where

$$\lambda = \frac{\sigma_{p0}}{f_{ptk}} \tag{B.2}$$

and f_{ptk} is the characteristic tensile strength; η_t is a dimensionless coefficient depending upon the steel quality and length of the period $(t - t_0)$ (when $(t - t_0)$ is infinity, η_t becomes equal to η; where η is defined in Section 1.4). The minus sign in Equation (B.1) indicates that the relaxation is a reduction of stress; hence a negative increment.

When the tendon is employed to prestress a concrete member, with a stress at time t_0 equal to σ_{p0}, a commonly smaller amount of relaxation occurs during the period $(t - t_0)$ given by

$$\Delta\bar{\sigma}_{pr}(t) = \chi_r\Delta\sigma_{pr}(t) \tag{B.3}$$

where $\Delta\bar{\sigma}_{pr}(t)$ is the reduced relaxation value; χ_r is the relaxation reduction coefficient. It will be shown below that the relaxation reduction coefficient

$$\chi_r = \int_0^1 (1 - \Omega\xi) \left(\frac{\lambda(1 - \Omega\xi) - 0.4)}{\lambda - 0.4} \right)^2 d\xi \qquad (B.4)$$

where Ω is the ratio of (total change in prestress–intrinsic relaxation) to the initial prestress; that is

$$\Omega = - \frac{\Delta\sigma_{ps}(t) - \Delta\sigma_{pr}(t)}{\sigma_{p0}} \qquad (B.5)$$

where $\Delta\sigma_{ps}(t)$ is the change in stress in the prestressed steel during the period $(t - t_0)$ due to the combined effect of creep, shrinkage and relaxation; $\Delta\sigma_{pr}(t)$ is the intrinsic relaxation in the same period.

ξ is a dimensionless time function defining the shape of the stress time curve (Fig. B.1). The value ξ varies between 0 and 1 as τ varies between t_0 and

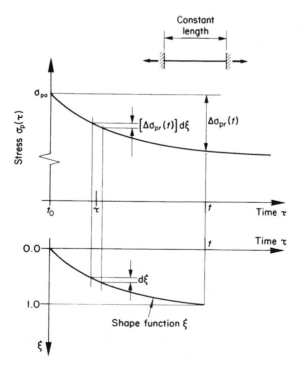

Figure B.1 Stress versus strain in a constant-length relaxation test. Definition of the shape function ξ.

t; where τ represents any intermediate instant. Thus, the intrinsic relaxation at the instant τ is

$$\Delta\sigma_{pr}(\tau) = [\Delta\sigma_{pr}(t)]\xi \qquad (B.6)$$

The prestress loss $\Delta\sigma_{ps}(\tau)$ due to the combined effects of creep, shrinkage and relaxation is assumed to vary during the period t_0 to t following the same shape function; thus,

$$\Delta\sigma_{ps}(\tau) = [\Delta\sigma_{ps}(t)]\xi \qquad (B.7)$$

At any instant τ, the absolute value $|\Delta\sigma_{ps}(\tau) - \Delta\sigma_{pr}(\tau)|$ represents a reduction in tension caused by shortening of the tendon. Thus, the elemental change in relaxation at the instant τ is the same as in a tendon with a reduced initial tension of value:

$$\bar{\sigma}_{p0}(\tau) = \sigma_{p0}(1 - \Omega\xi) \qquad (B.8)$$

Employing Equations (B.6) and (B.1), a differential value of the intrinsic relaxation is expressed as (see Fig. B.1):

$$d[\sigma_{pr}(\tau)] = [-\eta_t\sigma_{p0}(\lambda - 0.4)^2]d\xi \qquad (B.9)$$

Similarly, a differential value of the reduced relaxation

$$d[\bar{\sigma}_{pr}(\tau)] = \{-\eta_t\sigma_{p0}(1 - \Omega\xi)[\lambda(1 - \Omega\xi) - 0.4]^2\}d\xi$$
$$\text{where } \lambda(1 - \Omega\xi) \geq 0.4 \qquad (B.10)$$

The value between curly brackets, obtained by substitution of Equation (B.8) in (B.1) represents the intrinsic relaxation at time t for a tendon with a reduced initial tension $\bar{\sigma}_{p0}(\tau)$. The differential of the reduced relaxation is zero when $\lambda(1 - \Omega\xi) < 0.4$.

Integration of each of Equations (B.10) and (B.9) and then division gives Equation (B.4).

The graphs in Fig. 1.4 and Table 1.1 for the relaxation reduction coefficient χ_r are determined by expressing the integral in Equation (B.4) in a closed form and substitution of chosen values of λ and Ω, noting the above-mentioned restriction to Equation (B.10). The restriction is tantamount to replacing the upper limit of the integral in Equation (B.4) by the smaller of the two values 1 and $[(\lambda - 0.4)/(\lambda\Omega)]$.

The closed form expression for χ_r based on Equation (B.4) is rather lengthy. Instead, the following expression obtained by curve fitting from Fig. 1.4 may be employed as an approximation to the relaxation reduction coefficient:

$$\chi_r = \exp[(-6.7 + 5.3\,\lambda)\Omega] \tag{B.11}$$

In most cases Ω is positive and $\chi_r < 1$. In exceptional situations Ω is negative and $\chi_r > 1$.

The value of the intrinsic relaxation for any type of steel is commonly reported from tests in which a tendon is stretched between two fixed points for a period $(\tau - t_0) = 1000$ hours. The value of the intrinsic relaxation may be expressed as a function of the period $(\tau - t_0)$:

$$\Delta\sigma_{pr}(\tau - t_0) = \begin{cases} \Delta\sigma_{pr\infty}\left[\dfrac{1}{16}\ln\left(\dfrac{\tau - t_0}{10} + 1\right)\right] & \\ & 0 \le (\tau - t_0) \le 1000 \quad (B.12) \\ \Delta\sigma_{pr\infty}\left(\dfrac{\tau - t_0}{0.5 \times 10^6}\right)^{0.2} & \\ & 1000 < (\tau - t_0) \le 0.5 \times 10^6 \quad (B.13) \\ \Delta\sigma_{pr\infty} & (\tau - t_0) \le 0.5 \times 10^6 \quad (B.14) \end{cases}$$

The quantity $(\tau - (t_0)$ is the period in hours for which the tendon is stretched; $\Delta\sigma_{pr\infty}$ is the ultimate intrinsic relaxation. Equations (B.12–14) closely follow the requirements of MC90[1] and FIP report on prestressing steel.[2]

Notes

1 See reference mentioned in Note 2, page 19.
2 Fédération Internationale de la Précontrainte (1976), *Report on Prestressing Steel, Part 1, Types and Properties, FIP/5/3*, Aug., published by Cement and Concrete Association, Wexham Springs, Slough S13 6PL, England.

Appendix C

Elongation, end rotation and central deflection of a beam in terms of the values of axial strain and curvature at a number of sections

A number of geometry relations are given below to express the elongation, the end rotations and the deflection at the middle of a beam in terms of the axial strain ε_O at the centroid and the curvature ψ at a number of equally spaced sections. Fig. C.1(a) defines four coordinates at which the displacements are considered. ε_O and ψ are assumed to be known at three or five sections as shown in Fig. C.1(b) and (c). The variation of ε_O and ψ is assumed to be linear between each two consecutive sections. Parabolic variation is also considered for the three- and five-section systems of Fig. C.1(b) and (c).

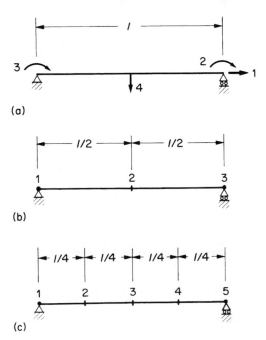

Figure C.1 Coordinates and station points referred to in Equations (C1–16): (a) coordinate system; (b) three sections (Equations (C.1–8)); (c) five sections (Equations (C.9–12) or (C.13–16)).

The equations presented in parts (a) to (e) of this appendix are not limited to the simple beam shown in Fig. C.1(a); they are applicable to any member of a framed structure. Fig. C.2 shows a straight member AB and its deflected

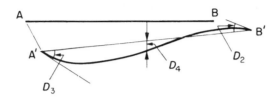

Figure C.2 Original and deformed shape of a member of a framed structure.

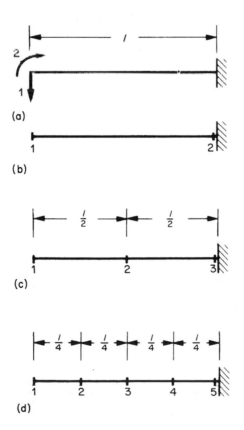

(a)

(b)

(c)

(d)

Figure C.3 Coordinates and station points referred to in Equations (C.17–22): (a) coordinate system; (b) two sections (Equations (C.17),(C.18)); (c) three sections (Equations (C.19),(C.20)); (d) five sections (Equations (C.21),(C.22)).

shape A′B′. The displacement D_1 in this case represents the elongation of the member while D_2 to D_4 are as indicated in the figure.

In parts (e) to (g) of this appendix, equations are given for deflection and rotation at the free end of a cantilever in terms of curvature at a number of sections (Fig. C.3).

Positive ε_O means elongation and positive curvature corresponds to elongation and shortening at the bottom and top fibres, respectively.

(a) Three sections, straight-line variation (Fig. C.1)

$$D_1 = \frac{l}{4}[1\ 2\ 1]\ \{\varepsilon_O\} \tag{C.1}$$

$$D_2 = -\frac{l}{24}[1\ 6\ 5]\ \{\psi\} \tag{C.2}$$

$$D_3 = \frac{l}{24}[5\ 6\ 1]\ \{\psi\} \tag{C.3}$$

$$D_4 = \frac{l^2}{48}[1\ 4\ 1]\ \{\psi\} \tag{C.4}$$

(b) Three sections, parabolic variation (Fig. C.1)

$$D_1 = \frac{l}{6}[1\ 4\ 1]\{\varepsilon_O\} \tag{C.5}$$

$$D_2 = -\frac{l}{6}[0\ 2\ 1]\{\psi\} \tag{C.6}$$

$$D_3 = \frac{l}{6}[1\ 2\ 0]\{\psi\} \tag{C.7}$$

$$D_4 = \frac{l^2}{96}[1\ 10\ 1]\{\psi\} \tag{C.8}$$

(c) Five sections, straight-line variation (Fig. C.1)

$$D_1 = \frac{l}{8}[1\ 2\ 2\ 2\ 1]\{\varepsilon_O\} \tag{C.9}$$

$$D_2 = -\frac{l}{96}[1\ 6\ 12\ 18\ 11]\{\psi\} \tag{C.10}$$

$$D_3 = \frac{l}{96}[11\ 18\ 12\ 6\ 1]\{\psi\} \tag{C.11}$$

$$D_4 = \frac{l^2}{192}[1\ 6\ 10\ 6\ 1]\{\psi\} \tag{C.12}$$

(d) Five sections, parabolic variation between sections 1, 2 and 3 and sections 3, 4 and 5

$$D_1 = \frac{l}{12}[1\ 4\ 2\ 4\ 1]\{\varepsilon_0\} \tag{C.13}$$

$$D_2 = -\frac{l}{12}[0\ 1\ 1\ 3\ 1]\{\psi\} \tag{C.14}$$

$$D_3 = \frac{l}{12}[1\ 3\ 1\ 1\ 0]\{\psi\} \tag{C.15}$$

$$D_4 = \frac{l^2}{24}[0\ 1\ 1\ 1\ 0]\{\psi\} \tag{C.16}$$

(e) Cantilever: two sections, straight-line variation (Fig. C.3)

$$D_1 = -\frac{l^2}{6}[1\ 2]\{\psi\} \tag{C.17}$$

$$D_2 = \frac{l}{2}[1\ 1]\{\psi\} \tag{C.18}$$

(f) Cantilever: three sections, straight-line variation (Fig. C.3)

$$D_1 = -\frac{l^2}{24}[1\ 6\ 5]\{\psi\} \tag{C.19}$$

$$D_2 = \frac{l}{4}[1\ 2\ 1]\{\psi\} \tag{C.20}$$

(g) Cantilever five sections, parabolic variation between sections 1, 2 and 3 and sections 3, 4 and 5

$$D_1 = -\frac{l^2}{12}[0\ 1\ 1\ 3\ 1]\{\psi\} \tag{C.21}$$

$$D_2 = \frac{l}{12}[1\ 4\ 2\ 4\ 1]\{\psi\} \tag{C.22}$$

Appendix D

Depth of compression zone in a fully cracked T section

Equation (7.20) can be solved to give the depth c of the compression zone of a T section (Fig. 7.2), subjected to an eccentric normal force N which produces compression and tension at the top and bottom fibres, respectively. The equation may be rewritten as a cubic polynomial:

$$c^3 + a_1 c^2 + a_2 c + a_3 = 0 \tag{D.1}$$

where

$$a_1 = -3(d_{ns} + e_s) \tag{D.2}$$

$$
\begin{aligned}
a_2 = -\frac{6}{b_w}[h_f(b - b_w)(d_{ns} + e_s - \tfrac{1}{2}h_f) + A'_{ns}(a_{ns} - 1)(d_{ns} + e_s - d'_{ns}) \\
+ a_{ps}A_{ps}(d_{ns} + e_s - d_{ps}) + a_{ns}A_{ns}e_s]
\end{aligned} \tag{D.3}
$$

$$
\begin{aligned}
a_3 = \frac{6}{b_w}[\tfrac{1}{2}h_f^2(b - b_w)(d_{ns} + e_s - \tfrac{2}{3}h_f) + A'_{ns}d'_{ns}(a_{ns} - 1)(d_{ns} + e_s - d'_{ns}) \\
+ a_{ps}A_{ps}d_{ps}(d_{ns} + e_s - d_{ps}) + a_{ns}A_{ns}d_{ns}e_s]
\end{aligned} \tag{D.4}
$$

The symbols related to the geometry of the cross-section and position of the normal force are defined in Fig. 7.2; $a_{ns} = E_{ns}/E_c$ and $a_{ps} = E_{ps}/E_c$ with E_{ns} and E_{ps} being moduli of elasticity of the non-prestressed and the prestressed steel and E_c is the modulus of elasticity of concrete at the time of application of the normal force. The limitations of Equation (7.20) mentioned in Section 7.4.2 also apply to Equations (D.1) to (D.4).

If the section has an additional steel layer of cross-section area A_{nsi} at a distance d_{nsi} below the top edge, additional terms $[a_{ns}A_{nsi}(d_{ns} + e_s - d_{nsi})]$ and $[a_{ns}A_{nsi}d_{nsi}(d_{ns} + e_s - d_{nsi})]$ should be included inside the square brackets in Equations (D.3) and (D.4), respectively. When the added layer is situated in the compression zone, a_{ns} should be substituted by $(a_{ns} - 1)$.

Solution of the cubic equation (D.1) is given by substitution in the following equations[1]:

$$a_4 = a_2 - \frac{a_1^2}{3} \qquad\qquad\qquad (D.5)$$

$$a_5 = 2\left(\frac{a_1}{3}\right)^3 - \frac{a_1 a_2}{3} + a_3 \qquad\qquad\qquad (D.6)$$

$$a_6 = \left(\frac{a_4}{3}\right)^3 + \left(\frac{a_5}{2}\right)^2 \qquad\qquad\qquad (D.7)$$

When a_6 is positive, the cubic equation has only one real solution:

$$c = \left(-\frac{a_5}{2} + \sqrt{a_6}\right)^{\frac{1}{3}} + \left(-\frac{a_5}{2} - \sqrt{a_6}\right)^{\frac{1}{3}} - \frac{a_1}{3} \qquad\qquad (D.8)$$

When any of the quantities between brackets in the first two terms on the right-hand side of this equation is negative, the term should be replaced by $[-$ (absolute value of the quantity) $^{\frac{1}{3}}]$.

When a_6 is zero or negative, the cubic equation has three real solutions, but only one is meaningful (with c between zero and d_{ns}). The three solutions are given by:

$$\cos\theta = -\frac{a_5}{2[-(a_4/3)^3]^{1/2}} \qquad\qquad\qquad (D.9)$$

$$c_1 = 2\left(-\frac{a_4}{3}\right)^{\frac{1}{2}} \cos\left(\frac{\theta}{3}\right) - \frac{a_1}{3} \qquad\qquad\qquad (D.10)$$

$$c_{2,3} = -2\left(-\frac{a_4}{3}\right)^{\frac{1}{2}} \cos\left(\frac{\theta}{3} \pm 60\right) - \frac{a_1}{3}. \qquad\qquad (D.11)$$

Note

1 Korn, G.A. and Korn, T.M. (1968), *Mathematical Handbook for Scientists and Engineers*, 2nd edn, McGraw-Hill, New York, see page 23.

Crack width and crack spacing

E.1 Introduction

Cracks in reinforced and partially prestressed concrete structures are expected to occur, but with adequate and well detailed reinforcement, it is possible to limit the width of cracks to a small value, such that appearance or performance of the structure are not hampered.

Accurate prediction of crack width is not possible. Many equations and methods have been suggested but most are merely empirical rules resulting from observations or testing. Furthermore, there is no agreement on what crack width should be permitted for different types of structures. This appendix discusses the main parameters which affect crack width and give equations which may be used in common situations.

External load applied on a concrete structure produces cracking when the tensile strength of concrete is exceeded. When the reinforcement is designed to provide ultimate strength in accordance with any of the existing codes, load-induced cracks rarely exceed a width of 0.5 mm (0.02 in). Cracks of larger width occur only when the structure is subjected to loads larger than what it is designed for or when there is a misconception of the statical behaviour of the structure which results in yielding of the reinforcement under service loads.

Internal forces and stresses develop due to temperature, shrinkage and settlements of supports only when the movement due to these effects is *restrained*. The magnitude of the forces produced by the *restraint* depend upon the stiffness of the members and hence the forces are much smaller in a cracked structure compared to a structure without cracks. When adequate reinforcement is provided, cracks caused by restraint are generally of small width and the number of cracks increases with the increase in the restrained movement. There is no generally accepted procedure for design of reinforcement necessary to control cracking caused by restraint. One approach is to provide reinforcement at all tension zones of a minimum ratio (See Section 11.6):

$$\rho_{min} = \frac{f_{ct}}{f_{sy}} \tag{E.1}$$

where f_{ct} is the tensile strength of concrete; f_{sy} is the yield strength of steel. This equation is based on the assumption that the tensile force carried by the concrete immediately before cracking is transmitted to the reinforcement causing stress which does not exceed its yield strength. With $f_{ct} = 2\,MPa$ and $f_{sy} = 400\,MPa$ (0.3 and 60 ksi), Equation (E.1) gives $\rho_{min} = 0.005$.[1]

Cracking can also occur due to causes other than what is discussed above. Much wider cracks can occur during the first few hours after placing of concrete, while it is in a plastic state. These are caused by shrinkage or by settlement of the plastic concrete in the forms. Cracks occur when movement of concrete is restrained by the reinforcement or by the formwork. Plastic cracking cannot be controlled by provision of reinforcement; it can only be achieved by attention to mix design and avoidance of conditions which produce rapid drying during the first hour after placing. This type of crack is not discussed any further below.

Permissible crack width varies with design codes. Acceptable values vary between 0.1 and 0.4 mm. The smaller value may be suitable for water-retaining structures and the larger value for structures in dry air or with protective membrane.

The width of cracks depends mainly on stress in steel after cracking. Other factors affecting crack width are thickness of concrete cover to reinforcement, diameter of bars, their spacing and the way they are arranged in the cross-section, bond properties of the bars, concrete strength and the shape of strain distribution. Load-induced cracks, unlike displacement-induced cracks (Section 11.3), increase in width with the duration of loading.

In Section 8.3–6 the following expression was derived for the average width of cracks which run in a direction perpendicular to the main reinforcement in members subjected to an axial force, bending moment or both (Equation (8.48):

$$w_m = s_{rm}\zeta\varepsilon_{s2} \tag{E.2}$$

where s_{rm} is the spacing between cracks; this will be discussed in the following section. ε_{s2} is the steel strain calculated for a transformed section in which the concrete in tension is ignored (state 2); ζ is a dimensionless coefficient between 0 and 1, representing the effect of the participation of concrete in the tension zone to the stiffness of the member, the so-called tension stiffening effect. The product $(\zeta\varepsilon_{s2})$ represents average excess in strain in the reinforcement relative to the surrounding concrete. (Further explanation of the meaning of the symbol ε_{s2} and its calculation are given in Section 8.6.1.)

The value of the coefficient ζ depends upon the ratio (N_r/N) or (M_r/M); where N and M are the values of the axial normal force or bending moment on the section; the subscript r refers to the value of N and M which produces tensile stress f_{ct} at the extreme fibre.

E.2 Crack spacing

A semi-empirical equation is presented below for prediction of spacing between transverse cracks in members subjected to axial force or bending.[2]

Fig. E.1 shows a reinforced concrete member subjected to an axial force of magnitude just sufficient to produce the first crack. At the cracked section the stress in concrete is zero (state 2) and the axial force is carried entirely by steel. At some distance s_{r0} from the crack, the cross-section is in uncracked state 1 and the stress in concrete is f_{ct}, the strength of concrete in tension; the force in steel at this section is only a fraction of the axial force. The remaining part of the force is transmitted to the concrete by bond stress over the length s_{r0}. Assuming f_{bm} is the average value of the bond stress, we can write:

$$A_c f_{ct} = s_{r0} f_{bm} \left(\frac{4A_s}{d_b} \right) \tag{E.3}$$

where A_c and A_s are the cross-section areas of concrete and steel; the quantity $(4A_s/d_b)$ is the sum of bar perimeters assuming that the bars have equal diameter d_b.

For a given type of reinforcement, the bond stress f_{bm} may be considered proportional to f_{ct}; thus

$$\kappa_1 = \frac{f_{ct}}{f_{bm}} \tag{E.4}$$

where κ_1 is a dimensionless coefficient depending upon bond properties of the reinforcing bars.

Substitution of Equation (E.4) into (E.3) gives

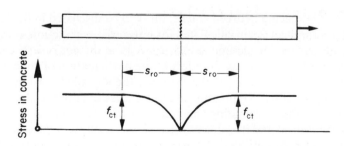

Figure E.1 Stress in concrete after first crack in a member subjected to axial force.

$$s_{r0} = \kappa_1 \frac{d_b}{4\rho} \qquad\qquad (E.5)$$

The symbol s_{r0} represents the distance between the first crack and the cross-section at which the concrete stress reaches f_{ct}. Subsequent small increase in applied force causes second and third cracks to occur at a distance s_{r0} on either side of the first crack, and so on until a so-called *stabilized crack pattern* is obtained. Further increase in load does not produce new cracks.

Restraint, which occurs when a member with fixed ends attempts to shorten due to shrinkage or temperature drop, may produce only few cracks, so that a stabilized state of cracking does not usually occur. This is because cracking is associated with a reduction in stiffness and hence alleviation of restraining forces.

Experiments indicate that crack spacing is affected by other parameters not included in Equation (E.5), namely the concrete cover and the spacing of bars. For this reason, Equation (E.5) is empirically modified in practice.

EC2–91 and MC-90[3] give equations for the characteristic maximum crack width, w_k. The two codes consider the value $w_k = 0.30$ mm (0.012 in) under quasi-permanent loading as satisfactory for reinforced concrete members (without prestressing). This limit may be relaxed when the exposure conditions are such that crack width has no influence on durability (for example the interior of buildings for habitation or offices). A lower limit for w_k should be specified in accordance with the client when de-icing agents are expected to be used on top of tensioned zones.

For prestressed concrete members, the two codes limit, in general, the value of the characteristic crack width to $w_k = 0.2$ mm (0.008 in); furthermore, for certain exposure conditions, it is required that under frequent load combinations, the prestressed tendons lie at least 25 mm (1 in) within concrete in compression, or no tension is allowed within the section.

The two codes differ in the equation to be used in calculation of w_k as given below. Provisions of ACI318-89 code are also discussed.

E.3 Eurocode 2–1991 (EC2–91)

The EC2–91 employs Equation (E.2) to calculate the average crack width, w_m; but the code defines the design or characteristic maximum crack width, w_k, as:

$$w_k = \beta w_m. \qquad\qquad (E.6)$$

For load-induced cracking, the value of the coefficient β to be used in Equation (E.6) is $\beta = 1.7$ or 1.3, respectively, for sections whose minimum dimension exceeds 800 mm (30 in) or is smaller than 300 mm (12 in).

According to EC2–91, the average crack spacing s_{rm} (mm), to be used in Equation (E.2) is:

$$s_{rm} = 50 + \kappa_1 \kappa_2 \frac{d_b}{4\rho_r} \tag{E.7}$$

where

d_b = bar diameter (mm)

κ_1 = coefficient depending upon bond quality; $\kappa_1 = 0.8$ for high bond bars and 1.6 for plain bars. When cracking is due to restraint of intrinsic imposed deformations (for example restraint of shrinkage), the coefficient κ_1 is to be replaced by $0.8\,\kappa_1$. The multiplier 0.8 should generally be used; but for rectangular sections of height h the multiplier should be equal to 0.8 for $h \leqslant 0.3\,\text{m}$ (12 in) and equal to 0.5 for $h \geqslant 0.8\,\text{m}$ (30 in)

κ_2 = coefficient depending upon the shape of the strain diagram; $\kappa_2 = 0.5$ in the case of bending without axial force; $\kappa_2 = 1.0$ in the case of axial tension. In the case of eccentric tension,

$$\kappa_2 = \frac{\varepsilon_1 + \varepsilon_2}{2\varepsilon_1} \tag{E.8}$$

where ε_1 is the greater and ε_2 the lesser tensile strain values (assessed on the basis of fully cracked section) at upper and lower boundaries of the effective tension area A_{cef} defined in Fig. E.2. The steel ratio ρ_r is defined as:

$$\rho_r = \frac{A_s}{A_{cef}} \tag{E.9}$$

The effective tension area is generally equal to 2.5 times the distance from the tension face of the section to the centroid of A_s (see Fig. E.2); but the height of the effective area should not be greater than $(h - c)/3$; where h is the height of the section and c is the depth of the compression zone.

E.4   CEB-FIP 1990 (MC-90)

MC-90 gives the following equation for calculation of the design crack width:

$$w_k = l_{s\,max}(\varepsilon_{s2} - \beta\varepsilon_{sr2} - \varepsilon_{cs}) \tag{E.10}$$

where

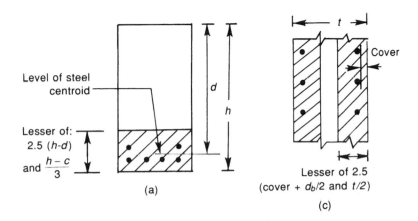

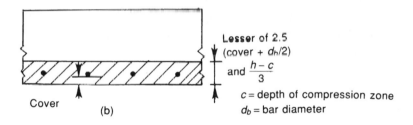

Figure E.2 Effective area, A_{cef} for use in Equations (E.9 and E.13): (a) beam; (b) slab;
(c) member in tension (reference MC-90 or EC–91).

ε_{cs} = the free shrinkage of concrete, generally a negative value.
ε_{sr2} = the steel strain at the crack, under a force causing stress equal to $f_{ctm}(t)$; within A_{cef}

$$\varepsilon_{sr2} = \frac{f_{ctm}(t)}{\rho_r E_s}(1 + a\rho_r)$$ (E.11)

$f_{ctm}(t)$ = the mean value of the tensile strength at time t at which the crack occurred (Equations (A.13) and (A.14))

$$a = E_s/E_c(t)$$ (E.12)

ρ_r and A_{cef} are defined in Equation (E.9) and Fig. E.2.

ε_{s2} = steel strain at the crack.
$l_{s\,max}$ = the length over which slip between steel and concrete occurs. This length is given by Equation (E.13) or Equation (E.14) for stabilized cracking and for single crack formation, respectively:

$$l_{s\,max} = \frac{d_b}{3.6\,\rho_r} \tag{E.13}$$

$$l_{s\,max} = \frac{\sigma_{s2}d_b}{2\tau_{bk}(1 + a\rho_r)} \tag{E.14}$$

d_b = diameter of reinforcing bar.

σ_{s2} = steel stress at the crack.
τ_{bk} = bond stress given in Table E.1 (assuming deformed bars are used).
β = empirical coefficient to assess the average strain within $l_{s\,max}$. The value of β is given in Table E.1 (assuming that deformed bars are used).

E.5   ACI318-89 and ACI318-99

The American Concrete Institute, Building Code Requirements for Reinforced Concrete, ACI318-89[4] controls flexural cracking by limiting the stress in steel at a cracked section due to service load to 60 per cent of the specified yield strength. Alternatively, the parameter z, defined by Equation (E.15) must not exceed 175 or 145 kip/in (30.6×10^6 or 25.4×10^6 N/m) for interior and exterior exposure, respectively. The parameter z is defined as:

$$z = f_s \sqrt[3]{d_c A} \quad \text{(force/length)} \tag{E.15}$$

where

f_s = stress in reinforcement at service load; this is to be calculated for a fully cracked section (state 2).
d_c = thickness of concrete cover measured from extreme tension fibre to centre of bar located closest thereto (Fig. E.3(a)).
A = effective tension area of concrete surrounding the flexural tension reinforcement and having the same centroid as that reinforcement, divided by the number of bars (Fig. E.3(a)). When the flexural reinforcement consists of different bar sizes, the number of bars is to be computed as the total area of reinforcement divided by the area of the largest bar or wire used.

Table E.1 Value of β and τ_{bk} for use in Equations (E.10) and (E.14)

	Single-crack formation		Stabilized cracking	
	β	τ_{bk}	β	τ_{bk}
Short-term or instantaneous loading	0.6	1.8 $f_{ctm}(t)$	0.6	1.8 $f_{ctm}(t)$
Long-term or repeated loading	0.6	1.35 $f_{ctm}(t)$	0.38	1.8 $f_{ctm}(t)$

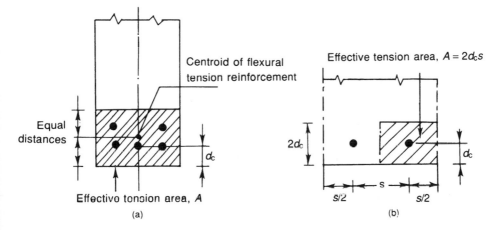

Figure E.3 Definitions of symbols A, d_c and s for use in Equations (E.15) and (E.19): (a) beam for 5 bars; (b) slab (references ACI 318-89 and ACI 224-86).

Equation (E. 15) is based on the Gergely–Lutz[5] expression for maximum crack width, w_{max} (Equation (E.16)), corresponding to limiting crack widths of 0.016 and 0.013 in (0.40 and 0.33 mm). The Gergely–Lutz equation predicts the maximum crack width as:

$$w_{max} = (76 \times 10^{-6})\, \beta z \text{ (kip-in units)} \tag{E.16}$$

$$w_{max} = (11 \times 10^{-12})\, \beta z \text{ (N-m units)} \tag{E.17}$$

where β is the ratio of the distances from the neutral axis to the extreme tension fibre and to the centroid of the main reinforcement. The limiting values for z and w_{max} given above are based on an average value $\beta = 1.2$ which applies for beams. For slabs, the average value of $\beta \approx 1.35$; thus for consistency, the maximum values for z are to be reduced by the ratio 1.2/1.35.

Derivation of Equation (E.16) involves the assumption that the maximum crack spacing is:[4]

$$S_{rm} = 4t_e \tag{E.18}$$

where t_e is an increased effective cover defined as:

$$t_e = d_c \sqrt{1 + \left(\frac{s}{4d_c}\right)^2} \tag{E.19}$$

where s is the bar spacing (Fig. E.3(b)).

The American Concrete Institute code ACI 318-99[6] replaces the requirement for the parameter z presented above by setting a limit to the spacing, s between bars in the zone of maximum tension in a cross-section as the smaller of:

$$s(\text{in}) = \frac{540}{\sigma_s\,(\text{ksi})} - 2.5\,(\text{cover (in.)}) \tag{E.20}$$

$$s(\text{mm}) = \frac{95}{\sigma_s\,(\text{MPa})} - 2.5\,(\text{cover (mm)}) \tag{E.21}$$

and

$$s(\text{in}) = \frac{12(36)}{\sigma_s\,(\text{ksi})} \tag{E.22}$$

$$s(\text{mm}) = \frac{76}{\sigma_s\,(\text{MPa})} \tag{E.23}$$

where σ_s is stress in the reinforcement at service, computed as the unfactored moment divided by the product of the steel area and the internal moment arm. The code permits to take the stress in steel as 60 percent of specified yield strength.

The parameter z in earlier ACI codes was based on empirical equations using a calculated crack width of 0.4 mm (0.016 in.). The ACI 318R-99 code commentary recognizes that crack widths are highly variable and adopts the Equations E.20 to E.23 that intend to control surface cracks to a width that is generally acceptable in practice, but may vary widely in a given structure. At the same time the ACI 318-99 code states that the bar spacing requirement is not sufficient and requires special investigations and precautions for structures subject to very aggressive exposure or designed to be watertight.

Similar to the commentary of the earlier code, ACI 318R-99 states that control of cracking is particularly important when reinforcement with a yield stress in excess of 40 ksi (300 MPa) is used. The commentary lists references to laboratory tests, involving deformed bars, that confirm that crack width at service load is proportional to steel stress.

E.6   British Standard BS 8110

British Standard BS 8110, Part 2, 1985[7] gives Equation (E.24) for 'design width' of flexural crack at a particular point on the surface of a member. The equation gives the design width of crack with acceptably small chance of being exceeded; actual cracks occasionally exceeding this width are

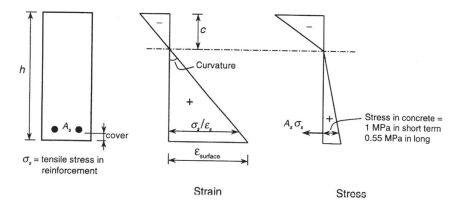

Figure E.4 Assumptions in calculation of strain and stress distributions according to British Standard BS 8110.

considered acceptable. Provided the strain in the tension reinforcement is less or equal to $0.8 f_y / E_s$, the design crack width at surface may be calculated by:

$$w_{\text{at service}} = \frac{3 a \varepsilon_{\text{surface}}}{1 + 2\left(\dfrac{a - \text{cover}}{h - c}\right)} \tag{E.24}$$

f_y = specified characteristic strength of reinforcement.
E_s = modulus of elasticity of reinforcement.
a = distance from point considered to the surface of the nearest longitudinal bar.
h = height of section.
c = depth of compression zone.
$\varepsilon_{\text{surface}}$ = strain at the tension face (Fig. E.4).

In assessing $\varepsilon_{\text{surface}}$ assume that the stress in the tension zone is as shown in Fig. E.4 and take the modulus of elasticity of concrete, E_c half the instantaneous value. Where shrinkage is abnormally high ($> 600 \times 10^{-6}$), $\varepsilon_{\text{surface}}$ should be increased by 50 per cent of the expected shrinkage; otherwise shrinkage may be ignored. This approach makes a notional allowance for long-term effects.

Notes

1 This steel ratio which may be adopted for water-retaining or structures exposed to weather is relatively high compared to the value of 0.002 or 0.0018 required by ACI 318-89 Code for shrinkage and temperature reinforcement at right angles to the

principal reinforcement in structural slabs. See *Building Code Requirements for Reinforced Concrete, ACI 318-89*, American Concrete Institute, Detroit, Michigan 48219 (Section 7.12).

2 See reference mentioned in Note 2, page 19.

3 See references mentioned in Notes 2 and 3, page 19.

4 See reference mentioned in Note 9, page 406.

5 See ACI Committee 224, *Cracking for Concrete Members in Direct Tension*, ACI224.2R-86, American Concrete Institute, Farmington Hills, Michigan 48333-9094, USA.

6 ACI 318 (1999), *Building Code Requirements for Structural Concrete* (318-99) and Commentary (318R-99), American Concrete Institute, Farmington Hills, Michigan 48333-9094.

7 See Note 3, page 533.

Appendix F

Values of curvature coefficients κ_s, κ_φ and κ_{cs}

Equations (9.1) to (9.4) give the instantaneous curvature at time t_0 and the changes in curvature during a period t_0 to t, caused by creep and shrinkage at a reinforced concrete section, without prestress subjected to a bending moment M applied at t_0. The equations include curvature coefficients κ_s, κ_φ and κ_{cs} which are evaluated in this appendix.

Figs F.1 to F.10 give values of κ_s, κ_φ and κ_{cs} for rectangular sections; the additional subscript 1 or 2 is used to refer to the two states of no cracking and full cracking, respectively.

For a general cross-section, the curvature coefficients may be calculated by the following expressions which can be derived from comparison of Equations (9.1) to (9.4) with Equations (2.16) and (3.16):

$$\kappa_s = \frac{I_g}{I} \tag{F.1}$$

$$\kappa_\varphi = \frac{I_c + A_c y_c \Delta y}{\bar{I}} \tag{F.2}$$

$$\kappa_{cs} = -\frac{A_c y_c d}{\bar{I}} \tag{F.3}$$

The above equations are applicable to uncracked and cracked sections (see Equation (6.27)): for this reason, the subscripts 1 and 2 are omitted. The symbols in the equations are defined below:

κ_s = coefficient smaller than unity which represents the stiffening effect of the presence of reinforcement on instantaneous curvature (Equation (9.1))

κ_φ = coefficient, smaller than unity, representing the restraining effect of reinforcement on creep curvature (Equation (9.2)).

κ_{cs} = coefficient to be used in Equation (9.3) for the curvature due to shrinkage

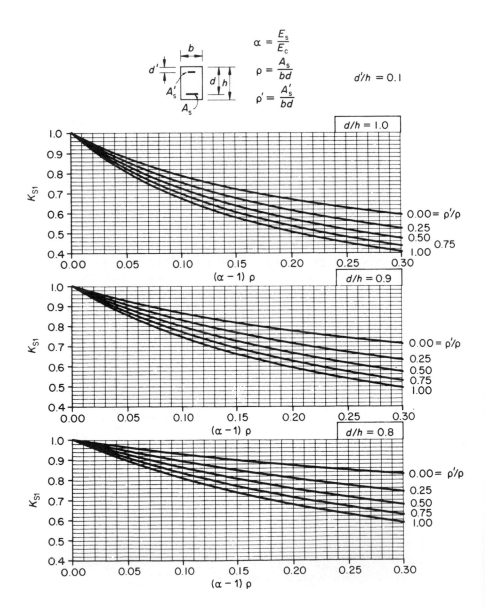

Figure F.1 Curvature coefficient K_{s1} for rectangular uncracked sections.

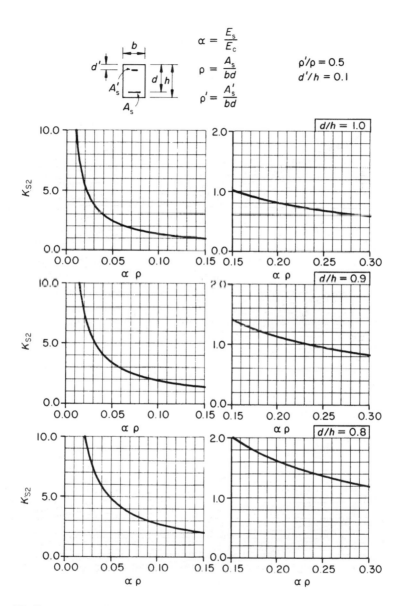

Figure F.2 Curvature coefficient K_{s2} for rectangular cracked sections.

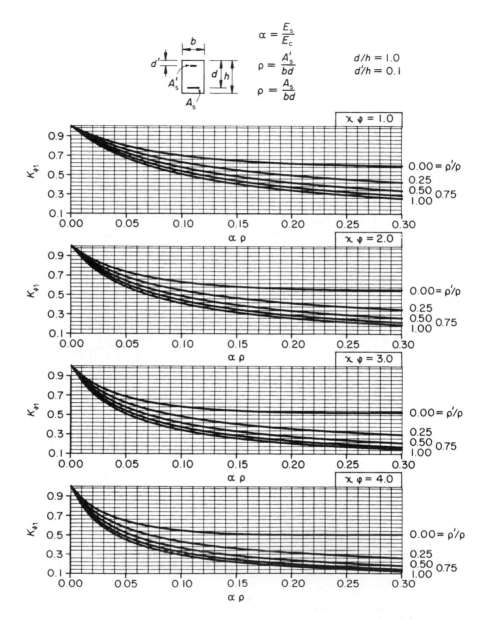

Figure F.3   Curvature coefficient $\kappa_{\varphi 1}$ for rectangular uncracked sections; $d/h = 1.0$.

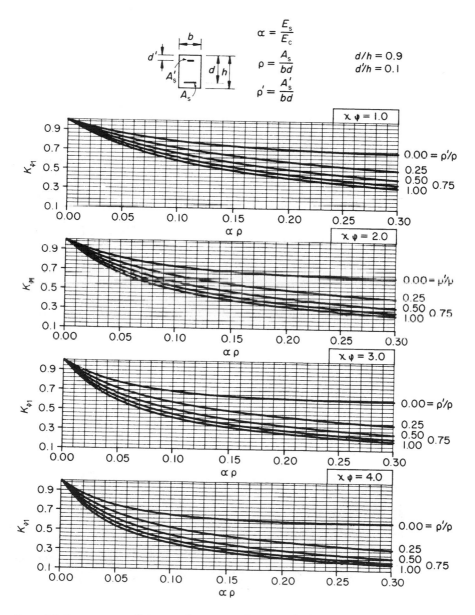

Figure F.4 Curvature coefficient $\kappa_{\varphi1}$ for rectangular uncracked sections; $d/h = 0.9$.

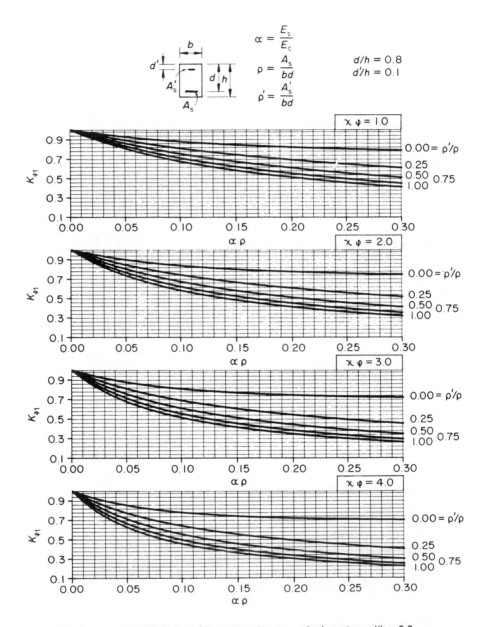

Figure F.5 Curvature coefficient $\kappa_{\varphi 1}$ for rectangular uncracked sections; $d/h = 0.8$.

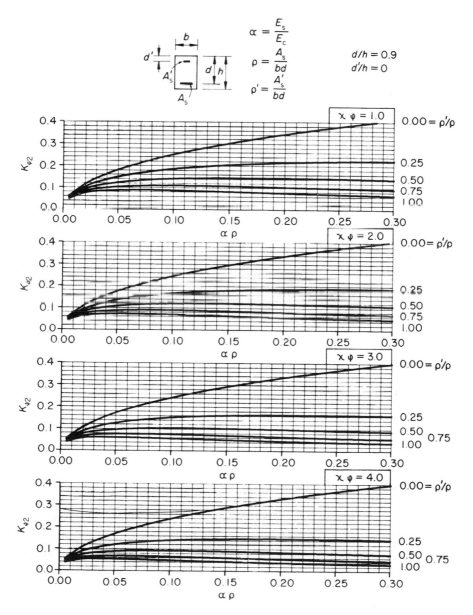

Figure F.6 Curvature coefficient $\kappa_{\varphi 2}$ for rectangular cracked sections; $d'/h = 0$.

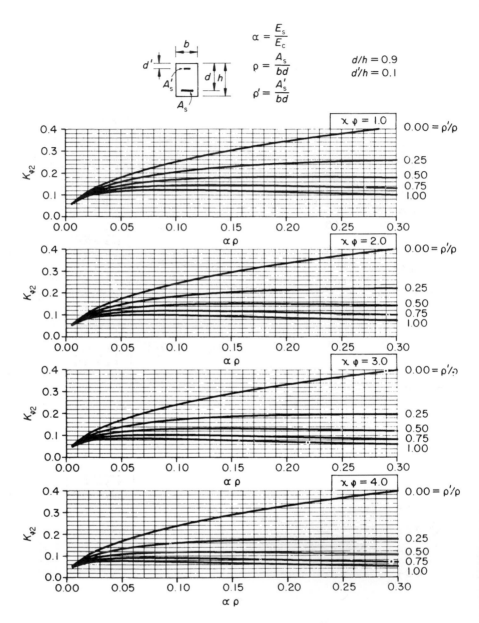

Figure F.7 Curvature coefficient $\kappa_{\varphi 2}$ for rectangular cracked sections; $d'/h = 0.1$.

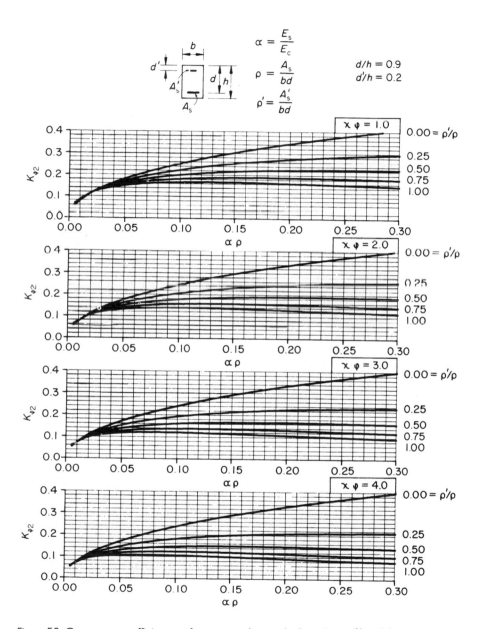

Figure F.8 Curvature coefficient $K_{\varphi2}$ for rectangular cracked sections; $d'/h = 0.2$.

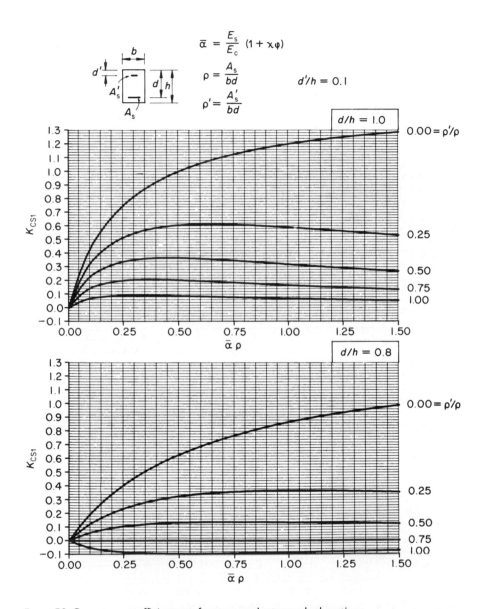

Figure F.9 Curvature coefficient K_{cs1} for rectangular uncracked sections.

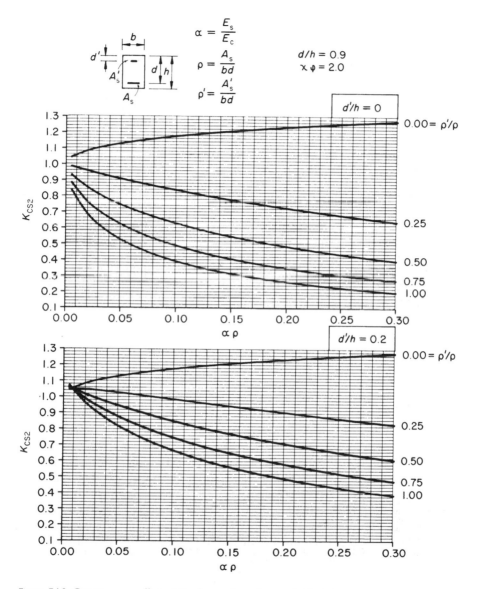

Figure F.10 Curvature coefficient K_{cs2} for rectangular cracked sections.

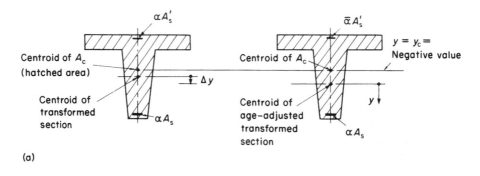

(a)

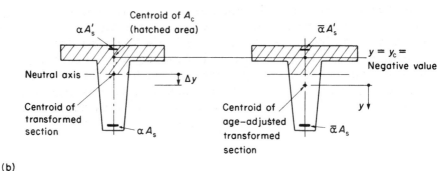

(b)

Figure F.11 Transformed section at t_0 and age-adjusted transformed section in states 1 and 2: (a) uncracked section; (b) cracked section, concrete in tension ignored.

Other symbols in Equations (F.1) to (F.3) are geometrical properties of the transformed section at time t_0 and the age-adjusted transformed section (Fig. F.11(a) and (b)). The first is composed of the concrete area plus a times the area of steel; where $a = a(t_0) = E_s/E_c(t_0)$; E_s is the modulus of elasticity of steel and $E_c(t_0)$ is the modulus of elasticity of concrete at t_0. The age-adjusted transformed section is composed of the area of concrete plus $\bar{a}$ times the area of reinforcement; $\bar{a} = \bar{a}(t, t_0) = E_s/\bar{E}_c(t, t_0)$; $\bar{E}_c(t, t_0)$ is the age-adjusted modulus of elasticity of concrete (Equation (1.31)). The geometrical section properties included in Equation F.1 to F.3 are:

I = moment of inertia of the transformed section at time t_0 about an axis through its centroid

I_g = moment of inertia of the gross concrete area about an axis through its centroid

$\bar{I}$ = moment of inertia of the age-adjusted transformed section about an axis through its centroid

I_c = moment of inertia of concrete area A_c about an axis through centroid of age-adjusted transformed section

A_c = area of concrete considered effective = entire concrete area in state 1, but only area of compression zone in state 2

y_c = the y-coordinate of the centroid of A_c, measured downwards from the centroid of the age-adjusted transformed section

Δy = the y-coordinate of the centroid of the age-adjusted transformed section, measured downwards from the centroid of the transformed section at t_0

d = distance between extreme compression fibre and centroid of tension steel.

In the usual case, when the tension steel A_s near the bottom fibre is larger than the compression steel and $\bar{a}$ is larger than a, the values of Δy and y_c are respectively positive and negative as shown in Fig. F.11(a) and (b).

When Equations F.1 to F.3 are used for a fully cracked section, the symbols refer to properties of transformed cross-sections for which the concrete in tension is ignored. The depth of the compression zone is determined by Equation (7.16), which is applicable for the case when the section is subjected to a bending moment, without normal force.

Description of computer programs provided at *www.sponpress.com/concretestructures*

The following password will be required to access the site:
CONCRETE
(NB: the password needs to be in capital letters)

G.1   Introduction

At the above web site, three computer programs are provided as optional companions of this book. The programs are for use on IBM personal computers or compatibles. One program gives creep and aging coefficient and relaxation function for concrete; the other two are for analyses of stress and strains in individual sections. A more comprehensive computer program that performs these anlyses for a number of sections and calculates deflections and rotations is:

RPM, "Reinforced and Prestressed members", Elbadry, M. and Ghali, A., American Concrete Institute, P.O. Box 9094, Farmington Hills, MI 48333–9094, USA.

The names of the three programs in the above-mentioned web site are:

- *CREEP*
- *SCS* (Stresses in Cracked Sections)
- *TDA* (Time-Dependent Analysis)

   The files are listed below separately for each program:

MANUAL.CRP	*MANUAL.SCS*	*MANUAL.TDA*
CREEP.IN	*SCS.IN*	*TDA.IN*
CREEP.EXE	*SCS.EXE*	*TDA.EXE*
CREEP.FOR		

The files in the web address should be copied in a directory of arbitrary name. The files CREEP.IN; SCS.IN and TDA.IN are input files of example problems. To generate an input file for a new problem, edit the relevant file with the ending 'IN', replacing the problem title, the integers and the real values by the data of the problem to be solved. Before editing any input file example, it should be copied – for future reference – in a file of arbitrary name. To run a program, type its name, while the computer is set on DOS prompt and press 'Enter'. The results will be written by the computer in a file named CREEP.OUT, SCS.OUT or TDA.OUT. Sections G.2 to G.4 describe the three programs.

The following are DOS commands that may be used, with the computer on DOS prompt. After typing each command, press 'Enter':

Command	*What the command achieves*
md JOE	Make a new directory named JOE.
cd JOE	Change directory by opening JOE.
copy a:*.*	Copy all files in drive A.
copy CREEP.IN CREEPIN.BAK	Copy an existing file in a new file.
edit CREEP.IN	Open a file to read it or edit it.
CREEP	Run the program named CREEP.
edit CREEP.OUT	Open a file to read it or edit it.

G.2   Computer program CREEP

The program CREEP calculates the creep coefficient for concrete, the relaxation function and the aging coefficient in accordance with CEB-FIP Model Code 1990[1] (See Section A.1). Use of the program CREEP gives results not much different from the answers calculated in accordance with Eurocode2 – 1991[2].

G.2.1   Input and output of CREEP

The input data file named CREEP.IN has three lines of data:

* Title of problem (less than 76 characters);
* Values of f_{ck}, h_o and RH in MPa, mm and per cent, respectively;
* Concrete ages t_0 and t in days.

The output file, CREEP.OUT includes $r(\tau, t_0)$, $\chi(\tau, t_0)$, $E_c(t_0)$, $\varphi(t, t_0)$, $\chi(t, t_0)$, $r(t, t_0)$; where:

f_{ck} (MPa) = characteristic compressive strength of cylinders 150 mm in diameter and 300 mm in height stored in water at $20 \pm 2\ °C$, and tested at the age of 28 days.

h_o (mm) = notional size = $2A_c/u$, with A_c and u being the area and the perimeter in contact with the atmosphere of the cross-section of the considered member.

RH (per cent) = relative humidity.

t_0 (days) = age of concrete at loading.

t (days) = age of concrete at the end of a period in which the load is sustained.

τ (days) = a time varying between t_0 and t.

$E_c(t_0)$ = modulus of elasticity of concrete at age t_0.

$\varphi(t, t_0)$ = ratio of creep to instantaneous strain.

$\chi(t, t_0)$ = aging coefficient of concrete.

$r(t, t_0)$ = relaxation function = concrete stress at time t due to a unit strain imposed at time t_0 and sustained to time t.

G.2.2   FORTRAN code

The file CREEP.FOR presents a listing of FORTRAN statements, which includes subroutine named *Phicoef* to calculate $\varphi(t, t_0)$ using equations of CEB-FIP Model Code 1990 (see Section A.1). This subroutine can be changed when use of other equations is required. A manual for quick reference is included in the web address in the file MANUAL.CRP.

G.2.3   Example input file for CREEP

The file CREEP.IN can generate the data to plot one of the two relaxation functions in Fig. A.3. The three lines of data for this problem are:

Title: Relaxation function Fig. A.3, for $t_0 = 3$ days.

30.0	400.0	50.0	f_{ck} (*MPa*), h_o (*mm*), *RH* (*per cent*)
3.0	30000.0		t_0, t (*days*)

G.3   Computer program SCS (Stresses in Cracked Sections)

The program SCS calculates stresses and strains in a reinforced concrete section subjected to a bending moment, M with or without a normal force, N. The section can be composed of any number of trapezoidal concrete layers and any number of reinforcement layers. The layers can have different elasticity moduli, E_c and E_s. Prestressed and non-prestressed reinforcement are treated in the same way. First the stresses are calculated for uncracked section. If stress in concrete at an extreme fibre exceeds the tensile strength, f_{ct}, the analysis is redone ignoring concrete in tension.

G.3.1   Input and output of SCS

The input and output files are named SCS.IN and SCS.OUT. Running the program must be preceded by preparation of the input file in which the data are presented as follows:

- Title of problem (less than 76 characters)
- Number of concrete and reinforcement layers, *NCL* and *NRL*, respectively

- A set of *NCL* lines; each line describes consecutively a trapezoidal concrete layer, starting by the top layer:
  Layer number, widths at top and at bottom, height and elasticity modulus, E_c
- A set of *NRL* lines; each line describes a reinforcement layer:
  Layer number, cross-sectional area, depth, d_s below top fibre, and elasticity modulus, E_s. When *NRL* = 0, skip this set of lines.
- Values of *M, N* and f_{ct}

The computer writes the results in file SCS.OUT, which includes the strain and stress parameters that define their distributions and area properties of the cross-section. When cracking occurs the output includes depth c of the compression zone.

G.3.2   Units and sign convention

The basic units used are: force unit and length unit. Any units for these two must be consistently used. As example, when Newton and metre are used for force and length, respectively, *M* must be in Newton-metre and E_c, E_s and f_{ct} in Newton per metre squared.

The reference point, *O* is at top fibre. When the resultant force on the section is a normal force, *N* at any position on vertical symmetry axis, it must be substituted by statical equivalent normal force, *N* at *O*, combined with a moment, *M*. The *y*-coordinate of any fibre and the depth, d_s of any reinforcement area are measured downward from the top fibre. A tensile stress and the associated strain are positive. A positive moment, *M* produces tensile stress at bottom fibre and induces positive curvature.

Prestressing duct: When it is required to deduct a cavity, such as a prestressing duct, from concrete area, enter it as a reinforcement layer having a negative cross-sectional area; a dummy real value, say a zero, should be entered for the modulus of elasticity.

G.3.4   Example input file for SCS

The following is file SCS.IN for analysis of the section in Example 7.6 in the cracking stage:

```
T-section, Example 7.6, cracking stage; N₂ = –327 kip; M₂ = 6692 kip. in
2    3            Number of concrete layers, number of reinforcement layers
1    80.  80. 4.      4000 Layer no., widths at top & bot., ht., modulus Ec
2    20.  20  36.     4000.
1    4.   2.  29000. Reinft. layer no., area, depth ds, modulus Es
```

2   3.   34. 27000.
3   10.   37. 29000.
6692. −327 0.0 Moment, M, Normal force, N and f_{ct}

G.4   Computer program TDA (Time-Dependent Analysis)

A section composed of any number of trapezoidal layers and any number of non-prestressed reinforcement layers is considered. All concrete layers have the same elasticity modulus. The section may have a single prestressed reinforcement layer, which can be pretensioned or post-tensioned. The prestressing is introduced simultaneously with a normal force, N at top fibre and moment, M about an axis at top fibre. After a period during which creep and shrinkage of concrete and relaxation of prestressed steel occur, additional normal force and moment are introduced, representing effect of live load. The purpose of this program is to calculate the strain and the stress immediately after prestressing, after occurrence of creep, shrinkage and relaxation and after application of the live load.

G.4.1   Input data for TDA

The input and output files have the names TDA.IN and TDA.OUT. Running the program must be preceded by preparation of the input file with the data presented as follows:

- Title of problem (less than 76 characters).
- Numbers of concrete and reinforcement layers, *NCL* and *NRL*, respectively.
- A set of *NCL* lines; each line describes a consecutive trapezoidal concrete layer, starting by the top layer: layer number, widths at top and bottom, height and elasticity modulus, E_c at the time of prestressing (first loading stage). The same value of E_c must be entered for all layers.
- A set of *NRL* lines; each line describes a reinforcement layer: layer number, cross-sectional area, depth d_s below top fibre and elasticity modulus, E_s. When *NRL* = 0, skip this set of lines.
- Values of M, N and f_{ct}. Value of f_{ct} is tensile strength at time of first stage of loading; M and N are values of moment and axial force introduced at first stage. No prestressing is included in values of M and N.
- *Iprestress, Ilayer*, prestress force, *Itda*; where *Iprestress* = 0, 1 or 2, meaning no prestress, pretensioning or post-tensioning, respectively. *Ilayer* is the number of the layer that is prestressed. *Itdata* = 0 or 1, meaning the time-dependent analysis is not required or required, respectively. When *Iprestress* = 0, enter 0 and 0.0 for the layer number and the prestressing force, respectively.

- Creep coefficient, aging coefficient, free shrinkage and reduced relaxation. Omit this line when *Itdata* = 0.
- Values of M, N, f_{ct}, E_c. Enter here magnitudes of moment and normal force introduced after the time-dependent changes; give also f_{ct} and E_c at this instant. Omit this line when *Itdata* = 0.

G.4.2   Units and sign convention

The references point, O is at top fibre. A normal force, N at any position on vertical symmetry axis is substituted by statical equivalent normal force, N at O, combined with a moment, M. The y-coordinate of any fibre and depth, d_s of any reinforcement layer are measured downwards from the top fibre. A tensile stress and the associated strain are positive. A positive moment, M produces a tensile stress at bottom fibre and induces a positive curvature. The free shrinkage is commonly a negative value, indicating shortening; the reduced relaxation is also negative, indicating loss of tension. Any basic units of force and length can be adopted; all parameters must be entered using the same basic units.

G.4.3   Prestressing duct

When it is required to deduct a cavity, such as a prestressing duct, from concrete area, it should be entered as a reinforcement layer having a negative cross-sectional area; a dummy real value, say a zero, should be entered for the modulus of elasticity. The prestressed steel in the duct must be entered on a separate line.

G.4.4   Example input file for TDA

The input file presented below is for solution of Examples 2.6 and 7.6. The T-section of a pretensioned beam (Fig. 2.15(a)) is to be analyzed for the time-dependent effects occurring between the time of prestress and a later instant. At this instant, a bending moment is applied, representing effect of live load. The immediate strain and stress due to live load are also required. The prestress transfer is accompanied by a given bending moment due to the self-weight. In this problem, basic units used for force and length are kip and in, respectively. The input data file is:

T-section of Examples 2.6 and 7.6 (Fig. 2.15)

2	3		No. of concrete layers, no. of reinforcement layers	
1	80. 80.	4	3600.	Layer no., widths at top & bot., ht., E_c
2	20. 20.	36.	3600.	
1	4. 2.	29000.		Reinft. layer no., area, depth d_s, E_s
2	3. 34.	27000.		

3   10. 37.   29000.
10560. 0.   0.0 M, N and f_{ct}.
1   2   600.   1   *Iprestress*, *Ilayer*, prestress force, *Itda*
3.   .8   −300. e-6 −13.   Creep coef., aging coef., fr. shrge., red. relaxn.
9600.   0.   0.   4000. M, N, f_{ct}, E_c.

Notes

1 See reference mentioned in Note 2, p. 19.
2 See reference mentioned in Note 5, p. 19.

Further reading

The following are selected relevant books. Extensive lists of references can be found in each of them.

Branson, D.E. (1977). *Deformation of Concrete Structures*. McGraw-Hill, New York.

Favre, R., Beeby, A.W., Falkner, H., Koprna, M. and Schiessl, P. (1985). *Cracking and Deformation*. Comité Euro-International de Béton (CEB), Federal Institute of Technology, Lausanne, Switzerland.

Favre, R., Koprna, M. and Radojicic, A. (1980). *Effects differés*. Fissuration et Déformations des Structures en Béton, Georgi Saint-Saphorin, VD, Switzerland.

Favre, R., Jaccoud, J.-P., Koprna, M. and Radojicic, A. (1990). *Dimensionnement des structures en béton*, volume 8 of *traité de Génie Civil*. Presses polytechniques et universitaires romandes, Lausanne, Switzerland.

Gilbert, R.I., (1988). *Time Effects in Concrete Structures*. Elsevier, Amsterdam.

Neville, A.M., Dilger, W.H. and Brooks J.J. (1983). *Creep of Plain and Structural Concrete*. Construction Press, London.

Index